Biological Control

Ecology and Applications

Biological control is the suppression of populations of pests, weeds and plant pathogens by living organisms. These organisms can reduce damage by invasive species and protect our environment by reducing the need for pesticides. However, they also pose possible environmental risks, so biological control interventions must be undertaken with great care. This book enhances our understanding of biological control interactions by combining theory and practical application. Using a combination of historical analyses, theoretical models and case studies, with explicit links to invasion biology, the authors cover the biological control of insects, weeds, plant pathogens and vertebrate animals.

This book reflects increasing recognition of risks over the past 20 years, and incorporates the latest technological advances and theoretical developments. It is ideal for researchers and students of biological control and invasion biology.

George E. Heimpel is a Distinguished McKnight University Professor of Entomology at the University of Minnesota. His research is focused on parasitoid biology and the use of parasitoids as biological control agents in agricultural and natural settings.

Nicholas J. Mills is a Professor of Entomology in the Department of Environmental Science, Policy and Management at the University of California, Berkeley. His research focuses on the biological control of arthropod pests using parasitoids and predatory insects in agricultural, urban and forest settings.

Biological Control

Ecology and Applications

GEORGE E. HEIMPEL
Professor
Department of Entomology
University of Minnesota

NICHOLAS J. MILLS
Professor of Entomology
Department of Environmental Science, Policy and Management
University of California, Berkeley

CAMBRIDGE
UNIVERSITY PRESS

University Printing House, Cambridge CB2 8BS, United Kingdom

Cambridge University Press is part of the University of Cambridge.

It furthers the University's mission by disseminating knowledge in the pursuit of education, learning and research at the highest international levels of excellence.

www.cambridge.org
Information on this title: www.cambridge.org/9780521845144

First published 2017

Printed in the United Kingdom by Clays, St Ives plc, March 2017

A catalogue record for this publication is available from the British Library

Library of Congress Cataloging-in-Publication Data
Names: Heimpel, George E., author. | Mills, Nicholas J., author.
Title: Biological control : ecology and applications / George E. Heimpel, Distinguished McKnight University Professor, Department of Entomology, University of Minnesota, Nicholas J. Mills, Professor of Insect Biology, Department of Environmental Science, Policy and Management, University of California, Berkeley.
Description: New York : Cambridge University Press, 2017. | Includes bibliographical references and index.
Identifiers: LCCN 2016047005 | ISBN 9780521845144 (alk. paper)
Subjects: LCSH: Pests – Biological control.
Classification: LCC SB975 .H45 2017 | DDC 628.9/6–dc23
LC record available at https://lccn.loc.gov/2016047005

ISBN 978-0-521-84514-4 Hardback

We dedicate this book to:
Mirella, Nicholas, Simon & Carmen
Alison, Rosie, Hester & Isobel

CONTENTS

FOREWORD

Today, controversies related to the use of chemical pesticides continue to increase social awareness of topics related to human health and the environment. As an ecologist, I consider biological control to be the best alternative to pesticides. Biological control is both effective at reducing populations of detrimental organisms and safe for our planet. The most obvious advantages relate to food security, human health and protection of the environment through the control of agricultural pests, disease vectors and invasive alien species, respectively. Notable indirect benefits also arise from reduced pesticide use by increasing farmer health, food safety, biodiversity conservation, maintenance of ecosystem services, as well as soil, water and air quality. All regions of the world can benefit from implementing biological control programs as they create and sustain public good.

Biological control has a long history of success in developing, emerging and well-developed countries. In classical (=importation) biological control almost 2 700 natural enemies have been introduced in 196 different countries or islands to control insects, weeds, plant pathogens and other pests. In addition, more than 440 species of natural enemies are currently produced and sold for augmentative biological control worldwide. Nevertheless, biological control is not a long, calm river. Researchers and practitioners have made some mistakes, the most obvious one being the introduction of a few exotic natural enemies into new areas where they became an ecological problem, mostly because they were not specific to the target pest. However, the science behind biological control has made remarkable progress during the past 50 years, changing from a trial and error method to a more predictive approach based on theories of predator-prey interactions and population dynamics. Such a rigorous research-based activity should help to prevent further errors from the past.

I agree with George Heimpel and Nicholas Mills that 'Biological control is at a crossroads in many ways'. Although biological control is progressing worldwide (e.g., in terms of numbers of biological control agents available from private markets), it remains a relatively small component of overall pest control in most settings. A number of factors have recently contributed to impeding the progress of biological control. For example, new national and international legal frameworks for the export, import and release of biological control agents have slowed the process of biological control. Right from the beginning of their book, George and Nick have attempted to unify definitions and concepts used by colleagues from different scientific disciplines that study various types of pests, weeds and plant pathogens. This is important because biological control has entered an era of globalization. These efforts should contribute to increased dialogue and networking among biological control scientists and practitioners from all backgrounds.

After decades of experience in modern biological control, it is also appropriate to reexamine the fundamental ecological principles of our discipline and their applications. Bradford Hawkins and Howard Cornell once wrote 'The scientific basis for successful biological control has been a relentlessly pursued but elusive goal[1]'. George and Nick have brilliantly succeeded in addressing this challenge. I find their book irresistible! From their expertise in ecology and evolutionary biology, their hands-on experience in the field and the laboratory, their teaching practice, and their dedication to biological control, George and Nick bring creativity to the field. Years of fundamental and applied studies have generated a wealth of information on all aspects of the biology of biological control agents, including complex interactions with biotic and abiotic elements of the environment. The authors have digested the literature and analyzed many pest-natural enemy associations in their quest for patterns and mechanisms. They not only summarize concepts and illustrate them with pertinent examples but also take the opportunity to express their own opinion about conventional definitions, paradigms, controversies and current challenges. This book emerges as a splendid illustration of the value of a comprehensive understanding of all ecological and evolutionary aspects of pest-natural enemy relationships when developing biological control programs. The more we understand the organisms themselves, as well as the nature and mechanisms of their interactions, the more confidently we can predict the potential of a biological control agent and the consequences of its introduction into a given habitat.

George and Nick introduce what is essential in biological control and their book can be read straight through with pleasure. Their contribution not only provides valuable information but is also a lively source of original insights and ideas. Writing a textbook is not an easy task and I know that this book represents several years of continuous efforts and dedication. It is an honour for me to be asked to write this foreword to *Biological Control – Ecology and Applications*. I thank George and Nick for assembling such a fine and sound piece of work. This book will rapidly become a foremost reference in biological control.

Jacques Brodeur
Canada Research Chair in Biological Control
Université de Montréal
February 2017

[1] Hawkins, B.A. & H.V. Cornell. 1999. *Theoretical Approaches to Biological Control*. Cambridge University Press. Cambridge, UK.

PREFACE & ACKNOWLEDGMENTS

Biological control is simultaneously a natural phenomenon, a pest management tactic, and a scientific discipline. It therefore can mean different things to different people. For example, a distinction is sometimes made between 'biological control scientists' and 'biological control practitioners'. In reality, many practitioners are able scientists and many scientists are heavily involved in implementing biological control solutions. Biological control also spans multiple taxa, levels of biological organization and fields of study. We therefore saw a need to present biological control as a unified set of interactions that apply equally well to all disciplines without the need to develop separate definitions, terminology or concepts in each field. To our knowledge this is the first time that biological control has been considered more broadly in the context of pests, diseases and weeds in an explicitly ecological context, and we hope that this will stimulate greater integration between disciplines.

The book has been written for a broad readership including scholars (researchers, graduate and undergraduate students) that are interested in biological control, invasion biology or environmental management, and environmental professionals that desire a deeper understanding of the underlying ecological processes that could facilitate or interfere with the management strategies that they are using or contemplating. This book is not a 'how-to' guide for biological control, and neither does it provide a comprehensive overview of the organisms that are prominent in biological control interactions. There are fortunately a number of other volumes that cover these important topics.

We are enthusiastic about the positive role that biological control has played in reducing negative impacts of invasive species and thus allowing for the production of a healthier and more sustainable food supply as well as the conservation of natural habitats. And we look forward to a future in which increasing knowledge of biological control interactions will allow for more effective manipulation of natural enemies in both managed and natural ecosystems. However, this is not a work of advocacy. We have tried our best to be dispassionate about biological control activities, presenting the complexities of biological control as objectively and scientifically as we are able. We recognize that biological control is not a panacea and that it is not always the best solution to a given environmental problem. Also, biological control is, of course, not without risks. In this way it is no different from other interventions or from the decision to do nothing about invasive species or pest organisms.

The book is organized into an introductory chapter on the scope and definition of biological control, followed by a series of chapters (2–7) that discuss different aspects of importation biological control. In Chapter 8 we consider augmentation of natural enemies, and in Chapters 9 and 10 we discuss the use of habitat manipulation and pesticide compatibility, respectively, as the two main components of conservation biological control. We use examples from biological control of arthropod pests, diseases and weeds wherever relevant throughout the book and these address a wide variety of managed landscapes including agricultural, forest, wildland and aquatic ecosystems.

Biological control is at a crossroads in many ways and we hope that this book will help the discipline to navigate new and changing scientific landscapes. Importation biological control in particular is emerging from a paradigm shift in which an era focusing only on the benefits of biological control was replaced by an era where risks dominated the conversation. Here, we see our goal as contributing to a new third era where benefits and risks are well understood so that solutions maximizing the former while minimizing the latter can be developed. Biological control is also expanding from an era in which it was seen as primarily relevant in an agricultural context to one in which it is being increasingly explored in arenas of natural systems and public health. In addition, biological control is increasingly taking advantage of modern tools in biology – particularly genomics. These new tools are driving novel approaches particularly in the biological control of plant pathogens. We are optimistic that these new tools can contribute to improve success rates in many areas of biological control.

Both of us teach biological control classes and we have benefitted enormously from discussions with a series of graduate and undergraduate student cohorts who have helped us to think more clearly about these topics over the years. We also thank a number of colleagues for their feedback on various chapters of the book including David Andow, Mariana Bulgarella, Daniel Cathay, Jeremy Chacon, Brett Couch, Camille Delebeque, Erin Donley, Jim Eckberg, Wolfgang Heimpel, Ruth Hufbauer, Yang Hu, Kathy Hughes, Mark Jervis, Joe Kaser, Christine Kulhanek, Eric LoPresti, Elias Marvinney, Georgiana May, Paul Ode, Sandy Olkowski, Jay Rosenheim, Tracy Schohr, Zeynep Sezen, Ruth Shaw, Peter Tiffin, Roy van Driesche, Rob Venette, Steve Welter, Rachel Wiggington, Alan Yip and Kate Zemenick. We also thank Keith Hopper for sharing unpublished data, Riccardo Bommarco, Nick Haddad, Richard Mack and Charles Mitchell for help with literature or data for figures, Tom Gao for help with graphics and Mary Marek-Spartz for drawing Figure 7.4. Lastly, we would like to thank Cambridge University Press, and in particular Lindsey Tate, for guidance and patience.

George Heimpel & Nicholas Mills
November 2016

1 Definitions and Interactions

The term *biological control* is broadly used to identify the suppression of populations of pests, weeds and disease-causing organisms by living organisms (referred to here as *biological control agents*). Humans have recognized the beneficial effects of biological control agents for thousands of years. For example, a puzzle provided in the 'Rhind Mathematical Papyrus', an Egyptian document that dates to 2000 BCE, indicates clearly that cats were appreciated as rodent hunters in ancient Egypt (Chace 1979):

> *In each of seven houses are seven cats, each cat kills seven mice, each mouse would have eaten seven ears of wheat, and each ear will produce seven measures of grain; how much grain is thereby saved?*

Cats or other predators may or may not have been *intentionally* used to suppress pests in early human history, but they certainly would have contributed to the control of pests in even the earliest days of agriculture. Remains of the parasitoid *Habrobracon hebetor*, which attacks stored-grain-infesting caterpillars, were isolated from Egyptian tombs dating to 2000 BCE as well (Chaddick & Leek 1972). By protecting the food supply of humankind from pests, these animals would have been among the first biological control agents.

The first credible cases of the intentional use of biological control agents to control pests of humans were deliberate movement and manipulation of predatory ant nests to control orchard and stored-grain pests in China 3,000 years ago (Olkowski & Zhang 1998; Figure 1.1). These practices continue to this day in various parts of the world (Perfecto & Castiñeras 1998; Van Mele et al. 2007), and are further discussed in Chapter 9 of this book. Histories

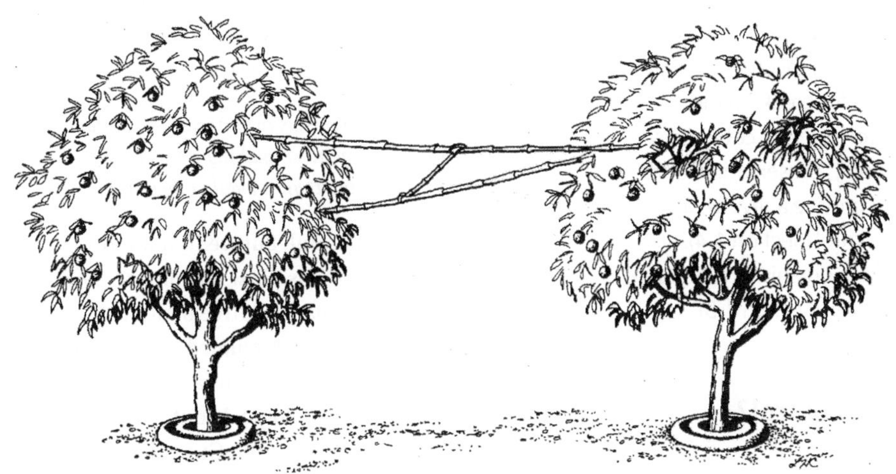

Figure 1.1. Illustration of a method used in China to conserve weaver ants on citrus trees and distribute them within groves. Weaver ant nests are collected in mountainous areas and placed on citrus trees outfitted with moats and ladders to keep the ants on the trees (from Olkowski & Zhang 1998).

of purposeful biological control have been pieced together by Steinhaus (1956), Doutt (1964), Simmonds and colleagues (1976), DeBach and Rosen (1991) and Legner (2008), among others, and Sawyer (1996) provides an account of the colorful beginnings of biological control in California starting in the 1880s. Generally speaking, the history of biological control can be characterized as a gradual increase in the understanding of the biology and ecology of pest–enemy interactions, as well as the development of increasingly sophisticated and creative methods of utilizing biological control agents to control pests. The result is that today, the term *biological control* can be simultaneously used to refer to a natural phenomenon, a method of pest control, and a scientific discipline.

In this introductory chapter, we will begin by defining biological control (Section 1.1). The definition of biological control has gone through some historical changes, and recent innovations in biologically based forms of pest management have blurred the divisions between activities that should and should not be classified as biological control. We introduce a new way of defining biological control that relies on ecological relationships among populations and the direct or indirect effects of these relationships on humans. This definition is then used to evaluate various pest control strategies that are at the fringes of traditional definitions of biological control (Section 1.2). We explain the rationale for breaking the discipline of biological control into four categories ('importation', 'augmentation', 'conservation' and 'natural') in Section 1.3 and provide a review of growth areas in biological control (Section 1.4). Finally, we briefly introduce the organisms that figure prominently in biological control (Section 1.5).

1.1 What Is Biological Control?

Definitions of biological control have varied in their inclusiveness of ecological processes and biological functions. For example, in an intentionally expansive definition by DeBach and Rosen (1991), any reduction of plant or animal populations by natural enemies (understood to be predators, herbivores, parasitoids or pathogens) occurring in natural or managed systems was called biological control. We favor definitions that are restricted to the control or suppression of organisms that have negative effects on humans (directly or indirectly). Another class of definitions that we find too expansive are those that include a broad range of nonchemical pest control methods. Depending on the authors, these definitions include host plant resistance, insect growth regulators, pheromone disruption, botanical insecticides, transgenic insecticidal crops and even vaccines of human and animal diseases (Pimentel 1980; US National Academy of Sciences 1988; Rechcigl & Rechcigl 1998). Definitions such as these were vigorously repudiated by Garcia and colleagues (1988), who objected in particular to the inclusion of recombinant genes or gene products in a definition of biological control. We favor restricting biological control to actions precipitated by living organisms or viruses against target organisms that cause harm to humans or their resources (Eilenberg et al. 2001; Hajek 2004; Van Driesche et al. 2008).

A definition of biological control from Eilenberg and colleagues (2001) that is used in Hajek's (2004) introductory text reads as follows:

> *The use of living organisms to suppress the population of a specific pest organism, making it less abundant or less damaging than it would otherwise be.*

This definition is useful, but we find it too restrictive for two reasons. First, the phrase '*The use of*...' emphasizes manipulative biological control, whereas the unmanipulated suppression of organisms that have the potential to attain pest status also constitutes biological control (e.g., Hornby 1983; Settle et al. 1996; Guretzky & Louda 1997; Rosenheim et al. 1997; Maron & Vilà 2001). Indeed, the global value of biological control as a naturally occurring 'ecosystem service' has been estimated at greater than $400 billion per year

Table 1.1. **Calculations that Losey and Vaughan (2006) used to arrive at the estimate of $4.5 billion in crop protection per year attributable to natural biological control of potential native insect pests by other insects**

Parameter	Estimate (US$ in Billions)
Crop losses due to native insect pests	$7.32
Proportion of potential native pests reaching damaging levels	0.35
Hypothetical damage by native pests in the absence of natural control	$7.32/0.35 = $20.91
Value of natural control	$20.91 – $7.32 = $13.59
Proportion of natural control attributable to insects	0.33
Crop protection attributable to insects	**$13.59*0.33 = $4.49**

(Costanza et al. 1997). Losey and Vaughan (2006) estimated that natural biological control by native insects in the United States alone provides $4.5 billion of savings per year in increased yield and reduced insecticide inputs in agricultural settings (Table 1.1). Second, the definition cited earlier technically excludes the action of viruses in pest control, since most if not all viruses are considered inanimate (Raoult et al. 2004; Villareal 2004). Having said this, viruses have been universally considered bona fide biological control agents – presumably because of their ability to replicate within host cells.

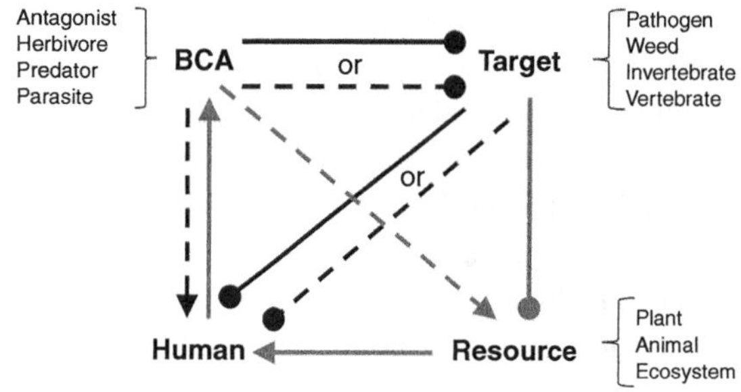

Figure 1.2. Diagram depicting biological control interactions among biological control agents (BCA), target organisms, resources and humans using the notation of Levins (1975). Solid lines indicate direct effects, dashed lines indicate indirect effects, arrowheads indicate positive effects and round heads indicate negative effects. The set of interactions in black shows the 'biological control triangle' – the minimum set of interactions that define biological control.

We seek a definition of biological control that recognizes the underlying ecological relationships that characterize biological control. These relationships include direct and indirect interactions between populations of target organisms (used in the broadest sense to include animal pests, weeds and disease-causing organisms), biological control agents, humans and their resources (Figure 1.2). Interactions are considered direct when they involve consumption or interference competition, but indirect when they are mediated by a third species and do not involve direct contact as can occur through processes such as resource competition, apparent competition or induced resistance (Wooton 1994). The effects of target organisms on humans are negative and can either be direct through nuisance, disease, injury or death, or be indirect through environmental impacts that reduce the economic, aesthetic or spiritual value of natural or managed resources. In this context, resources are externalities that are of direct benefit to human welfare, whether plant or animal, natural or managed. Direct negative effects of biological control agents on target organisms include herbivory, predation and parasitism, as well as antibiosis, a form of contest competition. Corresponding indirect negative effects include most forms of resource competition and induced resistance, as well as any other interaction in which the presence of the biological control agent results in harm to the target organism or disrupts its performance through the actions of one or more other species. Whether direct or indirect, effects of the biological control agents are of indirect benefit to humans or their resources. Finally, the strength of the various interactions depends not only on the individual species involved, but also on the suitability of the abiotic and biotic environment in which the interactions take place. Thus interactions that appear very promising under controlled conditions in the laboratory or greenhouse have often proved ineffective in the field due to unknown environmental constraints.

A subset of these interactions constitutes the core definition of biological control, shown in black in Figure 1.2. Biological control is the indirect positive effect of a biological control agent on humans that is mediated by direct or indirect negative effects of that agent on populations of one or more target species. In other words – 'The enemy of my enemy is my friend.' We call this set of minimum interactions the 'biological control triangle', and consider it the sole necessary criterion in defining biological control.

Further, we recognize three basic underlying ecological models of biological control using a simplified set of interactions (Figure 1.3). These models differ in the ways that humans affect the biological control agents of target organisms. In the first model, there is no effect of humans on the biological control agent so that biological control occurs naturally – this is 'natural biological control' (Figure 1.3a). The second includes direct human facilitation of the biological control agent (Figure 1.3b), which can be divided into importation and augmentative biological control methodologies, as discussed later in this chapter and in Chapters 2–8. In the final model, facilitation of the biological control agent is indirect (Figure 1.3c). This encompasses conservation biological control, which is discussed later in this chapter and in Chapters 9 and 10. Biological control can be further broken down by the type of interaction and by the kind of target organism and biological control agent. Examples of biological control interactions involving various organisms and mechanisms are shown in Figure 1.4.

(a)

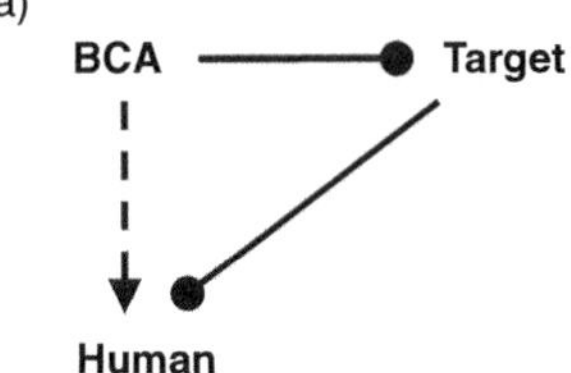

(b)

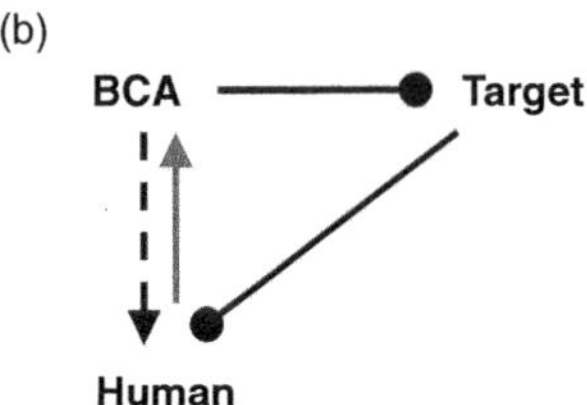

(c)

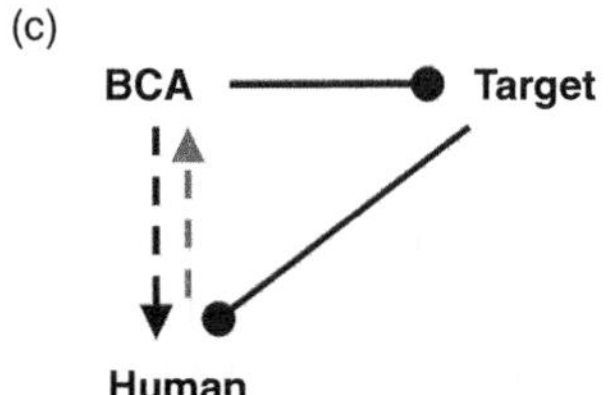

Figure 1.3. Interaction diagrams to characterize (a) naturally occurring biological control, (b) direct facilitation of biological control agents (importation or augmentative biological control) and (c) indirect facilitation of biological control agents (conservation biological control). See legend to Figure 1.2 for explanation of line types.

These models make clear that biological control interactions require a minimum of three interactors: humans, a target organism that is either directly or indirectly harmful to humans and a biological control agent. All of these protagonists are understood to be living organisms (or viruses). While the target organism and the biological control agent

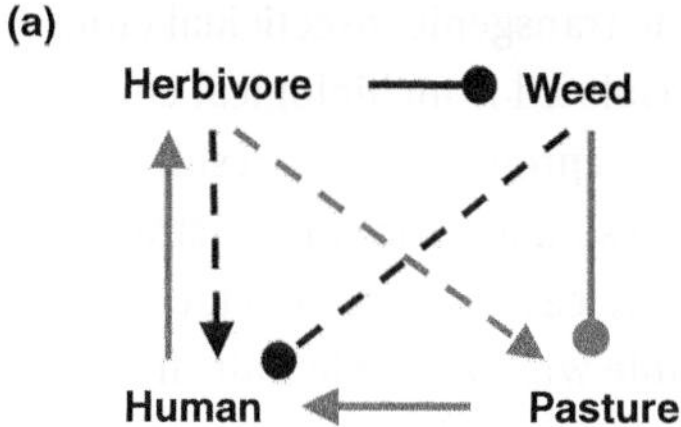

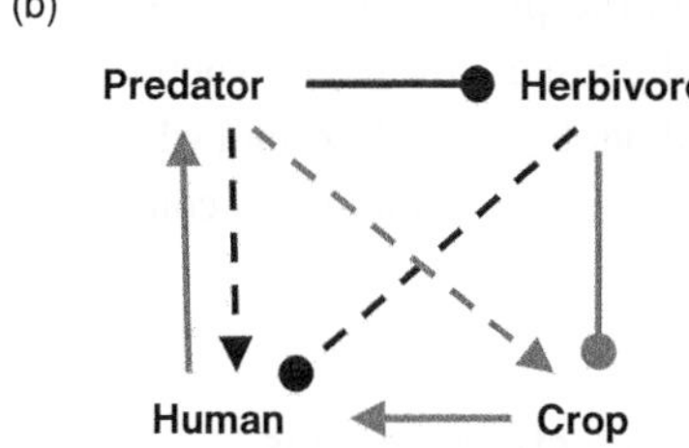

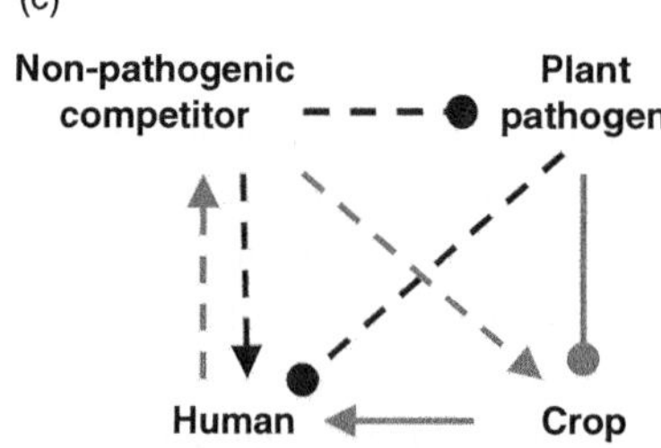

Figure 1.4. Interaction diagrams depicting three examples of human-facilitated biological control. (a) Importation or augmentative biological control of an agricultural weed by an herbivore or plant pathogen. (b) Importation or augmentative biological control of an arthropod crop pest by an arthropod natural enemy. (c) Conservation biological control of a plant pathogen by a competing microbe that is not itself a pest. See legend to Figure 1.2 for explanation of line types.

are usually distinct species, this need not necessarily be the case, as we illustrate later in this chapter and in Chapters 8 and 9.

Defining biological control on the basis of ecological interactions has a number of advantages. In addition to encompassing natural biological control, it allows for competitive as well as the more traditional trophic effects of biological control agents. The suppression of bush flies in Australia through the introduction of dung beetles from South Africa provides an interesting example of competition as an effective mechanism in biological control (Moon 1980; Tyndale-Biscoe & Vogt 1996). In this case, although bush flies that breed in cattle dung were the target organism, suppression was achieved by the introduction of a biological control agent that competed for the resource of the target organism rather than attacking the target organism itself. Other indirect effects are included as well. An interesting example of another indirect mechanism is that of induced resistance among spider mites in vineyards in California (Karban et al. 1997). In this case, the early season presence of Willamette mites, *Eotetranychus willametti*, induced resistance in the grapevines and resulted in smaller populations of the more damaging Pacific mite, *Tetranychus pacificus*, later in the season. While competition and induced resistance are unusual in the biological control of arthropod pests or weeds, they are frequent modes of action for the antagonists used in the suppression of plant diseases, as we discuss in Chapter 8 (Wilson 1997; Whipps 2001; Whipps & McQuilken 2009).

While our definition explicitly involves humans as the beneficiaries, defining biological control in terms of ecological interactions leads to the recognition that the ecological relationships underlying biological control have evolved to benefit nonhuman species as well. Indeed, so-called protection mutualisms, in which protection from enemies is provided in return for food and/or shelter (e.g., Bronstein 1998; Di Giusto et al. 2001), share the basic interaction structure with biological control. Ants are involved in some well-studied protection mutualisms. Examples include plants that produce specialized domiciles and food resources for ants that protect them from herbivores (Mayer et al. 2014), aphids that feed ants with honeydew and gain protection from parasitoids (Stadler & Dixon 2005) and antifungal bacteria used by leaf-cutting ants to control pest fungi that infect their gardens (Currie et al. 1999). Another class of protection mutualisms involves defensive symbionts in plants and animals, in which endosymbiotic bacteria or fungi protect their hosts from a wide variety of consumers (Clay 2014).

1.2 Interactions That Challenge the Definition of Biological Control

We have already alluded to some processes that we do not consider to fall within the category of biological control: host plant resistance, the use of botanical insecticides, insect growth regulators, pheromone disruption and similar environmentally friendly modes of pest control. None of these cases fits the ecological interactions outlined earlier (Figure 1.2). We also argued that the suppression of non-harmful organisms should not be considered biological control. Upon close inspection, however, this latter argument can be problematic. One could maintain that virtually *any* species could become harmful to humans if it were not regulated by consumers or competitors. And while there may be some truth to this, it may also be difficult to prove that any given innocuous species is being kept from harmful status by a particular biotic interaction or set of interactions. And there is by no means agreement that innocuous species are 'kept in check' by top-down forces (the action of natural enemies) (Hunter & Price 1992; Polis & Strong 1996). Losey and Vaughan's (2006) calculations outlined in Table 1.1, for example, use Hawkins and colleagues' (1999) estimate that one-third of native insect herbivores are controlled by arthropod natural enemies with an unknown percentage controlled by pathogens, and the balance is likely kept at harmless levels by host plant resistance or abiotic factors. This suggests that a sizable fraction, possibly even the majority, of natural control occurs through mechanisms other than predation, herbivory, parasitism, induced defense or competition. We therefore maintain our restriction of biological control to those interactions leading to the suppression of undesired organisms.

Beyond these considerations, however, there are recently developed and as-yet-unrealized approaches to pest control that truly challenge our conception of what is and what is not biological control. These methodologies include transgenic insecticidal crops that use toxin genes isolated from biological control organisms, plant growth promotion as a result of microbial action in the soil, genetic control strategies including sterile male releases and related tactics that replace undesirable with desirable traits in pest populations using genetic drivers and certain forms of immunocontraception. In the following paragraphs, we briefly discuss the overlap between these and other methodologies and explain why or why not we consider them to be biological control interactions.

1.2.1 Transgenic Insecticidal Crops

Various crop plants have been genetically transformed to express toxins derived from the bacterium *Bacillus thuringiensis* (*Bt*). A number of formulations of this entomopathogen have been available in spray form for decades, and have provided outstanding biological control of various pest species (e.g., Lacey et al. 2001). Some of these sprays include live *Bt* spores that have the capability to reproduce in the field, and some contain only the insecticidal toxin. It has been argued that this latter class of sprays should not be considered biological control because the agent does not reproduce in the field (Garcia et al. 1988; Hajek 2004), and we are in agreement with this position. What of the case, however, when the toxins are incorporated within crop plants? One could argue that the toxin reproduces or replicates as a component of its host plant and that its use therefore constitutes biological control. However, since the insecticide-encoding gene is now incorporated into the plant, it is no longer part of a third interactor that negatively impacts the pest. It thus does not conform to any of the ecological models of biological control outlined earlier. We therefore suggest that the use of *Bt* and other toxins within transgenic plants are best viewed as a form of host plant resistance that derives from biological control, and not as a form of biological control.

1.2.2 Plant Growth Promotion

Interactions among host plants, pathogens and biological control agents are complex, and it is well known that antagonists can exhibit multiple modes of action (Santoyo et al. 2012). While some modes of action such as parasitism, antibiosis, competition and induced resistance are clearly encompassed by biological control (Figure 1.5), others, such as solubilization of micronutrients and release of phytohormones, can promote plant growth in the absence of any pathogens and should not be considered biological control. In response to this dilemma it has been suggested that rhizobacteria be reclassified as either biocontrol plant growth-promoting bacteria (Biocontrol-PGPB) or simply as PGPB to distinguish between these two different mechanisms for increased plant growth (Bashan & Holguin 1998). Similarly, while the major mode

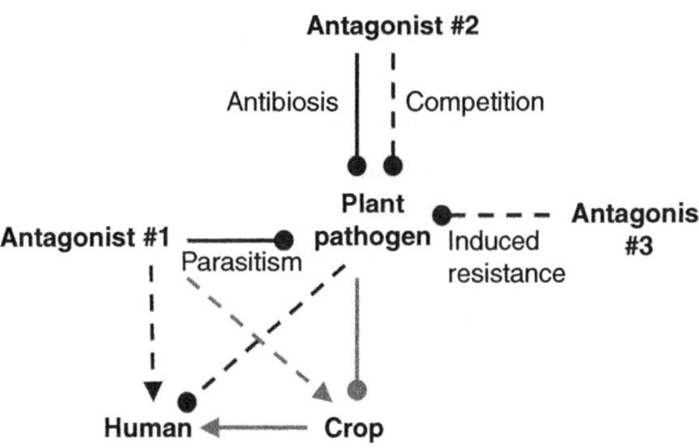

Figure 1.5. Biological control relationships involving three classes of microbial antagonists of plant pathogens that are crop pests. Antagonist #1 class are parasites that consume the pest pathogen, producing a direct negative effect on the plant pathogen. Antagonist #2 class engage in direct competition via antibiosis or indirect competition for space or resources with the pest pathogen and produce a net positive indirect effect on humans because its negative effect on the crop is less than that of the pest pathogen. The antagonist #3 class produces the same net effect via induced resistance. See legend to Figure 1.2 for explanation of line types; not all indirect links are shown in the figure.

of action of mycorrhizal fungi is improved plant nutritional status, they also function secondarily as biological control agents through induced resistance and competition for space and nutrients (Whipps 2004). Although factorial experiments can be used to separate the effects of plant growth promotion from those of biological control (Larkin & Fravel 1999), the potential for microbial control agents to have multiple modes of action that can vary with plant genotype, microbiota and environmental conditions makes such studies particularly challenging.

1.2.3 Resource Dilution Using Habitat Diversification

As we explain later and in Chapter 9, habitat diversification is one of the main strategies used in conservation biological control. This strategy is aimed mainly at arthropod pests and diseases of crops, and involves the establishment of polycultures that favor biological control agents. It is well known, however, that these manipulations can lead to pest and disease suppression in the absence of any benefit to biological control agents, and that a reduction in arthropod pest pressure can come from a dilution of the resource (the crop) for pest organisms – often because the pests have trouble actually locating and colonizing the crop within more diversified agroecosystems (Root 1973; Andow 1991). This scenario is not traditionally considered biological control, but the non-crop vegetation itself might be considered a biological control agent of the herbivorous pest by making its favored host plant less accessible or apparent – in essence acting as a foil and resulting in an indirect positive effect on the crop plant (Figure 1.6; see also Parolin et al. 2014). A definition of biological control that stresses ecological relationships suggests that the inclusion of pest and disease reduction through resource dilution should be classified as biological control.

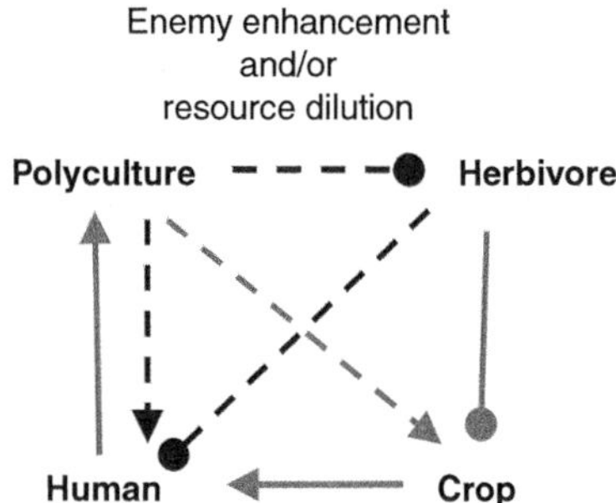

Figure 1.6. Interactions leading to biological control of herbivores through an increase in vegetational diversity by polyculture establishment within a cropping system. An indirect positive effect on the crop can occur through natural enemy enhancement and/or through decreased apparency of crop plants for herbivores, as can occur through resource concentration. See legend to Figure 1.2 for explanation of line types.

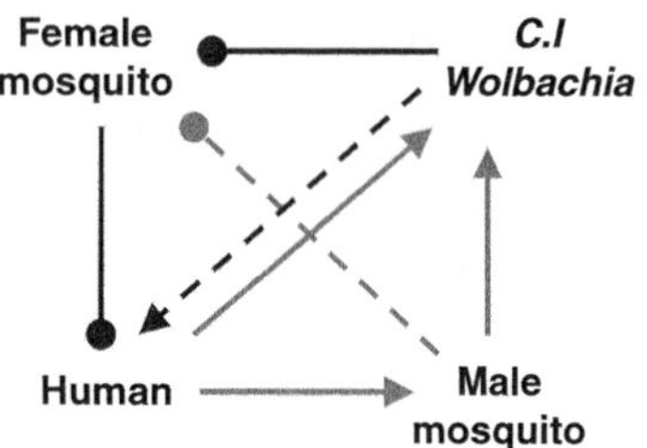

Figure 1.7. Interactions illustrating genetic biological control using the release of sterile males of a pest mosquito species. In this case sterility is achieved by *Wolbachia* bacteria that induce cytoplasmic incompatibility (C.I.). The C.I. *Wolbachia* can be either naturally occurring within particular mosquito strains or they can be purposefully incorporated (as illustrated in this diagram). The text describes extensions of this method that involve strain replacement. See legend to Figure 1.2 for explanation of line types.

1.2.4 Genetic Control

The best-known form of genetic control is the release of sterile male insects, which is also called the 'sterile insect technique' or 'autocidal control'. In traditional sterile male release programs, mass-reared and sterilized males of a pest species are released to mate with wild females, rendering them infertile. This mode of pest management was first used with great success to eradicate populations of the screwworm, *Cochliomyia hominivorax*, from the island of Curaçao and large areas of continental North America (Baumhover 2002). Sterile male releases have also been used with success against various fruit fly species (Krafsur 1998) and are under consideration for control of the highly invasive and destructive sea lamprey in the North American Great Lakes (Bergstedt et al. 2003; Twohey et al. 2003). Cytoplasmic incompatibility (C.I.) induced by bacterial endosymbionts in the genus *Wolbachia* can also form the basis of a sterile male release method. The incompatibility occurs when *Wolbachia*-carrying males mate with females that are either *Wolbachia*-free or harbor a different strain of *Wolbachia*. Females become effectively sterile as a result of these matings. Local eradication of the mosquito *Culex pipiens fatigans* was achieved by this method in the 1960s before the role of *Wolbachia* in causing cytoplasmic incompatibility was understood (Laven 1967).

These and other genetic methods of pest control have not been traditionally viewed as biological control tactics, but they do constitute biological control by our definition, because the sterile males have an indirect positive effect on humans by having a negative effect (indirect if *Wolbachia* are used) on female pests (Figure 1.7). Our ecologically based definition leads to the classification of these interactions as biological control despite the fact that sterile males belong to the same species as the target females, and despite the fact that the traditional mechanisms of biological control – predation, herbivory, parasitism, induced defense and competition – are not involved.

In some extensions of these genetic control programs, the goal is to replace the pest species (or strain) with a harmless one (Dobson et al. 2002; Gould & Schliekelman 2004) and these also constitute biological control by our criteria. Here, the released agents may include females, and they are expected to mate and reproduce successfully with pest species. Control comes about through the introduction of genetic material that compromises the fitness of hybrid individuals and favors the introduced (more benign) genotype at the expense of the harmful one. This method can involve various genetic constructs and it forms the basis for the goal of driving desirable genes into pest populations using transgenesis of the organisms themselves or their symbionts (Gould & Schliekelman 2004; Xi et al. 2005). This general approach was field-tested in northeastern Australia in 2011 using *Aedes aegypti* mosquitoes that contained a *Drosophila melanogaster*–derived strain of C.I. *Wolbachia* that greatly limits the transmission of dengue fever from mosquitoes to human hosts (Hoffmann et al. 2011). The frequency of *A. aegypti* harboring this *Wolbachia* strain reached near fixation within five months of the beginning of releases, proving that strain replacement using cytoplasmic incompatibility is possible. A regional reduction of cases of dengue fever is likely as a result of these releases (Hoffmann 2014). Despite the great promise of genetic control methods, implementation presents a number of challenges. These include identification of useful genes to be driven into pest populations, stability of the transgenes, negative fitness consequences of the genetically modified strains, economic feasibility, the possibility of ecological harm and a potentially negative public perception associated with the field releases of transgenic animals (e.g., Curtis 2000; Dobson 2003; Rasgon et al. 2003; Gould & Schliekelman 2004; Irvin et al. 2004; Turelli 2010; David et al. 2013; Alphey 2014).

1.2.5 Immunocontraception

Immunocontraception is being increasingly considered as a method for controlling vertebrate pests. The method consists of deploying one or more proteins from the pest's reproductive system as an antigen to induce an immune response that disables reproduction. In Australia, immunocontraception has been called the 'holy grail of vertebrate pest control' (McCallum 1996). Beyond injecting the pest organisms directly, the antigens can be delivered by one of two methods: baits or self-replicating and spreading vectors such as viruses. We follow Barlow (2000) in considering only vectored immunocontraception as true biological control (Figure 1.8). More details on these systems are discussed in the following section on disease organisms used in vertebrate pest control.

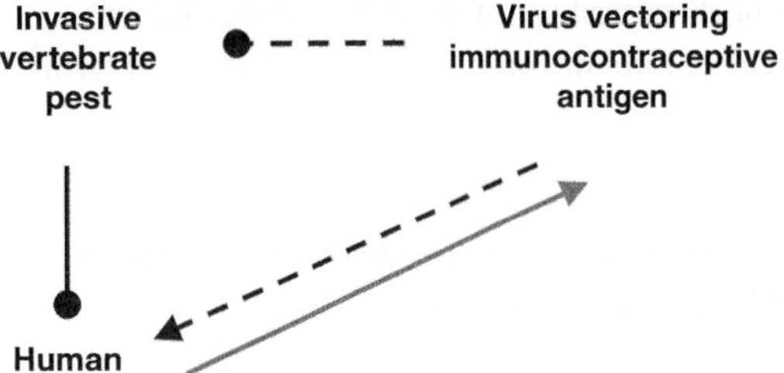

Figure 1.8. Interactions involved in the potential release of viruses that are genetically engineered to vector immunocontraceptive antigens against invasive vertebrate pests. See legend to Figure 1.2 for explanation of line types.

1.3 Categories of Biological Control

Manipulative biological control tactics have traditionally been classified as 'classical', 'augmentative' or 'conservation' biological control. These strategies can be used either singly or in combination (Newman 1998; Gurr & Wratten 1999). There has been some movement to erect the term 'systems management approach' to avoid perpetuating artificial borders between these strategies in weed biological control (Charudattan et al. 2002). For consistency, however, we retain the three terms and recognize both that multiple strategies can be used, and that these terms do not encompass all facets of biological control. In addition, we stress the importance of natural biological control, in which no human manipulation is involved. All of these forms of biological control will be discussed throughout this book, and we provide a brief introduction and definition of terms here.

1.3.1 Classical (= Importation) Biological Control (Chapters 2–5)

In classical or importation biological control, a biological control agent from the native geographic range of an exotic pest is imported and established to provide long-term control. This form of biological control has been used extensively for the suppression of invasive arthropods and weeds, occasionally for the suppression of invasive vertebrate pests, but not at all for the suppression of plant pathogens or nematodes. The biological control agent is typically collected from the pest species itself, but may also be collected from closely related species. Although the term *classical biological control* has been used to describe this practice, we prefer the term *importation biological control*, which has been used as well and is more descriptive and less jargonny – we therefore will mainly use the latter term throughout this book. Lockwood (1993) suggested that the term *neoclassical biological control* be used to describe cases in which exotic biological control agents are imported to suppress native pests. More broadly, *new associations* (or *novel associations*) *biological control* refers to use of biological control agents that do not have an evolutionary history with the pest without regard to whether the pest or biological control agent is native (Hokkanen & Pimentel 1989; Wiedenmann & Smith 1997). This includes importation biological control projects in which the biological control agent was collected from pests other than the target pest for logistic reasons.

1.3.2 Augmentative Biological Control (Chapter 8)

In *augmentative biological control*, biological control agents are mass-reared and released against pests with the aim of relatively short-term suppression. This is the primary form of biological control against plant pathogens and nematodes, and has been used extensively for the suppression of arthropod pests, but has not been developed for suppression of vertebrate pests. Augmentative control of weeds is achieved through the application of living mycoherbicides, but not yet via application of arthropods. The temporary control of weeds using management of grazing animals such as goats could be considered augmentative biological control (Wardle 1987), and mass-reared, sterilized grass-carp have been released to control aquatic weeds in parts of the United States (Hanlon et al. 2000). A critical difference between importation and augmentative biological control is that importation biological control is ideally permanent with no need to re-apply the biological control agent after it has been established. Despite this, augmentative releases of already established biological control agents that had been established through importation biological control are sometimes utilized if the agents fail to provide complete control. For example, importation biological control of the California red scale, *Aonidiella aurantii*, by the parasitoid *Aphytis melinus* is considered a success in some parts of California, but augmentation of *A. melinus* is still widely practiced in these areas (Hare & Morgan 1997).

Biological control agents used in augmentation are typically commercially mass-produced and the taxa utilized range from microbes and nematodes to arthropods. The common practice of propagation of introduced importation weed biological control agents for redistribution might be considered augmentation, but should more correctly be viewed as facilitation of the natural rate of dispersal of a colonizing biological control agent population. The terms *inundative* and *inoculative biological control* are used to differentiate between augmentative biological control releases in which either no reproduction or a few generations of reproduction is expected, respectively. Inoculative biological control programs are often used seasonally and are common in greenhouse settings, where augmentative biological control of arthropod pests has been particularly successful (Heinz et al. 2005).

1.3.3 Conservation Biological Control (Chapters 9, 10)

In *conservation biological control*, tactics are sought that will improve existing levels of biological control. These tactics include selection of pesticides that do not harm biological control agents, and habitat manipulation aimed at increasing biological control agent activity through increased retention, attraction, survival or reproduction. Conservation practices have played an important role in the biological control of arthropod pests, plant pathogens and nematodes, but have seldom been explored for weed and vertebrate suppression. The deployment of bird nesting boxes to increase the capacity for arthropod or rodent control is a form of conservation biological control (e.g., Mols & Visser 2007; Jedlicka et al. 2011; Paz et al. 2013). Many cases of conservation biological control involve the establishment of some kind of polyculture (e.g., intercropping, cover cropping, border vegetation) (Andow 1991; Landis et al. 2000). In these cases, reduced pest pressure may be achieved in polycultures by reduced colonization or retention of herbivorous pests (Root's [1973] 'resource concentration' hypothesis) and/or through the increased action of biological control agents (Root's 'enemies' hypothesis) (see Figure 1.6). Of these two mechanisms, only the enemies hypothesis has traditionally been considered to constitute conservation biological control as we have discussed previously.

1.3.4 Natural Biological Control

While conservation biological control takes advantage of resident biological control agents, it still involves management to more effectively utilize the services of these enemies. We have already pointed out that naturally occurring biological control plays an important role in pest suppression. One interesting way in which pest managers have taken advantage of natural biological control has been to modify recommendations for pesticide applications based on the naturally occurring densities of biological control agents (e.g., Hoffmann et al. 1990; Flint & Dreistadt 1998; Walker et al. 2010; Hallett et al. 2014). In some cases, the recommendation is not to spray at all if the density of a particular biological control agent is high enough (regardless of pest density), and in other cases, the threshold pest density at which spraying is recommended is higher when biological control agents are present.

1.4 Some Trends in the Science of Biological Control

Biological control is in a growth phase as a scientific discipline. Three major international scientific journals currently focus exclusively on biological control, two of which were launched in the early 1990s. The profile of biological control in the ecological sciences has been growing as well, with an increasing number of articles on biological control appearing in ecology-oriented journals (Figure 1.9b). Interest in biological control has also been growing in the field of plant pathology. The fraction of articles in the three major biological control journals (*Entomophaga/BioControl, Biological Control* and *Biocontrol Science and Technology*) devoted to biological control of plant pathogens or the use of plant pathogens to control weeds has increased to more than 20% over the past 20 years (Figure 1.9b). Other previously underrepresented areas in which interest is growing include biological control of marine pests (Lafferty & Kuris 1996; Secord 2003; Teplitski & Ritchie 2009) and biological control of mammals using disease organisms (Hoddle 1999).

How can we explain this general increase in interest in biological control? The continued

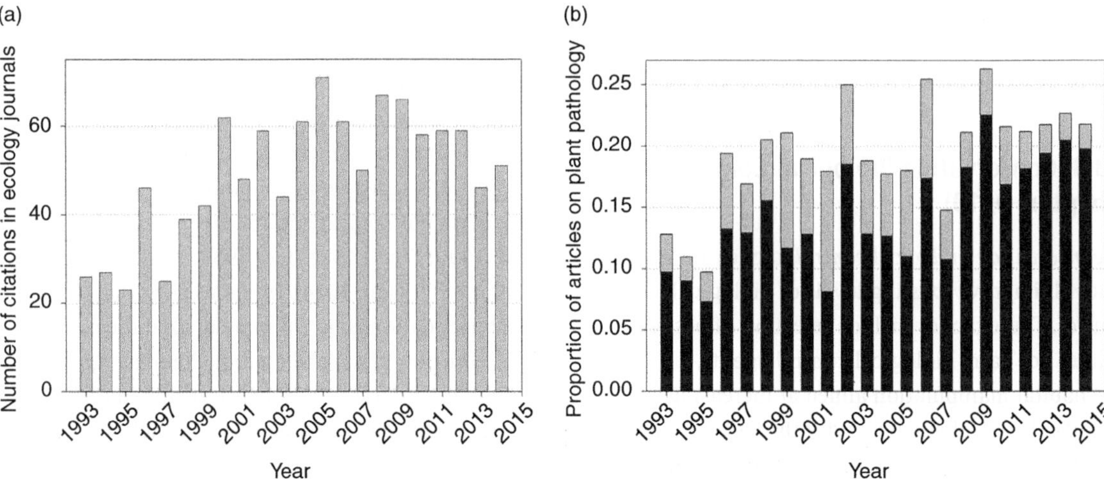

Figure 1.9. (a) The number of articles in ecology journals for which 'biological control' appears as a topic term in the Web of Science between 1993 and 2014. Journals consulted: *American Naturalist, Biological Conservation, Conservation Biology, Ecological Applications, Ecological Entomology, Ecological Modelling, Ecology, Functional Ecology, Journal of Animal Ecology, Journal of Applied Ecology, Journal of Chemical Ecology, Journal of Ecology, Molecular Ecology, Oecologia, Oikos, Researches on Population Ecology/Population Ecology* and *Trends in Ecology and Evolution.* (b) The proportion of articles published in biological control journals that focused on either biological control of plant pathogens (black portion of bars) or the biological control of weeds using plant pathogens (gray portion of bars) between 1993 and 2014. Journals consulted: *Entomophaga/BioControl, Biological Control* and *Biocontrol Science and Technology.*

relevance of pest, disease and weed problems in general, and the increase in invasive pests over recent decades in particular, clearly constitutes one class of reasons. For stemming the invasion of plants and animals into pristine or natural areas, biological control is often seen as the only viable option, and there has therefore been an increase in biological control of invasive species of conservation concern (Van Driesche et al. 2010). Also, while some effort has been made to develop chemical pest control methods that pose less risk to humans and the environment (this topic is covered in Chapter 10), highly toxic pesticides have remained the rule rather than the exception, at least into the early 21st century (Johnston 2001). The promise of reducing pesticide use remains one of the main incentives for studying and developing biological control.

The importance of invasive species in affecting not only human health and economies, but also natural ecosystems, has grown over the past decade as well, so that invasion biology is now recognized as a subdiscipline within ecology. Since biological control is one of the main strategies to combat invasive species, it has figured prominently in the field of invasion biology, as we discuss in the next chapter. Biological control introductions also constitute biological invasions themselves, and share some of the risks associated with unintentional biological invasions. And thus, some of the increased attention garnered by biological control is related to *negative* impacts of biological control introductions, and how to reduce them (Follett & Duan 2000; Wajnberg et al. 2001; Bigler et al. 2006). This topic is developed more fully in Chapters 4 and 5.

New technologies have also played a role in increasing interest in biological control over the past decade. Advances in molecular genetics have greatly increased our ability to distinguish between strains and cryptic species of biological control agents. This has been a boon particularly for the biological control of plant pathogens, where pest and beneficial organisms may belong to the same species, and where advances in molecular biology have aided elucidation of the mechanisms of systemic acquired resistance (Durrant & Dong 2004) among other processes. There is also considerable interest in developing transgenic biological control agents (Ashburner et al. 1998; Lacey et al. 2001; Robinson et al. 2004; Snow et al. 2005; St. Leger & Wang 2010), including the genetic control techniques discussed earlier in Section 1.2.4. Transgenic biological control organisms that have been released in the field include predatory mites and entomopathogenic nematodes with marker genes added for experimental purposes only, and baculoviruses engineered to express insect-specific neurotoxins (Harrison & Bonning 2000; Hoy 2000a). A variant of this approach is to use genetic engineering to transform nonpathogenic organisms into pathogenic ones, as has been done with the movement of *Bacillus thuringiensis* (*Bt*) toxin genes into otherwise non-toxic endophytic bacteria that colonize crop plants (e.g. Tomasino et al. 1995). Other examples of this general strategy are discussed in Chapters 5 and 8. While genetic engineering of biological control agents is an active research area, the widespread use of such agents may in the end be determined as much by issues related to economics, public acceptability and regulatory policy as by scientific discoveries.

The development of transgenic plants, particularly those modified to be resistant to pest insects, has also spurred research in biological control. This may seem paradoxical because these crops are designed to decrease pest problems and could therefore reduce the need for biological control. However, the ecological implications of the transition from traditional pesticide applications to genetically modified crops for biological control need to be understood. Thus, the advent of transgenic insecticidal crops has spurred research on effects of the transgenes on the abundance, diversity and fitness of biological control agents (Andow & Hilbeck 2004; Sisterson & Tabashnik 2005; Wolfenbarger et al. 2008; Lovei et al. 2009; Lundgren et al. 2009; Duan et al. 2010), as well as effects that biological control agents will have on the evolution of pest resistance to transgenic crops (Gould 1994; Heimpel et al. 2005; Onstad et al. 2013; Liu et al. 2014). In addition, decreases in insecticide use that accompany transgenic crops should lead to the opportunity to implement biological control against pests that are not targeted by the insecticidal transgene but that were previously controlled by broad-spectrum insecticides (Sisterson et al. 2007; Lu et al. 2012).

Another growth area in both ecology and biological control is the study of indirect interactions (e.g., Wajnberg et al. 2001; Persello-Cartieux et al. 2003; Harmon & Andow 2004; Schmitz et al. 2004; Brodeur & Boivin 2006), which is a theme throughout this book, but will receive particular attention in Chapters 2, 3, 4, 5 and 9. The successful biological control of herbivores, in which a biological control agent increases the fitness of the herbivore's host plant by reducing herbivore pressure, is an example of a trophic cascade. More complicated and interesting interactions are possible, however, and some of these are just now beginning to be studied and documented in field settings (Schmitz et al. 2004). Systemic acquired resistance is one of the most striking examples of indirect interactions in biological control. Here, the presence of one plant pathogen or herbivore can induce resistance to other species of plant pathogens or herbivores feeding on the same plant. This can lead to very effective biological control if the inducer is a non-pest and the species subject to the resistance is a pest (Harman et al. 2004). It has become increasingly clear that complex indirect interactions of this and many other kinds are widespread in biological control, a fact that should come as no surprise since many biological control interactions occur within a broad ecological context.

There has also been a tentative increase in the cross-fertilization between biological control and evolutionary biology over the past decade or so. Until relatively recently, the only robust investigation of evolutionary dynamics occurring within the realm of biological control involved interactions between invasive rabbits and myxoma viruses (Fenner & Fantini 1999; see Chapter 7). There is, however, great scope for evolutionary inquiries in many other areas of biological control (Ehler et al. 2004; Hufbauer & Roderick 2005; Roderick & Navajas 2008; Roderick et al. 2012). For example, biological control agents have often been presumed to be impervious to resistance evolution in the pests that they attack due to their presumed ability to counter resistance through coevolutionary mechanisms. This assumption, however, has begun to be questioned, on both theoretical and empirical grounds (e.g., Holt & Hochberg 1997; Hufbauer & Via 1999). The evolutionary dynamics of invading species has great relevance for biological control as well (Roderick 1992; Roderick & Navajas 2003, 2008), both because invasive pests may evolve responses to resistance from native biota, and because biological control agents themselves may undergo evolutionary shifts of various kinds (Ehler et al. 2004).

And last, the discoveries beginning in the 1990s that many biological control agents and their associated pests are involved in complex interactions with endosymbiotic microorganisms that manipulate sex ratio and reproduction has spurred research into the implications of these dynamics for biological control (e.g., Stouthamer 1993, 2004; Hsiao 1996; Majerus & Hurst 1997; Hunter 1999; Silva et al. 2000; Mochiah et al. 2002; Oliver et al. 2008). The discovery of defensive symbionts in arthropod pests as well as in crop plants is particularly important as it can help us explain the mechanisms by which resistance to biological control agents can occur and how this form of resistance can evolve (Hsiao 1996; Oliver et al. 2014; Panaccione et al. 2014).

1.5 Biological Control Agents

Unlike some other authors of books on biological control (DeBach 1974; DeBach & Rosen 1991; Van Driesche & Bellows 1996; Hajek 2004; Van Driesche et al. 2008), we do not provide a review of the biology of biological control agents. We feel that this material is well covered in these texts and also in more specialized sources cited later. In the following paragraphs, however, we provide a very brief summary of the organisms that figure prominently in biological control.

1.5.1 Predators

Predators that act as biological control agents include predatory arthropods (Hagen et al. 1999; Symondson et al. 2002), insectivorous and carnivorous birds (Mols & Visser 2007; Paz et al. 2013), fish that consume mosquito larvae and other aquatic pests (Garcia and Legner 1999), vertebrate predators of vertebrate pests (Hickling 2000) and predatory snails (Port et al. 2000; Rondelaud et al. 2006). Predatory nematodes may also be important biological control agents of plant-parasitic nematodes and other soil microfauna (Yeates & Wardle 1996). The presence of predators can have a strong suppressive effect on pest populations, and most or all of this suppression is presumed to be due to direct consumption. However, predators can also induce antipredator behavior in prey, which can similarly lead to decreased pest activity or abundance (Schmitz et al. 2004). For example, predators may induce escape behavior in their prey that reduces their local density and may lead to an increased risk of mortality from other causes (e.g., Nakasuji et al. 1973; Losey & Denno 1998; Nelson et al. 2004; Kaplan & Thaler 2010).

Not only strict predators engage in predation. For example, many arthropod biological control agents are omnivorous, consuming both herbivorous prey and plant material (Lundgren 2009). Plant consumption by these species can reduce their effectiveness as biological control agents (by reducing the predation rate or causing direct plant damage), or it can increase their effectiveness as biological control agents (by sustaining them during times of low prey availability) (Coll & Guershon 2002). A second group of biological control agents that engage in predation despite not being strict predators are parasitoids, which are discussed later. Adult females of many parasitoid species prey upon the same host species (or even the same individuals) that they parasitize. This behavior is known as 'host-feeding', and it can rival parasitism as a source of herbivore mortality, while supplying parasitoids with nutrients that support additional parasitism (Jervis & Kidd 1986; Heimpel & Collier 1996).

Predation can also interfere with biological control, as occurs when predators feed on biological control agents (Goeden & Louda 1976; Rosenheim 1998). Many predator species consume pests as well as other predator species, an interaction known as intraguild predation if both predators consume a shared prey (pest) species (Rosenheim et al. 1995). The net effect of intraguild predators on biological control of herbivorous pests depends on factors such as preference for and encounter rates with the herbivorous versus predatory–prey (Rosenheim & Corbett 2003) and the presence of additional links in the food web (Janssen et al. 2006; Rosenheim & Harmon 2006).

1.5.2 Herbivores

Herbivorous biological control agents include arthropods for the control of terrestrial weeds, and arthropods and fish for freshwater aquatic weeds. In addition, sea slugs have been contemplated for the biological control of algae in marine environments (Coquillard et al. 2000). While victims of predation are almost invariably killed, plants and algae that herbivores feed on often survive the encounter, and in some cases may even benefit from herbivory through compensatory growth or induced resistance (Whitham et al. 1991; Karban & Baldwin 2000). Myers and colleagues (1989) have suggested that compensatory growth and/or reproduction in response to herbivorous biological control agents may be especially common in invasive weeds that are limited by intraspecific competition (see also Ortega et al. 2012). On the other hand, herbivory does not necessarily need to kill plants for biological control to be successful. Weeds that are reduced in vigor by herbivory are often weaker competitors and may suffer disproportionately large effects from herbivory when growing in competition with other plants (e.g., McEvoy & Coombs 1999).

1.5.3 Parasitoids

All parasitoids are holometabolous insects, but beyond this they are bound together as a group by a unique suite of life history characteristics rather than by common ancestry. Adult females are free-living, and deposit one or more eggs in, on or near host individuals. The parasitoid larvae then develop on and consume the host organism, almost invariably killing it in the process (Godfray 1994; Quicke 1997). The parasitoid life history combines aspects of biology seen in some predatory, herbivorous and parasitic insects. Like many predators and herbivores, adult female parasitoids seek out resources that will be consumed. Unlike most predators, however, only the immature stages of most parasitoid species consume pest organisms (the exceptions are host-feeding species as noted earlier). Also unlike predators, at most a single host individual is consumed by an individual parasitoid larva, and parasitoid larvae are sessile during the feeding stage. And unlike ectoparasites such as ticks and lice, parasitoids almost invariably kill their host.

Most insect herbivore species are attacked by one or more parasitoid species, but some major arthropod pest groups appear to be entirely free of parasitoid attack. For instance, we know of no parasitoids that attack mosquitoes or other biting flies with aquatic larval stages, and no parasitoids that attack phytophagous mites or adelgids. Many social insects appear to be free of parasitoids as well – we know of no parasitoids of termites, and only a few parasitoid taxa that attack ants. There are some cases of parasitoid attacks on non-arthropods, and one of these is of relevance to biological control. Some parasitoid flies in the family Sciomyzidae attack aquatic pest snails, including species that transmit schistosomiasis to humans (Garcia and Legner 1999).

As is the case for predators, not all parasitoid species are beneficial biological control agents. Predators, herbivores and other parasitoids providing biological control can themselves be attacked by parasitoids, potentially compromising the suppression of pest populations (Goeden & Louda 1976; Rosenheim 1998). Hyperparasitoids (also known as 'secondary' parasitoids) are parasitoids that attack other, 'primary' parasitoid species, and there is some anecdotal (but little experimental) evidence that hyperparasitoids can interfere with biological control programs by attacking otherwise effective parasitoids of pests. Some hyperparasitoids are capable of attacking both the primary parasitoids and the host(s) of these primary parasitoids and thus act in ways functionally identical to intraguild predation. The effect of these facultative hyperparasitoids on biological control is variable, and likely depends on factors such as the attack rate of the facultative hyperparasitoid on the pest versus the primary parasitoid (Rosenheim et al. 1995; Pedersen & Mills 2004). In a subset of hyperparasitoids known as heteronomous hyperparasitoids, female larvae develop on pest hosts, while males develop on other parasitoids including conspecific females (Hunter & Woolley 2001). When heteronomous hyperparasitism is confined to conspecific females (autoparasitism), it can provide effective control, but when directed toward other parasitoid species, it is more likely disruptive (Mills & Gutierrez 1996).

1.5.4 Entomopathogens

Most pathogens of insects reproduce within their host by causing breakdown of its tissues and incorporating the breakdown products into additional pathogen biomass. This general form of consumption is common to viruses, bacteria, fungi, protozoans and nematodes. Many entomopathogenic bacteria and fungi used in biological control release toxins into the host to facilitate the breakdown process (Tanada and Kaya 1993). In the case of entomopathogenic nematodes in the families Steinernematidae and Heterorhabditidae (the families containing the most important biological control agents), mutualistic bacteria are the agents of host decomposition and the nematodes use the breakdown products and accumulated bacterial biomass for growth and reproduction.

The relatively short span of time separating infection and death that these toxins and mutualistic bacteria achieve has facilitated the use of some entomopathogens as augmentative biological control agents in ways that approximate insecticide use. Along these lines, attempts have been made to incorporate toxin-producing genes into the genome of baculoviruses for the express reason of hastening the time-to-death of their insect hosts (Bonning et al. 2002). Sub-lethal levels of host consumption that negatively impact various fitness components are caused by a number of entomopathogenic viruses, fungi, protozoans and nematodes. While these agents cannot be used to achieve a rapid host kill, they can be very effective at decreasing average or equilibrium population densities below pest levels (e.g., Lacey et al. 2001).

1.5.5 Plant Pathogens Used in Weed Control

Plant-parasitic fungi and to a lesser extent bacteria and nematodes are used as biological control agents of weeds. Many of these agents are formulated as sprays (bioherbicides) and used as inundative or inoculative agents, but there are also cases of successful importation biological control of weeds using rust fungi (Charudattan et al. 2002). As we have noted earlier, infection by pathogens can activate resistance responses within plants – both to the infection itself and to other pathogens and herbivores. Infection by plant pathogens can put weeds at a competitive disadvantage with respect to other plants so that biological control can be particularly effective when weeds are subject to strong interspecific competition (Cousens & Croft 2000; Abu-Dieyeh & Watson 2007).

1.5.6 Plant Pathogen Antagonists

Antagonists used in biological control of plant pathogens include bacteria, fungi, yeasts and viruses. While members of all of these groups are active in natural biological control of plant pathogens, their commercial development as products for biological control augmentation has been confined to bacteria and fungi, with one example of a virus product (Whipps & McQuilken 2009). As already discussed (Figure 1.5), agents can act by any (or a combination) of four mechanisms: (i) direct consumption such as actinomycete parasitism of fungal spores (El-Tarabily et al. 1997; Bolwerk et al. 2003; Steyaert et al. 2003), (ii) competition between fungal or bacterial pathogens and nonpathogenic fungi or bacteria for colonization sites and/or nutrients (Janisiewicz et al. 2000; Lindow & Leveau 2002; Olivain et al. 2006), (iii) antibiosis or the production of antibiotic compounds primarily by bacteria (Raaijmakers et al. 2002) or (iv) systemic induced or acquired resistance to other plant pathogens that can result from the action of all types of antagonists (Kloepper et al. 1992; Bakker et al. 2003). The application of molecular tools has contributed to an enhanced understanding of the modes of action of antagonists and the importance of environmental conditions and plant genotype in mediating the action of antagonists. New developments in the use of antagonists have focused on formulation, mixtures of biological control agents and integration with cultural and chemical control methods (Whipps 2001; Whipps & McQuilken 2009).

1.5.7 Diseases of Vertebrates

Disease organisms used or contemplated as biological control agents of vertebrate animals include viruses, bacteria, protozoans and nematodes (Dobson 1988; Fenner & Fantini 1999; Hoddle 1999). Viruses have played the most important role and examples include the myxoma virus and rabbit hemorrhagic disease against rabbits and feline parvo virus against cats (Di Giallonardo & Holmes 2015b). The case of the myxoma virus is one of the most important in the annals of biological control (Fenner & Fantini 1999). This program has been viewed largely as a success, leading to sizable reductions in rabbit populations in Australia and Chile (Jaksić & Yáñez 1983; Fenner & Fantini 1999; Saunders et al. 2010), despite the evolution of resistance to myxoma in Australia. The interaction between resistance by rabbits and virulence by the virus is a textbook example of the evolution of intermediate-level virulence in disease organisms, as we discuss in Chapter 7.

The most advanced form of biological vertebrate control is virus-vectored immunocontraception (VVIC), in which species-specific viruses are genetically engineered to carry a gene encoding for antigenic proteins associated with either host eggs or sperm. The infection of the target pest with this protein is designed to trigger an autoimmune response against its own gametes, blocking fertilization (Seamark 2001; Courchamp et al. 2003). This technique has been tested for use against rodents and other small mammals, but we are not aware of any releases of VVIC agents (Massei & Cowan 2014). In Australia, the primary targets are the house mouse, European rabbits and the European red fox, and viruses that have been studied as vectors for these three pests include murine cytomegalovirus, myxoma virus and canine herpes virus, respectively (Seamark 2001). A number of modeling studies have shown that VVIC has great potential in vertebrate pest control when used judiciously (e.g., Barlow 1994; Courchamp & Cornell 2000; Hood et al. 2000). Disadvantages that have been identified include irreversibility, development of host resistance, slow response and the need for engineering genetically modified vectors (Barlow 2000; Courchamp & Cornell 2000). One of the arguments put forth in favor of VVIC agents is that they can be designed to result in less animal suffering than alternative control measures such as trapping, shooting, poisoning or the dissemination of pathogens, all of which are aimed at reducing existing populations rather than reducing birth rates. However, as with genetically modified biological control organisms of arthropods discussed previously, the use in the field of these agents, once developed, may rest as much on economics and public acceptance as on scientific advances.

1.6 Conclusions

Biological control encompasses naturally occurring and human-aided decreases in the abundance or damage caused by pest organisms (including weeds) that is attributable to living organisms or viruses. Mechanisms by which biological control can function include but are not limited to predation, herbivory, parasitism, disease, competition, vectoring of disease agents, toxicity and sterility induction. Biological control is not restricted to interspecific interactions so that strategies involving inter-strain antagonism or replacement are included. Important biological control agents can be found within various animal groups as well as within the fungi, bacteria and viruses. Plants can be biological control agents via mechanisms of competition or by serving as a foil for pests of associated desirable plants. We have developed a definition of biological control based on ecological relationships. Biological control requires that an organism or virus have a negative effect on humans in some way (i.e., acts as a pest), and that an enemy (again – a living organism or virus) reduces this negative effect, resulting in an indirect positive effect on humans.

Biological control exists at the confluence of the fundamental natural sciences and applied sciences, and researchers have come to realize that the interactions that make up biological control are similar to ones that structure biological communities and reflect (as well as initiate) evolutionary processes, but also that biological control interactions can differ from natural interactions in important ways (Hawkins et al. 1999). There is therefore potential for a rich and reciprocal interplay between efforts to improve the efficacy of biological control and efforts to elucidate fundamental principles of ecology and evolution.

It has been remarked in the past that success in biological control is due in part to art and in part to science (Sheppard & Raghu 2005). While it is certainly true that success in collecting and rearing biological control agents can have an elusive quality in which experience and intuition figure prominently, predictability and firm scientific standing has always been the ultimate goal in biological control. In the chapters ahead, we attempt to address the major scientific issues and hypotheses that underlie biological control interactions. We employ an ecological perspective because of the obvious links and overlap between biological control and the ecological sciences.

I

Biological Control as Intentional Invasions

2 Biological Control and Invasion Biology

Invasion biology was launched as a scientific discipline with Charles Elton's 1958 classic, *The Ecology of Invasions by Animals and Plants*. In this book, Elton outlined in vivid detail the extent to which human activities have aided animal and plant invasions and put forth a research agenda for the study of invasive species. Prominent in this agenda was a quest to understand what ecological factors facilitate and impede biological invasions at various stages of the invasion process. Elton recognized the importance that importation biological control could play both in illustrating some of the principles of invasion biology and in reducing the negative impact of biological invasions. He called biological control agents 'counterpests', and noted that 'counterpests have done splendid work in ameliorating disastrous situations', but also that 'many [counterpests] become residents of the new country without necessarily producing control of the pest.' He went on to discuss some Hawaiian biological control projects for which he suspected that the agents were doing more harm than good. Thus, Elton recognized both the benefits and the risks of importation biological control in the context of invasion biology.

Much has been written about invasion biology since Charles Elton brought it to the attention of ecologists and the general public, and invasive species are now seen as one of the most important threats to natural and managed ecosystems in many countries (e.g., Williamson 1996; Shigesada & Kawasaki 1997; Mack et al. 2000; Pimentel et al. 2000; Sakai et al. 2001; NRC 2002; Ruiz & Carlton 2003; Davis 2009). We do not attempt to cover all aspects of invasion biology in this chapter, but rather we focus on areas of clear overlap between biological invasions and biological control. We begin with a brief discussion of the intersections of biological control and invasion biology as scientific disciplines (Section 2.1), and this is followed by treatments of how natural biological control can impede invasions via 'biotic resistance' (Section 2.2) and how the absence of biological control can facilitate invasions ('enemy release'; Section 2.3). A scenario of coupled biotic resistance and enemy release dynamics is discussed in Section 2.4, and we end this chapter with a discussion of ways that conceptual and theoretical developments in invasion biology can improve success in biological control (Section 2.5).

2.1 The Overlap between Invasion Biology and Biological Control

The disciplines of biological control and invasion biology are interconnected in numerous ways. The presence of naturally occurring biological control can interfere with invasions ('biotic resistance'), or its absence can facilitate invasions ('enemy release'). Biotic resistance can take the form of consumption of or competition with invaders. In either case, it constitutes natural biological control when the introduction, spread or impact of pests or potential pests is attenuated. Understanding the relative importance of biotic resistance and other factors that can limit biological invasions remains an important goal in invasion biology (Maron & Vilá 2001; Levine et al. 2004; Parker et al. 2006; Kimbro et al. 2013), and we review the evidence for biotic resistance against terrestrial plants and arthropods later in this chapter (Section 2.2.3). A counterpoint to biotic resistance is enemy release – the idea that exotic species are at an advantage in their introduced range because they are separated from enemies in their native range (Keane &

Crawley 2002). This 'enemy release hypothesis' is a foundational concept in biological control since it provides the theoretical justification for importation biological control.

The biotic resistance hypothesis is an explanation for how invasions fail, while the enemy release hypothesis is an explanation for how invasions succeed, and the two hypotheses are in essence mutually exclusive, although they can act differently on different stages of an invasion (Liu et al. 2007). Understanding the relative importance of biotic resistance and enemy release is an important part of the invasion biology agenda, and we discuss the evidence for and against both hypotheses later in this chapter (Sections 2.2 and 2.3). The relative importance of these two hypotheses also has important implications for biological control, as we illustrate in Figure 2.1. Strong biotic resistance implies strong natural biological control while strong enemy release suggests an opportunity for successful importation biological control. Weak biotic resistance and unimportant enemy release is certainly also possible, and this scenario implies that invasiveness is determined by factors other than enemy release. This scenario is less promising for the effectiveness of importation biological control (Figure 2.1).

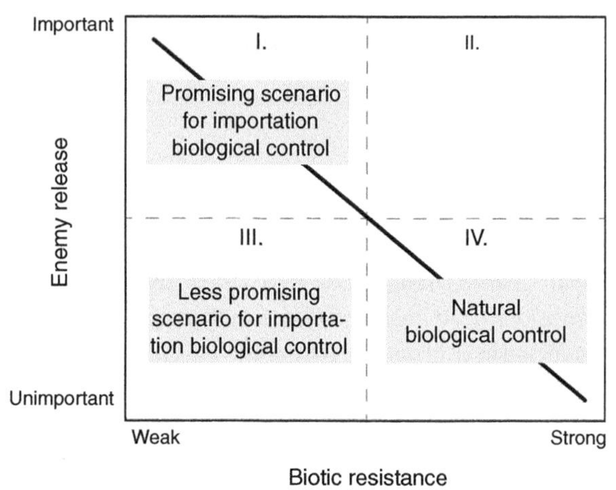

Figure 2.1. The relationship between biotic resistance and enemy release in determining the outcome of biological introductions. The diagonal line shows the effect of biotic resistance on the importance of enemy release in explaining the outcome of introductions. The quadrants in the figure highlight biological control scenarios. In quadrant I, biotic resistance is weak and enemy release is important, which is a promising scenario for importation biological control (see Section 2.3). Quadrant II (strong biotic resistance with important enemy release) is considered an unlikely scenario. In quadrant III, biotic resistance is weak but enemy release unimportant, suggesting that invasiveness is due to factors other than enemy release. This scenario is not promising for importation biological control since natural enemies may not be important in the native range (see Section 2.3, Chapter 3). In quadrant IV, biotic resistance is strong and enemy release is unimportant, a signature of natural biological control.

Biotic resistance and enemy release are not the only links between invasion biology and biological control. As Elton (1958) and many others have noted, manipulative forms of biological control can greatly diminish the negative ecological and economic impacts of invasive species. This is the very impetus of biological control as a management strategy, and the patterns underlying successful importation, augmentative and conservation biological control are explored in Chapters 3, 8, 9 and 10. Biological control organisms can also 'escape' from their intended function and become damaging invasive species themselves, as we discuss in Chapter 4. And since importation biological control projects are themselves intentional invasions that are typically more controlled than accidental invasions are, they can serve as 'model systems' to address critical questions in invasion biology (e.g., Strong et al. 1984; Simberloff 1986; Ehler 1998; Hawkins et al. 1999; Marsico et al. 2010; Fauvergue et al. 2012; Di Giallonardo & Homes 2015a, b). Conversely, insights gained through the study of invasive species more generally can guide practices in biological control (Fagan et al. 2002), a topic that we turn to at the end of this chapter (Section 2.5).

2.2 Biotic Resistance and Biological Control

Biotic resistance is a unifying theme in biological control because it can constitute naturally occurring biological control (when it constrains the establishment, spread or population growth of potential or actual pest organisms) or it can interfere with importation biological control (when biological control agents themselves are impeded). As we noted previously, biotic resistance is a hypothesis explaining how invasions fail. However, many readers will be aware that it is but one of many hypotheses, many of them non-exclusive. Other hypotheses explaining the lack of establishment, spread or severity of invasive species include low propagule pressure (Simberloff 2009), inappropriate traits in the invaders (Levine et al. 2003; Ortega & Pearson 2005) and inappropriate abiotic conditions such as climate in the recipient habitat (Venette et al. 2010). The primacy of biotic resistance among these hypotheses is by no means established, and it is not our intention to champion this hypothesis over others. Rather, we focus on biotic resistance because of its clear relationship to biological control. In the following paragraphs, we discuss general patterns for the failures of invasions (Sections 2.2.1 and 2.2.2) and then focus on evidence for (and against) biotic resistance against terrestrial plants and arthropods (Section 2.2.3).

2.2.1 The 'Tens Rule' and the 'Threes Rule'

A surprisingly robust trend in invasion biology is the so-called tens rule, which posits that each of the three transitions that characterize the invasion process – importation to introduction, introduction to establishment and establishment to pest status, occur with a frequency of approximately 1 in 10 (Williamson & Fitter 1996). Here, *imported* is defined as being brought into a new geographical area, *introduced* is defined as being found in the wild in the new area, *established* is defined as having a self-sustaining population in the area and *pest* is defined as an organism that has negative environmental and/or economic effects (Table 2.1). Each of these probabilities is in reality viewed as falling between 5% and 20% (the approximate

Table 2.1. **Elements of the tens rule for the success of biological invasions. See text for definition of terms. Adapted from Williamson (1996).**

Status of Potential Invader	Process	Probability of Process
Imported		
↓	Escaping	0.05–0.20
Introduced		
↓	Establishing	0.05–0.20
Established		
↓	Becoming a pest	0.05–0.20
Pest		
Imported → Pest		0.000125–0.008

bounds of a 95% confidence interval around 10% assuming a binomial distribution) (Williamson 1996). Ehler (1998) considered the data supporting the tens rule to be incomplete and suggested that the 'rule' should be viewed as a hypothesis. We agree with this sentiment, but remain impressed that any trend is discernible at all, given the myriad of factors that must influence the success of invasions. In addition, at least seven data sets supporting the tens rule have been published since Ehler's comments: plants in California, Florida and Tennessee (Lockwood et al. 2001), planarians in the United Kingdom (Boag & Yeates 2001), macrophytes in the Mediterranean Sea (Boudouresque & Verlaque 2002) and birds and vascular land plants in New Zealand (Vietch & Clout 2001).

Of course, exceptions to the tens rule have been documented, and biological control is one of these (Williamson & Fitter 1996; Courchamp et al. 2003). Not surprising, the rate of successful establishment of biological control agents, as well as the rate of becoming a 'pest' (equivalent to having an impact on the target pest species in this context) is higher for biological control than for organisms that are not intentionally introduced by humans. These rates are approximately three times as high for biological control organisms as they are for organisms that conform to the tens rule. Williamson & Fitter (1996) therefore referred to biological control projects loosely following a 'threes' rule (we discuss patterns underlying the success rate of biological control in Chapter 3).

For both accidental and intentional introductions, one of the key questions is: What prevents the majority of invaders from establishing, and the majority of established invaders from becoming pests? Clearly, climate and other abiotic factors play a large role, as can the absence of critical resources or enabling organisms such as hosts or pollinators. In many cases, however, introduced organisms fail to establish or to attain pest status in seemingly appropriate habitats with ample resources (NRC 2002). Biotic resistance, that is, interference via predation, parasitism, disease and/or competition, could conceivably account for the balance of the reduction in establishment and impact of invaders that are observed in the tens and threes rules.

2.2.2 Geography and Biotic Resistance

In this section we discuss two potential geographic determinants of biotic resistance – latitude and island versus mainland status.

2.2.2.1 Latitude

The general observation of highest biodiversity in the tropics holds true for native but not introduced species. Instead, the diversity of many groups of introduced species declines rather than increases from the mid-latitudes to the Equator (Figure 2.2, and see Rejmanek 1996; Lonsdale 1999; Sax 2001; Fine 2002; Pysek & Richardson 2006). One of the hypotheses to explain this trend is that biotic resistance is stronger in the tropics than in temperate zones due to higher native biodiversity (Freestone et al. 2013). Indeed, a number of studies have found that predation and herbivory are stronger in tropical than in temperate regions (Schemske et al.

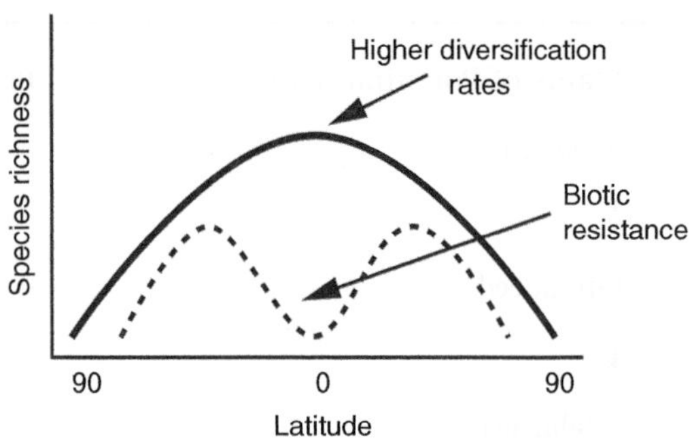

Figure 2.2. Diagram illustrating richness of native species (solid) and introduced species (dashed line) across a latitudinal gradient. Biotic resistance is one hypothesis for reduced species richness of introduced species in the tropics. From Freestone and colleagues (2013).

2009). Of particular relevance to biological control interactions, Hawkins and colleagues (1997) used a meta-analysis of life-table studies to show that predation on insect herbivore eggs and larvae, as well as parasitism of herbivore eggs, was higher in tropical than in temperate latitudes (another study, however, found no such effect; Stireman et al. 2005).

The dual observations of lower diversity of introduced species in the tropics coupled with higher predation in the topics are suggestive of stronger biotic resistance in the tropics. And in further support of this view, biotic resistance against biological control agents was deemed higher in the tropics (29%) than in temperate regions (15%) in a review of perceptions of biological control interactions (Stiling 1993). However, other hypotheses have been posed, and these include biases in the climatic tolerances and location of establishment of introduced species, and aspects of recent global climate change (Guo et al. 2012). And important data sets refute the trend. In particular, an assessment of 24 nature reserves throughout the world did not find evidence for higher invasibility at higher latitudes for continental sites (Usher 1988). All of these sites (excepting Antarctica) were invaded and the main determinant of invasion was the frequency of human visitation. Thus, propagule pressure may be a main determinant of invasion into relatively pristine areas (see also Simberloff 2009).

To summarize – there is some evidence for declining diversity of introduced species toward the tropics, and some evidence that biotic resistance can be involved in causing such a trend. However, neither the latitudinal trend in diversity nor biotic resistance as an explanation is universal by any means.

2.2.2.2 Islands versus Continents

Islands are notably more vulnerable to invasions than mainland habitats are (D'Antonio & Dudley 1995; MacDonald & Cooper 1995), and one of the potential explanations is weaker biotic resistance on islands than on mainlands (MacArthur & Wilson 1967; Sailer 1978). The biological control literature in general was seen as giving support to this hypothesis. Early commentators noted successes on islands and suspected that success in biological control was especially likely on islands in part because of reduced biotic resistance. For example, Imms (1937, pg. 393) wrote that, on islands,

> *Introduced parasites meet with relatively little competition from indigenous forms, and the parasitic element is but poorly developed.*

Similarly, Taylor (1955) stated that

> *The principal reason suggested for the greater success of introduced natural enemies in island conditions is that the relatively small fauna contains few or no species likely to attack any beneficial insects that may be introduced.*

Is this true? Few data are available to critically address this hypothesis, but in the first explicit comparison of islands and continents as sites for importation biological control, Hall and Ehler (1979) documented a significantly higher establishment rate of biological control on islands than on continents (40% versus 30%). While this supported the general hypothesis of greater biotic resistance on mainlands, the mechanism(s) explaining the difference were not addressed. Also, a subsequent analysis by Greathead and Greathead (1992) could not corroborate Hall and Ehler's finding. Furthermore, the *success* rates of biological control on islands and continents turned out not to differ significantly between islands and mainlands (DeBach 1965; Huffaker et al. 1976; Hall et al. 1980). This argues against decreased biotic resistance on islands (at least in the context of biological control introductions), since biotic resistance should affect both establishment and success of biological control agents.

Simberloff (1986) was interested in the phenomenon of biotic resistance in general, and also attempted to use the database on biological control introductions to determine whether biotic resistance was stronger on mainlands than on islands. To minimize phylogenetic bias, he examined patterns of establishment of six genera of biological control agents, of which various species had been moved to both islands and continents, and compared establishment records between the two source areas. The study included the coccinellid beetle genus, *Scymnus*, and five parasitoid wasp genera: *Apanteles, Aphytis, Bracon, Coccophagus* and *Opius*. He reasoned that if biotic resistance was weaker on islands, and mainland species were stronger competitors, establishment rates should be highest for movement of mainland species to islands, and lowest for movement of island species to mainlands. For the 71 cases of movement from mainland to islands involving these 6 genera (with Australia counted as a continent), 35 (49%) resulted in establishment, while for 15 cases of movement from island to mainland, only 5 (33%) resulted in establishment. While this fits the predicted pattern, the difference was not statistically significant. An analysis by Stiling (1993) was entirely inconsistent with higher biotic resistance on continents though: a higher fraction of biological control failures was attributed to native fauna on islands than continents, although this conclusion was from a rather small data set that was uncorrected for potentially confounding factors.

Taken together, these data sets from the biological control literature suggest that biotic resistance is not lower on islands *despite* lower species diversity on islands. Simberloff (1995) has made the point that biological control introductions may not offer a sensitive test case for biotic resistance because the agents are typically released into agricultural settings that may have unnaturally low levels of competition and predation. Differences in invasibility of the two habitats will obviously be harder to detect if disturbed habitats are more similar between islands and continents than pristine areas are.

2.2.3 Biotic Resistance as Biological Control

As we have noted, biotic resistance constitutes biological control when it impedes the introduction, establishment, spread or impact of pest or potential pest species. The fact that approximately 90% of imported species do not become introduced, 90% of introduced species do not become established and 90% of established species do not become pests (Table 2.1) means that there is a great potential for biotic resistance to limit invasions or their effects. It is not our goal here to determine what proportion of these failures are due to biotic resistance, but in the following paragraphs we review the evidence for biotic resistance against terrestrial or freshwater aquatic plants (Section 2.2.3.1) and terrestrial arthropods (Section 2.2.3.2). Readers interested in biotic resistance against marine invaders are referred to Kimbro and colleagues (2013).

2.2.3.1 Biotic Resistance against Invasive Plants

It is often assumed that interspecific competition can prevent or limit plant invasions by reducing resources available for invaders (Tilman 2004). Patterns consistent with this view can be found in studies showing higher rates of establishment and performance of introduced plants in ephemeral rather than stable habitats, and in habitats with less plant cover or lower plant diversity (e.g., Crawley 1986, Tilman 1997; Hector et al. 2001; Kennedy et al. 2002; Levine et al. 2004; Maron & Marler 2008). Two interesting caveats accompany these generalizations, however. First, there are a number of documented cases in which more diverse systems offer *less*, rather than more biotic resistance than do less diverse systems (Levine & D'Antonio 1999; Stohlgren et al. 2003; Fridley et al. 2007). This may come about if higher plant diversity generates more ecological niches (as is likely to occur if diversity is measured over a large geographic scale) or if higher plant diversity increases the chance that 'facilitator species' (such as nitrogen-fixing plants) will be present that aid the invasion process (Lonsdale 1999; Richardson et al. 2000).

Second, while disturbed sites may offer less biotic resistance via interspecific competition than do undisturbed sites, plants colonizing disturbed sites may be subject to more intense biotic resistance via herbivory. For example, seed predation by native grouse and quail experienced by scotch broom, an invasive shrub in northern California, was more intense in disturbed than in undisturbed sites (Bossard 1991). Similarly, while removal of the understory of a ponderosa pine forest in the northwestern United States improved conditions for germination and growth of invasive cheatgrass, it also made it more vulnerable to herbivory by small mammals (Pierson & Mack 1990). The net effect in this case, however, was that disturbed sites supported higher populations of cheatgrass despite enhanced herbivory.

Herbivory on introduced plants can be intense (Maron & Vilà 2001) and tends to be stronger when attributed to native rather than to introduced herbivores (Parker et al. 2006). Experimental field trials have revealed maximum seed predation of more than 50% for *Acacia* spp. in South Africa, scotch broom in California, *Opuntia* spp. in Spain and pines in Patagonia (Holmes 1990; Bossard 1991; Vilà & Gimeno 2003; Nuñez et al. 2008). In all of these cases, seed predation was by native rodents and/or birds. The extent to which seed predation actually limits invasions depends largely on the level of seed limitation experienced by populations of invasive plants (Hoffman & Moran 1998; Parker 2000). Despite this caveat, numerous experimental studies have demonstrated that herbivory can limit the establishment, spread and impact of introduced plant species. Cases range from the classic study done by Brown and Heske (1990) in which 13 years of fenced enclosures of kangaroo rats in the Arizona desert led to a 20-fold increase in the abundance of the South African grass *Eragrostis lehmanniana*, to the studies of Eckberg and colleagues (2012, 2014) in the Nebraskan prairie, where insecticide exclusion studies showed that native herbivorous insects were limiting the spread of the Eurasian bull thistle. These and other cases in which native herbivores have been shown to limit the establishment, spread or impact of introduced plants are listed in Table 2.2. All are examples of naturally occurring biological control of potential or actual weeds.

The effect of native plant pathogens on invading plants can be strong as well, as has been demonstrated by the extreme difficulty in establishing some non-native crop plants in the face of native pathogens (Hokkanen 1985a). The role of plant pathogens in limiting invasions of weeds has been documented in some cases as well. For example, the spread of multiflora rose, an invasive weed in North America, is apparently being restricted by a native but unidentified mite-vectored rose pathogen (Epstein & Hill 1999). More generally, Mitchell and Power (2003) have found that the exotic weeds that had acquired the greatest number of native pathogen species tended to be the least 'noxious' (Figure 2.3). Mitchell and Power's data set is important because it puts us in the position to address a question posed at the beginning of this chapter, namely how important biotic resistance is in preventing introduced organisms from becoming pests. Their data suggest that plant pathogens play an important role in preventing at least some introduced plants from becoming noxious weeds. It is also clear from their data set that many species were not noxious despite not having accumulated pathogens. This shows that factors other than biotic resistance can also limit the impact of introduced weeds.

Suspected cases in which establishment of non-cultivated exotic plants was actually *prevented* by herbivory or disease do exist, but their documentation remains imperfect. Examples include termites excluding invasion of *Eucalyptus* spp. in various tropical countries and a root rot with an exceptionally broad host range routinely excluding exotic trees and shrubs from the deserts of North America (Mack 1996, 2002). Levine and colleagues (2004) found no clear evidence of

Table 2.2. **Selected cases of biotic resistance against invading plants by herbivores. Entries represent studies in which evidence of decreased establishment, spread or density of an invasive plant was presented.**

Exotic Plant	Location	Herbivore	Specialist/ Generalist	Effect	Reference
Lehmann lovegrass (*Eragrostis lehmanniana*)	Arizona, USA	Rodents	Generalists	Population suppression	Brown & Heske 1990
Acacia spp.	S. Africa	Striped field mouse	Generalist	Reduction of seed density	Holmes 1990
Cheatgrass (*Bromus tectorum*)	Forests in western USA	Small mammals	Generalists	Recruitment reduced in disturbed areas	Pierson & Mack 1990
Scotch broom (*Cytisus scoparius*)	Coastal California, USA	California Quail, Blue Grouse	Generalists	Reduction of seed density in disturbed sites	Bossard 1991
Ice plant (*Carpobrotus edulis*)	Coastal California, USA	Mammalian herbivores	Generalists	Establishment prevented in some habitats	D'Antonio 1993
Hollyleaved barberry (*Mahonia aquifolium*)	Germany	*Rhagoletis meigenii* (Diptera: Tephritidae)	Specialist	Reduction of seed density, dispersal	Soldaat & Auge 1998
Guernsey fleabane (*Conyza sumatrensis*)	Southern England	Rabbits	Generalist	Establishment prevented in some areas	Case & Crawley 2000
Eurasian water milfoil (*Myriophyllum spicatum*)	North American ponds and lakes	*Euhrychiopsis lecontei* (Coleoptera: Curculionidae)	Specialist	Weed status prevented in certain lakes	Newman 2004
Western salsify (*Tragopogon dubius*)	Montana, USA	Rodents	Generalists	Population suppression	Pearson et al. 2012
Bull thistle (*Cirsium vulgare*)	Eastern Nebraska, USA	Various insect herbivores	unknown	Weed status, spread prevented	Eckberg et al. 2014

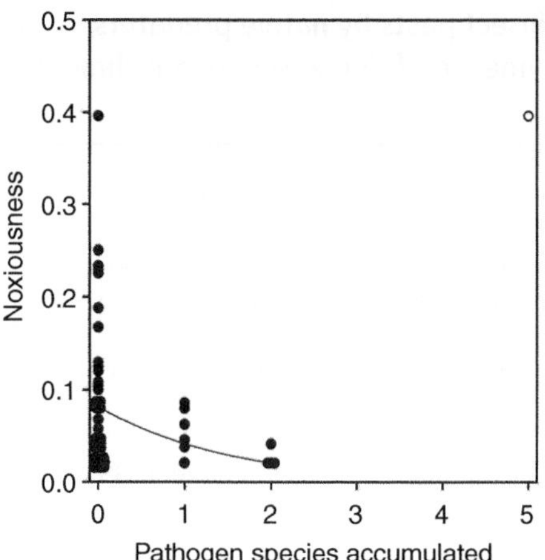

Figure 2.3. The impact of pathogen species accumulation on noxiousness of weeds. Noxiousness was determined using plants' status as noxious weeds; each dot represents one weed species, and the negative relationship between pathogen species accumulated and noxiousness is significant provided that the outlier (open circle, *Sorghum halapense* [Johnsongrass]) is excluded. From Mitchell and Power (2003).

plant invasions that had been prevented by biotic resistance, and argued that biotic resistance was unlikely to be able to prevent the establishment of introduced plants outright provided that propagule pressure is high enough (see also Simberloff 2009).

2.2.3.2 Biotic Resistance against Invading Arthropods

As we noted in Chapter 1, Losey and Vaughan (2006) calculated the extent to which native insects reduce the need to apply insecticides against crop pest insects (see Table 1.1). In their analysis, native predators and parasitoids reduced economic damage attributable to crop pest insects by 65% in the United States. While this suggests a potentially important role of biotic resistance in limiting the effects of exotic pest species, their analysis did not distinguish between native and introduced pests. A separate analysis revealed a slightly higher level of attack on exotic rather than native herbivores, but in this case, no distinction was made between introduced and native enemies (Hawkins et al. 1997). Still, there are some reports of suppression of introduced pest (or potential pest) insects by native consumers (Table 2.3). In addition, a number of manipulative studies from natural systems have demonstrated strong biotic resistance against introduced arthropods, including birds on introduced spiders in Hawaii (Gruner 2005) and toads on invasive ants in Indonesia (Wanger et al. 2011).

While these case studies suggest that biotic resistance against arthropods may be important in some contexts, this topic has received less attention than biotic resistance against invading plants has. However, the role of native predators, parasitoids and pathogens in limiting the establishment and impact of introduced biological control agents has fascinated biological control researchers for many years. Indeed, much of what we know about biotic resistance against arthropods comes from analyses of biological control releases. In the following sections, we review the evidence for this form of biotic resistance. We begin with biotic resistance against biological control agents of weeds and then move on to biotic resistance against biological control agents of arthropods. In Chapter 4, we will discuss the special case of biological control agents released to control arthropods interfering with weed biological control agents.

Biotic Resistance against Weed Biological Control Agents. Goeden and Louda (1976) were the first to systematically review biotic resistance against arthropod weed biological control agents, and in doing so they introduced the term *biotic interference* into the literature to denote biotic resistance against biological control agents. More important, Goeden and Louda uncovered experimental or anecdotal cases of biotic resistance in approximately half of the cases they reviewed. One of the most striking patterns they uncovered is that most of the biotic resistance against insect biological control agents

Table 2.3. **Selected cases of biotic resistance against invasive insect pests by native predators, parasitoids or pathogens. Entries represent studies in which evidence of decreased establishment, spread or density of an invasive insect pest was presented.**

Insect Pest	Location	Enemy	Reference
Barberry whitefly (*Parabemisia myricae*)	Citrus groves in California, USA	The parasitoid *Eretmocerus debachi*	Rose & DeBach 1992
Cotton aphid (*Aphis gossypii*)	Cotton fields in California, USA	Coccinellid beetles	Rosenheim et al. 1997
Brown citrus aphid (*Toxoptera citricida*)	Citrus groves in Puerto Rico, Florida, USA	Various Coccinellidae, Syrphidae*	Michaud 1999
Gypsy moth (*Lymantria dispar*)	Woodlands in N. America	Small mammals*	Elkinton et al. 2004
Beet armyworm (*Spodoptera exigua*)	Alfalfa fields in California, USA	Generalist predators and parasitoids	Ehler 2007
Soft brown scale (*Coccus hesperidum*)	Citrus groves in interior of S. California, USA	The parasitoid *Metaphycus luteolus**	Kapranas et al. 2007
Coffee berry borer (*Hypothenemus hampei*)	Coffee plantations of S. Mexico	Various ant species	Gonthier et al. 2013
Light brown apple moth (*Epiphyas postvittana*)	Urban ornamentals in California, USA	Generalist parasitoids*	Bürgi & Mills 2014, Bürgi et al. 2015
Emerald ash borer (*Agrilus planipennis*)	Ash forests in Ohio, USA	Birds	Flower et al. 2014

* Assemblage of both native and introduced species, with native species reported as playing an important role

was attributed to generalist predators. Birds, ants and a number of other arthropod predators are mentioned frequently, and in South Africa, baboons were also invoked, apparently feasting on *Cactoblastis cactorum* caterpillars, which had been introduced to control the prickly pear cactus. Parasitoids and pathogens are implicated as well, but to a much lesser extent (see also Hill and Hulley 1995). A similar trend was borne out in another review of biological weed control done by Crawley (1986, 1987). As part of this review, Crawley tabulated the suspected reasons for failure of insects used in weed biological control that were cited by scientists involved in the work. Interference by native predators was cited as an important cause of reduced impact or failure in 22% of 278 cases compared with 11% and 8% for parasitoids and pathogens, respectively. Competition came in at 12% in this analysis – a value perhaps higher than expected given that weeds would appear to be a virtually limitless resource at the time of insect introduction.

A weed biological control agent that has recruited more than its fair share of native natural enemies (despite aposematic coloration) is the cinnabar moth, *Tyria jacobaeae*, which was introduced into many countries to control tansy ragwort, *Senecio jacobaea*. In Australia, cinnabar moth failed to establish despite a massive breeding and release program, and the blame for this non-establishment was placed primarily on the native generalist scorpionfly, *Harpobittacus nigriceps*. For some of the early releases, there was suspicion that the non-establishment may have been due to the co-introduction of a nuclear polyhedrosis virus disease along with cinnabar moth (Bucher & Harris 1961; Bornemissza 1966), but natural enemies also thwarted later releases in which great pains were taken to release disease-free stock (Schmidl 1972). The failure of establishment of cinnabar moth in New Zealand was at least in part blamed on predation by birds, which, 'engorged themselves [on cinnabar moth larvae] until they were unable to take wing' (Miller 1936). Slowed establishment or reduced control was linked to biotic resistance by a number of predators, parasitoids and pathogens in other countries as well (Table 2.4).

Much of the information on biotic resistance against biological weed control agents is observational rather than experimental (including what we know of the tansy ragwort case). Goeden & Louda (1976) found only two experimental evaluations of resistance, and Crawley's data set, while large, essentially reports impressions of researchers. Part of the reason for this lack of hard data is the difficulty in testing hypotheses for why

Table 2.4. **Biotic resistance against the cinnabar moth, *Tyria jacobaeae*, released to control tansy ragwort, in various countries.**

Location	Enemies Noted	Purported Outcome	References
Australia	*Harpobittacus nigriceps* (Mecoptera: Bittacidae), stink bugs, damsel bugs, ants, rove beetles, larval parasitoids, birds, microsporidian, other fungal and viral pathogens[1]	Non-establishment	Currie & Fyfe 1938; Bornemissza 1966; Schmidl 1972
New Zealand	Birds, larval and pupal parasitoids	Non-establishment	Cameron 1935; Miller 1936
Western Canada	Ground beetles[2], pathogens	Non-establishment at one site	Wilkinson 1965; Harris et al. 1971
Eastern Canada	Egg predators and parasitoids	Population reduction, lag in establishment	Harris et al. 1971
California, USA	Microsporidian pathogen (*Nosema* sp.)[3], rodents, earwigs, sow bugs, carabid beetles, pupal parasitoids	Population reduction	Hawkes 1973

[1] Some of the pathogens may have been introduced (see text)
[2] The primary suspected ground beetle species was *Pterostichus melanarius*, itself an (accidentally) introduced species.
[3] *Nosema* infection may have been introduced along with initial releases of cinnabar moth into California.

biological control fails. Also, the impulse from an applied perspective has often been to try to obtain a success (perhaps with other agents) rather than dwelling on agents that have failed.

More recent studies, however, have addressed biotic resistance against weed biological control agents head on. Pioneering research along these lines was done by Briese (1986a), who performed predator exclusion and life-table studies of the geometrid moth, *Anaitis efformata*, which was introduced into Australia to control St. John's wort, *Hypericum perforatum*. Failure of this agent to establish in the 1930s had been attributed to predation by ants, and Briese set out to determine whether these initial observations were correct following a second round of releases in the 1980s. His results indeed supported the conclusion that biotic resistance against *A. efformata* was strong (Figure 2.4). The major causes of biotic resistance included egg parasitism and a suite of generalist predators including hemipterans, spiders and ants. Briese concluded that predators would probably not prevent establishment of *A. efformata*, but that they could hinder its effectiveness as a biological control agent. Having said this, the agent is not currently established in Australia and biotic resistance must be maintained as one

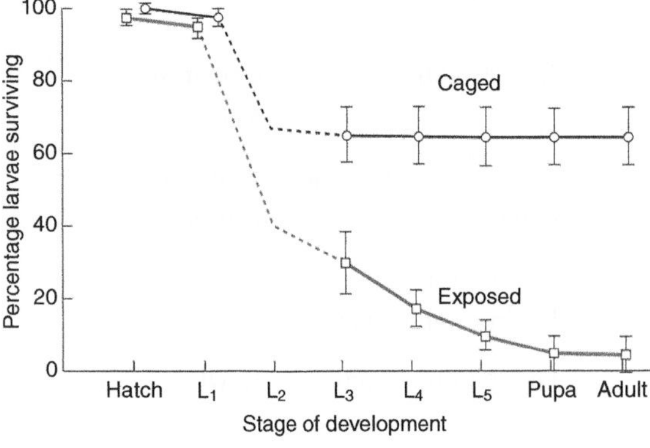

Figure 2.4. Survival rates of developmental stages of *Anaitis efformata*, a lepidopteran biological control agent of St. John's wort on plants that are caged to exclude predators (control) and exposed. From Briese (1986a).

of the hypotheses explaining non-establishment. Other biological control agents of St. John's wort, including the chrysomelid beetle, *Chrysolina quadrigemina*, have apparently escaped catastrophic biotic resistance and are providing partial control of this weed in Australia (Briese 1997).

Biotic resistance associated with biological control of purple loosestrife, *Lythrum salicaria*, by *Galerucella* leaf beetles in North America has been assessed experimentally as well, prompted by worries that generalist arthropod predators may limit establishment or control potential at some sites (Malecki et al. 1993; Nechols et al. 1996; Diehl et al. 1997; Hunt-Joshi et al. 2005). Laboratory studies showed that a number of native arthropods were indeed capable of feeding on *Galerucella* eggs (Sebolt & Landis 2002, 2004; Wiebe & Obrycki 2002). In experimental field studies done in the northern United States, Sebolt and Landis (2004) found substantial levels of biotic resistance when *Galerucella* eggs were at high densities, but no biotic resistance at low egg densities. Thus, they concluded that while predation was not likely to impede establishment of *Galerucella*, it was strong enough to slow population growth. Other studies have shown appreciable predation on *Galerucella* beetles in other parts of North America (Wiebe & Obrycki 2004; Denoth & Myers 2005).

Biotic resistance against arthropod biological control agents. Arthropod predators and parasitoids released for biological control can also face biotic resistance. Stiling (1993) conducted a survey similar to Crawley's in which the opinion of biological control researchers as to why certain biological control projects against arthropod pests failed was assessed. Biotic resistance (combined competition, predation, parasitism and hyperparasitism) accounted for 20% of all the suspected reasons for failure. We should not be surprised that predators and parasitoids can themselves be subject to interference, especially given the strong impact that 'higher order' predators (predators that include other predators in their diet) can have on predator and parasitoid populations in general (Rosenheim 1998). However, investigations of the impact of native higher-order predators, parasitoids and hyperparasitoids on invading biological control agents (especially at the time of introduction) are not as developed as similar studies on invading plants and weed biological control agents. Considerable work has gone into determining whether previously released biological control agents compete with, or otherwise interfere with subsequent species released against the same pest. Indeed, this is one of the most important topics in theoretical and applied biological control. We defer discussion of this issue to Chapter 3, however, and are primarily interested here in the effect of biotic resistance by native organisms. Here, we focus on two examples of biotic resistance against arthropod biological control agents: hyperparasitoids attacking parasitoids released as biological control agents, and intraguild predation among ladybird beetles.

Hyperparasitoids. Studies of native hyperparasitoids attacking introduced parasitoids would appear to offer good opportunities to investigate biotic resistance. While experimental demonstrations of hyperparasitoids (native or exotic) interfering with biological control are scarce (Luck et al. 1981; Rosenheim 1998; Schooler et al. 2011), there certainly are cases where native hyperparasitoids have inflicted severe mortality on populations of introduced parasitoids. For instance, *Cotesia melanoscela* (= *Apanteles melanoscelus*), first introduced to North America in 1911 against the gypsy moth, *Lymantria dispar*, suffered up to 97% mortality from a complex of 35 native hyperparasitoid species (Muesebeck & Dohanian 1927). Hyperparasitism of another *Cotesia* species – *C. rubecula* – may have led to non-establishment of this parasitoid at some sites in the eastern United States (McDonald & Kok 1991). Another interesting case involves hyperparasitism of two ichneumonid parasitoids introduced to North America against the alfalfa weevil, *Hypera postica*. One of the parasitoid species, *Bathyplectes curculionis*, overwinters as a larva within its cocoon, where it is subject to parasitism by approximately 20 species of native hyperparasitoids, leading to greater than 50% overwintering mortality in some areas. In the western United States, hyperparasitism has been widely suspected as limiting the effectiveness of this parasitoid (Pike & Burkhardt 1974; Simpson et al. 1979; Rethwisch & Manglitz 1986). The other species, *B. anurus*, overwinters as an adult within its cocoon and is subject to lower levels of hyperparasitism. In the larval stage, this species also has the unusual capability of causing its cocoon to 'jump' as far as 3 cm, a behavior that may provide some protection from hyperparasitism (Day 1970). Of the two species, *B. anurus* is considered the more effective parasitoid (Day 1981; Radcliffe & Flanders 1998).

High hyperparasitism rates can also accompany successful biological control. Indeed, some of our most famously successful cases of arthropod biological control have been achieved despite substantial hyperparasitism. These include biological control of the cassava mealybug and the mango mealybug in Africa, where native hyperparasitism of the highly successful agents *Anagyrus* (= *Epidinocarsis*) *lopezi* and *Gyranusoidea tebygi*, respectively, approached 90% in some areas (Agricola & Fischer 1991). Another example involves *Trioxys pallidus*, a successful parasitoid of walnut aphid in California, which was able to control this pest despite levels of hyperparasitism approaching 100% by the end of the growing season at some sites (Frazer & Van den Bosch 1973; Van den Bosch et al. 1979). One reason that hyperparasitoids may not interfere with biological control (and may even improve it!) is that they may help to stabilize the interactions between primary parasitoids and their hosts (Nicholson 1933; Beddington & Hammond 1977; Luck et al. 1981). More likely however, the extent of interference will depend on the level of specificity of the hyperparasitoids. Specialist hyperparasitoids may have a high capacity to interfere with biological control, but generalists are less coupled with the biological control system and therefore may not be able to prevent successful control. Biotic resistance from indigenous hyperparasitoids will typically involve generalist species and may be less damaging to control for this reason.

Interactions among ladybeetle species. A number of studies have examined interactions between species of native and introduced ladybeetles (Coleoptera: Coccinellidae). Since coccinellid species may compete and are known to engage in interspecific predation (Majerus 1994), one could reasonably ask whether native ladybeetles have interfered with the establishment, spread or abundance of introduced ladybeetles. For at least two well-studied species, the answer to this question appears to be an emphatic 'no', at least in North America. *Coccinella septempunctata* and *Harmonia axyridis* were introduced from Europe and Asia, respectively, to control various aphid species in North America (Day et al. 1994; Tedders & Schaeffer 1994; Wheeler & Stoops 1996). Numerous studies pairing *C. septempunctata* and *H. axyridis* with native ladybeetles have shown that the native species are affected more negatively than the introduced ones via both competition and predation (e.g., Cottrell & Yeargan 1998; Obrycki et al. 1998; Cottrell 2004, 2005; Hoogendoorn & Heimpel 2004; Snyder et al. 2004; Yasuda et al. 2004; Gagnon et al. 2011). Indeed, this asymmetry may be facilitating the spread of these species in North America, and there is considerable concern that *C. septempunctata* and *H. axyridis* may be displacing native ladybeetles from agricultural habitats in North America (Elliot et al. 1996; Michaud 2002; Turnock et al. 2003; Alyokhin & Sewell 2004; Evans 2004; Finlayson et al. 2008; but see Smith & Gardiner 2013) with similar trends being reported in Europe for *H. axyridis* (Raak-van Den Berg et al. 2012; Roy et al. 2012).

As a counterpoint to this, *Adalia bipunctata*, a coccinellid native to mainland Eurasia, may be encountering biotic resistance from native *H. axyridis* and *C. septempunctata* in Japan. The first Japanese record of *A. bipunctata* was from a park along Osaka Bay in 1993 and six years later, its distribution in Japan had not extended beyond this park. By 2004, it was still restricted to the city of Osaka (Toda & Sakuratani 2006). Kajita and colleagues (2000) showed that this species is highly vulnerable to predation by *H. axyridis* and *C. septempunctata* in the laboratory, and *H. axyridis* was observed feeding on *A. bipunctata* within Osaka Bay park (Sakuratani et al. 2000). Further laboratory and field studies showed that *A. bipunctata* eggs were more vulnerable to predation by *C. septempunctata* than the reverse (Sato & Dixon 2004; Kajita et al. 2006a; Toda & Sakuratani 2006; Sato et al. 2009) and also that *A. bipunctata* was reluctant to oviposit in the presence of *H. axyridis* or *C. septempunctata* (Kajita et al. 2006b). Taken together, these observations suggest that predation by *H. axyridis* and *C. septempunctata* may be impeding the spread of *A. bipunctata* in Japan.

2.2.4 Biotic Resistance: General Principles

While an assessment of the relative importance of biotic resistance to other factors in limiting invasions is beyond the scope of this book, we do conclude that biotic resistance can be an important force in reducing the establishment, spread and impact of invading organisms. This has been shown not only by sampling and observation, but by exclusion experiments as well, and the biological control literature has been a great source of information in addressing this question. Simberloff (2009) has argued that high propagule pressure can overcome biotic resistance, and indeed, biotic resistance is likely to be most important when propagule pressure is low.

In this section, we focus on two aspects of biotic resistance that have particular relevance for our understanding of biological control. The first involves the distinction between biotic resistance due to generalist versus specialist consumers, and the second involves the process of apparent competition.

2.2.4.1 Generalists versus Specialists

Two principles with clear implications for biological control involve the importance of generalist versus specialist enemies in conferring biotic resistance. The first principle is that generalist enemies are more likely to be agents of biotic resistance than are specialist enemies. The second principle constitutes an exception to the first principle, and it holds that specialist natural enemies can attack invaders that have close relatives in the introduced range.

Support for the hypothesis that generalist enemies are particularly important agents of biotic resistance comes from the preponderance of cases in which generalist vertebrate herbivores attack exotic plants (Table 2.2) and generalist predators attack biological control agents of weeds (Table 2.4). Introduced plants and insects support a disproportionately high fraction of generalist enemies with respect to their native counterparts (e.g., Strong et al. 1984; Cornell & Hawkins 1993; Fraser & Lawton 1994; Jobin et al. 1996; Novotny et al. 2003), and these trends underscore the importance of native generalist enemies as natural biological control agents of recently invading pests. They also support Ehler's (1998) recommendation to preserve or augment generalist natural enemies in areas prone to invasion. The dominance of generalists on invasive species can be transient, however. Andow and Imura (1994) found that communities of herbivorous insects attacking exotic crop plants in Japan have become increasingly specialized over the past 2,500 years. This is consistent with initial attack of invading plants by generalists followed by a gradual recruitment of specialized species.

The role of specialist enemies in biotic resistance is greatly enhanced when the invading species has close relatives in the introduced range (the second principle). As an illustration of this principle, a number of cases of host-range expansion come from insects attacking introduced tree species that have close relatives in their introduced range (e.g., Bush 1969; Connor et al. 1980; Auerbach & Simberloff 1988; Da Ros et al. 1993). Mack (1996) sought broader trends that could address the hypothesis that invaders with few relatives had a better chance of establishing and spreading. He reasoned that if this were the case, naturalized species would be disproportionately composed of species with no congeners in the introduced range. This prediction was upheld for the floras of California, Florida, Hawaii and Illinois (but not New York). Of a total of 1,774 naturalized plant genera in these first four states, 1,230 (69%) had no native members in the states in which they were naturalized. This trend is consistent with (but by no means proves) the hypothesis that plants with close relatives in the introduced range are excluded by herbivory (Mack 1996).

A good example of biotic resistance by a specialist herbivore involves natural biological control of Eurasian watermilfoil, *Myriophyllum spicatum*, introduced into North America in the 1940s. Native North American watermilfoils support at least two herbivorous insects that specialize on a subset of *Myriophyllum* species (Newman 2004). Among these is the milfoil weevil, *Euhrychiopsis lecontei*, which has expanded its host range to include Eurasian watermilfoil and has been responsible for a number of spectacular declines of this weed in North American lakes. The milfoil weevil develops faster and has higher fecundity when feeding on the exotic milfoil than on native milfoil species (Newman et al. 1997; Solarz & Newman 2001; Sheldon & Creed 2003). This result fits a general pattern in which host species that lack a historical association with a consumer species may be more suitable to consumers than closely related hosts that have had the time to evolve resistance to the consumer (Parker & Hay 2005). It has even been argued that such 'new associations' may be preferable to traditional importation biological control, where natural enemies with a long history of association with the pest are used (Hokkanen & Pimentel 1984, 1989; Wiedenmann & Smith 1997). The case of *E. lecontei* and Eurasian watermilfoil thus illustrates both the principle that specialist natural enemies can expand their host range to include invading hosts closely related to their ancestral host, and also that these invaders may be subject to higher rates of attack because of their lack of a historical association with the consumer.

2.2.4.2 Biotic Resistance and Apparent Competition

A concept that can unify many of the principles of biotic resistance is *apparent competition.* This term, first coined by Holt (1977), refers to the negative, indirect effect that one prey species can have on another by increasing the abundance or activity of a shared predator. The presence of native prey species in a given geographical area supports predators that are in a position to prey upon any species that may invade the area (the terms *predators* and *prey* are used loosely here to refer to any resource/consumer relationships). Thus, apparent competition is a potentially ubiquitous component of biotic resistance and plays an important role both in stemming harmful invasions and in interfering with biological control introductions.

However, apparent competition can also lead to other outcomes within the context of invasion biology. Predator reproduction may of course increase in the presence of an exotic species, and this can result in increased predation on native species. The trade-off between the 'positive' effects of biotic resistance in reducing the strength of invasions, and the 'negative' effects of increased predation of native species was modeled by Noonburg and Byers (2005). Their model included four species: a resource species fed upon by two competing prey species (one native and one exotic), both of which were in turn attacked by a single predator species. Extirpation of a native species through apparent competition with an exotic species was theoretically possible, but only for a relatively restricted combination of parameter values. Important parameters included the relative resource consumption rates of the exotic and native species, and the predator's preference for the native versus the exotic species. Predator-mediated coexistence is also possible if the exotic competitor is neither preferred by the predator nor a superior competitor. We illustrate some of the possible outcomes of the Noonburg and Byers model in Figure 2.5.

While the parameter space for predator-mediated displacement of a native species by an exotic competitor is relatively restricted in the Noonburg and Byers model, there is some evidence that this has occurred in California grasslands (Borer et al. 2007). Malmstrom and colleagues (2005) showed that introduced oats and bromes from Europe have a negative impact on native bunchgrasses by being a source for an aphid-vectored plant-pathogenic virus. The aphids prefer and perform better on the introduced grasses and then spread the virus to the native bunchgrasses. Virus loads on bunchgrasses are doubled by the presence of the introduced grasses, and separate studies showed that virus load can severely stunt bunchgrasses, especially under conditions of stress. This system therefore illustrates the scenario described in Figure 2.5c: an abundant exotic 'prey' species (the invasive grasses) greatly enhances the abundance of a 'predator' (the aphid-vectored virus), and this predator inflicts substantial damage on a native 'prey' species (the native bunchgrasses).

Within the context of this chapter, we are primarily concerned with the situation where the 'predator' in Figure 2.5 is a native enemy of a native and an exotic species. However, there is ample opportunity to consider these interactions with the view that the exotic prey species or the predator is a biological control agent (Holt & Hochberg 2001). Because such an interpretation would bring us beyond our current discussion of biotic resistance however, we defer a more detailed treatment of these interactions to Chapter 4.

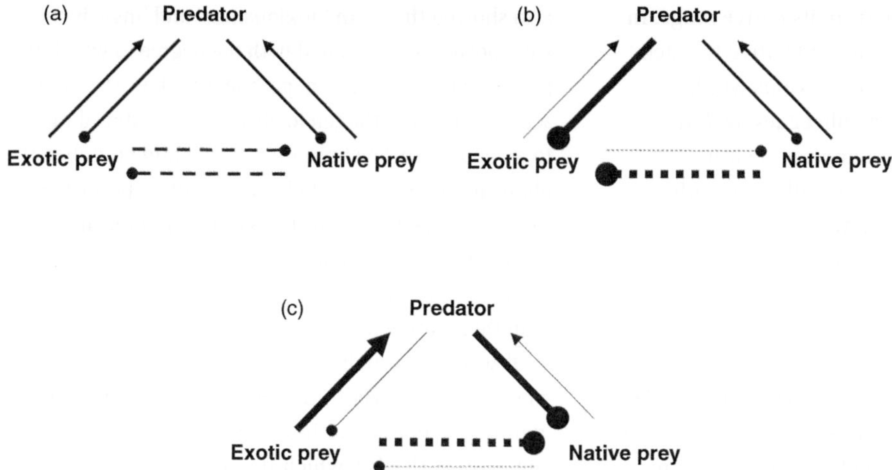

Figure 2.5. Modified Levins diagrams illustrating three possible outcomes of apparent competition when an exotic and a native prey species share one predator species. Pointed and rounded arrows indicate positive and negative effects on population abundances, respectively, and dashed lines indicate indirect interactions (apparent competition). Thickness of arrows is proportional to the strength of the interaction. In panel (a), apparent competition is symmetrical with both native and introduced prey having indirect negative effects on one another through their shared predator. Panel (b) illustrates the situation that occurs when an exotic species that is at low abundance is attacked by the predator such that its population declines with only negligible effects on the predator abundance. This is biotic resistance, and the apparent competition exerted by the native on the introduced prey population is stronger than vice versa. In panel (c), the presence of the exotic prey species greatly increases predator abundance and this affects the native prey more negatively than it does the exotic prey species. Apparent competition is stronger on the native than the exotic prey species in this scenario.

2.3 The Enemy Release Hypothesis

While biotic resistance may play an important role in limiting the effects of some invasions, it is also true that enemy attack against introduced species is often below that of either native species in the introduced range or the same species in its native range (Cornell & Hawkins 1993; Maron & Vilá 2001; Keane & Crawley 2002; Mitchell & Power 2003; Torchin et al. 2003; Tallamy 2004; Agrawal et al. 2005; Adams et al. 2009; Roy et al. 2011; Castells et al. 2013). In a few cases, there is circumstantial, experimental or comparative evidence that the invasive nature of some exotic species is directly due to relaxation of enemy attack. An example of strong circumstantial evidence comes from the white campion, *Silene latifolia*, which is an innocuous plant prized for its sweet-smelling flowers in its native Europe, but an invasive, noxious weed in North America. Damage from insect herbivores and fungal diseases is substantially and significantly higher in Europe than in North America, leading to a 17-fold higher probability of attack in Europe than in North America (Wolfe 2002). The damage these enemies inflicted was severe in many cases, leading to a strong likelihood that *S. latifolia* abundance was limited by these enemies in Europe, and that release from these enemies in North America contributed substantially to its invasive nature.

An experimental approach was taken by DeWalt and colleagues (2004), who worked with the invasion of the neotropical shrub *Clidemia hirta* (Koster's curse) into Hawaii. This species is native to lowland areas in Central and South America, but is absent from forest habitats in its native range. In Hawaii, conversely, it has invaded native forested areas, where it is considered a noxious weed. In the study by DeWalt and colleagues, *D. hirta* plants were grown in forest areas in their native Costa Rica and in Hawaii, and a subset of both groups of plants was treated with insecticides and fungicides to exclude natural enemies. The results were striking: application of both insecticides and fungicides tripled the survival of *D. hirta* plants in their native Costa Rica, but had no effect in Hawaii, where survival was uniformly high (Figure 2.6). The growth rate of surviving plants was not affected by these treatments, but the implications remain that: (1) insect herbivores and plant-pathogenic fungi suppress *D. hirta* in forested habitats in its native range, and (2) release from these enemies contributes to the spread of *D. hirta* in Hawaiian forests.

A comparative approach comes from a study of 473 plants introduced from Europe to North America by Mitchell and Power (2003). In addition to showing that the plant pathogen load of these plants was much lower in their introduced than their native range, they

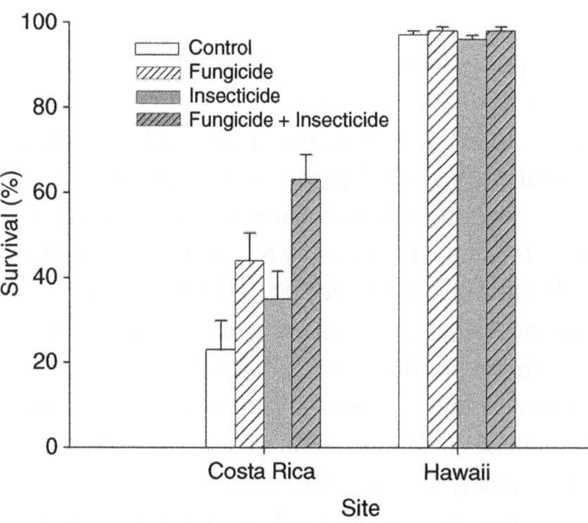

Figure 2.6. Survival of *Clidemia hirta* plants in forest understories of Costa Rica (the native range) and Hawaii (the introduced range) where plants were subjected to three spray treatments and a no-spray control. Data from DeWalt et al. (2004).

also showed that both 'noxiousness' and 'invasiveness' were positively correlated with the degree to which the plants were released from pathogens. Thus, their data set links lower pathogen loads in the introduced range to a population-level response over a large number of plant species. Cornell and Hawkins (1993) performed a similar analysis for parasitoids of herbivorous insects, finding both lower parasitoid species richness and level of attack on herbivore species in their introduced than their native range.

Observations such as these have led to the 'enemy release hypothesis' (ERH) (also known as the 'natural enemies hypothesis' or the 'escape-from-enemy hypothesis'), which states that invasiveness results from an organism being released from its native enemies upon being introduced into a new geographic setting. This hypothesis forms a primary justification for the practice of importation biological control, and conversely, successful importation biological control has been seen as evidence for the enemy release hypothesis (Strong et al. 1984). It is therefore important that the validity of the ERH itself be evaluated as rigorously as possible, and that the use of biological control projects as tests of the ERH be justified. While the studies cited earlier make clear that enemy release can occur and can be important, the generality of the ERH as an explanation for invasiveness is in question. It seems obvious that enemy release cannot be the sole factor explaining invasiveness of exotic species, and even before the debate about the ERH got into full swing, one of the most celebrated biological control scientists and advocates, C. B. Huffaker, warned against assuming that enemy release was a universal explanation for invasiveness (Huffaker 1957):

> *While aggressive and abundant alien pests are always good prospects for biological control, it does not follow and is ecologically unsound to postulate that such aggressiveness and abundance is due to an absence of natural enemies. There are far too many other reasons why a pest may become relatively innocuous in its native land yet aggressive and troublesome in environments new to it.*

In previous sections of this chapter, we discussed cases of strong biotic resistance against invaders (including examples listed in Tables 2.2, 2.3). Should these cases be seen as exceptions to the general rule of enemy release, or do they indicate that enemy release is itself an exception to biotic resistance? Colautti and colleagues (2004) reviewed the evidence for and against the ERH and found good support for the hypothesis that invasive species are attacked by a lower *diversity* of natural enemies in their introduced versus native ranges, but much less support for the hypothesis that the overall *effect* of natural enemies was lower in the introduced range. They considered eight hypotheses that might explain this difference, four of which are outlined in Table 2.5. From the review of Colautti and colleagues, as well as from the results of subsequent meta-analyses (Liu & Stiling 2006; Chun et al. 2010), it seems clear that while enemy release can be an important force allowing exotic species to invade and reach high population densities, it is not the whole story, and may not even be most of the story for many exotic species. For example, even in Mitchell and Power's (2003) study, which showed significant increases in 'noxiousness' and 'invasiveness' of plants as a function of release from plant pathogens, release explained only 11% and 4% of the variation in these variables, respectively. A number of other studies designed to look explicitly for evidence of enemy release have failed to find it (Agrawal & Kotanen 2003; Colautti et al. 2004; Parker & Hay 2005; Roy et al. 2011; Bürgi & Mills 2014). Clearly, factors other than release from enemies are important in determining which introduced species become noxious or invasive weeds. When exotic species are strongly invasive, and there is little support for ERH, other explanations for their invasiveness must be sought. Common alternative hypotheses include favorable abiotic conditions, plentiful resources or habitat, and facilitation by humans or other invaders (e.g., Williamson 1996; Simberloff & von Holle 1999; Mack et al. 2000; Richardson et al. 2000; Denslow 2003; Barlow & Kean 2004; Heimpel et al. 2010).

Table 2.5. **Hypotheses for the observation that introduced species tend to have a reduced *diversity* of natural enemy species in their introduced versus their native ranges, but do not necessarily experience decreased *effects* of natural enemies in the introduced ranges. Adapted from Colautti and colleagues 2004; see their table 2 for additional hypotheses and references.**

Hypothesis	Explanation
Compensatory versus regulatory release	While there may be fewer species of enemies in the introduced range, these species may have disproportionately strong effects or effects for which the invader has not evolved defenses.
Native enemies	Introduced species may be especially vulnerable to native enemies because of a lack of coevolved defenses.
Increased susceptibility of invaders	Introduced species may be especially vulnerable to native or introduced enemies due to genetic bottlenecks during invasion.
Sampling effort bias	More enemy species may be recorded from the native range due to higher sampling and research effort. Also, lists of enemies from the *entire native range* of an invader may overestimate diversity of enemies from the *source population.*

2.3.1 Is Successful Biological Control Evidence of the ERH?

In successful importation biological control, an introduced species that has reached pest status is controlled by one or more introduced natural enemies. In many cases, the pest was present at sub-economic levels in its native range, and the importation biological control agent is a specialist collected from the pest in its native range. This is certainly consistent with the hypothesis that the introduced pest had been controlled in its native range, and that it reached pest status in its introduced range due to the absence of specialist enemies – the enemy release hypothesis in a nutshell. Successful importation biological control does therefore constitute circumstantial evidence for enemy release (Strong et al. 1984). Keane and Crawley (2002) have argued, however, that importation biological may not be a proper test for ERH.

Focusing on exotic plants, Keane and Crawley noted that biological control agents are themselves introduced without *their* natural enemies, making them perhaps more effective in their introduced range than they were in their native range. The introduced biological control agent is also likely to find itself free from competitors that might limit its effect in its native range, and have other advantages associated with the act of biological control introduction (Colautti et al. 2004). Hawkins and colleagues (1999) used an analysis of life-table studies in natural versus biological control settings to ask whether importation biological control systems could be used to address long-standing questions concerning top-down versus bottom-up control of insect herbivores. They found that biological control systems were qualitatively different from natural systems – particularly exhibiting fewer food web links – and that it was therefore inappropriate to use cases of biological control to address hypotheses of how herbivores were regulated in natural systems.

And finally, what of the cases of failed biological control? More than half of the importation biological control agents that establish in their introduced range fail to control the pest they were introduced against, despite careful screening in many cases (see Chapter 3). While these cases do not necessarily invalidate the ERH, because it could be argued that these enemies were not solely responsible for controlling the pest in its native range, they also highlight the possibility that escape from enemies was perhaps not the only explanation for invasiveness of the pest in the first place.

2.3.2 Evolutionary Dynamics and Enemy Release

Blossey and Nötzold (1995) articulated an interesting potential implication of enemy release. They suggested that exotic plants that have been released from natural enemies will evolve to invest more in growth and reproduction, and less in costly defense mechanisms. Since this should lead to enhanced competitive ability in exotic plants, they called their hypothesis the 'evolution of increased competitive ability' (EICA) hypothesis. Of course, this should also lead to increased susceptibility to importation biological control agents, which would encounter less-defended genotypes of plants in their introduced ranges. Indeed, a general pattern of increased growth and biomass, along with increased susceptibility and colonization by some specialist herbivores, was found in comparisons of some populations of the invasive weed purple loosestrife, *Lythrum salicaria*, collected from introduced versus native ranges and evaluated in common garden settings (Blossey & Nötzold 1995; Blossey & Kamil 1996; Willis & Blossey 1999, but see Willis et al. 1999). For tansy ragwort, *Senecio jacobaea*, introduced populations also produced larger plants than native populations under conditions of competition, and suffered higher mortality from the specialist flea beetle, *Longitarsus jacobaeae* (Stastny et al. 2005). Similar results were found for the agricultural weed *Silene latifolia*, introduced from Europe to North America in the early 1800s (Blair & Wolfe 2004; Wolfe et al. 2004) and the Chinese tallow tree, *Sapium sebiferum*, growing in Texas in the United States (Siemann & Rogers 2003a, b; Rogers & Siemann 2004).

However, while the EICA hypothesis is intriguing, and may indeed help to explain invasiveness for some plants, a number of studies have cast doubt on its general applicability to invading plants (e.g., Van Kleunen & Schmid 2003; Vilá et al. 2003; Felker-Quinn et al. 2013). One of the underlying assumptions of the EICA hypothesis, namely that exotic plants are larger in their introduced range, has itself proven difficult to confirm (Willis et al. 2000; Thébaud & Simberloff 2001), leaving EICA with weak general support. Even plants that are indeed larger in their introduced range do not necessarily show decreased resistance to introduced herbivores (Joshi & Vrieling 2005). And in other cases, reduced resistance to herbivores has been found in introduced populations, but without concomitant increases in size, fecundity or competitive ability (Daehler & Strong 1997; Bossdorf et al. 2004a, b; Maron et al. 2004; Meyer et al. 2005). In these cases, reduced resistance may be due to a founder effect in which resistance traits are lost during introduction (Daehler & Strong 1997; Burdon & Thrall 2004; see also Chapter 7).

Thébaud and Simberloff (2001) feared that early and uncritical acceptance of the EICA hypothesis would lead to unwarranted optimism regarding biological control and an increase in the rate of biological control agent releases. In our view, however, the caution (and even skepticism) with which weed biological control scientists have viewed this hypothesis makes such a scenario unlikely.

2.4 Boom and Bust Invasion Dynamics: Enemy Release Followed by Biotic Resistance?

A number of well-established introduced species have suffered spectacular population crashes in the absence of any management intervention. These spontaneous collapses, which sometimes even result in local extinction, are somewhat of an enigma, as there appear to be few clear patterns governing their occurrence (Simberloff & Gibbons 2004). Examples of mysterious collapses of well-established introduced species include *Elodea* water weeds in Europe, cane toads in Australia, weasels in New Zealand, the Mediterranean fruit fly in Australia, the yellow fever mosquito in Europe, the giant African snail in the South Pacific and a number of others reviewed by Simberloff and Gibbons (2004). Explanations for the collapses include exhaustion of resources, interactions with other introduced species (including fortuitous biological control) and biotic resistance. In many cases, however, the cause of the declines is unknown, and even when explanations are given, they are largely anecdotal. The collapses may in some cases also be simply a low point in boom-and-bust cycles that have not yet rebounded.

One hypothesis that could explain these observations is that initial population increases are due to enemy release, with subsequent crashes attributable to biotic resistance – in other words, a biotic resistance lag. Growth in an enemy population sufficient to lead to a collapse of an invading species may take decades. And if the invader has a higher rate of spread than its enemies, the enemies may need to 'catch up' before they can have a strong impact on the invader (Fagan et al. 2002). In the case of evolutionary adaptation to an invading species, biotic resistance may take much longer than this (we return to this topic in Chapter 7).

The case of the giant African snail, *Achatina fulica*, may be illustrative in this context. This snail experienced explosive population growth following its introduction to many Pacific islands from East Africa. Populations tended to stay high for periods of approximately 10 years, after which time they crashed to very low levels, even becoming locally extinct in some areas (Waterhouse & Norris 1979; Simberloff & Gibbons 2004). These crashes could have been due to the introduction of the predatory snail *Euglandina rosea* (see also chapter 4; Civeyrel & Simberloff 1996), but they also coincided with disease symptoms associated with the bacterium

Aeromonas hydrophila, which is lethal when ingested by the snails and present in the soil of areas invaded by the giant snail (Mead 1979). Simberloff and Gibbons (2004) pointed out that many diseases require high host densities to cause epidemics, so it may be that the snails exceeded a critical density above which the disease caused an epidemic. In the case of *A. fulica*, this dynamic may also have been aggravated by increased susceptibility to the disease under stress, which can occur at high densities. The case of the giant African snail combines elements consistent with a hypothesis of enemy release early in the process of invasion, and biotic resistance later on. Or rather, if the bacterium was introduced along with the snails (as it well may have been), the scenario is more one of temporary enemy release. In either case, while the bacterium was present in the introduced range during the early phases of the snail invasion, its effect would have been minimal if it required high densities of its host to cause epidemics. The initial explosion of snail populations could therefore be said to have been facilitated by enemy release. Once at high densities though, the snails became vulnerable to increased rates of infection by the bacterium, leading to a precipitous crash in populations. On islands where these crashes led to local extinction, the story ends there, but on islands where low densities of snails persist, the cycle is poised to repeat itself. The extent to which these dynamics would interact with the presence of the introduced predatory snail (which we discuss further in Chapter 4) is unclear.

2.5 Improving Biological Control with Invasion Theory

Both Ehler (1998) and Fagan and colleagues (2002) reasoned that conceptual and theoretical advances in invasion biology should lead to insights that could improve success or prediction in biological control. In this section, we explore this idea by discussing two interrelated topics that illustrate strong links between invasion theory and success in biological control: Allee effects and dispersal rates.

2.5.1 Propagule Pressure and Allee Effects

Propagule pressure refers to the size of an introduced population as well as the number of introductions made. Unsurprising, the probability of establishment increases with propagule pressure, and this is due mainly to stochastic processes that occur at low population sizes to increase the chance of local extinction (Simberloff 2009). However, low population sizes can also lead to deterministic per capita fitness reductions that are unrelated to environmental or demographic stochasticity. These relationships are known as Allee effects, and they refer to increased per-capita population growth rates with increasing densities below a so-called Allee threshold. Above this threshold, population growth and per-capita fitness decline with density due to intraspecific competition (Courchamp et al. 2008). Typical causes of Allee effects include difficulties in mate finding, group defense of prey or limitations in resource processing at low densities.

The biological control literature is a valuable source of data that can be used to test hypotheses involving propagule pressure and Allee effects since the number of individuals released, and the frequency of releases, is often recorded. Comparative analyses have indeed confirmed that releases of high numbers of individual agents lead to a higher likelihood of establishment (Beirne 1975; Hall & Ehler 1979; Cameron et al. 1993; Hopper & Roush 1993; Fauvergue et al. 2012) as have most directed experimental studies (Campbell 1976; Memmott et al. 1998, 2005; Grevstad 1999a; but see Fauvergue et al. 2007). A retrospective analysis of parasitoid releases conducted by Hopper and Roush (1993) found that establishment rates from single releases were significantly higher when more than 100 individuals were released for parasitoids in the hymenopteran superfamily Ichneumonoidea or more than 1,000 individuals were released for parasitoids in the hymenopteran superfamily Chalcidoidea or the dipteran family Tachinidae.

Given the importance of release size and frequency on establishment success – what is the optimal strategy for release given limited numbers of individuals available for release and limited time to conduct releases in a given season? Memmott and colleagues (1998) used results from experimental releases of a thrips, *Sericothrips staphylinus*, to control invasive gorse bushes, *Ulex europaeus*, in New Zealand to address this question. They conducted releases of between 10 and 810 thrips per bush and found a strong positive relationship between numbers released and the establishment rate. They then used these results to show that an intermediate release size maximizes the total number of successful establishments given a limited number of thrips available for release. Models by Grevstad (1999b) and Shea and Possingham (2000) have extended these insights to more formally navigate the trade-off between conducting many small releases and fewer large releases. The main conclusion of these analyses was that the optimal strategy depends critically on the shape of the relationship between release size and establishment probability. Since this relationship is often not known, an adaptive management approach was advocated in which initial releases are done over a range of release rates so that information gained can be used to optimize later releases (Shea et al. 2002).

It is worth noting that despite these clear effects of propagule pressure on establishment of biological control agents, very small releases (< 10 individuals) have resulted in establishment in some cases (e.g., Cock 1986; Grevstad 1999a; Memmott et al. 2005). Hymenopteran parasitoids may be expected to be particularly amenable to establishing in small populations since unmated females can oviposit male eggs via haplodiploidy. Indeed, Fauvergue and colleagues (2007) achieved establishment by releasing single mated females of the parasitoid *Neodryinus typhlocybae* (Hymenoptera: Dryinidae) against the plant-sucking pest *Metcalfa pruinosa* in southern France as part of a formal Allee effects study. This finding, as well as a subsequent study in which the source of an Allee effect was difficult to identify (Fauvergue & Hopper 2009), has prompted the hypothesis that Allee effects may be less important in hymenopteran parasitoids than in other biological control agents. Specifically, Fauvergue and colleagues (2007) identified six potential explanations for weak Allee effects in parasitoids, including the following: (1) high intraspecific competition; (2) efficient host location; (3) adaptive patch use behavior; (4) haplodiploidy, which allows reproduction by unmated females and tends to dampen inbreeding depression (we return to this in Chapter 7); and (5) rescue of small populations by metapopulation dynamics. This is an intriguing hypothesis, but there are currently insufficient data to address its generality.

2.5.2 Dispersal Rates of Biological Control Agents

The dispersal rates of biological control agents relative to those of their target pest or weed has important implications for outcomes in biological control (Fagan et al. 2002; Cronin & Reeve 2005; Sivakoff et al. 2012). In particular, if the agent is a faster disperser than the target is, it would be expected to 'catch up' to the pest at a rate defined by the relative dispersal rates (Fagan et al. 2002). Beyond catching up, biological control agents with greater dispersal capabilities than their target species should be able to keep target populations from spreading further. In a potential example of this principle, Coll and colleagues (1994) report population genetic data consistent with much higher dispersal rates of the native ladybird *Coleomegilla maculata* than one of its prey species – the Colorado potato beetle, *Leptinotarsa decemlineata*. They argue that the high mobility of this predator contributes greatly to its effectiveness as a biological control agent. Similar arguments have been made for the southern pine bark beetle, *Dendroctonus frontalis*, and its predator, the checkered beetle, *Thanasimus dubius* (Cronin et al. 2000). Analogs can be found in natural systems as well. For example, a series of studies showed that various species of mobile parasitoids were able to 'pin' populations of less-mobile western tussock moths, *Orgyia vetusta*, within rather circumscribed patches of their host plant (Brodmann et al. 1997; Maron & Harrison 1997). Another example that is more analogous to weed biological control was found on the slopes of Mt. St. Helens in the northwestern United States in the years after a violent volcanic eruption that decimated all plant and animal life. The spatial structure of recolonization by prairie lupines, *Lupinus lepidus*, to this habitat was limited by herbivores with a high frequency of long-distance dispersal events (Fagan & Bishop 2000; Fagan et al. 2005).

High dispersal rates of biological control agents are not necessarily desirable, however. Strong dispersal by released biological control agents can lead to Allee effects at the edge of the dispersal front (Heimpel & Asplen 2011). Theoretical analyses have found that the probability of establishment decreases with dispersal rates of released biological control agents (Hopper & Roush 1993; Jonsen et al. 2007) and also that dispersal of established agents can trade off with local suppression (Kean & Barlow 2000b). These are specific results of the general finding that a greater population radius (i.e., propagule pressure) is needed for establishment as the diffusion coefficient (i.e., dispersal rate) of the invader increases (Lewis & Kareiva 1993). This relationship is not dependent on Allee effects, but it is strengthened by them (Figure 2.7). An added complication is sex-biased dispersal, which can alter rates of establishment and spread. In particular, either male- or female-biased dispersal can diminish the dispersal speed of invading organisms over that found in single-species models (Miller et al. 2011). Female-biased dispersal has been found in some parasitoid biological control agents (Vorley

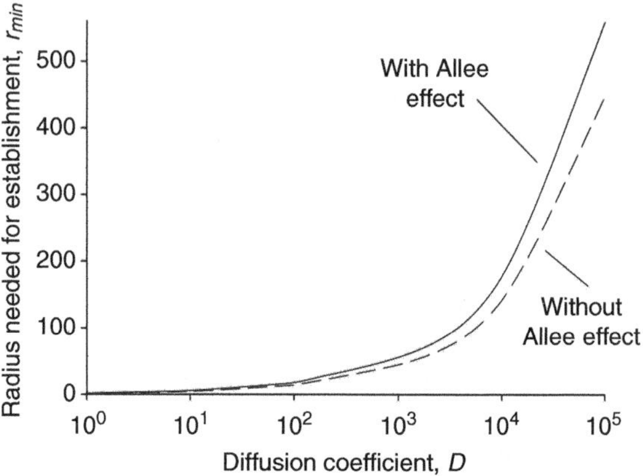

Figure 2.7. The relationship between the diffusion coefficient of an invader and the minimum radius of the population needed for establishment. The dotted line shows the relationship with no Allee effect and the solid line includes a mild Allee effect. The form of the relationship is from Lewis and Kareiva (1993) and figure and further explanations are from Heimpel and Asplen (2011).

& Wratten 1987; Ode et al. 1998; Bellamy & Byrne 2001; Wanner et al. 2006; Asplen et al. 2016). This dispersal pattern can exacerbate mate-finding Allee effects if females do not mate prior to dispersal.

Manipulating the dispersal of biological control agents is possible, and Heimpel and Asplen (2011) have discussed some strategies. One extreme example of this has been the breeding of flightless morphs of the ladybird *Harmonia axyridis* to limit dispersal from release sites (Nakayama et al. 2010). Short of this, Hougardy and Mills (2006) have shown how the post-release dispersal of the parasitoid *Mastrus ridens* (= *ridibundus*) against the codling moth, *Cydia pomonella*, can be manipulated by the kind of experience the parasitoid is given pre-release. Individual parasitoids held without hosts before being released tend to disperse much farther than those exposed to hosts. These authors envision scenarios where longer- or shorter-range dispersal might be desirable in different circumstances and that pre-release holding conditions could be varied accordingly.

2.6 Conclusions

Charles Elton launched the formal study of invasion biology in the middle of the 20th century, and it has since grown into an active subdiscipline of ecology. Many questions are still outstanding, but importation biological control has provided an important source of information to address hypotheses in this area. Examining the causes of biological control successes and failures has yielded important insights about the importance of biotic resistance and enemy release mechanisms in determining invasion dynamics.

We focused on biotic resistance and enemy release in this chapter as foundational concepts in invasion biology that have strong links to biological control. We found that biotic resistance can be strong and that it can act against accidentally introduced invaders (potential pests and weeds) and also against biological control agents themselves. In the former case, biotic resistance constitutes natural biological control and in the latter case it interferes with manipulative biological control. The enemy release hypothesis constitutes the underlying justification for the practice of biological control – if introduced species become invasive because they have 'left their enemies behind', then reuniting them with these enemies is a promising remedy. We found some evidence for the enemy release hypothesis in the literature, but it is clearly not a universal explanation for invasiveness. It is also tempting but naive to conclude that pests that are successfully controlled by importation biological control must have become invasive precisely because they were released from those enemies that have successfully controlled them in the introduced range.

Additional conceptual and theoretical advances in invasion biology, as well as more empirical information, will help to inform best practices in biological control, as has already begun.

3 Importation Biological Control – The Scope of Success

Although Charles Elton wrote his classic text *The Ecology of Invasion by Animals and Plants* in 1958, only within the past 15 years has the study of biological invasions grown to become a major focus in ecology (Davis 2009; Richardson 2011; Lockwood et al. 2013; Simberloff 2013). Invasive species are considered among the most important threats to global biodiversity and can impact ecosystems through influences on the fitness traits of individuals, their population growth rates, the structure and function of ecological communities and the ecosystem services they provide (Vitousek 1990; Ricciardi & Simberloff 2009; Simberloff 2013). Thus ecologists are rightly concerned about the likelihood of invasions of exotic species and the negative impacts that such invasions can have on indigenous plant and animal communities. From an alternative perspective, biological control practitioners have been concerned about the impact of invasive species on managed and natural ecosystems for more than a century, through the knowledge that a well-adapted specialist natural enemy can effectively mediate the detrimental effects of an invasive pest or weed (Caltagirone 1981; Ehler 1998; Hoddle 2004a; Müller-Schärer & Schaffner 2008; Van Driesche et al. 2010; Moran et al. 2011; Seastedt 2015). Thus biological control practitioners also focus their attention on the same process of invasion. Questions of importance in biological control are the likelihood of establishment of natural enemies as deliberate invasions (explored earlier in Chapter 2), the level of impact the introduced natural enemy will have on the targeted exotic pest (explored in this chapter) and the likelihood of a spillover effect onto indigenous plant and animal communities (discussed in Chapter 4).

Ecological processes are often studied in laboratory microcosms as this allows the exploration of mechanistic explanations through experimental manipulation of specific factors under controlled conditions. However, there is then the practical dilemma of how best to scale up from simplified laboratory experiments to verify that the same mechanistic explanations apply under field conditions. It is in this context that the practice of deliberate introductions through importation biological control has much to offer as biological control introductions represent very large-scale field experiments (Crawley 1986). The aim of such introductions is to suppress the abundance or activity of a pest on a regional scale, as occurred when the cassava mealybug, *Phenacoccus manihoti*, was famously controlled across the central belt of Africa through the deliberate introduction of the specialist parasitoid *Anagyrus lopezi* (Neuenschwander 2001). Thus, despite a lack of replication, the biological control record of natural, large-scale ecological experiments can inform our understanding of factors that influence both the establishment and the impact of invasive species.

In this chapter, we will begin by focusing on the scope of biological control importations and the various measures of success that have been used in documenting the outcome of such programs. We will move on to explore patterns of success in the biological control record and the factors most likely to influence the establishment and impact of introduced natural enemies in the management of invasive pest and weeds.

3.1 Scope of Importation Biological Control

Importation biological control, often referred to as *classical biological control*, is defined as the deliberate introduction of an exotic natural enemy for suppression of the abundance or activity of an undesirable species. The term *neoclassical biological control* (Lockwood 1993) has been applied to situations in which the natural enemy is exotic and the pest is indigenous, and *new associations biological control* (Hokannen & Pimentel 1989) refers to cases in which the exotic natural enemy is from a related host and thus has no coevolutionary history with the invasive pest. The first ever deliberate introduction of an exotic natural enemy is that of the mynah bird from India for control of red locusts in sugarcane on the island of Mauritius in 1762 (Legner 2008). More than 70 years later in 1795, the mealybug, *Dactylopius ceylonicus*, was imported into northern India from Brazil under the misapprehension that it was *D. coccus*, the cochineal insect that produces a wonderful carmine dye. It did not develop as expected on the cultivated *Opuntia ficus-indica*, but transferred to *O. vulgaris*, its natural host plant, which had become an invasive weed in India at that time. Thus in 1836, *D. ceylonicus* was used as an effective control agent for *O. vulgaris* in southern India, providing the first example of importation weed biological control (Goeden 1978).

Since the time of these pioneering introductions, numerous exotic natural enemies have been imported worldwide for the intentional control of a variety of target pests. Importation biological control has primarily been used for the management of arthropod pests and weeds using parasitoids, predators, arthropod herbivores and microbial pathogens. Notable among these was the introduction of the vedalia beetle, *Rodolia cardinalis*, from Australia for control of the cottony-cushion scale, *Icerya purchasi*, in California in 1888, as it has become a well-publicized and iconic example of the dramatic success that can be achieved (Caltagirone & Doutt 1989). Other examples of early successes in importation biological control are discussed in detail by DeBach and Rosen (1991). However, not all importations have lived up to these early successes. In fact, the biological control record includes a far greater number of failures than successes, and the challenge of predicting which pests and natural enemies are most likely to result in success is a topic explored in greater detail through the rest of this chapter.

In addition to arthropod pests and weeds, both terrestrial and aquatic mollusks (Cowie 2001; Coupland & Baker 2007; Pointier et al. 2011) and vertebrates (Hoddle 1999; Hickling 2000; Saunders et al. 2010; Di Giallonardo & Holmes 2015a) have occasionally been selected for importation biological control programs, and there has been some discussion of opportunities for marine invertebrates (Lafferty & Kuris 1996; Goddard et al. 2005). Three predatory snails, *Gonaxis kibweziensis* and *G. quadrilateralis* from East Africa and *Euglandina rosea* from Florida, were introduced to Hawaii in 1952–1957 to control the giant African snail, *Achatina fulica*, a terrestrial mollusk with a very broad host range (Cowie 2001). Subsequently, as many as 12 exotic predatory snails were released in Hawaii during the 1950s and 1960s, but only these first three became established. As the giant African snail is also an accidental introduction in many other tropical and subtropical islands, the same three predatory snails have also been imported, and some have established on islands in the Indian Ocean and French Polynesia (Cowie 2001; Gerlach 2001). The general consensus is that the introduced predatory snails failed to bring about any reduction in the abundance of the giant African snail, and that the pest subsequently declined in abundance due perhaps to bacterial disease as noted earlier in Chapter 2 (Civeyrel & Simberloff 1996; Cowie 2001, Simberloff & Gibbons 2004). An unintended consequence, however, was that the introduced predatory snails had an undesirable impact on the smaller indigenous tree snails in these regions (see Chapter 4 for further discussion). In contrast, deliberate introductions of predatory crayfish and snail competitors appear to have been more successful in the management of aquatic mollusks as intermediate hosts of schistosomes (Mkoji et al. 1999; Cowie 2001). For example, introductions of the competing snails, *Marisa cornuarietis* from northern South America and *Melanoides tuberculata* from the Old World, have succeeded in reducing populations of *Biomphalaria glabrata*, which is the intermediate host of *Schistosoma mansoni* in the Caribbean (Pointier et al. 2011).

For vertebrate pests there have also been few examples of importation biological control. Historical importations of vertebrate predators, such as domestic cats and the Indian mongoose, *Herpestes javanicus*, provided notorious examples of undesirable non-target impacts in importation biological control (see Chapter 4 for further discussion), and consequently the focus for importations of natural enemies shifted from vertebrate predators to pathogens and parasites (Saunders et al. 2010; Di Giallonardo & Holmes 2015a). While viral pathogens have been used successfully against rabbits and domestic cats, the case of rabbits in Australia provides ample evidence that pest suppression is only temporary due to the evolution of intermediate virulence on the part of the virus and resistance on the part of the rabbits (we return to this in Chapter 7). The use of competitors as natural enemies has also arisen on one occasion in the biological control of vertebrate pests. Sterilized red foxes, *Vulpes vulpes*, were introduced to the Aleutian Islands, where they were competitively superior to the introduced arctic fox, *V. lagopus*, and successfully eradicated the latter before being removed themselves from the islands (Bailey 1992).

Plant pathogens have also only rarely been targeted, with the introduction of a virus to control the invasive chestnut blight into North America (Milgroom & Cortesi 2004) being the only example of which we are aware. As far as we are aware, nematodes have never been targeted for biological control introductions. Of the many potential reasons for this, two stand out as particularly important constraints; the complexity of community structure and function in the rhizosphere and phylloplane, and the low host specificity of the natural enemies in many of these communities.

In a broader context, however, mollusks, vertebrates and plant pathogens represent only a very small proportion of the pests selected for deliberate importations of exotic natural enemies, and for the rest of this chapter we will focus our attention on the far greater range of importations used for the control of arthropod pests and weeds.

3.2 Measures of Success

One of the least satisfactory aspects of the historical record of biological control introductions is the documentation of the outcome of each project. For the majority of such projects there has been very little effort directed toward quantifying either the impact of the natural enemies that became established or the associated reduction in abundance of the target pest. This is in part a consequence of the reluctance of sponsors to continue to fund this final and open-ended phase of an importation biological control program (McFadyen 1998), but also stems from a lack of interest among practitioners once the project appears to be achieving the desired outcome. For example, while the benefits of continued monitoring were recognized early on (Huffaker & Kennett 1959), monitoring and evaluation were not included in a more recent guide to steps in importation arthropod biological control (Van Driesche & Bellows 1993), and little progress has been made in collecting quantitative data on the effects of introduced biological control agents (Gurr et al. 2000; Blossey 2004; Morin et al. 2009).

The global historical record of weed biological control has been documented by Winston and colleagues (2014), and that for arthropod pests has been recorded by Clausen and colleagues (1978) and Greathead and Greathead (1992). These publications and databases document the species of natural enemies introduced for each target pest, the date and country of introduction and a subjective measure of the degree of control achieved. The terminology used to describe the *biological success* of introductions against arthropod pests and weeds has been remarkably consistent (Van den Bosch et al. 1982; Hoffmann 1995):

- Failure – no long-term establishment of the introduced natural enemy
- Establishment – permanent establishment of the introduced natural enemy, but no obvious impact of the natural enemy
- Partial success – management is still dependent on the use of other control measures, despite an observable impact of the natural enemy
- Substantial success – the impact of the natural enemy is sufficient to reduce the frequency or severity with which other control measures are applied
- Complete success – other control measures are no longer needed to supplement the impact of the established natural enemy

These are clearly subjective criteria that can be influenced by a broad range of factors such as the type and extent of damage caused by the target species, the economic value of such damage and the degree to which it can be tolerated in the new environment. They are not based on any quantitative assessment of the degree of suppression of target populations and therefore provide no direct indication of the role the introduced natural enemy plays in the population dynamics of the target species. More recently, an alternative framework for classifying the effect of invasive species has been suggested by Blackburn and colleagues (2014), and we will return to this later on (Section 3.5.2).

Biological control introductions represent an example of pure public good from an economical perspective (Tisdell & Auld 1990), and are typically financed through government funding at a state, national or international level. As a result there has been an increasing need to evaluate the *economic success* and benefits of biological control introductions. *Ex post* analyses are carried out after a successful project has been completed to highlight the value of earlier natural enemy introductions (Hill & Greathead 2000; Culliney 2005), while *ex ante* analyses are carried out at the start of a project to justify investment in the development of a new program (Jetter 2005; Jarvis et al. 2006). Very large benefit-to-cost ratios have resulted from a number of biological control introductions that have shown complete success, such as the introduction of the parasitoid wasp, *Neodusmetia sangwani*, for control of the Rhodes grass scale, *Antonina graminis*, in Texas with a ratio of 1,285:1 (Gutierrez et al. 1999), and the introduction of the rust fungus, *Puccinia chondrillina*, for control of skeleton weeds in Australia with a ratio of 112:1 (Marsden et al. 1980). Even greater ratios have been estimated for repeat programs, where the same successful natural enemy is introduced against the same target species in another geographic region, such as the introduction of the parasitoid fly *Lixophaga diatraeae* for control of sugarcane borer in St. Kitts-Nevis with a ratio of 3,301:1 (Gutierrez et al. 1999), and the introduction of the weevil *Cyrtobagus salviniae* to Sri Lanka for control of salvinia with a ratio of 1,675:1 (McFadyen 1998). Benefit:cost ratios very much depend on the time period of assessment, and similarly suffer from uncertainties, such as variation in the types of costs and benefits included and the subjectivity of the values assigned to them (Tisdell 1990; Culliney 2005). Nonetheless, economic benefits and benefit:cost ratios will continue to play an important role in evaluating the success of biological control, and allowing comparisons to be drawn between the benefits of biological control in relation to other control strategies (Culliney 2005; Bale et al. 2008; McFadyen 2008).

In addition to considering biological and economic success, Anderson and colleagues (2000) and Myers and Bazely (2003) have suggested five other measures of success that can be applied to biological control introductions. *Ecological success* applies particularly to biological weed control, as the suppression of an invasive weed may or may not have ecological benefits in terms of land management. A frequent outcome of the successful suppression of rangeland weeds is the replacement of one invasive weed by another, and thus for land managers a biological success may not necessarily lead to ecological success in terms of the quality and productivity of the landscape (Reid et al. 2009). An additional aspect of ecological success is the persistence of the benefits gained from a natural enemy introduction. As discussed in detail in Chapter 7, the introduction of the myxoma virus to Australia for the control of European rabbits *Oryctolagus cuniculus* in 1950 led to complete, but transient biological success with tremendous temporary ecological benefits for land management.

A significant and less appreciated measure is *scientific success*, defined by the knowledge gained from a biological control introduction that can be used to improve the biological, economic or ecological success of future programs. The emphasis here is on scientifically rigorous programs that can make significant advances in our understanding of the mechanistic basis of biological control whether or not they achieve control of the target pest. The final three measures of success are often overlooked as a result of being less clearly defined goals of an introduction program, but nonetheless, are important incidental components of success: *political success* in generating greater stakeholder and funding agency support for future programs, *social success* in enhancing greater public awareness and appreciation of biological control introductions and *legal success* in developing effective legislation to reduce the incidence of invasions and to facilitate responsible implementation of biological control introductions. In this broader context, in a recent review of the benefits of biological weed control, Suckling (2013) discusses the criteria developed in New Zealand to evaluate environmental, economic and social benefits on a six-point scale from negligible to massive.

3.3 Quantifying the Biological Success of Natural Enemy Introductions

Although a lack of quantitative data from individual importation programs has often hampered our understanding of why a particular program was a success or failure, the extensive historical record of importations provides an opportunity to look for more general patterns associated with success and failure. As Crawley (1986) noted, such records are neither replicated nor random samples, but they do provide a valuable opportunity to try to relate differences between success and failure to biological traits associated with both pest or weed and natural enemy. In this section, we will focus on patterns in the historical record of biological control introductions and factors likely to influence biological success.

3.3.1 Approaches to Estimate the Biological Success of Introduced Natural Enemies

A number of different approaches have been used to provide evidence of success in importation biological control. For invasive weeds, photographic documentation of sites before and after release of introduced natural enemies has been used consistently to provide compelling evidence of declines in weed density following successful introductions (Huffaker & Kennett 1959; Room et al. 1981; McConnachie et al. 2004). However, such illustrations are only convincing when success is complete, and of course do not work in the case of arthropod pests due to their much smaller size and greater mobility. For some arthropod pests, detailed field sampling, to provide before and after estimates of the abundance of the target species and introduced natural enemy, can be sufficient to provide a very compelling argument. The successful control of the ash whitefly in California through the introduction of the parasitoid *Encarsia inaron (= partenopea)* provides a good example (Gould et al. 1992). Field sampling of whitefly populations at the original release site showed that densities declined by a factor of 1,000 over a period of a single year following the introduction of the parasitoid, with levels of parasitism building to and remaining around 90% (Figure 3.1a). Similarly, from field sampling at Loftus in northern California (Huffaker & Kennett 1959), the successful control of Klamath weed, *Hypericum perforatum*, clearly coincided with the initial increase and subsequent decline of the leaf beetle, *Chrysolina quadrigemina* (Figure 3.1b). Long-term field sampling of European rabbit *Oryctolagus cuniculus* populations at Koonamore and Witchitie

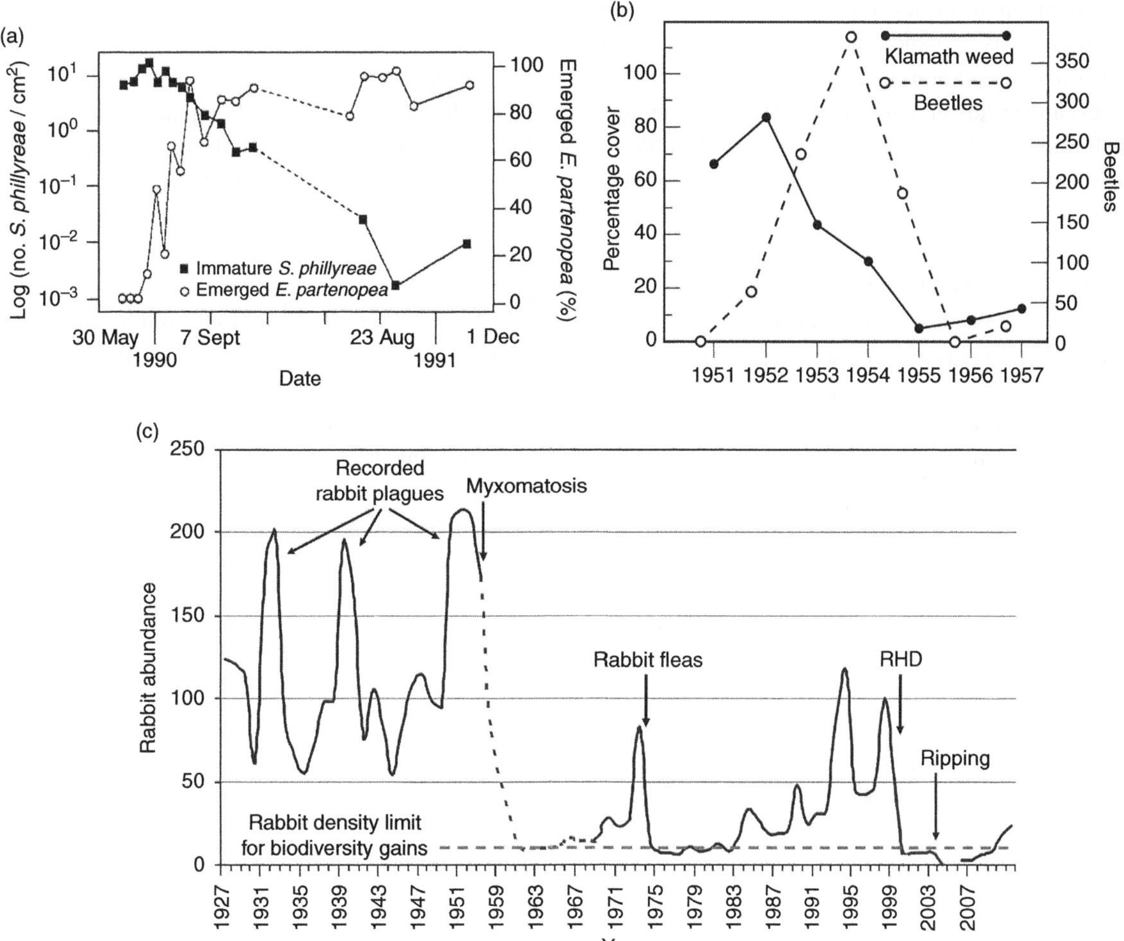

Figure 3.1. Examples in which field sampling has provided convincing evidence of the success of biological control introductions; (a) the parasitoid *Encarsia inaron (= partenopea)* for ash whitefly, *Siphoninus phillyreae*, in California (after Bellows et al. 1992), (b) the leaf beetle, *Chrysolina quadrigemina*, for Klamath weed, *Hypericum perforatum*, in California (after Huffaker & Kennett 1959) and (c) the myxoma and calici (RHD) viruses for European rabbit *Oryctolagus cuniculus* in eastern Australia (after Saunders et al. 2010).

sheep stations in northeastern Australia (Saunders et al. 2010) has also shown sharp declines in abundance coinciding with introductions of myxomatosis and rabbit hemorrhagic disease viruses (Figure 3.1c). However, despite these clear indications of cause-and-effect relationships, few biological control programs have been monitored and evaluated in sufficient detail to confirm such relationships.

Confirmation that an introduced natural enemy is responsible for at least some degree of suppression of a particular pest species can be verified only through life-table analysis, experimental methods comparing either release and non-release sites or natural enemy access and exclusion plots, or population modeling (Bellows & Van Driesche 1999; Carson et al. 2008; Morin et al. 2009). While life-table analysis provides a valuable quantitative approach to evaluating biological control agents (Bellows & Van Driesche 1999), it has seldom been used in biological control programs. This is undoubtedly due to the extensive data collection required over a sufficient period of time to be able to clearly evaluate the roles of the introduced natural enemies relative to other limitations on the survivorship and reproduction of the pest. One notable exception is the evaluation of introduced parasitoids for the biological control of larch casebearer, a defoliator of larch forests in Oregon (Ryan 1990). Data collected over a period of 18 years, spanning both before and after introduction of the two exotic parasitoids, *Agathis pumila* and *Chrysocharis laricinellae*, shows the change in casebearer larval densities and percent parasitism. A detailed life-table analysis was carried out in three separate sites, and *A. pumila* was shown to be the key factor at all three sites for the time period after parasitoid introduction from 1980–1988 (Figure 3.2; Table 3.1). This analysis confirms that parasitism by *A. pumila*, which reached as high as 80% in 1984, was the factor responsible for the decline in larch casebearer densities at these three sites.

Experimental assessment is a more generally applicable approach to quantifying the impact of natural enemies either through comparison of replicated release versus non-release sites or replicated exclusion versus access plots for the released natural enemies. The first of these two methods has been advocated for arthropod programs by Bellows and Van Driesche (1999) and for weed programs by Carson and colleagues (2008). The effectiveness of this method often depends on the proximity of the release and non-release sites and the dispersal ability of the natural enemy, but can provide at least a short-term comparison. Natural enemy exclusion techniques include the use of exclusion cages, insecticide removal and hand removal (Luck et al. 1999; Dhileepan 2003; Morin et al. 2009). A comparison of exclusion cages to open cages or

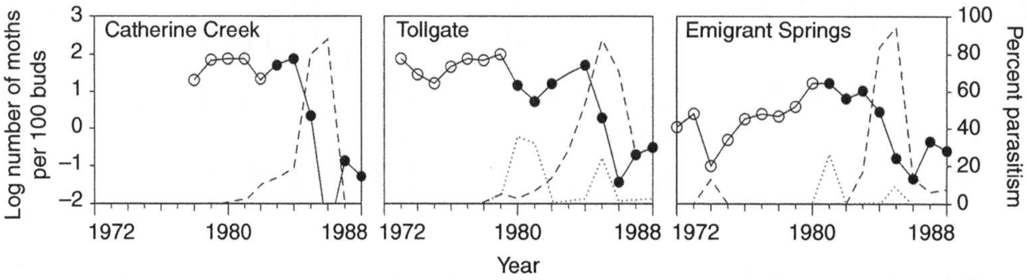

Figure 3.2. Life-table analysis was used to evaluate the success of importation biological control for suppression of larch casebearer, *Coleophora laricinella*, at three field sites in Oregon (after Ryan 1990). Extensive parasitism by *Agathis pumila* (dashed line), but not that by *Chrysocharis laricinellae* (dotted line), coincided with suppression of casebearer abundance (solid line) at the three different field sites.

Table 3.1. **Regression coefficients representing the contribution of individual mortality factors, k_i, to generational mortality from 1980-1988 identify parasitism by *Agathis pumila* as the key mortality factor responsible for the decline in larch casebearer abundance at all three sites shown in Figure 3.2. Modified from Ryan (1990).**

	Site (from Figure 3.2)		
k value	Catherine Creek	Emigrant Springs	Tollgate
k_1: egg infertility	−0.005	−0.002	0.008
k_2: egg predation	0.034	−0.009	0.044
k_3: mortality of advanced embryos	0.268	0.058	0.194
k_4: parasitism by *Agathis pumila*	**0.359**	**0.577**	**0.288**
k_5: mortality of leaf-mining larvae	0.108	0.134	0.086
k_6: mortality of case-bearing larvae (autumn)	0.111	−0.011	0.268
k_7: mortality of case-bearing larvae (winter and spring)	0.283	0.080	0.263
k_8: parasitism by species other than *A. pumila*	0.299	−0.132	−0.078
k_0: adult mortality, reduced fecundity and emigration	−0.466	0.311	−0.073

plots has frequently been used to evaluate biological control introductions including *Anagyrus lopezi* for control of cassava mealybug, *Phenacoccus manihoti*, in Africa (Neuenschwander et al. 1986), *Gyranusoidea tebygi* for control of mango mealybug, *Rastrococcus invadens*, in Africa (Boavida et al. 1995) and *Diadegma semiclausum* for control of diamondback moth, *Plutella xylostella*, in Australia (Wang et al. 2004) and Kenya (Momanyi et al. 2006). Insecticide removal was first used for evaluating the success of *Aphytis paramaculicornis* and *Coccophagoides utilis* in controlling olive scale, *Parlatoria oleae*, in California (Huffaker & Kennett 1966), but has more recently become a favored approach for evaluating weed biological control programs, including the impact of *Hypurus bertrandi* and *Schizocerella pilicornis* on common purslane, *Portulaca oleracea*, in California (Norris 1997), *Lixus cardui* on *Onopordum* thistles in Australia (Briese et al. 2004) and *Oxyops vitiosa* on *Melaleuca quinquenervia* in Florida (Tipping et al. 2009). Such experimental manipulations are not without drawbacks, however, as the limitations of each approach must be appreciated and the results from them carefully interpreted (Luck et al. 1999; Morin et al. 2009).

Another valuable approach that can be used to assess the outcome of biological control introductions is the use of population models. There have been numerous instances of modeling the impact of introduced natural enemies for both arthropod and weed biological control (Gutierrez 1996; Barlow 1999; Kriticos 2003; Gutierrez et al. 2005, 2011; Shea et al. 2005, 2010; Morin et al. 2009; Swope & Satterthwaite 2012; Gutierrez & Ponti 2013), and a few prospective models have been developed to explore the potential for vertebrate pests (Courchamp & Sugihara 1999; Barlow et al. 2002). As discussed in greater detail in Chapter 6, a number of tactical models have also been developed to aid the evaluation of biological control programs. The success of the parasitoid *Microctonus aethiopoides* in reducing the abundance of lucerne weevil, *Sitona discoideus*, by 75% in New Zealand, while achieving negligible reduction in Australia, provides an excellent example of this approach (Barlow & Goldson 1993; Kean & Barlow 2001). Typically *M. aethiopoides* has two generations per year compared to a single generation for its host, and first instar larvae enter diapause in their pre-reproductive adult hosts in summer. However, in New Zealand, a small proportion of parasitoid individuals did not go into summer diapause, but remained in the crop to complete two or more additional generations on late emerging hosts (Goldson et al. 1990). In addition, cooler autumn temperatures delay most weevil oviposition until spring. Neither atypical parasitoid development nor cooler autumn temperatures occur in Australia, and using different model formulations Kean and Barlow (2001) were able to confirm the critical importance of these biological traits in distinguishing between the success of *M. aethiopoides* in New Zealand and its failure in Australia. Similarly, models have been used very effectively in biological weed control to not only confirm the impact of natural enemy introductions, but also to predict how long it would take to achieve a certain level of reduction in weed density (Lonsdale et al. 1995; Buckley et al. 2004), and whether additional integrated tactics were needed to supplement the biological control (Rees & Paynter 1997; Kriticos et al. 2004).

3.3.2 Quantitative Estimates of Biological Success in the Historical Record of Natural Enemy Introductions

Williamson (1996) was among the first ecologists to recognize that the invasion process consists of a series of more or less discrete steps and that for a species to become a successful invader it must succeed in overcoming the barriers between each of the successive steps (see Chapter 2). There are typically three steps to the process: introduction or transportation of an organism to a new environment outside of its natural geographic range, establishment or persistence of a viable local population in the new environment and spread or sufficient expansion in geographic range and abundance to impact the novel environment (Williamson & Fitter 1996; Richardson 2000; Blackburn et al. 2011). The recognition that there is much uncertainty in the transition between each of these steps is incorporated in the *tens rule*, which was discussed in Chapter 2. The deliberate nature of biological control introductions circumvents the first barrier to invasion success as the natural enemy is directly transported and introduced into a novel environment in which food resources are readily available. Thus the series of steps that a natural enemy must achieve to become a success in importation biological control is subsequently reduced to two – establishment and impact. As natural enemies for biological control introductions have been preselected to maximize success, it is no surprise that the tens rule is less applicable and that a threes rule appears to be a better fit (Williamson & Fitter 1996). Thus, following Mills (1994, 2000), we note that in looking for patterns of success in the historical record of biological control it remains critical to distinguish between establishment and impact, as the two processes may be quite unrelated for a wide range of invasive species (Ricciardi & Cohen 2007; Ricciardi et al. 2013). We define the rate of establishment as the proportion of introductions that result in permanent establishment, the rate of impact as the proportion of establishments that lead to at least partial success in suppressing pest abundance and the overall rate of success as the product of the two.

The global record of biological control introductions for weeds has been documented and updated in a series of publications by Julien (1982,

1987, 1992) and Julien and Griffiths (1998), with the fourth edition recording 949 introductions of exotic plant feeders and pathogens up to the end of 1996. More recently this has been updated to a fifth edition by Winston and colleagues (2014). The corresponding global record of biological control introductions of parasitoids and predators for arthropod pests has been documented by Clausen (1978) and Greathead and Greathead (1992). The BIOCAT database records 4,769 introductions of exotic predators and parasitoids up to 1990 (Greathead & Greathead 1992). In addition, Hajek and colleagues (2005, 2007) document 131 introductions of exotic pathogens and nematodes against arthropod pests.

As noted by Stiling (1990), Waage (1990) and McFadyen (2000), one of the most important features of the historical record of biological control is that it contains a number of repeat introductions of the same species of natural enemy. As the outcome for each natural enemy introduction is recorded separately, when a natural enemy proves a success in one country, it is frequently introduced into other countries around the world that have the same invasive pest or weed, and thus repeat introductions enter into the historical record. Examples include repeat introductions of the parasitoid *Aphelinus mali* for control of woolly apple aphid, *Eriosoma lanigerum*, in at least 40 countries (Stiling 1990), and repeat introductions of the lace bug, *Teleonemia scrupulosa*, for control of *Lantana camara* in 29 countries (Crawley 1989b). Consequently, the separate records of introductions are not independent, and repeat introductions of successful natural enemies tends to bias and overestimate the true rates of establishment and impact of natural enemy species (Stiling 1990). Another potential source of bias in the historical record of natural enemy introductions is the lack of phylogenetic independence that results from an uneven representation of families, genera and species of pests, weeds and natural enemies. While it is currently not possible to resolve the phylogenetic contrasts for the pests, weeds and their introduced natural enemies, the bias due to repeat introductions can be resolved.

Early estimates of rates of success in biological control importation ignored the problem of repeat introductions and treated all records as independent (DeBach 1971; Hall & Ehler 1979; Hall et al. 1980; Crawley 1989b). However, Stiling (1990) suggested that the fraction of introductions that lead to establishment for each unique linkage of natural enemy and pest species could be used to effectively remove the bias from repeat introductions. Subsequently, Mills (1994) suggested that the single best outcome from among repeat introductions for each unique linkage may be more informative of the true potential or biological success for both establishment and impact. Many factors can influence a natural enemy species' ability to overcome the barriers to successful establishment and impact in biological control. These include not only the ecological traits that are of particular interest in understanding patterns and determinants of success, but also the random environmental factors that can prevent the expression of ecological traits. For example, environmental stressors, biotic resistance, synchronization and the quality of natural enemy individuals introduced could all have a profound influence on the probability of natural enemy establishment. From this perspective, the fraction of repeat introductions may be influenced more by random environmental events, whereas the single best outcome may be more likely to capture the importance of ecological traits. The single best outcome has also been adopted more recently by Boughton and Pemberton (2008) to analyze rates of establishment of different natural enemy taxa in the biological control of weeds.

As these different approaches to handling repeated introductions in the biological control record have yet to be compared, we do so here, based on data from Greathead and Greathead (1992) for parasitoid and predator introductions against arthropod pests, from Hajek and colleagues (2005, 2007) for pathogen introductions against arthropod pests, and from Julien and Griffiths (1998) for herbivore introductions against weeds. First of all, we excluded the numerous introductions for which the outcome remains unknown, and removed all introductions for which either the natural enemy or the target pest or weed was defined only to family level. For the remaining records, we estimated the rates of establishment, impact and overall success separately for each approach (Table 3.2); total attempts (no correction); average fraction (correction based

Table 3.2. **Comparison of different approaches to quantitative estimation of the success of biological control, based on databases of introductions of parasitoids and predators against arthropod pests (Greathead & Greathead 1992), introductions of pathogens against arthropod pests (Hajek et al. 2005, 2007) and arthropod herbivores against weeds (Julien & Griffiths 1998).**

Approach to Evaluation	Rate of Establishment (n)	Rate of Impact (n)	Overall Rate of Success (n)
Arthropod pests			
Predators and parasitoids			
Total attempts including repeat introductions	39.1% (3,013)	44.0% (1,178)	17.2% (3,013)
Average fraction to correct for repeat introductions	31.7% (2,108)	35.8% (737)	12.5% (2,108)
Single best outcome to correct for repeat introductions	35.0% (2,108)	41.4% (737)	14.5% (2,108)
Pathogens			
Total attempts, including repeat introductions	68.2% (110)	93.3% (75)	63.6% (110)
Average fraction to correct for repeat introductions	63.5% (69)	91.1% (45)	59.4% (69)
Single best outcome to correct for repeat introductions	65.2% (69)	91.1% (45)	59.4% (69)
Weed species			
Arthropod herbivores			
Total attempts including repeat introductions	56.2% (957)	62.4% (538)	35.1% (957)
Average fraction to correct for repeat introductions	52.3% (423)	56.4% (259)	30.7% (423)
Single best outcome to correct for repeat introductions	61.2% (423)	64.1% (259)	39.2% (423)

on a fraction for each linkage); and single best outcome (correction based on a binary outcome for each linkage). As expected, the average fraction provides lower estimates than the total attempts for all three categories of success, while the single best outcome is intermediate for parasitoid and predator introductions and for pathogen introductions against arthropod pests. Surprising, however, the single best outcome gave higher estimates of success for herbivore introductions against weeds than the total attempts, which appears to result from a greater level of repeat failures in the weed database than in the two arthropod databases. Overall the rate of success drops from 59% for pathogen introductions against arthropods to 39% for arthropod introductions against weeds to 15% for parasitoid and predator introductions against arthropod pests (Table 3.2).

This comparison also serves to highlight that although the threes rule of Williamson and Fitter (1996) provides a reasonable fit for both the rate of establishment and rate of impact of parasitoids and predators introduced against arthropod pests, it does not fit for the other two categories of biological control introductions. For herbivore introductions against weeds, a twos rule appears more applicable, and for pathogen introductions against arthropod pests, no such rule applies as the rate of establishment exceeds 60% and the rate of impact exceeds 90%. It is tempting to suggest that the much shorter generation times or replication cycles of pathogens enhances both their ability to overcome founder effects during establishment and their ability to numerically dominate host populations following establishment. However, this would be too simplistic a view of the dynamics of pathogen–host interactions, which are dependent on many other traits of both pathogen and host populations (Anderson & May 1981 and Chapter 6).

Finally, from a perspective of weed biological control, McFadyen (1998, 2000) has argued that any evaluation of the establishment and impact of individual natural enemies is confusing, and that a more important measure of success is the proportion of biological control *programs* that have achieved partial to complete control. This is particularly pertinent for weed programs, which often involve the introduction of multiple natural enemy species (McFadyen 2000; Denoth et al. 2002), but less so for arthropod programs, which more typically involve single natural enemy species (Myers et al. 1989; Denoth et al. 2002; Mills 2006a). While no such analysis has been carried out for arthropod programs, several regional assessments have suggested much greater success rates for weed programs: 83% for 23 programs in South Africa (Hoffmann 1995), 50% for 21 programs in Hawaii (Gardner et al. 1995), 69% for 36 programs in Australia (Cullen et al. 2011) and 83% for 6 programs New Zealand (Fowler et al. 2000). That programs have greater success rates than individual natural enemies may simply reflect the fact that funding support has allowed programs to continue to seek out and introduce additional natural enemy species when the first ones fail. However, it may also reinforce the notion that weeds require multiple natural enemies to provide sufficient cumulative stress to impact their demographic performance, an issue that we will return to later in this chapter.

3.4 Empirical Patterns in the Biological Success of Natural Enemy Introductions

A variety of empirical patterns have been explored in the historical record of biological control introductions. Inevitably these have primarily focused on the taxonomic representation of both pests and natural enemies, but also on broader questions of whether success improves with time or in simpler island communities versus more complex mainland communities.

3.4.1 Pest and Weed Taxonomy

It has long been known that rates of establishment from natural enemy introductions have varied significantly among the different orders of insect pests (DeBach 1964; Hall & Ehler 1979;

Hall et al. 1980; Hokkanen 1985b; Stiling 1990; Greathead & Greathead 1992; Mills 1994, 2000; Kimberling 2004). Of particular note is that establishment has been considerably more successful against many hemipteran pests than it has for lepidopteran pests (Mills 2006b). While these two orders of pests have attracted the greatest number of biological control introductions, it is more informative to consider rates of establishment and impact among the different families of pests to be more inclusive of other orders. Mills (1994, 2000, 2006b) has examined the pattern of establishment, impact and overall success of parasitoid introductions for the dominant pest families from the BIOCAT database corrected for repeat introductions using the single best outcome for each natural enemy–pest species linkage. Here we extend this analysis to overall rates of success for both parasitoid and predator introductions (Figure 3.3a). This analysis not only confirms the

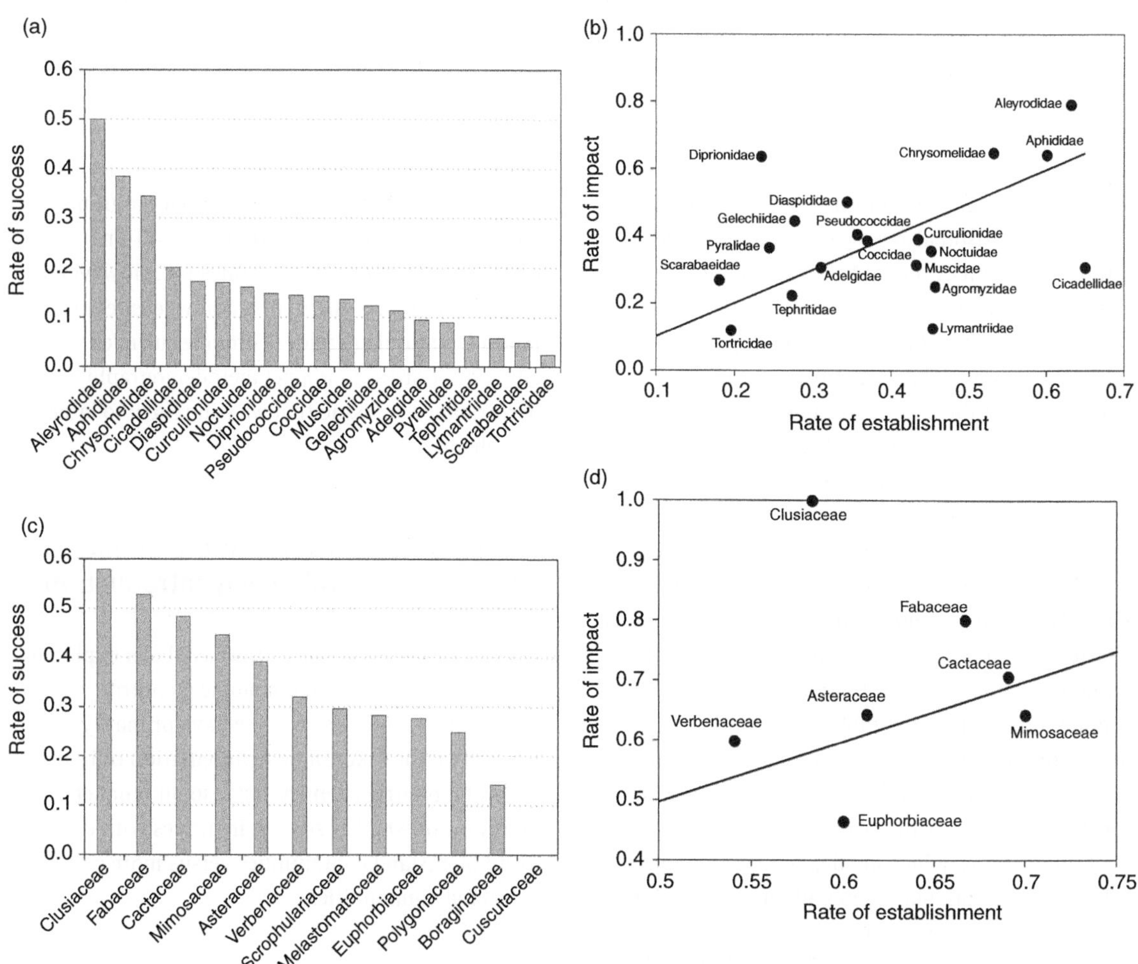

Figure 3.3. The overall rate of success of biological control introductions in relation to insect pest and weed families, based on the single best outcome to control for repeat introductions. Parasitoid and predator introductions for insect pest families with more than 30 unique link introductions (a) and 10 unique link establishments (b), and arthropod herbivore introductions for weed families with more than 6 unique link introductions (c) and 6 unique link establishments (d).

greater rates of success against hemipteran than lepidopteran families, but also points out other interesting anomalies such as the far greater rate of success against chrysomelid (leaf beetles; Coleoptera) than adelgid (woolly aphids; Hemiptera) pests. It is also valuable to compare rates of impact versus rates of establishment for these same pest families (Figure 3.3b). Those families that lie above the diagonal line of equal rates of establishment and impact are more difficult to establish as introductions, but have a greater impact when establishment is achieved, and vice versa for those families that lie below the diagonal line. From this comparison we also see that rates of impact have been far greater for certain families of pests, such as the conifer sawfly family Diprionidae, than rates of establishment. In contrast, although there have been high rates of establishment against Agromyzidae (leafmining flies; Diptera), Cicadellidae (leafhoppers; Hemiptera) and Lymantriidae (tussock moths; Lepidoptera), rates of impact have been disappointing.

As noted earlier, the biological control record for weeds includes fewer introductions than that for arthropod pests, and in particular includes far fewer target weed species. As a result, the only pattern that emerged from the biological control record is a greater overall rate of success for the Cactaceae, specifically *Opuntia* species, than for any other family of weeds (Crawley 1989b). Using an updated version of the historical record (Julien & Griffiths 1998) and the single best outcome approach to correct for repeat introductions, however, shows that weeds in the Clusiaceae (a tropical family of trees and shrubs) and Fabaceae (legumes) have had even higher overall rates of success than the Cactaceae (Figure 3.3c). In addition, the plant family with the greatest number of weed species targeted for natural enemy introductions, the Asteraceae (composites), has an intermediate overall rate of success. The rates of success for the Clusiaceae and Fabaceae are primarily driven by high rates of impact of established natural enemies, whereas rates of success for the Euphorbiaceae (spurges) have been limited by low rates of both establishment and impact (Figure 3.3d).

3.4.2 Natural Enemy Taxonomy

Greathead (1986) was the first to provide a more detailed comparison of the contribution of different families and even genera of parasitoids to the biological control record. Subsequently, Mills (2006b) examined the relative frequency of both parasitoid and predator introductions against the two most heavily targeted orders of insect pests, the Hemiptera and Lepidoptera. Using the BIOCAT database and the single best outcome to correct for repeat introductions, we extend this analysis to parasitoid and predator introductions for all orders of insect pests (Figure 3.4a). The overall rate of success has been greatest for the parasitoid families Aphelinidae, Encyrtidae, Eulophidae and Scelionidae, and notably lower for some predator families, such as Carabidae (ground beetles) and Histeridae (predatory beetles), and for the egg parasitoid family Trichogrammatidae. The Coccinellidae (ladybird beetles) and Tachinidae (parasitoid flies) appear to have been more difficult to establish, but have had a notable impact in cases where establishment has been achieved (Figure 3.4b). In contrast, Histeridae and Pteromalidae (parasitoid wasps) appear to have established more readily as introductions, but have had more limited impact once established.

For weed biological control, Crawley (1989b) compared the outcome of introductions (based on Julien [1987] and uncorrected for repeat introductions) of different insect herbivore families used for biological control of weeds, and found that cochineal insect (Dactylopiidae) introductions for control of *Opuntia* cacti (Cactaceae) had met with the greatest rate of success. More recently, Chrysomelidae (leaf beetles) and Curculionidae (weevils) have been favored as biological control introductions, and a recent meta-analysis has shown that these two families have been consistently more successful in reducing plant size following

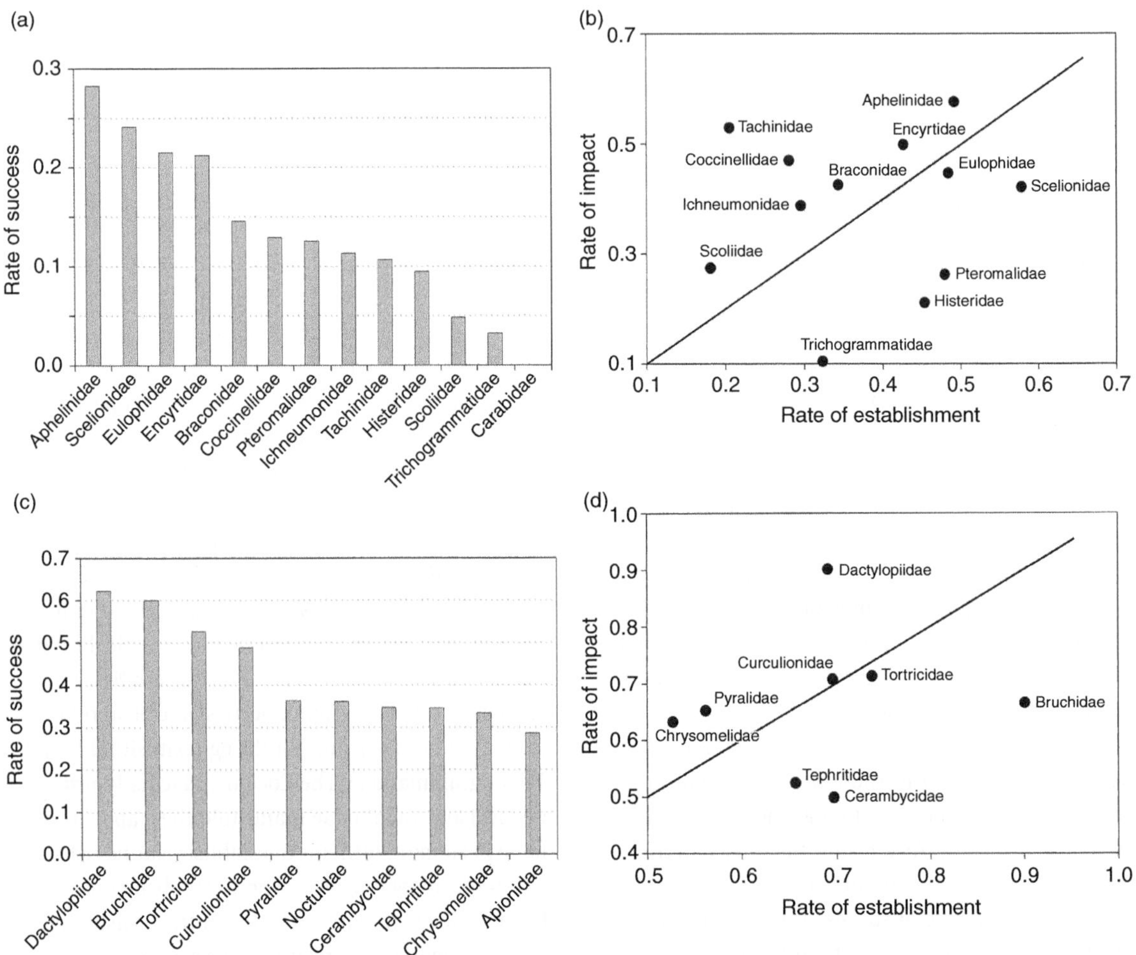

Figure 3.4. The overall rate of success rate of biological control introductions in relation to natural enemy family, based on the single best outcome to control for repeat introductions. Introductions of parasitoid and predator families with more than 30 unique link introductions (a) and 10 unique link establishments (b), and for introductions of arthropod herbivore families with more than 10 unique link introductions (c) and 9 unique link establishments (d).

introduction than others (Clewley et al. 2012). Boughton and Pemberton (2008) examined rates of establishment for different natural enemy orders and families of Lepidoptera (based on Julien & Griffiths [1998] and corrected for repeat introductions using the single best outcome approach). They found that the most frequently used orders are the Coleoptera, Lepidoptera and Diptera, with similar rates of establishment, and that the most frequently used families of Lepidoptera were Pyralidae (snout moths), Tortricidae (leafroller moths) and Noctuidae (owlet moths), with rates of establishment highest for Tortricidae and lowest for Pyralidae. Here we extend the analysis of natural enemy families to include all orders and find that the overall rates of success for Bruchidae (bean weevils), Curculionidae and Tortricidae are higher than for other families (Figure 3.4c). The success of the Dactylopiidae can be seen to be due to their high rate of impact, whereas the success of the Bruchidae has been driven primarily by a high rate of establishment (Figure 3.4d). Other patterns of note are that the Chrysomelidae have a low rate of establishment and that the Cerambycidae (longhorn beetles) and Tephritidae (true fruit flies) have not achieved as great a rate of impact as other families.

The success of the Chrysomelidae, Curculionidae and Dactylopiidae has often been attributed to reduced susceptibility to parasitism in comparison with other insect herbivore families (Goeden & Louda 1976; McFadyen & Spafford Jacob 2004). While the Dactylopiidae have no parasitoids, it is still not clear that vulnerability to parasitism limits the impact of introduced weed control agents. For example, the Cecidomyiidae (gall midges) have been surprisingly successful as introductions (rate of success 0.83) despite their infrequent use (only six unique introduction links) and high levels of parasitism by resident parasitoids (McFadyen & Spafford Jacob 2004). Similarly, despite supporting much richer parasitoid assemblages, Pyralidae and Tortricidae have achieved a greater rate of success than the Chrysomelidae (Figure 3.4c). However, parasitism by a resident parasitic mite *Pyemotes tritici* is considered to limit the impact of *Eustenopus villosus* (Curculionidae), a seed predator of yellow starthistle *Centaurea solstitialis*, in coastal California (Swope & Satterthwaite 2012). Overall the success of biological weed control seems more likely to be related to the extent of per capita damage that a natural enemy can inflict on its host plant than to its vulnerability to parasitism (Stiling & Cornelissen 2005; Clewley et al. 2012).

3.4.3 Functional Groups of Natural Enemies

The success of predator versus parasitoid introductions for the biological control of arthropod pests has been greatly influenced by repeat introductions of the predatory vedalia beetle *Rodolia cardinalis*, which have consistently provided complete control of cottony cushion scale, *Icerya purchasi* (Hall & Ehler 1979; Hall et al. 1980). However, using data from the BIOCAT database, corrected for repeat introductions by using the single best outcome, Hawkins and colleagues (1999) found a marginally significant difference in the overall rate of success of predators (12%, n = 299) compared to that of parasitoids (17%, n = 1244). Similarly, from a separate analysis of introductions into the United States only, Kimberling (2004) found parasitoids to be 9.5 times more successful than predators. As noted earlier, in comparison, the overall rate of success of pathogen introductions against arthropods has been much higher than either of these two groups (Table 3.2). Although ectoparasitoids have been suggested to be less successful as biological control introductions than endoparasitoids (Stiling 1990), Mills (1994) was unable to find a statistically significant difference in their rates of establishment or impact. In contrast, Mills (1994) found that egg parasitoids have a high rate of establishment, but a low rate of impact, and that there is a steady decline in rates of impact for other guilds of parasitoids that kill their target host at progressively later developmental stages. Thus larval parasitoids that kill their hosts before pupation have a greater rate of impact than those that kill their hosts as prepupae or pupae. These observations suggest: (1) that egg parasitism could be compensated for by a reduction in subsequent competition for resources (van Hamburg & Hassell 1984; see also Chapter 4); and (2) that parasitoids with a more protracted development strategy may be more constrained in their ability to realize the potential fecundity needed to balance the greater losses during their development from host mortality due to other causes (Jervis et al. 2012).

The two main functional groups of natural enemies used as introductions for the biological control of weeds are insect herbivores and fungal pathogens. Again using data provided by McFadyen (2000, 2003), there have been 40 successes from introductions of insects for control of 122 weed species, a rate of success of 33%. Similarly,

Charudattan (2005) reports six successes from introductions of fungal pathogens for control of 21 weed species, a very comparable rate of success of 29%. Thus in contrast to natural enemy introductions against arthropod pests, the two main functional groups of natural enemies introduced for biological weed control would appear to have been equally successful. However, from a meta-analysis of studies published between 1994 and 2003, Stiling and Cornelissen (2005) found reductions in plant biomass to be significantly more pronounced for pathogens than for either sap-sucking or leaf-chewing arthropods.

3.4.4 Other Empirical Patterns

Analyses by Hall and Ehler (1979), Greathead and Greathead (1992) and Gurr and colleagues (2000) suggest that rates of establishment and overall success of natural enemies introduced for control of arthropod pests declined by decade from the 'classical era' of biological control introductions (1880–1939) to a lower level during the 'chemical era' (1940–1959) and have since risen back to former levels during the 'integrated era' (1960–present) of pest management. While this pattern may well have been driven by the changes in insecticide usage, particularly the use of organochlorines, there have also been substantial changes in the protocols implemented before releasing introduced natural enemies in biological control programs that could also have played a role. Two other patterns that have been addressed several times in the literature are a comparison of the success of biological control introductions for control of arthropod pests on islands versus continents (Greathead 1971, 1986; Hall & Ehler 1979; Hall et al. 1980; Stiling 1993) and among different geographic regions (Hall & Ehler 1979; Hall et al. 1980; Greathead & Greathead 1992; Clewley et al. 2012). As discussed in Chapter 2, the potential effect of reduced biotic resistance on the success of biological control introductions in island environments has proved inconsistent, both for establishment and impact. Nonetheless, the greater rates of success of biological control programs in New Zealand against both weeds (Hayes et al. 2013; Suckling 2013) and pastoral insect pests (Goldson et al. 2014), for example, provide additional evidence that there could be a greater potential for stronger effects of natural enemies in island communities that are characterized by reduced community richness and connectedness. Although some variation in success among continents has been observed, as discussed in Chapter 2, other than a potential effect of biotic resistance on establishment via latitude, it is less clear why such patterns should occur.

3.5 Ecological Determinants of the Biological Success of Natural Enemy Introductions

While patterns in the biological control record have been examined on a number of occasions, rather less attention has been paid to the mechanisms or ecological factors that underlie these patterns and determine the outcome of deliberate natural enemy introductions. Again, the two key steps to biological control introductions are establishment and impact, and these represent the final two stages of a typical invasion process (Blackburn et al. 2011, Chapter 2). The initial step of transportation and introduction is deliberately manipulated by the biological control practitioner, but the subsequent steps of establishment and impact are less easily managed and frequently subject to the uncertainties and complexities of natural ecological processes.

3.5.1 Success of Establishment

The establishment step of a biological control introduction is a population process that depends on the ability of the introduced natural enemies to survive and reproduce in a novel environment. As a resource, an invasive pest or weed is abundant, but may not always be suitable for the establishment of an introduced natural enemy if the latter either fails to recognize the resource (mismatch in host location cues), to synchronize effectively with the susceptible stage of the resource (phenological mismatch), or to utilize the resource effectively for survival and reproduction (physiological mismatch). These barriers to natural enemy establishment can result from a lack of pre-adaptation or phenotypic plasticity to a novel environment, or from differences in the abiotic conditions experienced in a novel environment. From a biological control perspective, the most important influences on the establishment of exotic natural enemies are climate match, propagule pressure and habitat disturbance (D'Antonio et al. 1999; Wiens & Graham 2005; Simberloff 2009; Richardson & Pyšek 2012).

3.5.1.1 Climate Match and Species Distribution Models

As for all introductions, deliberate or accidental, climate suitability can have a profound effect on the potential for establishment of exotic species. For example, Stiling (1993) reported that climate is thought to be responsible for 34.5% of the failures of natural enemy introductions against arthropod pests. While Harley and Forno (1992) provide examples of natural enemy introductions against weeds that proved successful in climates that differ from those in which they were collected, climate matching has long been considered an important criterion for improving the success of establishment in biological control programs (Messenger 1970; DeBach & Rosen 1991; Hoelmer & Kirk 2005; Cullen et al. 2011).

It is generally accepted that natural species distributions result from an interaction between abiotic factors, biotic interactions and dispersal (Peterson et al. 2011). Thus there has been much debate regarding the extent to which the observed geographic range of a species in its native region is shaped by each of these contributing factors. It has been argued that as abiotic variables operate over much larger spatial scales than biotic interactions, they play a much greater role in shaping species distributions (Peterson et al. 2011; Petitpierre et al. 2012; Godsoe et al. 2015). Dispersal can also play an important role, however, and thus it is essential in modeling species distributions to carefully select a region of study in which the species of interest has had an opportunity to disperse (Barve et al. 2011). Although a wide variety of species distribution modeling approaches have been developed over the past 20 years and used for predicting geographic ranges, caution is needed in interpreting the inherent uncertainties associated with such models (Peterson et al. 2011). The most commonly used approaches are correlative rather than mechanistic, and based on statistical relationships between the occurrence of a species and a variety of bioclimatic or other environmental variables (Dormann et al. 2012). Consequently such models have limited biological realism and may not extrapolate accurately to novel environments. In contrast to correlative species distribution models, mechanistic niche models are based on environmental drivers of physiological or ecological processes and have been suggested as an alternative approach to understanding species distributions (Kearney & Porter 2009). While the inclusion of mechanistic processes is appealing, the models tend to be more complex and have yet to be widely adopted.

In the context of biological control, the climate matching model Climex has been most widely used to select collection sites and release sites for natural enemy introductions (Wood et al. 2004; Rafter et al. 2008; Ceballo et al. 2010; Fisher et al. 2011; Mausel et al. 2011; Dhileepan et al. 2013; Smith 2014), and can also be used to assess non-target effects (see Chapter 5; Wyckhuys et al. 2009). Climex has the advantage of consisting of two different applications; one a hybrid mechanistic model that can include known details of biological responses of species to climate variables, and the other a simple correlative climate matching model in the absence of any biological data (Sutherst 2005). In addition, Maxent, a purely correlative model, has gained wide appeal for species distribution modeling and has also been used to source natural enemy collections for biological control introductions (Mukherjee et al. 2011; Manrique et al. 2014). The use of species distribution models to source natural enemy populations for introduction provides biological control practitioners with a very valuable tool to maximize the climate match and to enhance the probability of successful establishment. However, due to the uncertainty in using a species distribution model to predict the potential geographic range of an exotic species in a novel environment, the most effective strategy is to base decisions on a comparison of multiple models (Lozier & Mills 2011).

Climate match may not always be the limiting factor influencing the success of establishment, however, as illustrated from the introduction of three different populations of the flowerhead weevil, *Rhinocyllus conicus*, for control of musk thistle, *Carduus nutans*, in Australia (Cullen & Sheppard 2012). The sources of the three populations were southern France, which is climatically matched to the southern part of the range of *C. nutans* in Australia; northern Italy, which is matched to the northern range; and eastern France via Canada and New Zealand with no close match to any part of the range. All three source populations of the natural enemy were released in separate locations in the northern, central and southern part of the range of *C. nutans* in the tablelands of New South Wales in the late 1980s. The population from eastern France established throughout the range of musk thistle, the population from southern France became established only in the southern (best matched) part of the range, and the population from northern Italy failed to establish. One explanation for the greater success of the population from eastern France is that prior biological control introductions in both Canada and New Zealand resulted in genetic change that allowed it to become better adapted as a colonizer. Alternatively, however, as Australia experiences a higher occurrence of summer rainfall than either southern France or northern Italy, perhaps the population from eastern France was in fact better matched in the context of summer rainfall. This population also had a partial second generation, unlike the other two, and consequently may have been better able to exploit the extended flowering period of *C. nutans* in Australia. While climate matching is often based on average temperatures and rainfall, this example suggests that to better predict the potential for natural enemy establishment, we may need to focus greater attention on critical periods during the season that drive key patterns in the phenology or activity of the pest or weed.

3.5.1.2 Propagule Pressure and Allee Effects

Propagule pressure is represented by the number of individuals per introduction event and the frequency of introduction events, and is a factor known for its consistent influence on the establishment of invasive species (Lockwood et al. 2005; Hayes & Barry 2008; Simberloff 2009). Propagule pressure has also been shown to have an important influence on the establishment of deliberate introductions, such as insect parasitoids (Hopper & Roush 1993) and passerine birds (Blackburn et al. 2013). More generally, from the biological control record, the number of natural enemies released has been shown to have a strong influence on the probability of establishment of natural enemies (Beirne 1975; Hall & Ehler 1979; Cameron et al. 1993; Hopper & Roush 1993; Memmott et al. 2005; Yeates et al. 2012) as noted in Chapter 2. Individual release events for natural enemies in biological control programs have varied from 1 to more than 1,000 individuals (Cock 1986; Hopper & Roush 1993), but releases of from 50 to 200 individuals are probably more typical. The prominence of propagule pressure as a factor influencing the success of establishment of founder populations likely depends on the interaction between demographic and environmental stochasticity and dispersal.

From a theoretical perspective, the relationship between founder population size and the success of establishment can be influenced by all three factors (see Chapter 2). For example, demographic stochasticity or Allee effects can occur as a result of failure to find mates, failure to escape natural enemies, inbreeding depression or insufficient individuals for cooperative feeding (Liebhold & Tobin 2008), and has been documented for a number of terrestrial arthropods (Kramer et al. 2009). However, the magnitude of such effects is likely to vary among species, due to extensive variation in life history traits. Experimental verification that a larger number of individuals leads to greater success of establishment has been provided from releases of gorse thrips *Sericothrips staphylinus* (Memmott et al. 1998), purple loosestrife chrysomelids *Galerucella calmariensis* and *G. pusilla* (Grevstad 1999b) and broom psyllid *Arytainilla spartiophila* (Memmott et al. 2005). Similarly, the historical record for success of establishment of three different parasitoid taxa (Chalcidoidea, Ichneumonoidea and Tachinidae) against lepidopteran pests was shown to increase with both the number of individuals per release event and the total number released (Hopper & Roush 1993). Allee effects have been documented from releases of the parasitoids *Aphytis melinus* (Kramer et al. 2009) and *Aphelinus asychis* (Fauvergue & Hopper 2009), and from the purple loosestrife chrysomelids *Galerucella calmariensis* and *G. pusilla* (Grevstad 1999a), but appeared not to have influenced releases of the parasitoid *Neodryinus typhlocybae* (Fauvergue et al. 2007) or the broom psyllid *Arytainilla spartiophila* (Memmott et al. 2005). Nonetheless, the potential importance of founder effects prompted Hopper and Roush (1993) to suggest a minimum threshold of ~1000 individuals per release site to ensure establishment from biological control introductions. Similarly, Ireson and colleagues (2008) estimated that 250 *Sericothrips staphylinus* per gorse bush was the minimum number of release individuals to maximize the subsequent population growth rate for this natural enemy.

Dispersal can also interact with Allee effects in founder populations. From a theoretical perspective, low rates of natural enemy dispersal could enhance the success of establishment by mitigating Allee effects (Lewis & Kareiva 1993; Jonsen et al. 2007), while greater rates of dispersal could exacerbate the problem of Allee effects (Heimpel & Asplen 2011). As Heimpel and Asplen (2011) pointed out, low rates of dispersal can also be detrimental for natural enemies prone to inbreeding depression, such as parasitoids with complementary sex determination (Heimpel & de Boer 2008; and see Chapter 7). While little is known of the dispersal capabilities of biological control agents, Heimpel and Asplen (2011) suggest that greater attention should be paid to screening this natural enemy trait and to experimentally manipulating dispersal tendency prior to introduction.

Environmental stochasticity or the chance occurrence of abiotic disturbance events can also influence the probability of establishment irrespective of founder population size, and has prompted the suggestion that the number of release sites or events could be more important than the number of individuals released per site in less stable environments (Grevstad 1999b). In this context, the success of establishment of small chalcidoid parasitoids against lepidopteran pests has been shown to be influenced by the number of release events, but this was not the case for larger ichneumonoid and tachinid parasitoids (Hopper & Roush 1993). In the absence of prior information on the likely effects of demographic and environmental stochasticity and dispersal on the probability for natural enemy establishment, Shea and Possingham (2000) suggest that an optimal strategy might be to release different-sized founder populations at a series of different sites.

3.5.1.3 Habitat Disturbance

That more stable habitats are more conducive to biological control introductions was noted early on by both Varley (1959) and DeBach (1964). From a global perspective, however, Hall and Ehler (1979) found rates of establishment against arthropod pests for habitats differing in levels of management disturbance to be 28% for annual crops, 32% for orchard crops and 36% for forests, suggesting that disturbance may play less of a role than originally thought. In contrast, for introductions into New Zealand against arthropod pests, Cameron and colleagues (1993) found a stronger influence of habitat disturbance on rates of establishment: 21% for field crops, 41% for fruit crops and 63% for forests. Since disturbance from management practices such as insecticide applications, planting and harvesting has a dramatic effect on pest densities, it seems likely that it should also affect the establishment of natural enemies, although most introductions probably take place at sites in which disturbance can be minimized at least at the time of release.

While it is well known that disturbance can have a strong positive influence on the establishment of invasive plants (Richardson & Pyšek 2012, Taylor & Cruzon 2015), far less is known about the influence of habitat disturbance on the establishment of natural enemies introduced for the management of invasive weeds. Peschken and McClay (1995) suggest that weeds of cultivated land are more difficult to control than those occurring in less disturbed habitats, and success in the biological control of water hyacinth *Eichhornia crassipes* can be compromised by disturbance from flooding (Coetzee et al. 2011) and herbicide use (Center et al. 1999). In a more detailed study, Yeates and colleagues (2012) examined the influence of disturbance and propagule pressure on the colonization success of natural enemies of purple loosestrife, *Lythrum salicaria*, a wetland weed along the Columbia River estuary in the Pacific Northwest of the United States. In this example, disturbance was represented by the level of exposure of introduced natural enemies to tidal inundation in the estuary, and had a significant negative effect on colonization by both leaf beetles, *Galerucella* species, and a seed weevil, *Nanophyes marmoratus*. In fact, tidal inundation had a stronger and more consistent influence on the colonization of the natural enemies than propagule pressure and is suggested to have had a direct effect through dislodgement of the insects themselves.

3.5.2 Success of Impact

The effects of introduced species have received far less attention than the process of colonization in invasion ecology. In an important first step, Parker and colleagues (1999) suggested that the impact of an invasive species can be quantified as the product of the area occupied, abundance and per capita effect. It has since become clear that threshold effects of abundance and nonlinearities between abundance and per capita effect may also need to be taken into consideration when quantifying impact. It is only more recently, however, that a more unified framework for classifying the impacts of invasive species has been suggested (Blackburn et al. 2014), that potential mechanisms for variation in ecological impacts have been considered (Ricciardi et al. 2013), and that approaches to quantifying impacts have been explored (Kumschick et al. 2015).

While these more recent developments address the impacts of invasive species in general, they include components that are of considerable interest for biological control introductions, and provide a valuable update to the criteria by which impacts have traditionally been evaluated in biological control. For example, Suckling (2013) has applied these criteria to a recent assessment of the benefits of biological control of weeds in New Zealand, updating the traditional measures of biological control success (see Section 3.2) to a more modern standard (negligible, minimal, minor, moderate, major and massive). In addition, while the impacts of biological control introductions have often been limited to measurements of economic costs and benefits (Culliney 2005; McFadyen 2008; Naranjo et al. 2015), Suckling has expanded this perspective to consider environmental and societal benefits in addition to economic benefits.

For the rest of this section, we begin by discussing three old debates that have centered on the impacts of biological control introductions, and subsequently consider some of the more recent developments and mechanisms that can cause variation in the impact of invasive species.

3.5.2.1 Single versus Multiple Introductions

One of the oldest debates regarding biological control introductions is focused on single versus multiple introductions of natural enemy species (Pedersen & Mills 2004; Stephens et al. 2013). The traditional approach to natural enemy introductions, referred to as the *lottery model*, was based on the notion that introductions of multiple species would elevate the chances that at least one effective species would eventually be established (Denoth et al. 2002), and has been implemented more extensively for biological control of weeds than for arthropod pests. This was perhaps bolstered by Harris (1981, 1991) who argued that, as plants have a modular structure, cumulative stress from multiple natural enemy species may be needed to generate sufficient damage to reduce seed production and suppress weed populations. In contrast, the unitary structure of arthropods ensures that death is the inevitable result of attack by a single natural enemy species.

While Myers (1985) initially argued that successful biological weed control often results from the action of a single effective natural enemy species, subsequent, more extensive analyses have revealed that in general success has increased with the number of natural enemy species introduced (Denoth et al. 2002; Paynter et al. 2012). A notable exception to this general pattern is *Lantana camara* in South Africa, where success remains elusive despite the introduction of 19 different natural enemy species (Cilliers & Nesser 1991). In a meta-analysis of complementarity between natural enemies of plants (both herbivorous arthropods and plant pathogens), Stephens and colleagues (2013) found that only 14% of 74 unique interactions were antagonistic, while 86% showed either additive (independent) or synergistic impacts on plant performance. In addition, they found that antagonistic effects were primarily restricted to direct competitors, natural enemies that attack the same part of the plant at the same time and natural enemies that attack reproductive parts of a plant. Consequently there is greater evidence in support of multiple introductions for biological control of weeds than there is for biological control of arthropod pests for which there have been no detailed studies of complementarity. A good example of a synergistic impact from multiple introduced natural enemies comes from the biological control of tansy ragwort *Senecio jacobaea* in the Pacific Northwest of the United States. Using a manipulative field experiment, James and colleagues (1992) were able to show that the ragwort flea beetle, *Longitarsus jacobaeae*, alone primarily reduced plant biomass and density and cinnabar moth, *Tyria jacobaeae*, alone primarily reduced seed output, but that the two together reduced the ability of plants to compensate for herbivory and resulted in negligible seed production (Figure 3.5).

In contrast, although the available evidence remains limited, it would appear that competition among natural enemies from multiple introductions is more common for arthropod pests than for weeds. This is supported by the well-known examples of competitive displacement that have resulted from multiple introductions of parasitoids against insect pests (Murdoch et al. 1996; Reitz & Trumble 2002; Mills 2006a), and yet an absence of any examples associated with multiple introductions of insect herbivores against weeds (Reitz & Trumble 2002). Despite the evidence for competitive displacement among introduced parasitoids, such as the classic examples for biological control of California red scale (DeBach & Sundby 1963) and tephritid fruit flies in Hawaii (Bess et al. 1961), the outcome historically has always been greater suppression of the insect pest (Mills 2006a). Similarly, multiple introductions for arthropod pests have been found to increase pest mortality by 12.9% and to reduce

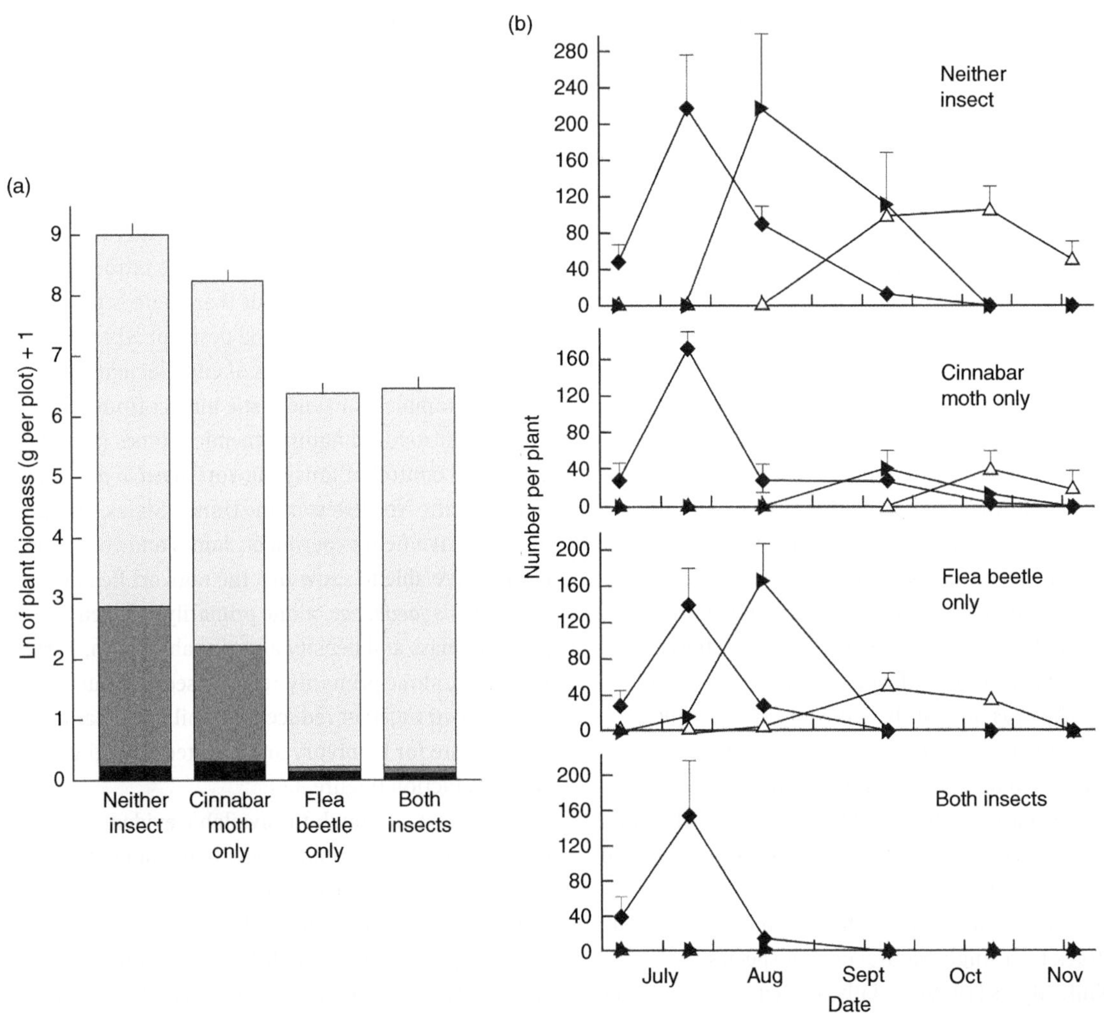

Figure 3.5. Single versus multiple natural enemies in the biological control of tansy ragwort, *Senecio jacobaea* (from James et al. 1992), showing (a) the dominant effect of flea beetles on the plant biomass of small (black) and medium-sized (dark gray) vegetative plants, but not of flowering plants (light gray); and (b) the dominant effect of cinnabar moth in reducing the number of capitula per plant when they were flowering (solid triangle) and fruiting (open triangle), but not during the flower bud stage (solid diamond). Note also that in the presence of both insects, none of the flower buds opened for flowering or seed set.

pest abundance by 27.2% in comparison to single introductions (Stiling & Cornelissen 2005).

These observations contrast, however, with more recent experimental and theoretical evidence that suggests that a superior competitor may not always provide the greatest levels of suppression of a pest or weed population (Story et al. 1991; Briggs 1993; Briese 1997; Kidd & Amarasekare 2012). The difference is almost certainly dependent on the extent of niche separation that occurs among species within a natural enemy assemblage. If there is broad overlap in resource use by the natural enemies, competitive displacement is more likely to result from multiple introductions with the single most effective species prevailing and providing greater host suppression. In contrast, if there is greater niche separation, as often occurs for weeds as modular resources, then coexistence and additive or synergistic effects can result from multiple introductions.

From a theoretical perspective, refuge breaking can be a powerful strategy for greater suppression of arthropod pests through biological control introductions (Hochberg & Hawkins 1994; Pedersen & Mills 2004). While no attempt has been made to look for evidence of complementarity with respect to multiple introductions in the biological control record for arthropod pests, some notable cases provide anecdotal evidence. For example, the effective suppression of olive scale, *Parlatoria oleae*, in California appears to result from a temperature-driven seasonal switch in the dominance of two introduced parasitoids, *Aphytis paramaculicornis* and *Coccophagoides utilis* (Rochat & Gutierrez 2001). This suggests that multiple introductions were successful in avoiding a temporal refuge from parasitoid attack for olive scale. Similarly, while the success of cottony cushion scale, *Icerya purchasi*, control in California has often been attributed to predation by the vedalia beetle, *Rodolia cardinalis*, alone, as a particularly effective species in the warmer interior region, *Cryptochaetum iceryae* is an equally effective parasitoid in the cooler coastal areas (Caltagirone & Doutt 1989). In this case, multiple introductions were successful in avoiding a regional spatial refuge (coast versus interior) from natural enemy attack. Refuges play an important role in the dynamics of consumer–resource relationships (Berryman & Hawkins 2006) and can arise through many other aspects of resource partitioning among consumers, such as spatial separation within plants, spatial separation between plant genotypes and separation between levels of resource abundance. Mills (2006a) provides anecdotal examples of coexistence among natural enemies in the biological control of arthropod pests, although the mechanisms involved are poorly understood.

An emerging consensus in the debate over single versus multiple introductions in biological control is that not only should pests and weeds be carefully prioritized according to severity of impacts and chances of success, but that candidate natural enemy species should also be carefully selected to ensure that they are necessary, effective and safe (McEvoy & Coombs 1999; Raghu & Klinken 2006; Barrett et al. 2010; see Chapter 5). This does not preclude the use of multiple introductions, but rather suggests that they should be carefully justified, based on clear evidence of niche separation among natural enemy species and the likelihood of refuge breaking for greater suppression of a pest or weed.

3.5.2.2 Density Dependence and Pest Suppression

A widely prevailing view of importation biological control is that an introduced natural enemy can control an invasive pest or weed by reducing its abundance to a lower and locally stable equilibrium (Smith & Van den Bosch 1967; Murdoch 1990). Dramatic successes in biological control were considered examples of large-scale ecological experiments in need of a sound theoretical framework to explain the features of the natural enemy–host interaction that could account for the local stability and persistence of the two populations. Consequently, the theory of biological control became a quest to find life history, behavioral, spatial or temporal components of natural enemy–host interactions that would generate sufficient density dependence for the natural enemy to be able to regulate (or at least provide some boundedness to) the abundance of its host around a low stable equilibrium (Briggs et al. 1999; Briggs 2009; Chapter 6). The focus on local stability as a key characteristic of biological control has been misleading, as it is by no means clear that the consumer–resource interaction alone is responsible for the persistence of the long-term suppression that can result from a natural enemy introduction. In general, local stability can be achieved only at the expense of increased host abundance, the so-called paradox of enrichment (or biological control), which highlights a general mismatch between theory and observation in biological control (Murdoch 2009).

Murdoch and colleagues (1985) were the first to question whether importation biological control is always characterized by local stability of parasitoid–host interactions, citing the winter moth, olive scale, larch sawfly and walnut aphid as examples of success for which there is no good evidence for the existence of a stable low equilibrium, and in some cases, even some evidence for local extinction. Waage (1990) also argued that failure in biological control more commonly results from lack of establishment or insufficient impact of a natural enemy rather than from lack of stability. In fact, the successful control of California red scale, *Aonidiella aurantii*, remains the only case study in which there is sufficient evidence to suggest that introduced parasitoids do contribute to local stability of the reduced equilibrium abundance of a pest (Murdoch et al. 2005). For other cases of successful biological control, the introduced natural enemies may contribute solely to the suppression of pest or weed abundance with other components of the environment contributing to persistence and stability. It is often considered that persistence occurs at a metapopulation rather than local scale and is mediated by asynchronous dynamics of independent local populations, differential dispersal rates of hosts and natural enemies and the connectedness of a fragmented landscape. However, the contribution of movement to the success of biological control remains poorly understood due to a scarcity of data on the dispersal characteristics of hosts and their natural enemies under field conditions (Cronin & Reeve 2005) and to the challenge of integrating species traits, spatial processes and landscape characteristics to predict the potential for pest and weed suppression (Cronin & Reeve 2014; Schellhorn et al. 2014).

The challenge of integration appears to be even greater for weeds than for arthropod pests, as multiple stressors contribute to the successful suppression of invasive species, and both individual plants and populations of plants often have a considerable capacity to compensate for herbivory (Crawley 1989a; Seastedt 2015). Thus the extent to which natural enemies contribute to density dependent mortality in weed populations is questionable, particularly when the dynamics of invasive weeds can vary so much with environmental context, as has been observed for musk thistle, *C. nutans*, in Australia and New Zealand (Shea et al. 2005). The multiple stressors on weed populations can include one or more introduced natural enemies that cause damage to the weed through herbivory, resident vegetation that competes with the weed for the resources available, and abiotic conditions such as climate and nutrient availability. While additive or synergistic effects of herbivory and plant competition have been documented for a number of weed biological control programs, such as tansy ragwort, *Senecio jacobaea*, in the western United States (McEvoy et al. (1993), Patterson's curse, *Echium plantagineum*, in Australia (Sheppard et al. 2001) and mile-a-minute weed, *Persicaria perfoliata*, in the mid-Atlantic region of the United States (Cutting & Hough-Goldstein 2013), this may not always be the case (Ferrero-Serrano et al. 2008; Weed & Schwarzlander 2014), and even antagonistic effects have been observed (Callaway et al. 1999). Thus it is more difficult to imagine how (let alone test whether) herbivory generates the density dependence that was traditionally thought necessary for the successful regulation of an invasive weed population.

3.5.2.3 New Associations

In a set of classic papers, Hokkanen and Pimentel (1984, 1989) promoted the idea that novel interactions can be characterized by greater natural enemy virulence and consequently greater impact on host populations than coevolved interactions. From an analysis of the historical biological control record, they concluded that there was a 75% greater chance of a successful outcome using new associations rather than old associations. By avoiding the evolutionary homeostasis associated with old interactions, the new association approach to selection of natural enemies was considered a superior one and recommended for use against native (known as neoclassical biological control) as well as invasive pests and weeds. Subsequently, Goeden and Kok (1986) argued that the analysis of new associations in the historical record for weed programs was biased toward repeat introductions of cactus-feeding insects and that the approach is not appropriate for biological control of weeds. Further, in a more detailed analysis of the biological control record for arthropod pests, corrected for repeat introductions, Waage (1990) found that the probability of establishment of new association natural enemies was only half that of old association natural enemies, and that there was no evidence for any greater impact of the new association natural enemies that did become established. Despite this lack of differentiation in effectiveness of new and old associations, the idea served to highlight both the potential importance of evolutionary processes in biological control and the lack of independence among introduction events in the historical record of biological control. However, the added risk of exotic new-association natural enemies, due to an inherently greater host range, has prevented the more widespread consideration of this approach for selection of candidates to introduce against native pests and weeds.

Despite these reservations, the fact that new associations can result in significant impacts of natural enemies on host populations is supported not only by the historical record of biological control introductions, but also by many of the examples of biotic resistance to invasive species that are discussed in detail in Chapter 2. One of the best-documented examples of success of neoclassical biological control, the suppression of a native pest by an exotic new-association parasitoid, is that of the tarnished plant bug, *Lygus lineolaris*, in North America by the European parasitoid *Peristenus digoneutis*. Damage to apples in New Hampshire has been reduced by 63% to less than 1% (Day et al. 2003) and plant bug nymphal densities have been reduced by 75% in alfalfa in the northeastern United States (Figure 3.6) and have remained low for more than 19 years (Day 2005). Similarly, an interesting new example of biotic resistance to an invasive pest by a native new-association parasitoid is that of the emerald ash borer, *Agrilus planipennis*, in North America by the larval parasitoids *Atanycolus* spp. (particularly *A. cappaerti*) and *Phasgonophora sulcata*, in spite of an

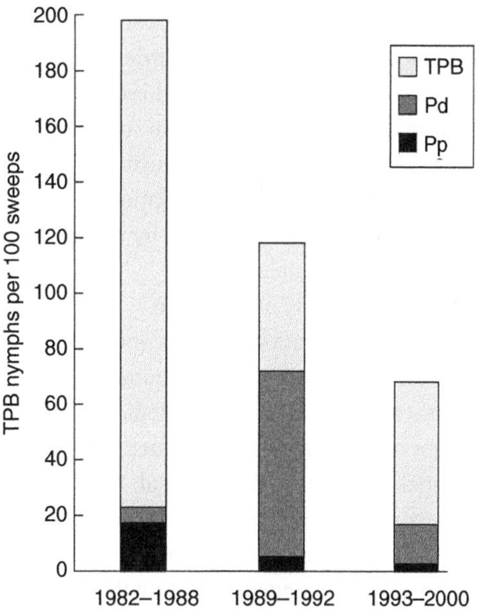

Figure 3.6. The abundance of native tarnished plant bugs (TBP; *Lygus lineolaris*) was initially high in alfalfa fields in New Jersey in the United States (1982–1988), with a low level of parasitism by the native parasitoid *Peristenus pallipes* (Pp). Following the establishment of the exotic new-association parasitoid *P. digoneutis* (Pd), parasitism initially increased substantially (1989–1992) and plant bug abundance continued to fall over a period of 12 years (1989–1992 and 1993–2000) (from Day 2005).

ongoing program of introduction of old-association parasitoids (Duan et al. 2015). Whether parasitism can reduce the spread of this invasive pest remains questionable, however, and many other tactics are currently being used to slow its spread and to reduce tree mortality (McCullough et al. 2015).

3.5.2.4 Population Structure and Microevolution

The process of colonization has long been considered detrimental to the success of introduced species due to the effects of genetic bottlenecks associated with small population size. These effects include a reduction in genetic diversity, drift-induced changes in allele frequencies and exposure of deleterious recessive alleles through inbreeding (Handley et al. 2011; Bock et al. 2015). However, it is now known that following successful establishment, many introduced species experience little loss of neutral genetic diversity. The apparent escape or recovery from bottleneck effects can in some cases be explained by admixture between introductions from genetically distinct source populations, by hybridization with resident species or by rapid adaptation following establishment.

Despite a growing body of literature, the importance of population genetics and microevolutionary processes for understanding variation in ecological impacts of biological control introductions remains unclear (Hufbauer & Roderick 2005; Roderick et al. 2012; Vorsino et al. 2012). Allelic diversity has been shown to be reduced for a few biological control agents in introduced regions relative to their native ranges, including the aphid parasitoids *Diaeretiella rapae* (Baker et al. 2003) and *Aphidius ervi* (Hufbauer et al. 2004), the gall midge, *Spurgia capitigena* (Lloyd et al. 2005), and the psyllid *Boreioglycaspis melaleucae* (Franks et al. 2011). This seems noteworthy and expected, but less clear is how important the loss of genetic variability has been in influencing the impacts of these biological control introductions. For example, Hufbauer (2002b) has documented a reduction in virulence, and Fauvergue and Hopper (2009) a reduction in population growth rate, of introduced aphid parasitoid populations, but the role of genetic diversity and microevolution in mediating these effects remains unresolved. In contrast, Facon and colleagues (2011) have argued that repeated introductions of the harlequin ladybird beetle, *Harmonia axyridis*, to North America has allowed them to avoid inbreeding depression through the purging of deleterious genes and contributed to emergent populations that have subsequently become globally invasive.

Genetic diversity in biological control introductions can be maximized by including as much variation from the native range as possible and by allowing natural enemies to avoid some of the effects of genetic bottlenecks through intraspecific hybridization or admixture (see also Chapter 7). Historically, one of the recommendations for biological control introductions was to allow distinct source populations of natural enemies to mix either immediately before or after release (Messenger et al. 1976b; Hopper et al. 1993). More recently, however, concerns regarding the safety of biological control have led to introductions of single-source populations of natural enemies to minimize the potential for unexpected host shifts or non-target interactions. While the outcome of admixture between populations can often be beneficial through acquisition of novel trait combinations or purging of genetic load, it can also be detrimental due to heterozygote disadvantage or to outbreeding depression (see Chapter 7). So far, evidence from short-term studies of admixture on the F1 generation of natural enemies is mixed, generating negative effects for both the mealybug, *Dactylopius opuntiae* (Hoffmann et al. 2002), and for the aphid parasitoid *Trioxys pallidus* (Messing & AliNiazee 1988), and yet positive effects for the flea beetle, *Longitarsus jacobaeae* (Szűcs et al. 2012a), and the harlequin ladybird beetle, *Harmonia axyridis* (Facon et al. 2010). While it is difficult to predict the longer-term consequences of admixture for subsequent generations, at least some evidence exists to suggest that short-term negative effects can translate into longer-term benefits (Hwang et al. 2011).

The extent to which rapid adaptation contributes to the success of biological control has seldom been considered, although we now know that evolutionary change occurs more commonly among invasive species than previously thought, and that adaptation is not limited by genetic variability (Bock et al. 2015). One pioneering study to test for the occurrence and consequences of adaptive evolution in biological control is that of Phillips and colleagues (2008), based on *Microctonus hyperodae*, a parasitoid introduced to New Zealand for control of the invasive pasture weevil, *Listronotus bonariensis*. In this study, two genetically and morphologically distinct parasitoid populations were introduced from South America and monitored for a period of 10 years post release. It was observed that the relative frequency of one population, originally collected from east of the Andes, increased progressively at 12 of 14 sites in New Zealand at the expense of the other. Using a simulation model to generate expected relative frequencies of the two populations in response to the random effects of genetic drift and sample size, it is argued that the observed change in relative frequencies was significantly greater than expected and thus indicative of adaptive evolution. In addition, the rapid adaptation led to increased levels of weevil mortality with rates of parasitism that often exceeded 60%. Similarly, other recent evidence also points to the ability of established natural enemies to become better adapted to the climatic environment through rapid adaptive evolution, as shown in warmer climates for the leaf beetle, *Diorhabda carinulata* (Bean et al. 2012) and in cooler climates for the cinnabar moth, *Tyria jacobaeae* (McEvoy et al. 2012).

With next-generation genetic tools offering new approaches to gain further insights into the importance of genetic variation and microevolution in biological control, we can expect to see significant advances in our current understanding of such processes in the near future.

3.5.2.5 Ecological Traits

Aided by the rapid development of consumer-resource models during the 1970s–1990s, the difference between success and failure in biological control was often attributed to the right combination of ecological traits (DeBach 1972; Hassell 1978; Hawkins et al. 1993; Kindlmann & Dixon 2001; Murdoch et al. 2003). All other things being equal, an effective natural enemy was expected to be specialized, with a high search rate, high fecundity, short generation time, low handling time relative to searching life time, pronounced mutual interference, an ability to aggregate in patches of high host density and an ability to avoid host refuges from natural enemy attack. However, in practice, few of these ecological traits have correlated well with success in the biological control record.

In analyzing traits correlated with the success of arthropod biological control, Kimberling (2004) found that specificity and number of generations a year of the natural enemy were associated with greater success. Similarly, Stiling and Cornelissen (2005) found natural enemy specificity to be an important trait in influencing abundance in the context of arthropod pests and extent of damage in the context of weeds. That specificity can be an important trait in influencing the success of biological control comes as no surprise, but Kimberling's finding of the importance of the number of generations per year is more interesting and supports the theoretical notion that generation time ratios can be an important determinant of the dynamics of consumer–resource interactions (Kindlmann & Dixon 2001). Based on a simple parasitoid-host model, Mills (2006b) has also shown that both a shorter generation time than the pest and gregarious development can provide an introduced specialist parasitoid with a significant reproductive numerical advantage. From an analysis of the biological control record, he further showed that a shorter generation time occurred more than twice as frequently among successes for hemipteran pests than among failures, and that gregarious development occurred almost twice as frequently among successes for lepidopteran pests than among failures. Similarly, the importance of parasitoid fecundity in influencing the degree of pest suppression has been highlighted by Lane and colleagues (1999) with supporting evidence from the biological control record of introductions against lepidopteran pests.

In the context of biological control of weeds, analyses of ecological traits have primarily focused on those associated with the weed rather than the natural enemy. Burdon and Marshall (1981) identified a pattern of greater success among asexually reproducing weeds and argued that weed populations with asexual reproduction are less likely to be genetically diverse and consequently would have greater vulnerability to natural enemy impact. However, in re-analyzing this pattern, Chaboudez and Sheppard (1995) found little supporting evidence and concluded that the success of biological control is more likely to be determined by a variety of factors rather than by mode of reproduction of the weed alone. While molecular tools now provide the opportunity to investigate whether mode of reproduction really does influence genetic diversity in weedy plants (Gaskin et al. 2011), few studies have been conducted so far (but see Kettenring & Mock 2012). More recently, Paynter and colleagues (2012) found that three plant traits, aquatic habit, asexuality and non-weed status in the region of origin, account for significantly higher rates of success. While the pattern for greater success in suppression of asexual weeds has resurfaced, its mechanistic basis has yet to be elucidated. The greater success against plants that are not considered weeds in their region of origin is perhaps not surprising, as this provides some evidence that natural enemies may be contributing to the lower level of abundance of such plants. This latter trait undoubtedly applies to arthropod pests as well, although it has yet to be formally documented, and is an important component of the enemy release hypothesis, which was discussed in Chapter 2.

One of the clearest patterns arising from the record of biological weed control is that the efficiency of natural enemies in suppressing plant density is greater for aquatic and wetland plants than for terrestrial plants (Paynter et al. 2012; Reeves & Lorch 2012). While this does not appear to be influenced by ecological traits per se, the difference in success between habitat types is interesting. One of the key differences between aquatic and terrestrial plants for insect herbivores is that nutrients are often more limiting in aquatic systems, with indirect effects on herbivore abundance and plant compensation for herbivore damage (Room 1990; Raghu et al. 2013). This is particularly evident for water hyacinth, *Eichhornia crassipes*, an aquatic weed of South American origin that has proved less predictable in terms of biological control success using the weevils *Neochetina bruchi* and *N. eichhorniae* (the dominant players among six introduced herbivore species). While control has been spectacular in some cases, such as in Lake Victoria (Wilson et al. 2007), it has been far less so in other regions such as Florida (Center et al. 1999) and South Africa (Coetzee et al. 2011). Weevil reproduction is directly related to the nutritional quality of plants in Florida (Center et al. 1999) and to the level of fertilizer applied to experimental tanks (Center & Dray 2010). Plants can also compensate for simulated herbivory, and allocate more resources to flowering at low nutrient concentrations, but to vegetative reproduction at higher nutrient concentrations (Soti & Volin 2010). Thus biological control of water hyacinth may be ineffective at the lowest nutrient concentrations, but most effective under limiting nutrient concentrations rather than eutrophic conditions, due to a greater impact of herbivory that results from a balance between effects on plant compensation and weevil reproduction.

From the evidence available so far, it might appear that ecological traits of the resource are most important in the context of weed biological control, whereas those of the consumer are more important for arthropod biological control. In reality, however, the apparent difference is more likely to be an artifact, with ecological traits of both consumers and resources of importance in both cases. Nonetheless, note that in any consideration of ecological traits, all other things are rarely equal, and consequently there are often negative trade-offs that result from a variety of physiological or genetic constraints during life history evolution (Rose et al. 1996; Zera & Harshman 2001). Thus the extent to which individual or combinations of ecological traits can be more broadly associated with success in biological control remains a challenge with many similarities to that of identifying the determinants of success for biological invasions.

In this context, Godfray and Waage (1991) recommended the integration of ecological traits into models of intermediate complexity to aid in the selection of those candidate natural enemies that are likely to have the greatest impact in suppressing pest populations. They illustrated this more holistic approach by using a stage-structured population model parameterized for the mango mealybug, *Rastrococcus invadens*, and two potential candidate parasitoids for introduction, *Gyranusoidea tebygi* and *Anagyrus mangicola* (Figure 3.7a). One important difference in the ecological traits of these parasitoids that was explored in the model is that *G. tebygi* attacks first and second instar nymphs of both sexes of its host, whereas the majority of parasitism by *A. mangicola* occurs in third instar and pre-reproductive female hosts. Through numerical solution of equilibrium densities, the model predicted that parasitism by *G. tebygi* alone resulted in a lower host equilibrium density than parasitism by *A. mangicola* alone, and that this prediction was robust to a sensitivity analysis of all parameters in the model (see Figure 3.7b for the analysis of parasitoid search rate and density dependence). Not only did the introduction of *G. tebygi* to West Africa provide successful suppression of mango mealybug (Bokonon-Ganta & Neuenschwander 1995), but the success of the introduction was estimated to have a benefit:cost ratio of 145:1 for Benin alone (Bokonon-Ganta et al. 2002). Thus the prospective model of biological control for mango mealybug provides a valuable example of how a more holistic demographic approach can be used to facilitate selection of the single best natural enemy for introduction.

Another modeling approach used to guide the selection of natural enemies that are likely to have greater impact as biological control introductions involves the analysis of vulnerability in the life cycles of pests and weeds. Matrix models have been used extensively to examine life history transitions and population viability analysis in the context of conservation biology (Mills et al. 1999; Crone et al. 2011), but have been applied to biological control only more recently (McEvoy & Coombs 1999; Mills 2005, 2008; Davis et al. 2006; Ramula et al. 2008; Dauer et al. 2012; Evans et al. 2012). Stage-structured matrix models can be used to determine at which stage in the life cycle of a pest or weed added mortality from an introduced natural enemy would contribute most to a reduction in population growth. Using this approach, Davis and colleagues (2006) and Evans and colleagues (2012) found that the most vulnerable components in the life cycle of growing populations of garlic mustard, *Alliaria petiolata*, are rosette survival, seed output and seed germination. From this analysis they were able to prioritize a sequence of natural enemy introductions, selecting a root crown weevil, *Ceutorhynchus scrobicollis*, first for its impact on rosette survival and a seed weevil, *C. alliariae*, second for its impact on seed output. From an analysis of the demographics of 21 invasive and 179 native plants, Ramula and colleagues (2008) suggested that a reduction in either growth or fecundity would be sufficient, more generally, to suppress the populations of short-lived invasive plants, such as garlic mustard, while simultaneous reductions in either survival and growth or survival and fecundity would be needed to suppress populations of long-lived invasive plants. Similarly, Mills (2005) used a stage-structured matrix model to investigate vulnerability in the life cycle of codling

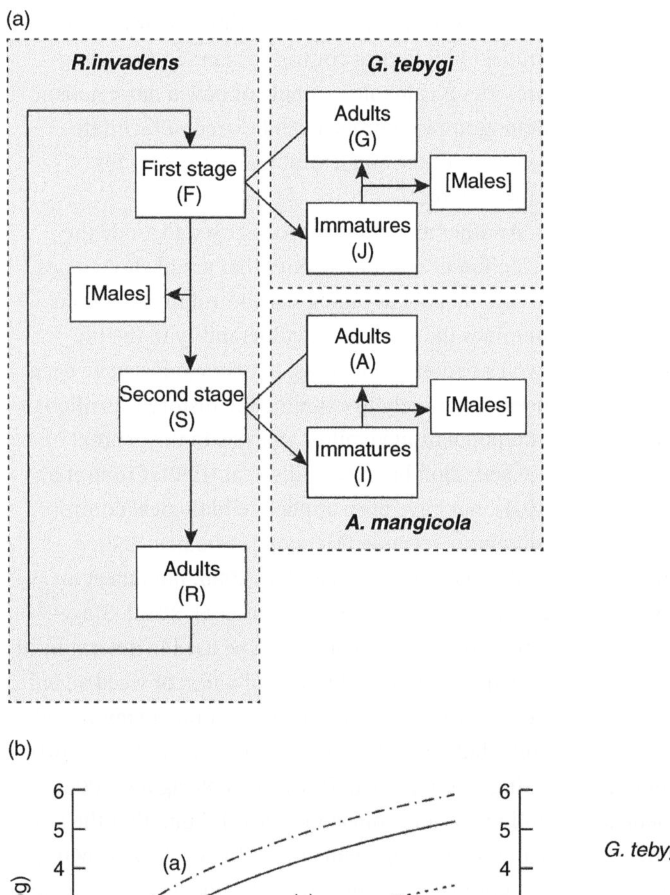

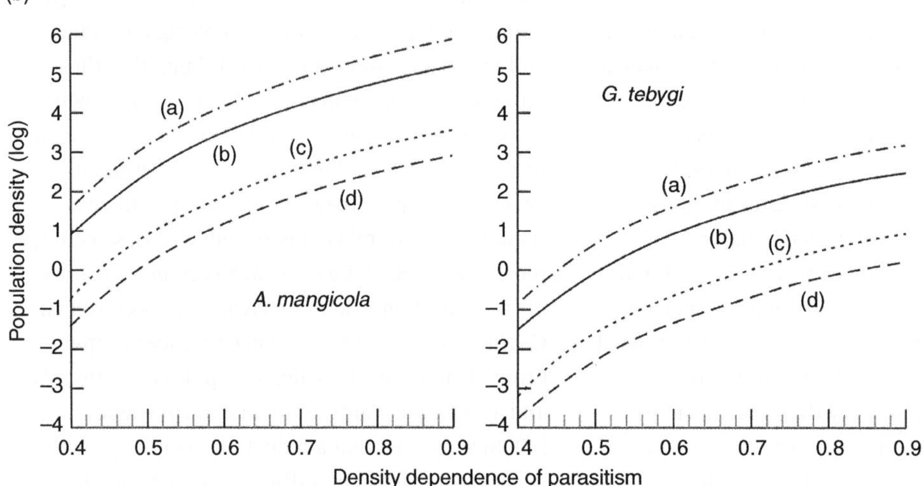

Figure 3.7. Prospective model used to guide the selection of natural enemy species for importation biological control of the mango mealybug, *Rastrococcus invadens* (from Godfray & Waage 1991), showing the stages in the life cycles of the mealybug and two candidate parasitoids, *Gyranusoidea tebygi* and *Anagyrus mangicola*, that were considered (a), and a sensitivity analysis of mealybug equilibrium density to parasitoid search rate (a, 0.0005; b, 0.001; c, 0.05; d, 0.01) and density dependence (b).

moth, *Cydia pomonella*, finding that added parasitism at the fifth instar larval and cocoon stages would contribute most to population suppression. Together with other ecological traits such as a positive response to host density, a shorter generation time than its host and gregarious development, parasitism of the cocoon stage by *Mastrus ridens* (= *ridibundus*) helps to explain the greater impact of this introduced parasitoid in western North America than achieved by earlier parasitoid introductions.

3.5.2.6 Temporal Synchronization

Recent research on the effects of climate change has provided many examples of advancement in the timing of seasonal phenological events for many different types of organisms. As organisms can be expected to vary in their sensitivity to climate change, this could lead to a phenological mismatch between interacting species with consequent effects on their population dynamics (Hance et al. 2007; Miller-Rushing et al. 2010). Differences between the climates of ancestral and introduced geographic regions could also result in a mismatch between the timing of the flight period of an introduced natural enemy and the presence of the susceptible stage of an invasive pest or weed. In an analysis of potential causes for the failure of established natural enemies to suppress insect pests, Stiling (1993) noted that climate was considered the most important factor in 24% of the 148 cases examined, followed by lack of alternative hosts (14.9%) and inappropriate strain of the natural enemy (11.5%). Lack of synchronization, however, was suggested to be important in only 9.5% of the cases examined. While no comparable analysis has been made for biological control introductions for weeds, we suspect that in both cases the importance of phenological mismatch in limiting impacts of the natural enemies has frequently been underestimated.

In a timely review, Welch and Harwood (2014) considered a variety of aspects of the temporal synchrony between natural enemies and their hosts, including differences in life cycle phenology, diel activity patterns and seasonal events in agroecosystems, and called for a greater focus on temporal dynamics in biological control. In many cases, temporal synchrony is thought to be driven by responses to temperature, but responses to day length (Bean et al. 2013) and rainfall (Chavalle et al. 2015) can also affect the synchronization of natural enemies and their hosts. For example, the impact of tamarisk beetles, *Diorhabda carinulata*, on biological control of tamarisk *Tamarix* spp. in the western United States has varied greatly between regions (Bean et al. 2012, 2013; Hultine et al. 2015). Due to the influence of latitude on the critical day length for induction of diapause in *D. carinulata*, adult tamarisk beetles in the southern part of the United States ceased reproducing and entered diapause in early August, and subsequently experienced low survivorship through to resumption of activity the following spring. This lack of synchrony in seasonal activity allowed tamarisk plants to compensate for the impact of early-season herbivory and resulted in ineffective suppression of weed populations. Since the initial introductions of *D. carinulata* in 2001, two additional species, *D. elongata* collected from Crete and *D. sublineata* from North Africa, have now established in the southern part of the United States and remain active until late in the season. In addition, the original introductions of *D. carinulata* have shown evidence of rapid adaptation with as much as a 54-minute reduction in the critical day length for diapause induction in field populations at the expanding southern edge of their range.

The influence of temperature on the synchronization and successful suppression of lucerne weevil in New Zealand, but lack thereof in Australia, has already been discussed in Section 3.3.1. Another example of the influence of temperature on temporal synchrony involves the specialized larval parasitoid *Tetrastichus julis*, which has been credited with providing effective importation biological control of the cereal leaf beetle, *Oulema melanopus*, in western North America

(Evans et al. 2006). A recent study of parasitism rates over a 10-year period in relation to spring temperatures suggested that the beetle may be gaining a greater temporal refuge from parasitism in warmer years (Evans et al. 2013). While the timing of cereal leaf beetle activity appears to respond to the more rapid degree-day accumulation in warmer springs, the timing of parasitoid emergence from the soil does not, creating a mismatch in the temporal synchronization of the susceptible host stage and parasitoid flight.

Even when there does appear to be broad overlap in the flight period of a natural enemy and the susceptible stage of its host in the region of origin of an invasive pest, that synchrony can be lost in a novel environment. In Japan, the elongate hemlock scale, *Fiorinia externa*, has two full generations each year and adult emergence of the parasitoid *Encarsia citrina* is perfectly synchronized with the peak abundance of second instar nymphs that experience up to 90% parasitism (McClure 1986; Figure 3.8a). Since its discovery in the United States in 1908, the scale has spread from Massachusetts in the north, where it may have as little as one generation each year, to South Carolina in the south, where it has two complete generations. Throughout its range in the United States, the elongate hemlock scale is also parasitized by the cosmopolitan *E. citrina*, and McClure (1986) suggested that the lack of suppression of the scale by parasitism in Massachusetts is due to the absence of a complete second generation. However, Abell and Van Driesche (2012) failed to find improved synchrony between second instar scales and emergence of *E. citrina* along a latitudinal gradient from Massachusetts in the north to North Carolina in the south (Figures 3.8b–d). Thus number of scale generations did not improve the synchrony of the interaction, nor did it increase the low levels of parasitism or reduce scale densities. Potential differences in host plant quality of the hemlock species present in Japan and the eastern United States were suggested to be a factor that could perhaps account for the differences in synchrony and impact of *E. citrina* between the two countries.

For pests and weeds that are continuous breeders, overlapping generations ensure that stages susceptible to natural enemy attack are likely to be present throughout the season or year and consequently, this minimizes any limitation due to synchronization. In such cases, the impact of a biological control introduction may be determined more by the temporal dynamics associated with generation time ratios or other forms of numerical reproductive advantage than by synchronization. In contrast, however, for pests and weeds that are discrete breeders with non-overlapping generations, synchronization is likely to play a dominant role in the temporal dynamics and impact of biological control introductions.

3.5.2.7 Higher-Order and Indirect Interactions

While there has been some debate in the literature as to whether higher-order interactions can disrupt biological control (Lawton 1985; Rosenheim 1998), this issue has yet to be fully resolved due to a paucity of experimental studies. It could be argued that whenever an exotic natural enemy is deliberately introduced to a new environment there is always the possibility that it could experience enemy release and realize a greater impact on an invasive pest or weed than in its region of origin. However, any such release would apply only to specialist enemies, as it is likely that at least some generalists would be able to colonize the novel host unless it had sufficient phylogenetic or functional distinctiveness for there to be no ecological analogues as a source of generalist enemies in the resident community (Paynter et al. 2010).

As invaders, insect herbivores generally accumulate a parasitoid complex of lower richness than as natives in their region of origin, and the complex is generally composed of a greater proportion of generalists (Hawkins & Cornell 1993). Nonetheless, some invaders can accumulate the same

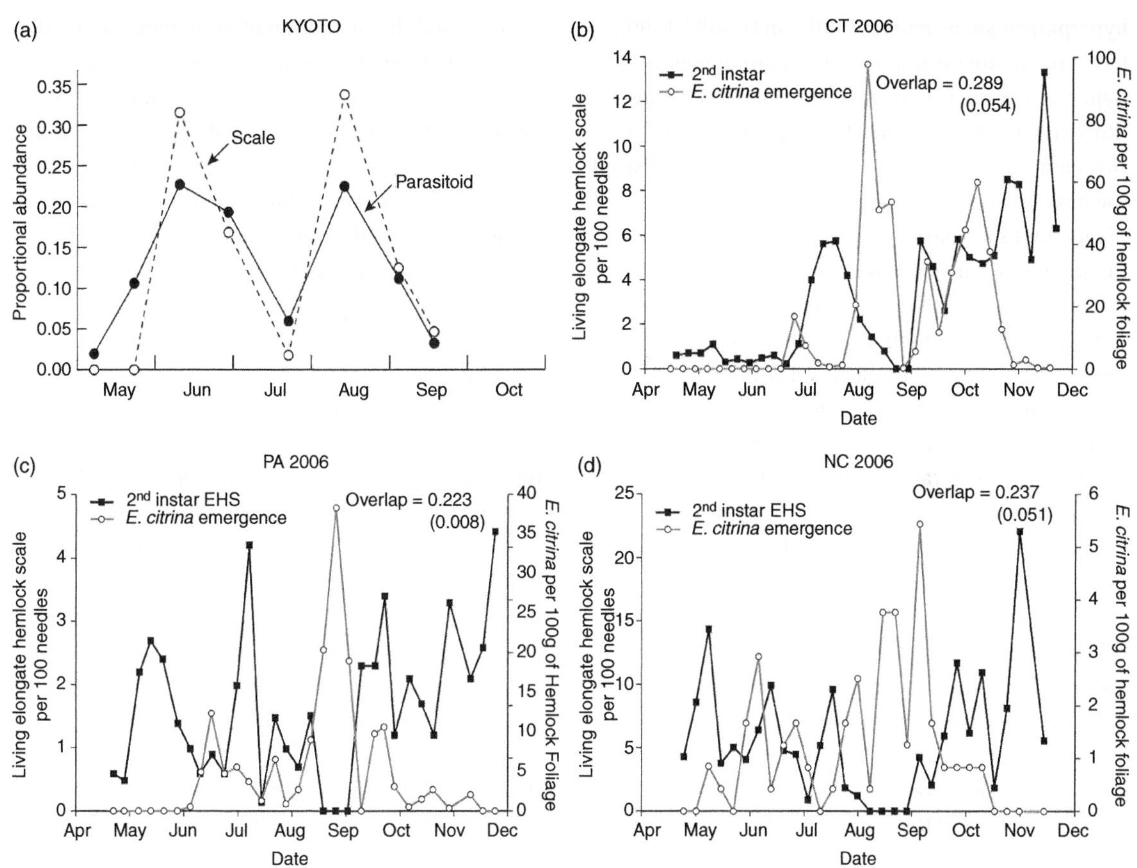

Figure 3.8. Lack of temporal synchrony between susceptible second instars of the invasive elongate hemlock scale, *Fiorinia externa*, and emergence of adult *Encarsia citrina* parasitoids in the eastern United States, showing (a) the temporal match in Kyoto, Japan in 1984 (from McClure 1986), but the temporal mismatch in (b) Connecticut, (c) Pennsylvania and (d) North Carolina in 2006 (Abell & Van Driesche 2012), despite two full generations of the scale in North Carolina, as in Kyoto.

level of parasitoid richness that they experienced as a native, within 10 years of introduction. Despite this, parasitism rates of insect herbivores as invaders were almost always much lower than those experienced as natives. As discussed earlier in Chapter 2 and Section 3.4.2 however, parasitism continues to be considered an important concern for the impact of natural enemies in weed biological control programs (Goeden & Louda 1976; McFadyen & Spafford Jacob 2004). For example, recent evidence from field surveys has demonstrated that parasitism is significantly associated with the failure of introduced natural enemies of weeds in New Zealand (Paynter et al. 2010). Some of the failed natural enemies appear to have rapidly accumulated native parasitoids from ecological analogues in the resident community with a median level of parasitism of 49%. While we know of no similar studies for invasive insect pests, a similar pattern is likely to hold true for introductions of exotic primary parasitoids and the accumulation of native hyperparasitoid species (see Chapter 2). For example, aphid parasitoids and microgastrine parasitoids of Lepidoptera, two groups of primary parasitoids that experience high rates of hyperparasitism as natives (Sullivan & Volkl 1999), have frequently been found to support extensive hyperparasitism as biological control introductions (Rosenheim 1998). The level of hyperparasitism was sufficient in the case of *Asaphes suspensus* to disrupt the biological control of pea aphid, *Acyrthosiphon pisum*, by its primary parasitoid, *Aphidius ervi*, in glasshouse cages (Schooler et al. 2011). Thus maximum levels of higher-order parasitism appear to be greater for introduced parasitoids than insect herbivores and may be associated with a lower level of specialization among hyperparasitoids than primary parasitoids, a greater likelihood of the presence of suitable ecological analogues as a source of generalist enemies in the introduced range or a diminished influence of sequestered plant chemical defenses. However, the fact that some parasitoid introductions can still be successful despite greater than 90% hyperparasitism (see Chapter 2) is surprising, and this may result from a numerical reproductive advantage from shorter generation times and greater fecundity of the primary parasitoids, at least in the case of aphid hyperparasitoids (Mackauer & Volkl 1993).

In addition to the direct effects of higher-order parasitism, indirect effects may also play a role in limiting the impact of biological control introductions. Both apparent competition and intraguild predation have the potential to disrupt biological control, although both have been better documented in short-term laboratory studies than in the field. Aphid parasitoids are often subject to asymmetrical intraguild predation as they develop within aphids that are also prey for insect predators (Brodeur & Rosenheim 2000). The outcome of intraguild predation can vary with habitat complexity (Denno & Finke 2006), and the extent of intraguild predation is generally reduced with increasing abundance of the shared prey (Daugherty et al. 2007). However, in an experimental field study of intraguild predation of mummies of *Binodoxys communis*, an introduced parasitoid of the invasive soybean aphid *Aphis glycines*, Chacon and Heimpel (2010) found that predation of mummies was greater on potted plants stocked with more aphids due to density dependent aggregation of aphid predators. Thus not only was there was no dilution effect of an increased availability of aphids as shared prey for the generalist predators, but predator aggregation on plants with higher aphid abundance also led to increased intraguild predation, disrupting the biological control potential of *B. communis*.

Due to the more frequent use of multiple introductions of natural enemies, indirect effects may also be a common occurrence in the biological control of weeds. An interesting example of the complexity of direct and indirect effects of introduced natural enemies has been documented for yellow starthistle, *Centaurea solstitialis*, in California (Swope & Stein 2012). Plant biomass, and consequently the number of inflorescences per plant, was reduced independently both by infection with the introduced fungal pathogen *Puccinia jaceae solstitialis* and by growth in serpentine versus non-serpentine soils (Figure 3.9a). Similarly, the number of viable seed per inflorescence was reduced independently through larval seed predation by introduced seed predators, notably the weevil *Eustenopus villosus*, and by growth in serpentine soils. However, the impact of the seed weevils was less effective for fungal-infected than non-infected plants due to an indirect effect of the pathogen on the seed weevil (Figure 3.9b). The net effect of the direct and indirect interactions of the two introduced natural enemies on viable seed output per plant was additive for yellow starthistle plants growing in serpentine soil, but not for those growing in non-serpentine soil (Figure 3.9c). The authors speculate that the difference in outcome for the two soil types may be due to changes in seed quality or systematic acquired resistance. In a further field study, it was found that uninfected plants growing on north-facing slopes with greater soil moisture experienced only 10% pollen limitation when open to visitation by pollinators (Swope 2014). Plants infected with the introduced fungal pathogen, however, experienced

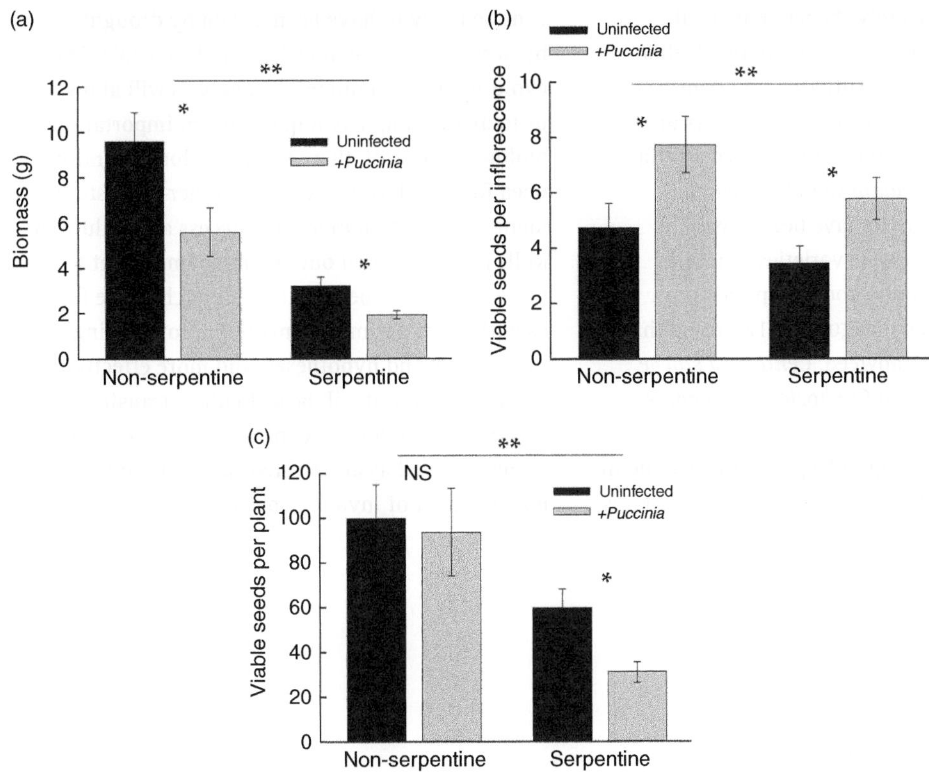

Figure 3.9. Indirect interactions influence the impact of biological control introductions for management of yellow starthistle, *Centaurea solstitialis* (from Swope & Stein 2012). In the absence of seed weevils there were direct effects of both serpentine soils and infection by the fungal pathogen *Puccinia jaceae solstitialis* on plant biomass (a). In the presence of seed weevils, however, there was an indirect effect of fungal infection on the extent of seed predation per inflorescence on both soils (b). Ultimately, the net effects of both fungal infection and seed predation and on viable seed output per plant were additive on serpentine soils, but cancelled each other out on non-serpentine soils (c). Asterisks indicate statistical significance at the alpha = 0.05 level (*) and 0.01 level (**); 'NS' signifies no statistical significance.

72% pollen limitation. Thus infection by the fungal pathogen not only caused a direct reduction in seed set by 72%, but also caused an indirect reduction in seed set by increasing pollen limitation by a factor of seven.

3.6 Conclusions

The practice of importation biological control is under increasing pressure to be more predictive in the selection of invasive species that will be most amenable to success and natural enemies that will have the greatest probability of establishment and impact on host populations. The patterns of success that have been documented from the historical record of biological control introductions provide some insight, but provide no guarantees, as they are often based on poor documentation of the scope of success and the factors contributing to observed changes in the abundance of pests and weeds. Poor documentation has also made it very difficult to know when introduced natural enemies are really the drivers of change in biological control programs or merely the passengers of change caused by other unknown factors (Didham et al. 2005). For example, Gutierrez and Ponti (2013) have shown that the classical success of importation biological control for spotted alfalfa aphid *Therioaphis maculata* in California and elsewhere is more likely to have been driven by the introduction of resistant varieties of alfalfa than by parasitism from introduced parasitoids. Similarly, Ortega and colleagues (2012) found that mortality of spotted knapweed, *Centaurea stoebe*, caused by the root weevil *Cyphocleonus achates* can lead to compensation at a population level by increased recruitment, suggesting that the decline in abundance of this weed in the western United States is more likely to have been driven by drought than by herbivory. While novel methods of ecological, evolutionary and numerical analysis will allow us to answer important questions in importation biological control, there remains a long-standing need for population-level and longer-term studies to identify the determinants of success and failure. As Mills (2000) pointed out, it will be important to use a more rigorous experimental approach in the future, as without experimental manipulation, testing of concepts and hypotheses, and more effective documentation, it will be difficult to transform importation biological control into a predictive science as well as an effective strategy for the management of invasive species.

4 Negative Consequences of Biological Control

'Happy families are all alike; every unhappy family is unhappy in its own way.' Thus begins Tolstoy's classic novel, *Anna Karenina*. Tolstoy himself does not define clearly what he meant by this, but in one interpretation, it indicates that any number of a diverse set of factors can go awry and doom an enterprise as complex as a family to failure, while success can be expected only when everything falls into place (Diamond 1997). Diamond coined this interpretation the 'Anna Karenina Principle', and applied it to the domestication of animals, which has occurred with success for only a relatively small fraction of potential candidates. Can the Anna Karenina Principle also be extended to biological control? Success in importation biological control does appear to require the confluence of a diverse array of factors, and the absence of any one of these factors could indeed doom the entire enterprise to failure, or worse, negative unintended consequences (McClay & Balciunas 2005). We devoted the previous chapter to successes in biological control, and in this chapter we focus on failures in the sense of negative unintended consequences. In the next chapter, we turn our attention to risk assessment in biological control and also attempt to reconcile Chapters 3 and 4 by considering risk-benefit analysis as a means of assessing the net outcome of biological control introductions.

4.1 The History of Concern over Negative Effects of Biological Control

The possibility of adverse ecological effects of biological control introductions was raised at the very inception of the practice (Perkins 1897). Perkins wrote of the period when Alfred Koebele, fresh from his successes controlling cottony cushion scale in California, turned his attention to the Hawaiian Islands. After extolling the virtues of biological control in Hawaii, Perkins (1897) went on to discuss possible effects on the native fauna:

> *In conclusion, I cannot help turning to the darker side of the picture. What will be the result of all these importations on the endemic fauna? The introduction of many other species – parasitic and predaceous – is contemplated, and will be performed. That success, from an economic point of view, will be attained there is little doubt, and while industries are threatened, or even the gratification of æsthetic tastes, it is certain that no consideration will be given to the native fauna. When even now the ladybirds are affecting the latter, what will be the result of the introduction of more widely predaceous species? The effect of the former is not imaginary but proven. In June 1895, in a lovely forest in Hawaii – 5000 feet above sea level – I found the native trees much affected by a black Aphis. By beating these trees the blight came down in abundance, and amongst them various fine species of endemic Chrysopa and Hemerobius, predatory creatures. One or two introduced ladybirds were also noticed. By September the ladybirds were in thousands, the blight and native insects in small numbers. In August 1896 not an Aphis was to be found, and only one or two stray specimens of ladybirds, as one may find anywhere throughout the forests. They had done their work and disappeared. This is a high testimonial as to the capabilities of the beetles, and as the existence or non-existence of Hawaiian Chrysopa is not likely to be regarded by people at large, and seeing that sooner or later the greater part of this most interesting native fauna is, under any circumstances, in all probability doomed to extinction, it only remains to wish Mr. Koebele a success in the future equal to that which he has already attained.*

This quote shows that many of the issues that the biological control community is struggling with today were on the mind of at least one observer 120 years ago. Perkins realized, for instance, that biological control agents were capable of colonizing habitats far from where they were released, and grappled with the balance between risks and benefits of biological control.

As we noted at the beginning of Chapter 2, Charles Elton also alluded to ecological risks of biological control in his landmark book on invasion biology (Elton 1958). Here again the focus was on Hawaii – Elton cited warnings by Zimmerman (1948) that parasitoids introduced into agricultural areas of the Hawaiian islands were invading upland forests and attacking native moths, with knock-on effects on hunting wasps that relied on the native moths for food. With nearly 60 years of hindsight, we can now see just how prescient these admonitions were. More recent studies by Hennemann and Memmott (2001) have confirmed that parasitoids introduced into Hawaii before 1945 have indeed infiltrated pristine habitats and are attacking native lepidopterans, and a number of other studies also demonstrate non-target effects of biological control in Hawaii (Howarth 1991; Follett et al. 2000; Johnson et al. 2005; King & Rubinoff 2008).

However, despite these early efforts and others from the 1980s (Howarth 1983; Pimentel et al. 1984; Turner 1985; Harris 1988; Ehler 1990), non-target effects of biological control were largely ignored until the 1990s. Broader recognition of the environmental risks of biological control was spurred by a review published by Howarth (1991), which highlighted information suggesting that various biological control projects had brought about the extinction or severe population reduction of non-target organisms, mostly on tropical islands, and that biological control agents could themselves become economic pests. Lockwood (1993) also argued strongly against using exotic natural enemies to control native pests (so-called neoclassical biological control), and he used biological control of native grasshoppers in the central United States as a case study.

Another very influential set of reviews of the risks of biological control was published by Simberloff and Stiling in 1996 (1996a, b). Like Howarth, Simberloff and Stiling argued that evidence existed for catastrophic environmental effects, and that many other such effects likely remained undocumented. Simberloff and Stiling went farther than Howarth did though, in seeming to condemn most if not all importation biological control as irresponsible and dangerous with the provocative statement in the abstract of one of their papers that biological control agents should be considered 'guilty until proven innocent' (Simberloff & Stiling 1996a). This, along with an emphasis on the axiom that an absence of evidence for negative impacts does not constitute evidence of an absence of such effects, could be interpreted as suggesting that importation biological control was too environmentally dangerous to pursue and should be abandoned altogether (Ehler 2000). It should be noted, however, that Simberloff and Stiling did not actually advocate an end to importation biological control, and conceded that the environmental benefits of controlling invasive species using biological control may outweigh the risks when other control options are not available or effective.

A number of empirical studies on negative consequences followed these reviews, some of them carefully executed. The most influential of these was work on the flowerhead weevil, *Rhinocyllus conicus*, which was released against musk thistle, *Carduus nutans*, in the United States in the 1960s. Louda and coworkers noted that *R. conicus* had moved onto non-target thistles in the genus *Cirsium* in nature preserves and national parks in the central United States by the 1990s (Louda et al. 1997; Louda 1998). This constituted a geographical and host-range expansion for the weevil, although previous host range tests had shown that various *Cirsium* spp. could be used as hosts (Gassmann & Louda 2001).

Follow-up studies have shown that *R. conicus* has indeed led to population declines of native thistles (Rose et al. 2005). We will return to various aspects of this project later in this chapter and in Chapter 5.

Following on the heels of the reviews of Howarth, Lockwood and Simberloff and Stiling, the publication of Louda and colleagues' (1997) results and accompanying commentary with the witty title 'Fear no Weevil?' (Strong 1997) spurred a great deal of interest in non-target impacts of biological control. Scientific publications dealing with biological control and non-target impacts or indirect effects went from a few per year in the early 1990s to more than one per week beginning in the mid-2000s (Figure 4.1). Thus, studying the risks of biological control has been transformed from a scientific backwater to an area receiving abundant and rather detailed attention. As a reflection of this, a number of edited volumes and review articles have been devoted to the topic (e.g., Hokkanen & Lynch 1995; Follett & Duan 2000; Goettel et al. 2001; Wajnberg et al. 2001; Hokkanen & Hajek 2003; Louda et al. 2003; Secord 2003; Carruthers & D'Antonio 2005; Bigler et al. 2006; Messing & Wright 2006; Parry

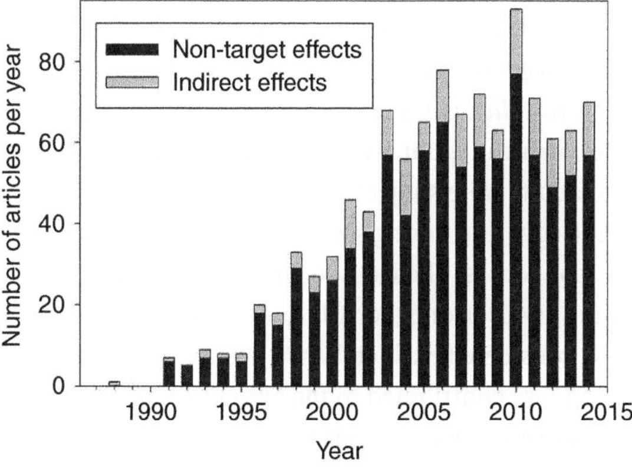

Figure 4.1. Number of scientific journal articles per year published with topic terms 'biological control' and ('non-target' or 'nontarget') or 'biological control' and 'indirect effects' between 1987 and 2014, as determined by the Web of Science.

2009; Simberloff 2012). Partly for this reason, we do not attempt a comprehensive review of negative non-target and indirect impacts of biological control in this chapter. Instead, we discuss three key topics in this general area. First, we will evaluate the evidence that biological control has caused global or local extinction of non-target species. Second, we evaluate the evidence that importation biological control has led to substantial declines in the abundance of non-target species. Third, we discuss negative indirect effects in biological control, in which effects accrue not through consumption, but through ecological processes involving at least three species. We defer discussion of risk analysis related to negative consequences to the next chapter of this book.

4.2 Extinction by Biological Control

Clear evidence of extinction of native organisms via biological control is hard to come by. Indeed, Stiling and Simberloff (2000) stated that 'biological control agents rarely, if ever, cause extinction of their target, so it is probably equally unlikely that they would cause the extinction of a non-target species.' For the case of insect extinctions in general, Dunn (2005) has pointed out that the mere documentation of such extinctions is difficult enough, let alone documenting the *reason* for extinctions. Factors that complicate the documentation of insect extinctions include difficulty of sampling some taxa as well as boom-bust population dynamics that can lead to very low densities of some species for years or even decades. Even when local or global extinction can be documented, assigning causality to one factor such as a biological control agent can be very problematic in the face of other, often simultaneous stressors such as habitat degradation and other invaders. Despite these difficulties, the literature contains enough data sets to suggest that a number of extinctions, both global and local, are the direct result of importation biological control projects. In the following paragraphs, we discuss possible extinctions of native fauna caused by introductions of vertebrate predators, predatory snails, parasitoids and the myxoma virus.

4.2.1 Vertebrate Predators

Vertebrate predators were almost exclusively used in the 18th and 19th centuries (an exception includes the continued use of mosquitofish, *Gambusia* spp.). Vertebrate predators have mainly been used in attempts to control rats and rabbits, and they include mongooses, ferrets, monitor lizards and cats. Other vertebrates introduced as biological control agents include insectivorous birds and the infamous cane toad, *Bufo marinus*. While some of these releases have shown some success (Hickling 2000), most have instead had disastrous negative effects on native species and ecosystems. Here we discuss mongooses and mosquitofish, focusing in particular on evidence that introduction of these species has caused local or global extinction of non-target species.

The small Indian mongoose. Local and possibly global extinctions almost certainly occurred as a result of the introduction of generalist mammal predators onto islands. Most notably, the small Indian mongoose, *Herpestes auropunctatus*, was introduced onto a number of tropical and subtropical islands to control rats and poisonous snakes in the 19th and early 20th centuries. While there is some indication that introduced mongooses prey upon rats on some islands (Lever 1985; Cavallini & Serafini 1995), the main effect of these introductions seems to have been widespread declines of ground-nesting birds and reptiles, almost certainly including a number of extinctions (Seaman & Randall 1962; Honegger 1981; Wodzicki 1981). An 1898 quote from the US Department of Agriculture cited by Lever (1985) summarized the situation in Jamaica:

> *Its omnivorous habits became more and more apparent as the rats diminished. It destroyed young pigs, kids, lambs, kittens, puppies, the native 'cony' or capromys, poultry, game, birds which nest on or near the ground, eggs, snakes, ground lizards, frogs, turtles' eggs and land crabs.... Toward the close of the second decade the mongoose, originally considered very beneficial, came to be regarded as the greatest pest ever introduced into the island.*

We could find no conclusive evidence of mongoose-caused extinctions, but reports from various archipelagoes do suggest that the diversity of native fauna is higher on islands onto which mongooses were not introduced than on islands where they were introduced. This was apparently the case in the West Indies, as well as in Hawaii, where mongooses were established on all of the major islands with the exception of Kauai prior to 1972, which at that time had a notably higher diversity of native birds than did the mongoose-infested islands (Lever 1985). The mongoose then did become established in Kauai between 1972 and 1985, negatively affecting some native bird populations, but we know of no Kauian extinctions attributed to the mongoose. A thorough and rigorous comparison of extinction rates on islands with and without mongoose introductions would be a fruitful way to assess the role of mongooses in causing extinctions. In the absence of such analyses, it will be difficult to separate effects of mongooses from those of habitat destruction, direct human interference or other introduced animals. The case of the Jamaican iguana, *Cyclura collei*, can serve as an example of this point. The Jamaican iguana was endemic to Jamaica and adults were subject to hunting by humans, while eggs and young were subject to predation by mongooses. The last specimens were reported in the 1940s (Honegger 1981), and the species is now considered extinct. While it is clear that the mongoose contributed to the extinction of the Jamaican iguana, it seems possible that this species would have gone extinct due to human activities even without the introduction of the mongoose. Lever (1985) also noted that effects of mongooses on Caribbean bird species was confounded with potential effects of introduced rats and native opossums.

Mosquitofish. Two main species of mosquitofish, *Gambusia affinis* and *G. holbrooki*, have been used extensively in importation biological control. *G. affinis* is known as the western mosquitofish and is native to southern North America, and *G. holbrooki*, the eastern mosquitofish, is native to the southeastern United States. Both species have been introduced all over the world, including novel habitats in North America, with some notable success in controlling mosquito populations (Murdoch & Bence 1987; Garcia & Legner 1999). However, *Gambusia* are generalist feeders with great invasive potential (Rehage & Sih 2004) and have been implicated in a number of non-target effects, including effects on rare or threatened fish, amphibian and invertebrate species (Minckley & Deacon 1968; Galat & Robertson 1992; Gamradt & Kats 1996; Goodsell & Kats 1999; Pyke & White 2000; Hamer et al. 2002; Leyse et al. 2004; Mills et al. 2004; Goldsworthy & Bettoli 2006; Laha & Mattingly 2007; Olden et al. 2008). The World Health Organization does not recommend the use of *Gambusia* for mosquito control programs for these reasons despite some apparent successes in malaria control in Europe and Asia (Garcia & Legner 1999). Courtenay and Meffe (1989) concluded that the negative impacts of *Gambusia* introductions have far outweighed the positive impacts.

The specter of local extinction due to *G. affinis* has been raised for the Gila topminnow, *Poeciliopsis occidentalis*, in the Southwestern United States since the 1960s (Minckley & Deacon 1968; Meffe et al. 1983; Meffe 1985; Minckley 1999). The Gila topminnow was placed on the US Federal Endangered Species List in 1973 due both to habitat loss and predation by *G. affinis* (Galat & Robertson 1992). As of the late 1980s, Gila topminnows persisted only in remnant populations in Arizona, New Mexico and parts of northern Mexico. Gila topminnows are better able to withstand flash-flooding events than are mosquitofish, and this appears to be a major contributor to their survival in these areas (Meffe 1984; Galat & Robertson 1992). Meffe (1983) attempted to protect topminnows in a spring that makes up part of the Gila River drainage basin in Arizona by poisoning *G. affinis*. A sample of Gila topminnows was removed from the spring and held in coolers while the spring was treated with Antimycin A, which appeared to kill all fish in the spring. Once the water was no longer toxic, the topminnows were reintroduced and thrived, although *G. affinis* were later found in the spring, indicating that they had not been completely eradicated by the poison. Redistribution of Gila topminnows to parts of their ancestral range has been suggested as well, with Childs (2006) proposing that these fish may actually be better biological control agents of mosquitoes than *G. affinis* in some areas!

4.2.2 The Predatory Snail *Euglandina rosea*

The best-documented cases of extinction via biological control probably come from the predatory land snail *Euglandina rosea*. *E. rosea* is known as the rosy wolf snail, and it is a generalist predator that feeds by sucking out the contents of snail shells, or swallowing small snails whole. This snail was introduced from Florida onto Oahu, Hawaii, in the mid-1950s to control the notorious giant African snail, *Achatina fulica*, which had itself been introduced as a food source, but had developed into an agricultural pest and a human health threat as an intermediate host of the nematode parasite causing eosinophilic meningitis in humans (Civeyrel & Simberloff 1996). Once *E. rosea* was established in Hawaii, many releases of *E. rosea* were made onto other Pacific islands. Spectacular declines of the giant African snail often followed introductions by *E. rosea*, but a number of authors have suggested that these declines were in many cases not attributable to *E. rosea* (Mead 1979; Waterhouse & Norris 1987; Murray 1993; Civeyrel & Simberloff 1996; Cowie 2001; Simberloff & Gibbons 2004). As we noted in Chapters 2 and 3, population crashes of the giant African snail could have been due to density-dependent disease outbreaks rather than introduced predatory snails.

***Euglandina* in Hawaii.** Negative impacts on native snail species were evident in Hawaii by the 1970s. In particular, 12 species within the genus *Achatinella* disappeared from Oahu closely following the introduction of *E. rosea* (Hadfield et al. 1993; Civeyrel & Simberloff 1996), and additional extinctions and population declines are suspected on other Hawaiian islands and for other snail taxa (Kinzie 1992; Civeyrel & Simberloff 1996). The main evidence implicating *E. rosea* in these extinctions in Hawaii was an increased rate of extinction of *Achatinella* spp. after the introduction of *E. rosea* coupled with evident vulnerability of native tree snails in Hawaii to predation by *E. rosea*. Traits such as low fecundity and late reproduction also put many Hawaiian *Achatinella* at a high risk of extinction, no matter the final cause (Hadfield et al. 1993). However, despite strong circumstantial evidence that *E. rosea* brought about the extinction of *Achatinella* species, the demise of these native snails had been noted since early in the 20th century, before the introduction of *E. rosea*. Stokes (1917; cited in Elton 1958) noted that shells of *Achatinella* and *Amastra* snails showed characteristic chew marks of introduced rats, and Hadfield and colleagues (1993) noted that habitat loss and degradation also likely contributed to the decline of native snails in Hawaii.

Hadfield and colleagues (1993) conducted a detailed 10-year field study (from 1983–1992) on the demographics and population dynamics of two populations of the native tree snail *Achatinella mustelina* on Oahu. This study came a decade after a different study in which *E. rosea* had apparently decimated a population of *A. mustelina* elsewhere on the island (Hadfield & Mountain 1980). One of the goals of the new, longer-term study was to compare the role of introduced rats and *E. rosea* in causing population declines of *A. mustelina*. One of the main differences between these two predators is the life stage of the snail that they attack. While *E. rosea* tends to concentrate on smaller individuals, rats have a preference for larger individuals. Hadfield and colleagues (1993) argued that predation on larger size classes of snails had a stronger negative effect on long-term population viability because of the disproportionate loss in reproductive value. Their census data showed a spike of predation of *E. rosea* followed closely by rat predation (Figure 4.2), which was halted by poisoning the rats. Hadfield and colleagues (1993) described the relative roles of the predatory species as follows:

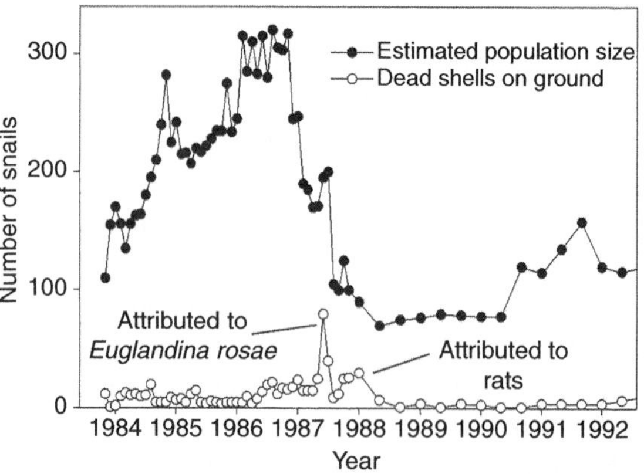

Figure 4.2. The estimated numbers of live *Achatinella mustelina* snails on trees and dead *A. mustelina* at the base of the same trees in Hadfield and colleagues' (1993) Pahole site on Oahu, Hawaii. The two spikes in snail mortality indicated by increased numbers of shells were attributed to the introduced predatory snail *Euglandina rosea* and introduced rats as indicated. Redrawn from Hadfield and colleagues (1993).

> *While the total population [of* A. mustelina*] declined during the period of intense predation attributed to* Euglandina rosea*, a small reproductive capacity remained. Under intense rat predation, adult snails were selected so severely that nearly all reproductive and near-reproductive snails were eaten, leaving the population without reproductive potential for several years. Recovery of the population began only after the predators had been removed [by poisoning] and remaining juvenile snails grew to adulthood.*

The end of the episode of intense predation by *E. rosea* coincided with the beginning of rat predation, suggesting that rats preyed upon *E. rosea* itself.

Hadfield and colleagues (1993) suspected that the population of *A. mustelina* would have been driven extinct had it not been for rat control.

Taken together, these observations do not allow a clear conclusion as to whether *E. rosea*, rats or habitat degradation is most responsible for population decline of *A. mustelina*, and by extension, extinction of other *Achatinella* spp. It seems likely that all three factors were important. Hart (1978) also noted that over-collecting decimated populations of tree snails in Hawaii, particularly in the late 19th century.

***Euglandina* in Moorea.** Despite the documentation of negative effects of *E. rosea* on native snails in Hawaii, introductions were made in the 1970s in Tahiti, Moorea and other islands in the Society Islands chain in the Pacific Ocean following the introduction of the giant African snail. At the time, Moorea had a very notable and storied snail fauna, comprising in particular nine species of *Partula*, which had been subjects of a number of noteworthy genetic and evolutionary studies beginning in the 19th century and continuing through the 1980s (Murray 1993). This exciting body of work was brought to an abrupt halt by the extinction of all nine of these species on Moorea within a 10-year period from 1977 to 1986. The species extirpated from Moorea were not endemic to that island, so the extinctions were local and not global, but at least two species have gone globally extinct following extirpation on other islands.

The evidence linking *E. rosea* to these extinctions is stronger than it is for *Achatinella* species in Hawaii. Clarke and colleagues (1984) and Murray and colleagues (1988) sampled *Partula* species and *E. rosea* extensively during and immediately after the introduction of *E. rosea*, and the timing of extinction of *Partula* species coincided with the spread of *E. rosea* into the habitats used by these species. Alternative hypotheses, such as habitat degradation or predation by other invaders, were not mentioned by Clarke and colleagues (1984) or Murray and colleagues (1988), which probably indicates that no notable differences in these factors were evident over the time period of *Partula* extinction. One could argue that the giant African snail itself could have played a role, perhaps via competition, in the demise of *Partula* spp., but this seems unlikely since it is reported almost exclusively from farm and garden areas. Furthermore, a natural 'control' island, Huahine, which is near Moorea, did not receive *E. rosea* despite having been invaded by the giant African snail, and suffered no decline in *Partula* species. Another nearby island, Tahiti, which did receive *E. rosea* introductions, suffered *Partula* declines very similar to the ones documented on Moorea (Murray et al. 1988).

Similar effects were seen on other Pacific islands, and on Mauritius in the Indian Ocean. In a study done on Mauritius, gut contents were examined to determine the feeding patterns of *E. rosea*. More than 70% of the gut contents of 57 *E. rosea* from Mauritius that were dissected consisted of native snail species, values that exceeded the proportion of those snails in the habitat, yet included no identifiable remains of the giant African snail (Griffiths et al. 1993). *E. rosea* was not abundant on these islands, but based on the gut contents analysis, Griffiths and colleagues concluded that *E. rosea*, along with habitat destruction and collecting, had likely contributed to the decline and possibly the extinction of native snails on Mauritius and other islands in the Indian Ocean.

4.2.3 Parasitoids

Parasitoids used in importation biological control have been implicated in a number of cases of non-target impacts (e.g., Barratt et al. 1997, 2007; Boettner et al. 2000; Follett et al. 2000; Hennemann & Memmott 2001), and, in a few cases, the claim has been made that introduced parasitoids have driven non-target species to extinction (Howarth 1991). We review two of these cases here that serve as hopeful cautionary tales in the sense that species thought to have gone extinct were either rediscovered later or are suspected to have survived after all.

Parasitoids of Hawaiian Lepidoptera. The Hawaiian Islands support an ecologically diverse and speciose endemic lepidopteran fauna that was well-documented by Perkins and Zimmerman beginning in the late 19th century (Gagne & Howarth 1985). The collections of Perkins, Zimmerman and others provided a baseline against which to compare modern lepidopteran faunas, allowing the hypothesis that certain species have gone extinct. Gagne and Howarth (1985) concluded that 27 species of Lepidoptera had been driven to extinction on the Hawaiian Islands, and that biological control was involved in at least 16 of these cases. Other suspected reasons for the extinctions that were cited included habitat loss, effects of exotic mammals, host plant loss and hybridization with exotic species. Gagne and Howarth acknowledged that these factors were not mutually exclusive to biological control, but suspected that biological control introduction played a dominant role in the extinctions.

The genus *Omiodes* suffered a disproportionate share of extinctions among the endemic lepidopteran fauna of Hawaii. *Omiodes* contains species that were agricultural pests in Hawaii and targets of biological control from the late 19th to the mid-20th centuries (*O. accepta* was known as the sugarcane leafroller and *O. blackburni* was known as the coconut leafroller). At least two parasitoid species were released against these native pest species and attacked these as well as native non-pest *Omiodes* and other Lepidoptera (Funasaki et al. 1988). By the early 21st century, 14 *Omiodes* species were suspected to be extinct (Evenhuis 2002), with biological control introductions the prime suspect for the cause of these extinctions (Gagne & Howarth 1985). However, later research confirmed that five of these species were not extinct after all, although their distribution and abundance had apparently been drastically reduced (King & Rubinoff 2008). *O. continuatalis*, for instance, had been known from all of the Hawaiian Islands in the early 20th century before being deemed extinct, but was later found on only three islands. King and Rubinoff (2008) conducted a series of fascinating observations that shed light on potential mechanisms by which this species may have escaped extinction. They placed *O. continuatalis* larvae onto potted sugarcane plants and placed the pots into a field setting for four days. Upon retrieving the larvae, they noted that some had buried themselves within the soil at the base of the plant. This was the first observation of non-pupation-related fossorial behavior of a Hawaiian *Omiodes*, and the behavior may provide a refuge from parasitoids and predators.

Levuana moth. Our second case of reported extinction due to parasitoids involves biological control of the Levuana moth, *Levuana iridescens*, on Viti Levu, one of the islands in the Fijian archipelago. This species was a serious pest of coconuts, and J. D. Tothill undertook a biological control program against it in the 1920s. A tachinid parasitoid, *Bessa remota*, was imported from Malaysia and led to rapid declines in the Levuana moth to the point where it was declared likely extinct in Fiji by the 1940s. Because of the thorough documentation of the Levuana moth biological control project by Tothill and colleagues (Tothill et al. 1930), Howarth (1991) called this case 'probably the best documented study of extinction among the insects'. However, two recent re-analyses of this case have cast doubt on the extinction hypothesis (Kuris 2003; Hoddle 2006). These new analyses have uncovered previously overlooked reports and collections that put the most recent reports of the Levuana moth at 1956, after which time research on coconut insects in Fiji declined dramatically. Kuris (2003) and Hoddle (2006) have also pointed out that Tothill and colleagues (1930) originally suspected that the Levuana moth was not native to Fiji but had invaded from islands west of Fiji. Thus, the Levuana moth may persist in its native range even if it was indeed extirpated from Viti Levu.

Howarth (1991) also suggested that *B. remota* drove non-pestiferous moths to extinction in Fiji, particularly *Heteropan dolens*, which is in the family Zygaenidae, as is the Levuana moth. However, Hoddle (2006) has noted that this species is known from other Pacific islands and that there is no evidence that it is a suitable host for *B. remota*. Other lines of reasoning casting doubt on the hypothesis that *B. remota* drove either the Levuana moth or *H. dolens* extinct have been articulated by Sands (1997), Kuris (2003), Tarmann (2004) and Hoddle (2006).

4.2.4 The Myxoma Virus

The extinction of the British population of the 'large blue' butterfly, *Maculinea* (= *Phengaris*) *arion* in the 1970s has been linked to biological control in a particularly fascinating way. The demise of the large blue in Britain has been traced to the illegal introduction of the myxoma virus, which is lethal to rabbits (see Chapter 7). Why would a butterfly go extinct because of the introduction of a mammalian virus? Like many other lycaenids, this butterfly has a very specialized life history that involves ants. The larvae feed on thyme in their early instars and for the last larval stage are transported by workers of the ant *Myrmica sabuleti* into their nest, where the ants feed on an exudate produced by the caterpillars (the caterpillars return the favor by feeding on ant larvae). This species of butterfly appears to be entirely dependent on these ants (as well as on thyme as a host plant) for its survival. The ants, for their part, need short grass in order to construct their underground nests, and the short grass was apparently supplied by rabbit grazing. The myxoma virus, which had been introduced in a series of unsanctioned releases against rabbits in the early 1950s, greatly reduced rabbit populations along with the conditions that had allowed the large blue to thrive (Thomas 1980, 1995; Moore 1987).

The large blue went extinct in Britain despite vigorous attempts to stave off extinction (Thomas 1980), and although myxoma certainly must have played a role in its demise, a number of other factors, such as land use practices, likely contributed. It is also worth mentioning that the European rabbit is not native to the British Isles, having been brought there from France sometime after (or during) the Norman Invasion of 1066 (Moore 1987). The association between rabbits and the large blue (via ants) is therefore relatively novel and was likely precipitated by land use practices in Britain before the 1950s that restricted grazing by animals other than rabbits.

Populations of both the large blue and its ant associate can be found on mainland Europe, where the large blue is listed as endangered for a series of reasons not related to rabbits or the myxoma virus (Mouquet et al. 2005). To the extent that British populations of the ant *M. sabuleti* depended on the large blue, it would seem likely that these populations would have gone extinct as well, but they did not. The large blue was successfully introduced from Sweden to the United Kingdom beginning in the 1980s in large part by the creation of conditions favorable for *M. sabuleti*, which was present but rare (Thomas 1995). This indicates either that the ant species is not as specialized as previously supposed (Pech et al. 2007) or that populations of the large blue persisted unrecognized.

4.2.5 Conclusions about Biological Control and Extinction

Let us now return to the statement by Stiling and Simberloff (2000) asserting that the fact that biological control agents do not drive their intended targets to extinction suggests that extinction of non-targets is similarly unlikely. It is indeed difficult to find clear evidence that biological control agents drive non-target species extinct, which supports their claim. However, differences between target and non-target species may have a bearing on this question. One difference is that target species are, by definition, present at high densities in the introduced range where they are considered pests or weeds. In contrast, non-target species may be present at vanishingly low densities, making them more vulnerable to local or global extinction. In this case, biological control may be the last stressor to 'push them over the edge' to extinction. Thus, endangered or threatened species may be at a relatively high risk of extinction even by natural enemies that are not able to drive their intended targets to extinction. In fact, this is the reasoning behind the fear that the moth *Cactoblastis cactorum* will lead to the demise of the endangered semaphore cactus, *Opuntia spinosissima*, in the Florida Keys (Johnson & Stiling 1996), and that the weevil *R. conicus* will cause the extinction of the endangered Pitcher's thistle, *Cirsium pitcheri* (Louda et al. 2005a).

4.3 Population Reduction of Non-target Species via Direct Effects

In this section, we consider evidence for population reduction of non-target species attributable to biological control agents by direct consumptive effects. The discussion focuses almost exclusively on importation biological control because almost all of the documentation of non-target impacts comes from importation biological control. Despite this, there is increasing recognition that augmentative biological control can pose risks to non-target species as well, and this is covered in more detail in Chapter 5.

4.3.1 Demonstrating Population Reduction

While a number of studies have documented *attack* of non-target species by biological control agents, fewer have demonstrated actual *population reduction* attributable to biological control, and our main task in this section will be to evaluate the evidence for, or likelihood of, population reduction of non-target species as a result of biological control. The problem of demonstrating population reduction of non-target organisms is analogous to the problem of demonstrating efficacy in biological control as discussed in the previous chapter. We consider two main mechanisms by which the attack of non-target species can fail to cause population reduction. The first is redundant mortality, or mortality of individuals that would have died anyway. This happens for simultaneous mortality sources such as occurs when predators feed on parasitized hosts (Carey 1989; Elkinton et al. 1992) and also when life stages are attacked that do not limit fitness of the non-target species. The most transparent example of this latter phenomenon is perhaps seed mortality of plants that are not seed-limited. If plants reproduce vegetatively or if seedling establishment is more limited by germination sites than by seed production, even very high seed mortality can have only negligible effects on plant population size (Hoffmann & Moran 1998; Parker 2000; Turnbull et al. 2000).

A less transparent example of this phenomenon occurs with the attack of life stages of a non-target species that occur prior to stages that experience density-dependent mortality. Van Hamburg and Hassell (1984) considered this scenario within the context of successful versus unsuccessful biological control of pest lepidopterans by *Trichogramma* egg parasitoids. They modeled the situation that occurs when larvae suffer from density-dependent, competition-mediated mortality. In this case, egg mortality (from *Trichogramma*) is compensated by larval mortality at high densities to the point where egg parasitism does not contribute to population reduction of the host. Like May and colleagues (1981) before them, Van Hamburg and Hassell (1984) found that under strong, overcompensating, density-dependent mortality of a given life stage, attack of a previous life stage can even *increase* the host population. Aside from explaining why egg parasitoids sometimes fail as biological control agents (see also Chapter 3 and Way et al. 1992; Collier & Van Steenwyk 2004), Van Hamburg and Hassell identified a general mechanism by which attack of non-target organisms may not lead to population reduction of that species. It is important to note that this mechanism will only protect relatively abundant non-target species from negative effects of biological control agents because competition will only compensate for mortality at relatively high densities. The densities of populations of rare or endangered species would likely be too low for this compensation to play a role.

A second way that attack of a non-target species may not lead to population reduction involves transient dynamics. If non-target attack occurs briefly, perhaps in the context of spillover onto only non-preferred or even unsuitable species in the vicinity of the target species, then populations of the non-target species may be affected only minimally or in a small portion of their geographic range (e.g., Blossey et al. 2001; Paynter et al. 2008b). This is not to say that all spillover effects in biological control produce only negligible non-target impacts. Spillover effects can be important when the non-target species is at low densities and spatially correlated with the target species (Raghu et al. 2007).

In the following sections, we will discuss a number of cases of biological control for which attack of non-target organisms has been documented and consider the evidence for, or likelihood of, population reduction of the non-target species. We provide an overview of these cases in Table 4.1 and discuss selected cases in more detail later. We finish this section with a treatment of the special case of biological control agents themselves as non-target species, which occurs when arthropod biological control agents attack weed biological control agents. Throughout these case studies, we hope to illustrate that a number of creative methodologies have been used to ask whether non-target impacts are having population-level impacts. These methods

Table 4.1. **Selected cases of non-target impacts of weed and insect biological control.**

BC Agent	Target Species	Main Non-target Impact Species	Substantial Population Reduction of Non-target Species?
Weed biological control			
Rhinocyllus conicus (Coleoptera: Curculionidae)	Musk thistle (*Carduus nutans*)	22 native *Cirsium* spp. attacked in N. America	Yes, for *Cirsium canescens*[1]
Larinus planus (Coleoptera: Curculionidae)	Canada thistle (*Cirsium arvense*)	4 native *Cirsium* spp. attacked in N. America	Unclear[2]
Cactoblastis cactorum (Lepidoptera: Pyralidae)	Weedy prickly pear (*Opuntia* spp.)	7 native *Opuntia* species attacked in Caribbean and S.E. United States	Unclear[1]
Insect biological control			
Compsilura concinnata (Diptera: Tachinidae)	Gypsy moth (*Lymantria dispar*)	Many native lepidopterans attacked, including various giant silk moth species in eastern United States	Yes, for at least 3 giant silk moth species[3]

Table 4.1. (cont.)

BC Agent	Target Species	Main Non-target Impact Species	Substantial Population Reduction of Non-target Species?
Trichopoda pilipes (Diptera: Tachinidae)	Southern green stink bug (*Nezara viridula*)	Koa bug (*Coleotichus blackburniae*) attacked in Hawaii	Yes, but modest[3]
Microctonus aethiopoides (Hymenoptera: Braconidae)	An alfalfa-feeding weevil (*Sitona discoideus*)	6 native weevil species attacked in New Zealand	No[3]
Cotesia glomerata (Hymenoptera: Braconidae)	Imported cabbage white butterfly (*Pieris rapae*)	The green-veined white butterfly, *Pieris napi*, attacked in southern New England	Likely[3]
Pteromalus puparum (Hymenoptera: Pteromalidae)	Imported cabbage white butterfly (*Pieris rapae*)	The New Zealand red admiral butterfly (*Bassaris gonerilla*) attacked in New Zealand	No[4]
Trigonospila brevifacies (Diptera: Tachinidae)	Light brown apple moth (*Epiphyas postvittana*)	At least one native lepidopteran attacked in New Zealand	Unclear[5]
Lysiphlebus testaceipes (Hymenoptera: Braconidae)	Two exotic citrus aphid species	Various native aphids in Mediterranean Europe	Unclear[6]
Various parasitoid species	Various pest lepidopterans in Hawaii	Various native lepidopterans attacked in Hawaii	Unclear[7]
Various ladybird species	Various aphid species	Various native ladybird species attacked in N. America	Unclear[8]

[1] See Section 4.3.2.
[2] Louda & O'Brien (2002)
[3] See Section 4.3.3.
[4] Barron et al. 2003; Barron 2007
[5] Munro & Henderson 2002
[6] Costa & Stary 1988; Stary et al. 1988
[7] Henneman & Memmott 2001
[8] See Sections 4.3.3 and 4.4.2.

range from manipulative field studies including the outplanting of sentinel non-target organisms to life table analyses and theoretical modeling approaches. Collectively, they reveal significant population-level effects in some cases and modest to undetectable levels in others.

4.3.2 Non-target Effects of Biological Control Agents of Weeds

***Rhinocyllus conicus* and the Platte thistle.** Beginning in the 1960s, the Eurasian flowerhead weevil, *Rhinocyllus conicus*, was introduced to North America, New Zealand, Australia and Argentina to control non-native thistles in rangelands (Andres & Rees 1995; Cullen & Sheppard 2012; de Briano et al. 2013). The success of *R. conicus* against its main target, the musk thistle, *Carduus nutans*, in North America has been inconsistent (Andres & Rees 1995), and in New Zealand, field-level declines have been minimal (Groenteman et al. 2011). More recent analyses suggest, however, that *R. conicus* has impeded the spread of musk thistle in both of those countries, and that suppression was likely underestimated (Marchetto et al. 2014).

Concerning non-target effects, releases in Australia, New Zealand and Argentina were considered risk-free because native thistles do not occur there. However, host-specificity studies suggested that some native thistles in North America would likely be attacked (Zwölfer & Harris 1984 – see also Section 5.2.2.1). And it was the finding that *R. conicus* attacked some of these native thistles that has made this insect iconic in biological control. As we have discussed previously, the finding by Louda and colleagues (1997) that *R. conicus* was feeding within the seed heads of native thistle species in the central United States ushered in a period of investigations of non-target impacts of importation biological control that continues to this day (see Figure 4.1). This case is particularly instructive from the standpoint of developing a sound scientific estimate of population-level non-target effects.

In particular, Louda and colleagues have made a very compelling case for serious population-level effects of *R. conicus* on the North American native Platte thistle, *Cirsium canescens*, in the Sandhills of Nebraska. The initial discovery of *R. conicus* feeding on Platte thistle seeds was not accompanied by data on the effect of this feeding on actual population sizes of Platte thistles. However, Louda and colleagues could rely on previous studies that had demonstrated seed limitation in field settings to predict strong population-level responses from seed feeding by *R. conicus*. In particular, experiments done before the establishment of *R. conicus* showed that inflorescence-feeding insects decreased not only seed survival, but subsequent seedling and adult plant recruitment (Louda et al. 1990; Louda & Potvin 1995) (Figure 4.3). These studies showed that neither physiological compensation for inflorescence feeding nor density-dependent mortality of plants canceled the negative effects of seed mortality.

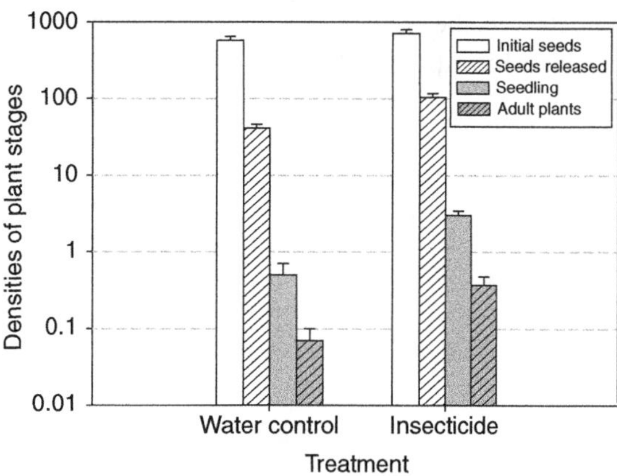

Figure 4.3. Effect of insecticides on demographics of the Platte thistle, *Cirsium canescens*, in experimental sites at Arapaho Prairie, Nebraska, the United States. The study was initiated in 1984 and 1985 and adult plants were counted in 1988. The herbivores controlled by the insecticides included two flies and one moth, all of which feed on *C. canescens* inflorescences in their larval stage. From Louda and colleagues 1990; effect of treatment significant for seeds, seedlings and adult plants.

While these studies showed that seed feeding in Platte thistle could produce dramatic population-level consequences, they did not involve the biological control agent *R. conicus*. Instead, they quantified effects of native herbivores. *R. conicus* was found on Platte thistle after these studies were completed (Louda et al. 1997), and a demonstration of population-level impacts came as a result of demographic studies done from 1991 through 2002 in plots where *R. conicus* was first detected in 1993 (Rose et al. 2005). No manipulation of these plots was done to exclude *R. conicus* or other insects from some of the plants, but *R. conicus* feeding was estimated and demographic modeling was used to estimate the role of *R. conicus* herbivory on Platte thistle population growth and size. These analyses indicated that *R. conicus* has been responsible for population declines of Platte thistle during the 1990s (Figure 4.4).

A potential caveat to the conclusions reached by this body of work is that *R. conicus* may impact Platte thistle only at sites adjacent to its intended target weed, the musk thistle. A series of geographic analyses have dispensed with this caveat, however (Rand & Louda 2006). These analyses show clearly that *R. conicus* attacks on Platte thistle are not spatially correlated with attacks on musk thistle at a regional scale and that populations on Platte thistle are not dependent on immigration from populations on musk thistle. Indeed, *R. conicus* is present throughout the range of Platte thistle, including in areas devoid of musk thistle.

In conclusion, a series of carefully conducted ecological studies that began before the establishment of *R. conicus* in the Sandhills of Nebraska demonstrated strong population-level impacts of this insect on the native Platte thistle. These studies have excluded two important mechanisms that could attenuate the long-term effects of biological control agents on non-target species: compensation and limited spillover dynamics. Given the demographic trajectory of Platte thistle in the presence of *R. conicus* (Rose et al. 2005) and the regional scope of the establishment of this insect (Rand & Louda 2006), it is not unreasonable to suppose that *R. conicus* may eventually drive Platte thistle to extinction (Rose et al. 2005).

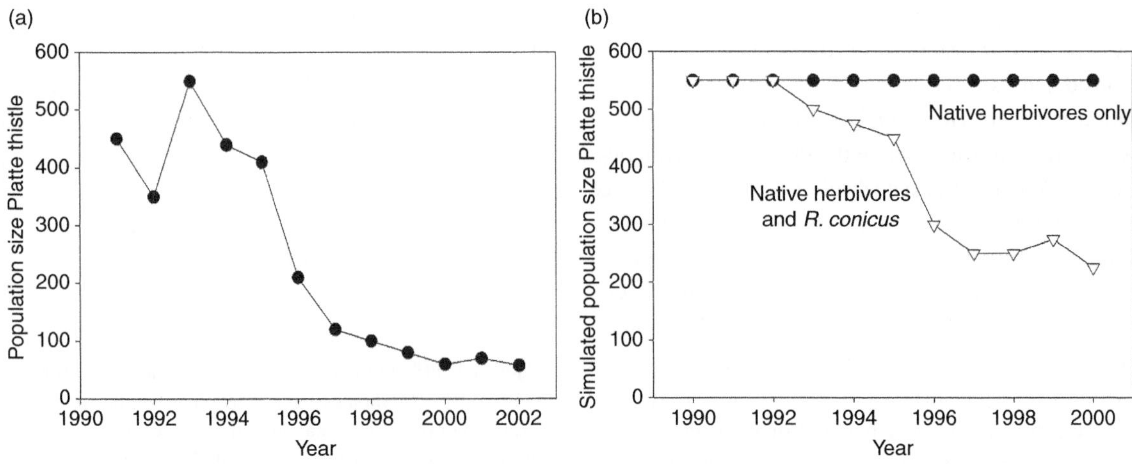

Figure 4.4. Platte thistle populations at Arapaho Prairie, Nebraska, the United States, where *Rhinocyllus conicus* was first recorded in 1993. (a) Population sizes of Platte thistle from 1991 to 2002. (b) Simulated population sizes of Platte thistle under assumptions of herbivory by native insects only (closed circles) and *R. conicus* in addition to native insects (open triangles). From Rose and colleagues (2005).

Cactoblastis cactorum and North American *Opuntia*. The Argentinean cactus moth, *Cactoblastis cactorum*, has been celebrated as a highly successful control agent of prickly pear cactus, *Opuntia* spp., in Australia and a moderately successful control agent of prickly pears in South Africa (Raghu & Walton 2007). Females of this moth lay eggs directly onto cactus spines and larvae feed within pads, killing plants outright at high larval densities and providing for the entry of plant-pathogenic fungi and bacteria with their feeding wounds. Since there are no native *Opuntia* in Australia or Africa, and *C. cactorum* is an *Opuntia* specialist, no non-target effects occurred on these continents that we are aware of.

Things began to unravel, however, when *C. cactorum* was released onto Nevis and other Caribbean islands to control Spanish lady cactus, *O. tricantha*, a native but weedy prickly pear. These releases were initiated in 1957 and were largely successful in controlling *O. tricantha* (Simmonds & Bennett 1966; Pemberton & Liu 2007). However, the releases also led to the establishment of *C. cactorum* in North America for the first time, putting at risk dozens of native *Opuntia* species there (Zimmerman et al. 2000; Stiling 2002). Laboratory and field studies confirmed the suspicion that *C. cactorum* oviposited and was able to complete development on a number of *Opuntia* species native to the Caribbean, the southeastern United States and Mexico (Johnson & Stiling 1996; Vigueras & Portillo 2001; Mafokoane et al. 2007), and sampling on various Caribbean islands 45 years after the introduction of *C. cactorum* revealed feeding on three *Opuntia* species (all native), including *O. tricantha* (Pemberton & Liu 2007).

Cactoblastis cactorum was first documented on the North American mainland in 1989, when it was found attacking native *O. stricta*, the erect prickly pear cactus, on Pine Key in Florida. This insect had presumably 'island hopped' from the Caribbean, either naturally or with inadvertent human assistance (Zimmerman et al. 2000). It then spread throughout the state of Florida, along coastal Georgia and South Carolina to the north, and into Alabama to the west, attacking at least five *Opuntia* species on the way (Johnson & Stiling 1996; Pemberton & Liu 2007). Of special concern is the endangered semaphore cactus, *O. corallicola* (= *O. spinosissima*), which is known only from the southern coast of Jamaica and two sites in Florida (Stiling et al. 2004).

To what extent are these attacks on native *Opuntia* in North America having population-level impacts? The scope for population-level impacts is certainly great, given severe population reductions of exotic *Opuntia* in Australia and native *Opuntia* in the Caribbean during biological control efforts. In addition, *C. cactorum* kill adult (albeit small- to medium-sized) *Opuntia* individuals outright. This averts the delay in population-level effects that can accompany seed feeding and can in principle lead to 'instant' population reduction. However, such effects could still be transient, very minor or localized, in which case population-level effects could be negligible. Johnson and Stiling (1998) collected data on the impact of *C. cactorum* on the widespread and abundant *O. stricta* in Florida between 1991 and 1993 and found a high incidence of attack – 90% of large plants showed evidence of damage. They addressed the question of population-level impacts by following individual cacti with feeding damage to estimate the impact of *C. cactorum* on net population growth rate. The results showed that *C. cactorum* killed some of the smallest plants during the first two years of the study, but larger plants withstood strong effects of feeding. The increased vulnerability of smaller plants is balanced in part by the fact that they are attacked less frequently, but it could eventually lead to a demographic shift in which larger plants predominate.

Of greater concern are potential population-level impacts of *C. cactorum* on the endangered semaphore cactus, *O. corallicola*. As we mentioned earlier, only three populations of this species are known, one in Jamaica, and two in Florida. Of these two Florida sites, one (Pine Key) had only 12 individual plants as of 1989 and a population of *C. cactorum* on *O. stricta*. To make matters worse, laboratory tests suggest that *C. cactorum* prefers *O. corallicola* over *O. stricta* (Johnson & Stiling 1996). Obviously, any mortality imposed upon any of the remaining semaphore cacti at this site would constitute severe population reduction. Such population reduction would almost certainly have occurred had it not been for the valiant efforts of a group of volunteers who protected the remaining *O. corallicola* on Pine Key by caging plants and manually picking off *C. cactorum* eggs and larvae (Stiling et al. 2004).

4.3.3 Non-target Effects of Biological Control Agents of Arthropods

4.3.3.1 Native Arthropods as Non-target Organisms

In this section, we discuss five cases of importation biological control in which native non-target arthropods have been subjected to unintended attack by a biological control agent. Other cases are listed in Table 4.1, but we focus here on cases that illuminate the relationships between attack and population reduction of non-target species particularly well. We reserve for the next section consideration of cases in which arthropod biological control agents have had a negative impact on weed biological control agents.

Compsilura concinnata and giant silkworm moths in North America. The parasitoid fly *C. concinnata* is notoriously polyphagous and was released into the United States from Europe beginning in 1906. It has been reared from more than 180 native species in North America, including mainly lepidopterans, but hymenopterans as well (Clausen 1978). *C. concinnata* is a gregarious, larval endoparasitoid, and females deposit fully formed parasitoid larvae into host larvae (Culver 1919). Depending on the host species and the stage attacked, *C. concinnata* may emerge either from host larvae or pupae.

Releases of *C. concinnata* were made primarily against the gypsy moth, *Lymantria dispar*, and it was recognized from the beginning that non-target hosts would likely be used (Culver 1919; Clausen 1978; Weseloh 1982, 1984). There is some evidence that *C. concinnata* contributes to preventing outbreaks of gypsy moth in some areas (Liebhold & Elkinton 1989; Gould et al. 1990; Gray et al. 2008; Elkinton & Boettner 2012), but it has by no means eliminated gypsy moth as a pest in North America. On the other hand, it is likely that *C. concinnata* has provided very effective and nearly complete biological control of the browntail moth, *Euproctis chrysorrhoea*, which was introduced from Europe at around the same time as the gypsy moth and was a co-target with the gypsy moth of some of the early releases (Elkinton et al. 2006; Elkinton & Boettner 2012).

Concerns about attacks of native non-target insects began in the 1980s, when sampling of native giant silkworm moths (family Saturniidae) revealed parasitism by *C. concinnata* (Stamp & Bowers 1990). Subsequent work by Boettner and colleagues (2000) and Kellogg and colleagues (2003) followed up on these observations by doing detailed studies with the express aim of testing the hypothesis that long-term population declines of saturniids in eastern North America were due to attack by *C. concinnata*. The main method both of these research groups employed was the outplant study, in which unparasitized hosts are placed into the field and recollected to estimate natural levels of parasitism. These experiments showed high levels of parasitism of three saturniid species (the Cecropia moth, *Hyalophora cecropia*; the Prometheus moth, *Callosamia promethea*; and the Luna moth, *Actias luna*) by *C. concinnata* (Figure 4.5), and opportunistic sampling of a fourth species, the Buck moth, *Hemileuca maia maia*, also suggested high parasitism. Similar studies in New York State and on Cape Cod in Massachusetts found lower rates of parasitism of some of these saturniids (Parry 2009).

The studies by Boettner and colleagues (2000) and Kellogg and colleagues (2003) present very strong evidence that *C. concinnata* is capable of reducing population sizes of these saturniid species. If only

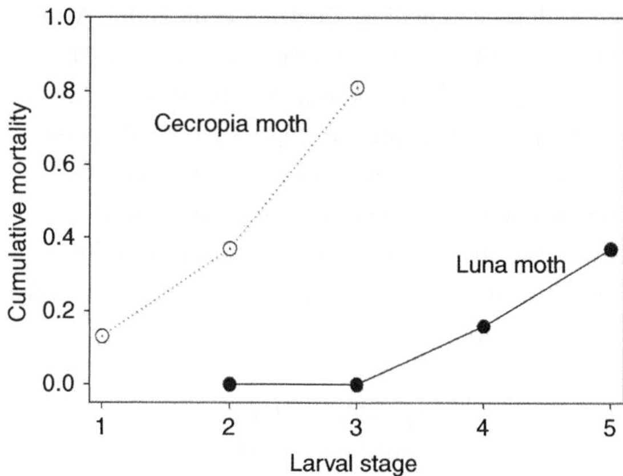

Figure 4.5. Mortality imposed by *Compsilura concinnata* parasitism of Cecropia moth and Luna moth in outplant studies in Massachusetts and Virginia in the United States, respectively. Data from Boettner and colleagues (2000) and Kellogg and colleagues (2003) with cumulative mortality calculated as outlined by Boettner and colleagues (2000). Data for larval stages 4 and 5 of Cecropia moth and for stage 1 of Luna moth not available.

for the sake of argument, however, we explore here potential scenarios under which the impact of *C. concinnata* on saturniid populations could be less than the experimental results suggest. The first such scenario is redundant mortality, including in particular compensatory density-dependent mortality of saturniid stages following those killed by *C. concinnata*. This is the scenario outlined earlier that was modeled by May and colleagues (1981) and then applied to egg parasitoids by Van Hamburg and Hassell (1984) (see Section 4.3.1). In this scenario, early mortality can protect later stages from higher levels of mortality that are triggered by high densities. The first question to be addressed in such an analysis is the stage at which *C. concinnata* imposes saturniid mortality, and this differs for the species studied by Boettner and colleagues (2000) and Kellogg and colleagues (2003). *C. concinnata* emerges from overwintering pupae of the Prometheus moth (Culver 1919; Boettner et al. 2000), which greatly restricts the later life stages available for compensatory mortality. However, this overwintering strategy increases the scope for simultaneous mortality if parasitized pupae suffer high predation rates. While there is little information on this point, Culver (1919) observed that *C. concinnata* was reducing Prometheus moth populations even during the first 10 years after introduction. Also, Marsh (1937) showed for the Cecropia moth that attack by native parasitoids was the main reason moths did not emerge, rather than predation on overwintering pupae, and the fate of overwintering Prometheus moth pupae is likely similar. Such attack is less likely to be redundant with attack by *C. concinnata* than is attack by pupal predators.

For Cecropia and Luna moths, however, *C. concinnata* kills the mid-larval and late-larval stages, respectively. For these species, it is reasonable to ask whether density-dependent mortality in later larval and pupal stages may be strong enough to lessen the effect of *C. concinnata*-imposed mortality. Density-dependent mortality has been documented for larvae and overwintering pupae of other forest lepidopterans and sawflies (Holling 1959; Dempster 1983; Parry et al. 1997; Mason & Torgersen 1987), so this scenario is by no means far-fetched. In the most famous of the lepidopteran cases, overwintering pupae of the winter moth, *Operophtera brumata*, suffer from density-dependent predation in a woodland of their native England (Varley et al. 1973). This winter moth population is attacked during the larval stage at relatively high rates by the tachinid fly *Cyzenis albicans*, but the parasitoid has relatively little influence on winter moth population dynamics. Hassell (1980) used a modeling approach to argue that the reason for this lack of control can be traced to the compensation of density-dependent pupal mortality (as well as predation of parasitized pupae).

Can a similar mechanism be posited for *Compsilura concinnata* and Cecropia and/or Luna moths in the northeastern United States? Probably not. Marsh (1937, 1941) studied mortality of Cecropia moth pupae and found that substantial mortality was due to native parasitoids, rodents and birds. Boettner and colleagues (2000) compared the larval survival needed to compensate for pupal mortality found by Marsh (1937) and concluded that mortality imposed by *C. concinnata* was insufficient to allow for replacement-level reproduction of Cecropia moths, leading to a situation in which 'No *H. cecropia* population can persist for long with these levels of larval mortality.' One could still argue that, under density dependence, mortality levels could be much higher in years with higher Cecropia densities, and that this could compensate for additional larval mortality as experienced by *C. concinnata*. However, Marsh found no evidence for density-dependent mortality (at least within one year when comparing different spatial placement of pupae), and the densities described by Marsh were already quite high and deemed locally stable at the time.

A second potential caveat involves density-dependent parasitism by *C. concinnata* itself. If *C. concinnata* parasitism is itself density-dependent (as found by Liebhold & Elkinton 1989 and Gould et al. 1990) and the densities of non-target larvae outplanted by Boettner and colleagues (2000) and Kellogg and colleagues (2003) were unnaturally high, then the high levels of parasitism observed could be an overestimate of natural levels. This was not the case however, as the density of outplanted larvae did not exceed natural levels.

In summary, we conclude that caveats that could nullify or ameliorate negative population-level impacts of *C. concinnata* on native saturniids in North America are without support and that this parasitoid has had significant impacts on population densities of the Cecropia, Prometheus, Luna and, likely, Buck moths. The broader question of whether *C. concinnata* alone can explain saturniid declines in the northeastern United States is less clear because other potential causes, such as habitat loss, insecticide use (including *Bacillus thuringiensis*) and mating disruption due to streetlights, cannot be entirely ruled out (although none of these is particularly compelling; Boettner et al. 2000). Also, the Luna moth has not experienced documented population declines in Virginia as of 2001 despite the data of Kellogg and colleagues (2003) showing strong effects of *C. concinnata*. It seems likely to us that *C. concinnata* has played an important role in the observed declines of saturniids in the northeastern United States and that the same is likely to occur farther south.

Parasitoids and the koa bug in Hawaii. The koa bug is one of the most beautiful of the Hawaiian endemic insects. It is a large shield bug that gets its name from feeding primarily on the koa tree, *Acacia koa*. Unfortunately, the koa bug is related to the Southern green stink bug, *Nezara viridula*, an important agricultural pest that invaded the Hawaiian islands in the 1960s and has been the subject of at least three importation biological control releases there. Released agents include the egg parasitoid *Trissolcus basalis* (Hymenoptera: Scelionidae) and two tachinid flies that attack primarily adult bugs: *Trichopoda pilipes* and *T. pennipes*. This was considered a partially successful biological control project in the years immediately following releases, but subsequent study showed that generalist predators were probably more important in limiting *Nezara* populations than were introduced parasitoids (Jones 1995). More important for our discussion, both *Trissolcus basalis* and *Trichopoda pilipes* attacked the koa bug on a number of Hawaiian islands, including in non-agricultural areas (Follett et al. 2000). Howarth (1991) argued that a decline in koa bug populations in the 1960s and 1970s was likely due to these biological control releases. Follett and colleagues (2000), however, were skeptical of this conclusion, noting that the timing of declines of museum specimens was not consistent with attacks of koa bug by *Nezara* parasitoids, and also that urbanization on Oahu during this time period was a confounding factor. They also noted that observed parasitism of koa bug by *T. pilipes* overestimates mortality imposed by this parasitoid because attacks on adult bugs are non-lethal, and male bugs are preferentially attacked.

Johnson and colleagues (2005) performed detailed field studies in 1998 and 1999 to address the question of whether parasitism by *T. pilipes* or *Trissolcus basalis* had indeed led to declines of the koa bug. These authors used data on koa bug mortality at various field sites on three Hawaiian islands to construct life tables in order to evaluate the role of biological control agents in suppressing koa bug populations. One of the main findings of this study was that the egg parasitoid *T. basalis* was not having a major effect on koa bug populations, parasitizing less than 4% of egg masses. A similar amount of egg parasitism was recorded from a different non-native parasitoid that was not intentionally introduced into Hawaii. The role of the tachinid *Trichopoda pilipes* was more nuanced, and requires explanation. Parasitism levels of koa bug adults and nymphs exceeded 10% in 3 of 24 sites, but these sites had the highest bug densities, leading to a pattern of very strong density-dependent parasitism. This density dependence is likely due to the attraction of the parasitoid to a sex pheromone that male bugs produce, which also probably explains why more male than female bugs were attacked. Johnson and colleagues (2005) noted, however, that this density-dependent parasitism 'suggests that the impact *of T. pilipes* may be underestimated by averaging across sites as if they were equivalent'. Indeed, parasitism levels reached 100% for males and 70% for females with additional parasitism of nymphs at some of the largest koa bug aggregations. Such high levels of parasitism could clearly make it difficult for koa bugs to reach high densities.

While these high levels of parasitism are alarming, their effect is attenuated by the fact that parasitism by *T. pilipes* is not always lethal nor does it necessarily preclude reproduction by the host. The life table Johnson and colleagues (2005) constructed for koa bug incorporated a number of details of the *T. pilipes*–koa bug interactions, which were gleaned from various laboratory and field-based studies:

- 90% of parasitized first–fourth instar nymphs of koa bug escape death by molting
- 85% of parasitized fifth instar nymphs of koa bug die
- Adult female koa bugs produce 1.5 egg masses if parasitized early

These factors reflect the sub-lethal nature of parasitism produced by *T. pilipes*, and reduced the impact of this parasitoid on koa bug populations in simulations run by Johnson and colleagues (2005). The decrease in the koa bug population replacement rate due to *T. pilipes* was estimated to be approximately 30% given these considerations. Although this is certainly a non-trivial impact, Johnson and colleagues (2005) concluded that it was unlikely that *T. pilipes* and the egg parasitoid *Trissolcus basalis* were responsible for the koa bug declines documented since the 1960s.

Microctonus aethiopoides and native weevils in New Zealand. *Microctonus aethiopoides* is an Old World parasitoid of adult weevils. It has been used successfully against the alfalfa weevil, *Hypera postica*, in North America, and was introduced into New Zealand in the 1980s, where it controlled *Sitona discoideus*, a serious pest of alfalfa (also called lucerne) (Kean & Barlow 2000c). However, this parasitoid has also been found attacking six species of non-target weevils that are native to New Zealand (Barratt et al. 1997, 2007; Barlow et al. 2004; Barratt 2004). Parasitism rates of native weevils varied by species and location, but exceeded 20% for *Nicaeana fraudator* for one sampling date at one site (although average season-wide parasitism never exceeded 3% for any site) (Barratt et al. 2007). Barlow and colleagues (2004) developed a model specifically to explore the effect of parasitism levels on impacts of *M. aethiopoides* on *N. fraudator* and one other non-target species, and their results suggested that an average parasitism level of 15% would result in

8% population suppression. Application of Barlow and colleagues' model to supplemental sampling data gathered by Barratt and colleagues (2007) suggested a population reduction of *N. fraudator* of approximately 13%, but no evidence of population declines were detected over five years of sampling. We conclude that evidence for significant population reduction of native weevils by *M. aethiopoides* in New Zealand remains weak.

Cotesia glomerata and *Pieris napi* in New England. *Cotesia glomerata* is a gregarious endoparasitoid of *Pieris* butterfly larvae and it was the first parasitoid introduced to the United States as a biological control agent. It was introduced against the imported cabbage white butterfly, *Pieris rapae*, during the 1880s in Washington, DC. Its performance as a biological control agent has been unspectacular, but it did establish and has been a ubiquitous member of the parasitoid fauna attacking *P. rapae* throughout North America. Unfortunately, it was also found attacking the green-veined white butterfly, *P. napi*, which is native to the northeastern United States and eastern Canada, and has suffered a dramatic reduction in its geographic range within New England over the past century. Benson, Van Driesche and colleagues used a series of elegant studies to address the hypothesis that *C. glomerata* was responsible for this range reduction (Benson et al. 2003b; Van Driesche et al. 2003, 2004).

The research these authors conducted had to contend with two rather puzzling complications. The first was that *P. napi* had declined only in southern New England despite the fact that *C. glomerata* was abundant throughout the region. If *C. glomerata* had indeed suppressed *P. napi* in the south, why did it not do so farther north? The answer is linked to habitat specificity of the parasitoids. *Cotesia glomerata* females forage for hosts mainly in open habitats such as meadows and agricultural fields and do not venture into woodland areas even if suitable hosts are present (Benson et al. 2003a, b). And *P. napi* spends its first summer generation feeding on woodland host plants and its second generation on meadow plants. The summer is long enough in southern New England to allow a full second generation of *P. napi* in meadows, but the season is restricted farther north so that only a partial second generation can be supported on meadow plants (Van Driesche et al. 2004). Thus, *P. napi* is more available to parasitism by *C. glomerata* in southern than northern New England, and this is consistent with *C. glomerata* suppressing *P. napi* in the south but not the north. But this leads to the second complication: The partial second generation of *P. napi* in the north is heavily attacked by *C. glomerata* – almost at the same level as southern populations. How can this be consistent with *C. glomerata* not suppressing *P. napi* in the north? Van Driesche and colleagues (2004) noted that the partial second generation of *P. napi* is unable to overwinter so that parasitism of these larvae represents redundant mortality and does not increase overall mortality to *P. napi*. Northern populations of *P. napi* can persist because a high proportion of first-generation pupae overwinter in the north. The larvae of this first generation are not attacked by *C. glomerata*, however, because they occur in woody habitats, as we have already mentioned. Taken together, this scenario can explain, in principle at least, how the parasitoid *C. glomerata* could suppress populations of *P. napi* in southern but not northern New England (Van Driesche et al. 2004).

Let us return now to the question of how strong the evidence is for population suppression of *P. napi* by *C. glomerata* in southern New England. The studies of Benson, Van Driesche and colleagues demonstrated that *C. glomerata* is capable of excluding *P. napi* from part of its historical range. This was done by doing outplant studies in which sentinel *P. napi* larvae were placed into field settings and parasitism was observed. In areas where *P. napi* is currently extinct or very rare, mortality caused by *C. glomerata* consistently exceeded 50% in agricultural areas and averaged around 10% in meadow habitats in one study and 100% in another study (Benson et al. 2003b; Van Driesche et al. 2004). This suggests that *P. napi* may currently be excluded from re-colonizing southern New England by *C. glomerata*, and it supports the hypothesis that *C. glomerata* contributed strongly to the extirpation of *P. napi* in this area. Alternative hypotheses for the range contraction of *P. napi*, including competition with the invasive *P. rapae*, and a change in the distribution of suitable host plants, were both deemed unlikely by Benson and colleagues (2003b).

These studies have led to a likely scenario in which the parasitoid *C. glomerata* is supported by the presence of the pest lepidopteran, *P. rapae*, throughout agricultural and meadow habitats in New England. The native *P. napi* suffers from this association through apparent competition in these habitats, but has a refuge in wooded habitats where *C. glomerata* does not forage.

Introduced and native lady beetles. As we have already noted in Chapter 2, a number of data sets over the past 30 years have shown declines of native lady beetles in North America coincident with introduction of the ladybirds *Coccinella septempunctata* and *Harmonia axyridis*. And as we discuss in Section 4.4.2, one of the main hypotheses for these declines involves competition between the introduced and the native lady beetles. However, another possibility is intraguild predation. It is common knowledge that aphidophagous lady beetles (including the species just listed) feed on other lady beetle species (Dixon 2000), and a number of laboratory studies of *C. septempunctata* and *H. axyridis* have shown that these species are particularly aggressive against native North American lady beetle species (Obrycki 2000). Confirmation that such predation is pervasive enough to cause the documented declines in native species is lacking, but recent field studies have documented the appropriate trophic links using molecular gut contents analyses (Hautier et al. 2008; Gagnon et al. 2011).

4.3.3.2 Weed Biological Control Agents as Non-target Organisms

A special kind of non-target effect pits one biological control project against another. In particular, predators and parasitoids used in biological control of arthropods have the potential to attack weed biological control agents (Kuhlmann et al. 2006a). One example of this involves *Microctonus aethiopoides*, the parasitoid introduced into New Zealand to control forage weevils that we discussed earlier. This species was found to attack a weed biological control agent, namely the flowerhead weevil, *Rhinocyllus conicus* (Barratt et al. 1997; Barratt & Johnstone 2001). This case carries particular irony because while *R. conicus* is itself responsible for negative non-target impacts against native thistles in North America as discussed earlier, its release in New Zealand is considered safe because of the absence of native thistles there. And to add insult to injury, *M. aethiopoides* has been released in North America as well (Radcliffe & Flanders 1998), but has not been documented attacking *R. conicus* there. Other cases in which arthropod biological control agents have been found attacking weed biological control agents include parasitoids introduced into Hawaii against fruit-feeding tephritids attacking a gall-forming tephritid species previously released as a biological control agent of lantana (Duan & Messing 2000) and commercially reared predatory mites found feeding on *Tetranychus lintearius*, a biological control agent of gorse, a weed in the western United States (Pratt et al. 2003).

4.4 Negative Indirect Effects

Indirect effects of biological control include any effects other than those that directly affect target or non-target organisms. Lonsdale and colleagues (2001) pointed out that the main positive effects of biological control are indirect, and we have already discussed some of these interactions in previous chapters. They include simple trophic cascades in which the natural enemy of a pest arthropod improves the growth and reproduction of a crop plant by suppressing herbivores. Successful weed biological control can also be the result of an indirect interaction if, for example, increased growth and reproduction of crop or native plants is due to decreased interspecific competition when an herbivore or plant pathogen suppresses a weed. However, the subject of this chapter is negative effects of biological control and evidence for negative indirect effects of biological control agents has been growing. We have already discussed one such case in this chapter – the case of (unsanctioned) myxoma virus introductions in England that suppressed rabbits and led to extinction of British populations of the large blue butterfly by habitat alteration and knock-on effects on the ability of a certain ant species to maintain its nests. This is an almost absurdly complicated domino-effect scenario that would have been very difficult to foresee. In this sense, this early example of an indirect negative effect of biological control encapsulates the danger of such effects: their unpredictability. Another problem associated with indirect effects is that they can be caused by highly specialized biological control agents, which are considered safe from the standpoint of direct non-target impacts (see Chapter 5 for a discussion of specificity testing in risk assessment). We will begin this section by outlining the different types of negative indirect effects that have been identified, and discuss each in turn, giving examples where appropriate.

4.4.1 Overview of the Types of Negative Indirect Effects

A number of authors have catalogued the kinds of indirect interactions that can lead to negative effects in biological control (Holt & Hochberg 2001; Pearson & Callaway 2005; Messing et al. 2006). Our compilation of ecological indirect effects does not differ greatly from these, and the four main indirect effects that we recognize in this chapter are diagrammed in Figure 4.6. Two of these effects involve resource competition, one in which the biological control agent engages in competition with a non-target species that also attacks the target species (Figure 4.6a), and the other in which the agent engages in competition with a non-target species that attacks a different non-target species that is itself also attacked by the agent (Figure 4.6b). Beyond this, a number of potentially negative indirect indirections can flow from the increase in the abundance of species consuming the biological control agents. This has been termed *enrichment* and one of its most straightforward effects is apparent competition, in which population levels of a non-target species are depressed as a result of this enrichment process (Figure 4.6c). A less intuitive negative indirect effect can occur through compensatory responses to attack by the biological control agent in the target species if these responses negatively impact non-target species (Figure 4.9d). And last, biological control can be involved in negative indirect effects if the target of biological control itself becomes an important ecological resource (sometimes called ecological replacement). These are cases, however, where the negative effects would not be different for other forms of control. Essentially, the 'target' organism is no longer considered a pest and its control is viewed as ecologically detrimental. Such is the case, for example, for the negative effects of rabbit biological control on large blue butterflies in Britain discussed in Section 4.2.4, as well as in Chapter 7, and it also explains the initial reluctance of the US government to allow biological control of saltcedar, *Tamarix*, because it had become an important nesting site for the endangered willow flycatcher (Dudley & DeLoach 2004). Since these kinds of negative effects are not particular to biological control, we do not discuss them further here. We do however discuss the roles of resource competition, enrichment and compensation in

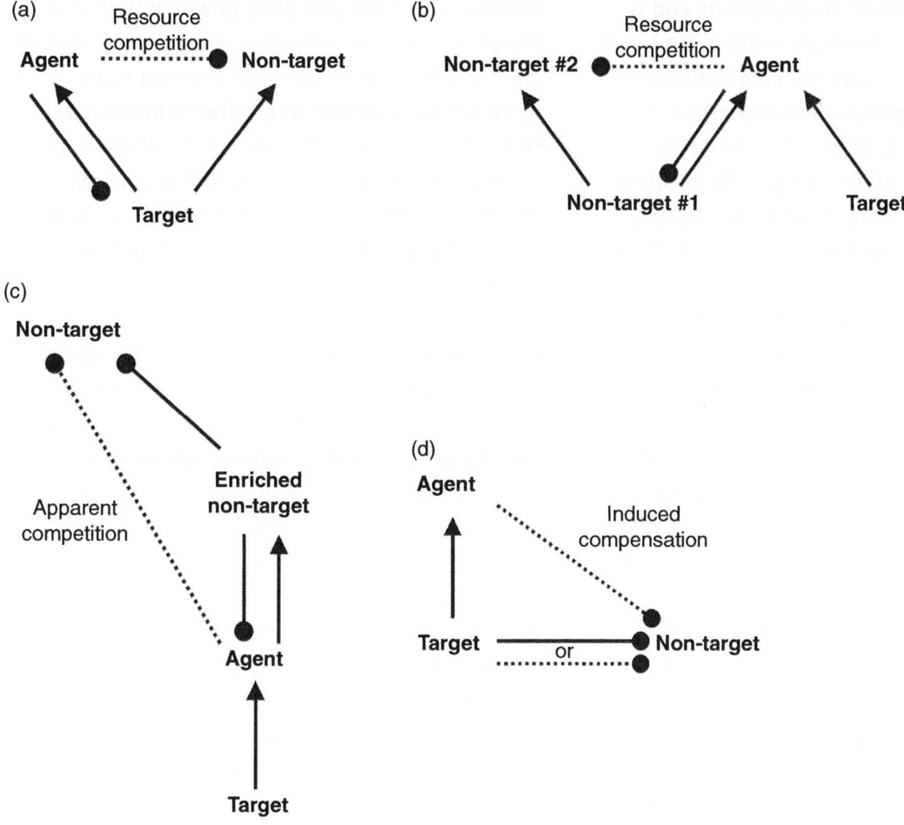

Figure 4.6. Diagrams illustrating some of the negative indirect interactions that can occur in biological control. For each diagram, solid lines indicate direct effects, with lines ending in arrows indicating a positive outcome in the direction of the arrow and lines ending in circles indicating negative outcomes in the direction of the circle. Dashed lines indicate indirect interactions. We show primarily those direct interactions that are critical to the outcome of the negative indirect action. (a) The agent negatively affects a non-target species through resource competition for the target species. (b) The agent negatively affects a non-target species by producing direct negative effects on a non-target species and engaging in resource competition with a consumer of that non-target species. (c) The agent supports the increased abundance (enrichment) of a non-target species, which itself has a negative direct effect on a different non-target species. The result is apparent competition between the agent and the second non-target species. Other, less predictable effects are possible if the enriched non-target species has positive effects on other non-target species (indicated on the right side of the diagram). (d) The agent induces a compensatory response in the target that has a negative (direct or indirect) effect on a non-target competitor of the target.

producing negative indirect effects of biological control in the following sections. While evidence for such effects remains relatively scant, at least one case study can be pointed to for each of these scenarios.

4.4.2 Resource Competition

Resource competition can be classified either as a direct or an indirect interaction between two species. Interference competition usually involves some kind of combat between competitors and is therefore classified as a direct interaction. However, exploitative competition is considered indirect when it is mediated through a third species – the resource. In the early stages of a biological control project, exploitative interspecific resource competition between the agent and a non-target species seems unlikely because of the very fact that a pest or weed species is in need of control. The high densities necessitating control imply that the pest is not a limiting resource for any consumer. However, as pest or weed densities decline, competition can come into play, as has been demonstrated in a number of cases in which introduced biological control agents engage in resource competition with each other (Reitz & Trumble 2002; Mills 2006b). Resource competition between biological control agents and native consumers of target pests is therefore not entirely unexpected, and should be more likely for highly effective biological control agents that drive the target pest or weed to low densities.

A number of cases of negative competitive effects of biological control agents are suspected, and while some or even many of these suspicions may have some validity, competition can be difficult to demonstrate in the field (Schoener 1983). For example, Howarth (1991) argued that native Hawaiian birds and predatory wasps were likely declining in part due to competition for caterpillars with parasitoids introduced as biological control agents, and that parasitoids released against the 'green looper' moth, *Chrysodeixis eriosoma*, in New Zealand displaced native parasitoids of this pest through competition. While these scenarios are plausible, they have not been subjected to competition studies in the field as far as we are aware. In this section, we discuss cases of competition between biological control agents and non-target organisms that have been evaluated experimentally. We differentiate between cases in which the biological control agent competes with a consumer of the target organism itself (Figure 4.6a) and cases in which the agent competes with the consumer of non-target species (Figure 4.6b). The former case is exemplified by competition between introduced and native consumers of aphids (Evans 2004) and the latter by competition between weed biological control agents and herbivores of non-target plants that the agents attack (Louda et al. 2005b). Both of these cases are discussed next.

Competition between biological control agents and other target consumers. As we noted earlier, if densities of target organisms are reduced to low enough levels, biological control agents may engage in competition with native consumers of these pests or weeds. This scenario has been proposed for a number of natural enemies of aphids (Mills 2006b). The sampling data alone paints a damning picture for lady beetles (coccinellids) in agricultural fields of North America, as mentioned in Section 4.3.3. While these data are suggestive of competitive effects of introduced coccinellids, demonstrating such competition in laboratory and field cage trials has produced more negative than positive results (Evans 1991, 2004; Obrycki 2000; Hoogendoorn & Heimpel 2004), and Evans (2000) failed to find a decline in the size of native lady beetles associated with the invasion of *C. septempunctata*. As we have alluded to previously, intraguild predation may be a more important mechanism by which introduced lady beetles displace native species.

Evidence is stronger for competition between a parasitoid released as a biological control agent of pea aphids and a native parasitoid of pea aphid. As is the case of the lady beetles studied by Evans, this work was done in alfalfa, but it involved the parasitoid *Aphidius ervi*, which is a highly successful biological control agent of the pea aphid, *Acyrthosiphon pisum*. The establishment of *A. ervi* in alfalfa ecosystems in North America coincided with declines of the native parasitoid *Praon pequodorum* in alfalfa, and Schellhorn and colleagues (2002) used a combination of laboratory studies and modeling to ask whether this decline was due to competition with *A. ervi*. Previous work had shown that *P. pequodorum* was a stronger direct interspecific competitor than *A. ervi* in the larval stage (Danyk & Mackauer 1996), leaving Schellhorn and colleagues with the task of showing that exploitation competition between adults favored *A. ervi* so strongly that it overrode their inferiority as larval competitors. Indeed, their laboratory experiments showed that *A. ervi* females were much more active searchers than *P. pequodorum* females were, resulting in a higher oviposition rate into pea aphids. Furthermore, a model simulating the trade-off between larval and adult competition showed that superiority in adult competition trumps superiority in larval competition when host populations are driven to extremely low values because this situation favors the efficient searcher. The pea aphid/alfalfa system is particularly susceptible to just these kinds of dynamics because of the fact that alfalfa is harvested multiple times per season, and Schellhorn and colleagues argued that this suggests that *A. ervi* could indeed have led to observed declines in *P. pequodorum*.

Evans (2004) articulated an interesting question that comes up when biological control agents competitively reduce populations of native natural enemies within agricultural or other managed ecosystems: Since these native natural enemies are feeding on introduced pests, which are themselves feeding on introduced crops, does this competition really represent ecological harm? Evans argued that it could represent harm if the competition within agricultural fields is strong enough to depress landscape- or region-wide abundances of natural enemies. In the particular system studied by Evans – *C. septempunctata* and native coccinellids in alfalfa field in Utah in the United States – there was little evidence for such a scenario, as alfalfa fields represented a relatively minor component of the landscape in his studies. It is certainly conceivable that in areas dominated by agricultural landscapes, competitive displacement could lead to lower abundances in the few remaining natural areas (Gardiner et al. 2009a). And it should go without saying that if introduced biological control agents infiltrate natural ecosystems, a number of negative direct and indirect consequences are possible. Unfortunately however, the activity of biological control agents suspected of displacing native natural enemies within agricultural settings is rarely studied in natural habitats despite the fact that such infiltration has been observed in the northeastern United States and Canada for introduced Coccinellidae (Acorn 2007; Finlayson et al. 2008).

Competition between biological control agents and consumers of non-target species. If a biological control agent consumes a non-target species, the likelihood that it will engage in competition with other consumers of that non-target species can be high, particularly if the non-target species is rare. A prime example of this kind of competition involves the thistle flowerhead weevil, *Rhinocyllus conicus*, that we discussed earlier (Section 4.3.2). The best-documented non-target host of *R. conicus* is the endangered Platte thistle, *C. canescens*, which is itself fed upon by the native picture-winged fly, *Paracantha culta*, a thistle specialist that feeds within thistle flowerheads as does *R. conicus*. Louda and colleagues have shown that the presence of *R. conicus* leads to suppression of *P. culta*, presumably in large part through exploitation competition (Louda et al. 1997, 2005a; Louda & Arnett 2000). The results of an experiment where *R. conicus* density was manipulated in the field are shown in Figure 4.7.

4.4.3 Enrichment and Apparent Competition

As we have discussed in Chapter 2, biological control agents are often attacked by natural enemies in their introduced range. These attacks form the basis of enrichment effects, in which the presence

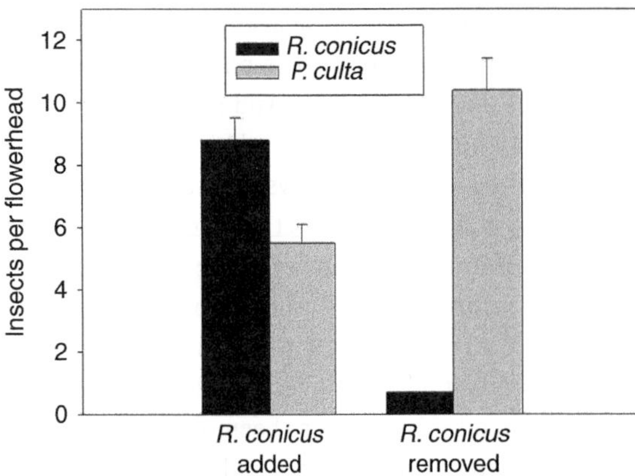

Figure 4.7. Results of an experiment in which the biological control agent *Rhinocyllus conicus* was added to or removed from flowerheads of the non-target Platte thistle. Effects on numbers of the native picture-wing fly *Paracantha culta* per flowerhead are shown along with numbers of *R. conicus*. The addition of *R. conicus* led to declines in *P. culta*, presumably due to competition. From Louda and colleagues (2005b).

of an introduced biological control agent leads to increases in abundance of some other species, typically a consumer of the agent. These enrichment effects have gone by other names in the literature, including 'food-web subsidies', 'food-web effects' or 'food-web interactions' (Pearson & Callaway 2003, 2005) and 'circuitous competition' (Messing et al. 2006). In all of these cases, the basic ecological mechanism involves an increase in abundance of a consumer of a biological control agent resulting in some sort of ecological change (Figure 4.6c). The enriched species can have negative or positive effects on other species in the community, making the outcome of enrichment difficult to predict. We discuss two relatively well-documented outcomes of enrichment linked to importation biological control projects. Both cases illustrate what can happen when a biological control agent becomes established and abundant without controlling its intended target organism.

The best-documented case of ecological changes due to enrichment through biological control involves biological control of spotted knapweed, *Centaurea maculosa*, in North America by two species of tephritid fruit flies in the genus *Urophora*. The *Urophora* species are host-specific flowerhead gallers that are capable of decreasing knapweed seed set, but have not effectively controlled knapweed populations or limited their spread in western North America (Maddox 1982), although the seed bank has been substantially reduced, which may ultimately lead to enhanced suppression (Story et al. 2008). Despite their poor initial showing as biological control agents, however, the *Urophora* flies have become widely established in western North America and can be locally superabundant (Myers & Harris 1980; Story & Nowierski 1984; Story 1995). The fly larvae overwinter within knapweed seedhead galls and it is the presence of high densities of these larvae during a time (winter) that is otherwise characterized by resource limitation that has led to significant ecological alterations. In particular, deer mice, *Peromyscus maniculatus*, have taken advantage of this new resource and experienced substantial population growth. Gut contents analyses and population sampling have shown that *Urophora* larvae make up more than 80% of the deer mouse diet in the winter and the presence of *Urophora* has led to a doubling of deer mouse populations as well as a shift of habitat preference of the mice from forests to meadows, which is where spotted knapweed grows (Pearson et al. 2000; Ortega et al. 2004). The ecological effects of such a shift in the size and spatial distribution of deer mouse populations are potentially wide-ranging, but most alarming is the fact that deer mice are a vector for the Sin Nombre hantavirus, which causes Hantavirus Pulmonary Syndrome, a disease with a 37% fatality rate in humans. Pearson and Callaway (2006) have shown that locations with higher knapweed (and therefore *Urophora*) abundance have higher populations of hantavirus-carrying deer mice. There was also a suggestion in their study that the *proportion* of hantavirus-infected mice was higher where

Urophora was more prevalent due to density-dependent horizontal transmission rates.

Perhaps the most intuitive and likely negative ecological outcome of biological control-related enrichment is apparent competition, in which the enriched species puts pressure on a native species as described in Chapter 2 (see Figure 2.5, Figure. 4.6c). This kind of an effect has been hypothesized and predicted by a number of authors (Holt & Hochberg 2001; Strong & Pemberton 2001; Heimpel et al. 2004a; Willis & Memmott 2005; Van Veen et al. 2006; Veldtman et al. 2011), but we know of only one well-documented case for importation biological control. This case involves biological control of Bitou, *Chrysanthemoides monilifera rotundata*, an invasive weed in Australia that is native to southern Africa. Biological control of Bitou bush is reminiscent of the efforts against spotted knapweed that were just mentioned – a seed-feeding tephritid fly, *Mesoclanis polana*, was introduced, and this agent established well but did not control the weed. Instead, it remained at high densities post-release and became a subsidy for a number of insect predators and parasitoids in its introduced range within Australia. In a food-web study of 17 Bitou plots that had been subjected to *M. polana* releases, Carvalheiro and colleagues (2008) found that 12% of the *M. polana* individuals were attacked by natural enemies, which included eight parasitoid species as well as a predatory midge species. This level of attack on *M. polana* led to a six-fold increase in overall abundance of these natural enemies. Variation among the 17 plots was used to construct multiple-regression models to estimate effects of *M. polana* abundance and other variables on native species abundances. These analyses showed that *M. polana* had a negative effect on some seed predators of native plants and also on parasitoids of hosts other than *M. polana*. Species richness of both dipteran seed predators and parasitoids was negatively correlated with *M. polana* abundance, particularly in coastal forested habitats (Figure 4.8). These results are non-experimental and therefore must be seen as

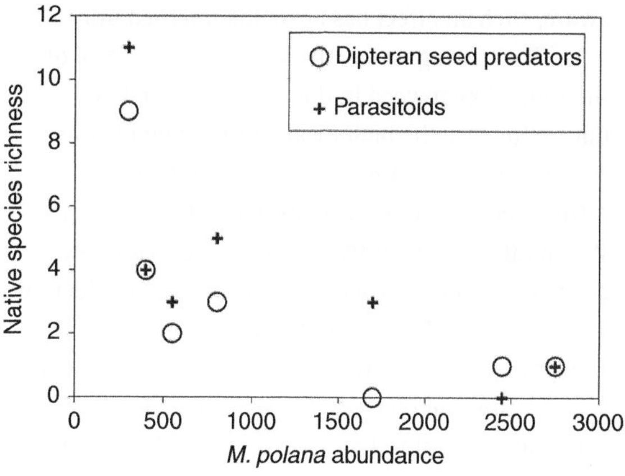

Figure 4.8. Correlation between the abundance of the dipteran seed feeder *Mesoclanis polana* (a biological control agent of the weed Bitou) and species richness of dipteran seed feeders and parasitoids in seven Bitou plots embedded within a coastal forest habitat in Australia. From Carvalheiro and colleagues (2008).

suggestive, but they are consistent with widespread ecological impacts of Bitou biological control. Since the agent, *M. polana*, did not produce effective biological control of the weed, this case seems to be a realization of the worst-case scenario envisioned by Holt and Hochberg (2001), in which an ineffective biological control agent becomes established and abundant, leading to strong apparent competition between itself and other species. This general scenario is discussed further from the standpoint of risk assessment in Chapter 5.

4.4.4 Compensation

Compensation to herbivory has been documented in a number of plant species, as we have already discussed in Chapter 1. Much less studied is the extent to which herbivore-induced compensation influences the soil and plants that surround the affected plant. To the extent that such effects are induced by biological control agents and have negative effects on non-target species, they constitute negative indirect effects of biological control (Figure 4.6d). As far as we are aware, such an effect has been documented only once. The case again involves biological control of the spotted knapweed in the western United States, but in this case the biological control agent is the root-boring moth *Agapeta zoegana*, which was introduced from Europe to Montana, Oregon and Washington (Story 1995). While the release of *A. zoegana* did not have a discernible negative effect on knapweed biomass (it did have a negative effect on flowering), reproductive capacity of a native fescue grass, *Festuca idahoensis*, declined significantly when growing adjacent to knapweed being fed upon by *A. zoegana* larvae (Callaway et al. 1999; Ridenour & Callaway 2003). The results of these studies showed that (1) spotted knapweed roots were able to mount an impressive compensatory response to herbivory (as had been shown previously by Müller-Schärer 1991; Steinger & Müller-Schärer 1992), and (2) the herbivory had an indirect negative impact on a neighboring plant of a different species. How could such a negative effect occur? Three non-exclusive hypotheses have been posed (Pearson & Callaway 2003): first, knapweed plants may have become more effective interspecific competitors in the presence of herbivores (the compensatory resource uptake hypothesis); second, herbivory may have altered mycorrhizal interactions between the two plants to the detriment of the fescue (the mycorrhizal hypothesis); and third, knapweed may have been induced to exude toxins into the soil that negatively impacted the fescue grass (the allelopathy hypothesis).

Finding evidence to support or refute any of these three hypotheses has been maddeningly difficult. There was little evidence for differential nutrient uptake by knapweed in the presence of herbivory in the studies done by Callaway and colleagues (Callaway et al. 1999; Ridenour & Callaway 2003), providing little support for the compensatory resource uptake hypothesis. However, it is possible that compensatory resource uptake is important under some soil conditions and not others. Turning to the mycorrhizal hypothesis, there is indeed some evidence for a mycorrhizal connection between fescue and knapweed (Ridenour & Callaway 2003), but again, it is not clear whether nutrient transfer from fescue to knapweed is increased by herbivory. Last, some evidence consistent with allelopathic effects of knapweed on fescue was initially found by Ridenour and Callaway (2001) and led to the hypothesis that effects of plant invasion can be intensified by allelopathic 'novel weapons' (Callaway & Ridenour 2004). There was no indication, however, that such affects are inducible by herbivory, and the allelopathy results were brought into question on methodological grounds (Blair et al. 2005, 2006, 2009; Lau et al. 2008; Chobot et al. 2009). The criticisms of the allelopathy data were directed mainly at claims of the production of a putative allelochemical catechin and the effects of this compound on the native fescue. As such, these studies do not exclude the possibility that allelopathy could be occurring via other chemicals (Blair et al. 2006). As far as we are aware, the mechanism(s) causing negative indirect effects of biological control via a compensatory response to herbivory remain unidentified.

Regardless of the mechanism, it is fair to ask how common such compensation effects are. As far as we are aware, the spotted knapweed/*Agapeta* case is the only such example known involving a biological control agent, but other herbivores are able to produce similar compensatory effects on neighboring plants. In one of these cases, Ramsell and colleagues (1993) found that perennial ryegrass, *Lolium perenne*, subject to root-grazing by larvae of the cranefly, *Tipula paludosa*, were stronger interspecific competitors than *L. perenne* plants free of such feeding. Other similar cases involving clipping as opposed to herbivory are reviewed by Pearson and Callaway (2005). Thomas and colleagues (2004) suggested that compensatory effects were more likely in weed rather than arthropod control, and this would certainly seem to be true for the kinds of mechanisms discussed earlier. However, Pearson and Callaway (2004) countered that one possible compensatory effect in arthropod biological control could be a host shift by an herbivore to escape its biological control agent. Such effects have been predicted since at least the 1980s and demonstrated in natural systems (Bernays & Graham 1988; Murphy 2004; Wiklund & Friberg 2008) and could occur in biological control systems as well.

4.4.5 Hybridization between Introduced Biological Control Agents and Native Species

Introduced biological control agents could in principle hybridize with closely related species, leading to a complex set of evolutionary and ecological outcomes (Hopper et al. 2006). The potential for hybridization has been investigated for a few parasitoids and a lacewing introduced into Japan as biological control agents (Naka et al. 2005; Davies et al. 2009; Yara et al. 2010). There is evidence of field hybridization between the Chinese parasitoid *Torymus sinensis*, which was introduced into Japan to control the chestnut gall wasp, *Dryocosmus kuriphilus*, and the native *T. beneficus* (Yara et al. 2010). This study showed that one of two temporally distinct populations of the native species was subject to hybridization by *T. sinensis*. Another case of hybridization involves biological control of the Asian hemlock woolly adelgid, *Adelges tsugae*, in North America. The derodontid beetle, *Laricobius nigrinus*, was introduced from the western to the eastern United States to control hemlock woolly adelgid with some success, and there it encountered the closely related *L. rubicus*. Interspecific copulations of these species were observed in the field (Mausel et al. 2008), and this prompted a thorough investigation of hybridization between the introduced *L. nigrinus* and the native *L. rubicus* in the eastern United States. Havill and colleagues (2012) used molecular and morphological markers to confirm the presence of hybrids between the two species, and also found evidence for F2 hybrids and backcrosses, showing that hybrids are fertile. The proportion of hybrids among all samples increased with time since the release of *L. nigrinus*, reaching almost 30% over three years. Havill and colleagues (2012) concluded that this hybridization could improve the establishment rate of *L. nigrinus* by alleviating Allee effects (see Section 2.5.1), but also that it could lead to the loss of *L. rubicus* as a pure species.

4.5 General Trends for Non-target and Other Negative Effects of Biological Control

In this section, we briefly review attempts to come to synthetic conclusions about trends in non-target and other negative effects of biological control using comparative methodologies. In a prescient attempt at a comparative analysis of non-target effect of parasitoids used for importation biological control, Hawkins and Marino (1997) looked for trends in the adoption of native hosts by parasitoids introduced as biological control agents. Their goal was to identify characteristics of either the parasitoids or their targets that would increase the likelihood of non-target impact. For this purpose, they used six classes of independent variables to predict whether non-target host use occurred:

(i) **Parasitoid biology**, incorporating various life history strategies of the parasitoids
(ii) **Parasitoid region of origin**
(iii) **Project outcome**; that is, whether the parasitoid established on the target host
(iv) **Feeding niche of the target host**
(v) **Target habitat**
(vi) **Time since introduction**

The analysis was restricted to parasitoids of holometabolous insects that were released in North America and incorporated 313 parasitoid species gleaned from the BIOCAT database that was discussed in Chapter 3. Various univariate and multivariate analyses were conducted to estimate the contribution of these six variables to the percentage of parasitoids using native host species. Overall, Hawkins and Marino found that 16% of the parasitoids used native host species, but that the independent variables had very little predictive power. The only variable that predicted use of native hosts was project outcome. As expected, parasitoids that established on their target host moved onto native hosts at a greater rate than parasitoids that did not. Note, however, that a non-trivial fraction (11%) of parasitoids that *did not* establish on their target host were reared from native hosts as well. Hawkins and Marino (1997) concluded rather pessimistically that

> *[W]e may never be able to predict with any certainty whether or not an exotic parasitoid will colonize native insects, just as the general problem of success of any particular invading species is believed to be inherently unpredictable.*

While it is certainly true that predicting non-target use of species that are not monophagous is difficult, some progress has been made since this analysis. In particular, the incorporation of two biological variables that Hawkins and Marino were not able to include in their analysis have proved useful: the specificity of the natural enemy and the degree to which non-target organisms are related to the target host. Pemberton (2000) tackled the influence of taxonomic relatedness of non-target species to target species and showed quite conclusively that, for weed biological control agents at least, non-targets were at much greater risk of attack by biological control agents if they were closely related to target species. We defer a more complete discussion of this important result to Chapter 5. Related to this, Kimberling (2004) found (as should be expected) that host-specific arthropod biological control agents were less likely to attack non-target organisms than were generalist biological control agents. Kimberling's analyses included a number of independent variables beyond specificity, but some of these were highly correlated with one another, making firm conclusions about the effect of other variables difficult. However, a consistent outcome of her models is that traits that increase efficacy of biological control agents also increase their safety. This was true not only for specificity, but also for voltinism (agents with multiple generations per pest generation were both more effective and safer) and sex ratio (agents with female-biased sex ratios are both more effective and safer) (Figure 4.9).

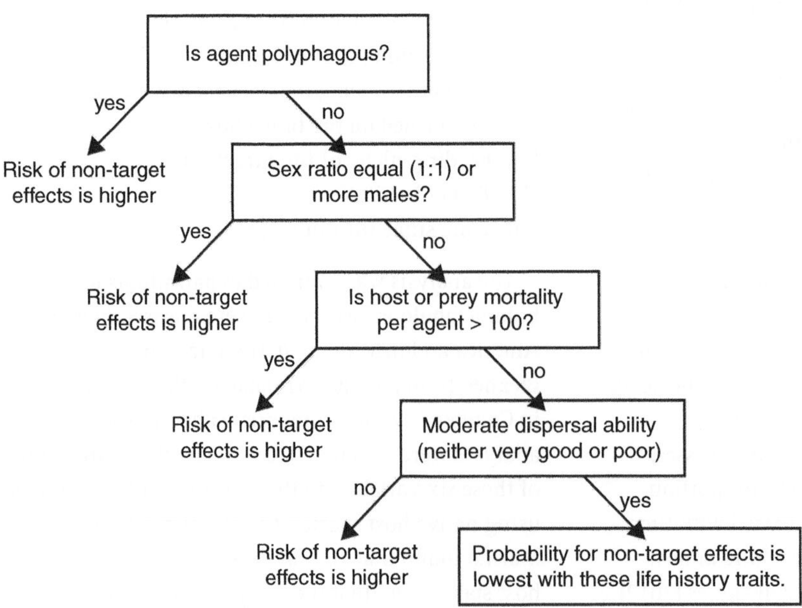

Figure 4.9. Statistical model evaluating non-target effects of biological control introductions. From Kimberling (2004).

These trends are consistent with predictions made by Holt and Hochberg (2001) that less effective agents will lead to more non-target effects because of their inability to drive pest populations (and therefore their own populations) to low levels. These predictions are discussed in more detail in Chapter 5. Last, Kimberling found that biological control has been both the least successful *and* the least safe in forest settings, and she interprets this as due to both the high habitat heterogeneity found in forest settings and the fact that generalist biological control agents have frequently been used against forest pests.

It is also possible that negative direct effects are more likely to occur in natural rather than managed ecosystems. This point was made by Sheppard and colleagues (2003), who also pointed out that biological control in natural ecosystems is sharply on the increase for weed biological control (see also Van Driesche et al. 2010). Natural ecosystems obviously have a higher diversity of non-target species, and their greater complexity should also provide opportunities for more classes of indirect effects (Hawkins et al. 1999; Holt & Hochberg 2001), generally increasing the risk of negative consequences (Simberloff 2012).

4.6 Conclusions

Biological control introductions carry a level of risk that other control methods do not carry simply because the agents are alive: they can disperse far from the areas where they are intended to control pests, and they can, at least in principle, evolve to use new plant or host species (Simberloff & Stiling 1996a; see Chapter 7). To this should be added the problem that they can engage in indirect interactions with native and other non-target species. And to compound things, biological control releases are virtually irreversible – agents cannot be recalled, and unlike with pesticides, a cessation of application is not a viable strategy to improve the situation in the case of established agents (but see Chapter 5). Despite this gloomy assessment, it is important to remember that the mere documentation of negative consequences does not imply that the net effect of the introduction is negative. Balancing negative effects against the positive effects of biological control discussed in Chapter 3 is needed to make a determination about the net effect of a program. An important component of such a reckoning is a comprehensive risk assessment process for biological control, which is the subject of the next chapter.

5 Ecological Risk Analysis in Biological Control

The father of medicine, Hippocrates, admonished physicians to 'help, or at least to do no harm'. In modern times, physicians take an oath that includes this sentiment before beginning their practice. As we have seen in the previous chapter, practitioners of biological control also need to be mindful of not doing harm. And just as physicians have to balance the benefits of curing patients with the potential risks of negative side effects of their interventions, biological control scientists need to be able to balance the benefits of biological control with its risks.

5.1 Introduction

The processes of environmental and ecological risk analysis have developed over the past half century from the assessment of risks of environmental contaminants on human health to the assessment of a broader range of stressors on nonhuman species and ecosystems more generally (US EPA 1998). Ecological risk analysis (ERA) encompasses risk assessment, risk management and risk communication. From the standpoint of importation biological control, the most important of these is risk assessment because information gained from risk assessment studies is used in deciding whether to do biological control releases in the first place (Sheppard et al. 2003; Van Lenteren et al. 2006a). Risk management and risk communication can play important roles as well, but these usually come into play post-release if the biological control agent is actually causing harm. An extension of ERA is risk-benefit analysis, in which risks are weighed against potential benefits (Delfosse 2005; Bigler & Kölliker-Ott 2006; Thomas & Reid 2007). This can be seen as comparing the risks of doing biological control with the risks of not doing biological control.

While ERA applies most transparently to importation biological control, it is relevant to augmentative biological control as well. For example, entomopathogenic nematodes released as biological control agents are thought to rarely persist in the field for more than a few months, yet non-target impacts have been examined in both the laboratory and the field (Bathon 1996; Barbercheck & Millar 2000; Millar & Barbercheck 2001; Ehlers 2003; De Nardo et al. 2006). And biological control agents released into greenhouses are sometimes found to be established in nearby field settings. Examples include the predatory mite *Neoseiulus* (= *Amblyseius*) *californicus* in the United Kingdom (Hart et al. 2002a) and the Asian coccinellid *Harmonia axyridis* in Europe (Brown et al. 2008). Indeed, many augmentative biological control agents are used in areas where they are not native, and the conventional wisdom that agents native to warm areas can be expected not to 'escape' from greenhouses in cooler climates has come into question, with a growing number of studies evaluating the risk that augmentative agents intended for greenhouse use will establish outdoors (Hart et al. 2002a, b; Hatherly et al. 2004, 2005; Tullett et al. 2004; Boivin et al. 2006). In sum, while there are some differences in the risk assessment procedure for importation and augmentative biological control, many of the overarching principles are the same (Van Lenteren & Loomans 2006). We therefore will not explicitly discuss ERA separately for these classes of biological control.

Conservation biological control is typically considered risk-free since it involves no releases of organisms into the environment. Even here, though, risk can come into play. For example, the use of habitat diversification in agricultural systems as a method to enhance biological control can have unintended side effects. Weedy cultures or cover crops can increase pest abundance by serving as alternative host plants for the pests (e.g., Shearer & Jones 1998) or by providing nectar to pests instead of natural enemies (e.g., Baggen et al. 1999; Romeis & Wäckers 2000). In one case, plantings of flowering buckwheat in New Zealand orchards for the purpose of attracting and feeding parasitoids of pest insects instead increased the density of a parasitoid of (beneficial) brown lacewings (Stephens et al. 1998). This general topic will be considered in more depth in Chapter 9.

Our focus in the first part of this chapter will be the three traditional aspects of risk analysis – risk assessment, risk management and risk communication – in the context of importation and augmentative biological control. This will be followed by an attempt to reconcile risks and benefits of importation biological control in order to arrive at a discussion of the net effect of biological control on ecosystems. A detailed review of the legal regulation of biological control agents and how this varies among countries is beyond the scope of this chapter, but we direct interested readers to the following treatments of this topic: Strong and Pemberton (2000, 2001), Sheppard and colleagues (2003), Babendreier and colleagues (2006), Moeed and colleagues (2006), Van Lenteren and colleagues (2006a) and Cock et al. (2009).

5.2 Risk Assessment

Ecological risk assessment has been defined by the United States Environmental Protection Agency (US EPA) as 'a process that evaluates the likelihood that adverse ecological effects may occur or are occurring as a result of exposure to one or more stressors' (US EPA 1998). For importation biological control, the stressors are the agents themselves, and the adverse effects include non-target impacts and negative indirect effects of the agents as outlined in the previous chapter. The goal of risk assessment in biological control is to estimate the probability that negative effects will occur as well as the magnitude of these effects. The primary process for evaluating the likelihood of non-target effects in biological control is host-specificity testing. There is no formal empirical process in use to assess the risk of negative indirect effects, although narrower diet breadth leads to a reduction in the *types* of indirect interactions that can occur (Holt & Hochberg 2001; Pearson & Callaway 2005; see Chapter 4), so that in this sense at least, host-specificity testing addresses both direct and indirect risks.

Before embarking upon discussions of host-specificity evaluations and predictions of negative indirect effects, we will consider briefly how these processes fit into the ERA framework more generally. And while we focus on the framework developed by the US EPA for illustrative purposes, it is important to recognize that a number of other frameworks have been introduced as well (Power & McCarty 2002). The US EPA framework for risk assessment consists of three components: problem formulation, analysis (which is itself divided into exposure and effects components) and risk characterization (Figure 5.1). This and other frameworks of risk assessment have been developed separately from the discipline of biological control, so assigning steps in the process of biological control research into the framework is somewhat artificial. Despite this, many linkages between established protocols in biological control and formal risk assessment are rather straightforward, suggesting that biological control research has been implicitly following a general risk-assessment model. As we describe later, however, research supporting weed biological control (by herbivores and pathogens) resembles formalized risk assessment much more closely than does research supporting arthropod biological control. In Figure 5.1, we diagram the main elements of the US EPA ERA framework, noting particular issues in the process of biological control that address these elements.

In the following paragraphs, we will explain how the three components of the US EPA risk assessment

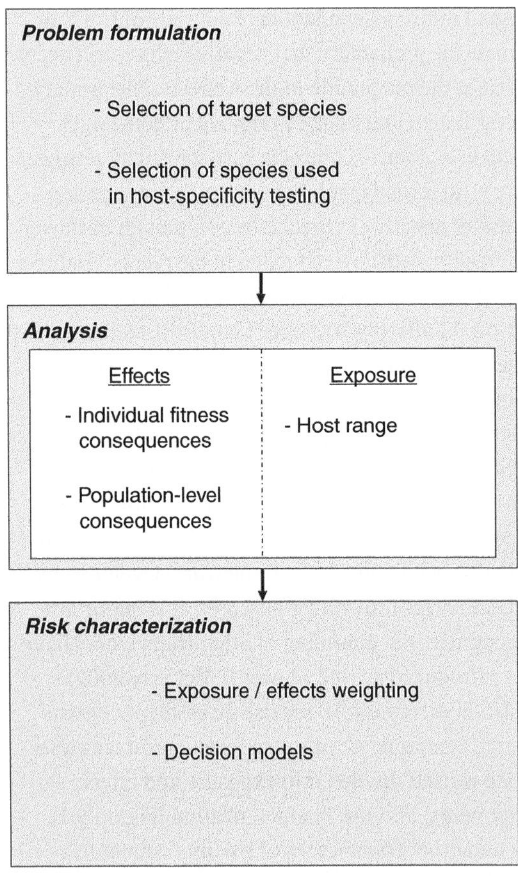

Figure 5.1. The major phases of ecological risk assessment as outlined by the US EPA (1998), along with activities and topic areas that are relevant to ecological risk assessment for biological control.

framework can be applied to biological control research (Section 5.2.1), after which we launch into a relatively detailed discussion of host-specificity testing (Section 5.2.2). We feel that close attention to this topic is appropriate since host-specificity testing is the mainstay of risk assessment for biological control. However, other classes of evaluation are important as well, and the section on host-specificity testing is followed by discussions of assessment of other kinds of risk, such as dispersal and indirect effects (Section 5.2.3). We finish our treatment of risk assessment with a discussion of the special case of transgenic biological control agents (Section 5.2.4).

5.2.1 Applying the Phases of Risk Assessment to Biological Control

Problem formulation. The first phase of risk assessment is problem formulation, in which objectives are refined and a plan for characterizing risk is developed. In terms of biological control, this phase includes both selection of target species for biological control and selection of potential non-target species to be used in host-specificity testing. The US EPA (1998) framework envisions three outcomes from the process of problem formulation, and these are:

(i) assessment endpoints that adequately reflect management goals and the ecosystem that they represent;
(ii) conceptual models that describe key relationships between a stressor and assessment endpoint; and
(iii) an analysis plan.

These outcomes are readily recognizable in biological control research as (i) identifying potential non-target species, ecosystem or human health properties that are to be protected during biological control; (ii) an understanding of potential non-target and indirect negative effects; and (iii) a set of experimental protocols designed to arrive at such an understanding.

Analysis: effects and exposure. The analysis phase within risk assessment includes characterization of exposure of non-target species or the environment in general to the stressor as well as the effect of that stressor on non-target species or the environment based on the level of exposure. In terms of importation biological control, the analysis of effects begins with host-specificity testing, as will be discussed shortly. Such tests can only determine the capability for a potential biological control agent to attack, feed on or develop on non-target host or prey species at the individual level. Further testing is needed to gain insights into the potential population-level consequences that released agents may have on non-target organisms. Characterization of exposure to the stressor is the second component of the analysis phase. Here, the contact or at least the co-occurrence of the agent and non-target species or habitat must be predicted. Ideally, the likely level of exposure is also assessed. While analyses of effects are used to determine which species the agent is capable of attacking, analyses of exposure assess the likelihood that the agent will actually come into contact with these species. Exposure characterization in biological control is analogous to an estimation of the 'ecological host range' – that portion of the potential host range that is expressed under field conditions, which we discuss later. For biological introductions in general, the analysis phase of risk assessment is not easily quantifiable because of the ability of introduced species to disperse, reproduce, interact with other species and evolve. Thus, qualitative estimates of both effects and exposure tend to be favored for risk assessment of introduced species (US EPA 1998).

Risk characterization. In the US EPA framework, risk characterization encompasses risk estimation, description and reporting. The most important of these for biological control is risk estimation, in which exposure and effects are integrated.

A number of authors have formulated particular conceptual frameworks that integrate exposure and effects for non-target effects of biological control agents. For example, the legal framework for biological control introductions in New Zealand begins with a qualitative scale for both effects and exposure ('likelihood of effect' in this framework) and then imposes a qualitative measure of the overall level of risk that balances exposure and effects (Table 5.1). Similar frameworks that are somewhat more quantitative in which assigned values of effects magnitude and likelihood are multiplied have been introduced as well (Van Lenteren & Loomans 2006). Flowchart decision models for ERA can be useful, and one such application that is relevant to both

Table 5.1. **Risk characterization framework used in biological control risk assessment by the Environmental Risk Management Authority of New Zealand as part of its Hazardous Substances and New Organisms Act. The matrix includes qualitative estimates of effects likelihoods and magnitudes, combinations of which produce overall risk of estimated qualitative levels. The overall risk estimates are depicted as shades of gray and black with lighter shades indicating lower risk. Modified from Moeed and colleagues (2006).**

Likelihood of Effect	Magnitude of Effect				
	Minimal	Minor	Moderate	Major	Massive
Highly improbable					
Improbable					
Very unlikely					
Unlikely					
Likely					
Very likely					
Extremely likely					

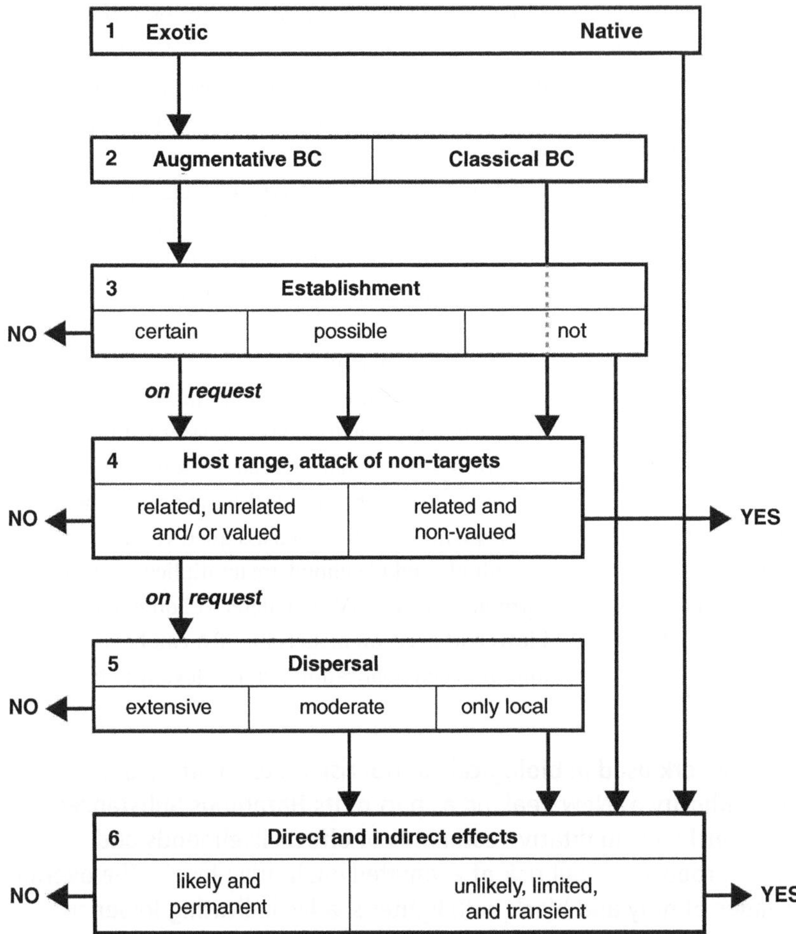

Figure 5.2. A flowchart decision model for augmentative and importation biological control in which 'Yes' means that release of a biological control agent is recommended and 'No' means it is not. 'On request' signifies that more information may be necessary to make a decision. From Van Lenteren and colleagues (2006).

importation and augmentative biological control is presented in Figure 5.2. We are not aware of such models that allow for separation or integration of exposure and effects components of risk, but this could in principle be easily incorporated. Andersen and colleagues (2005) have argued that meaningful risk characterization for biological control requires mathematical models that incorporate dispersal and other behavioral and life history information of agents within a population dynamics framework. They illustrate this with an individual-based model for a hypothetical weed biological control agent, stressing that movement patterns of the agent are likely to play an important role for both exposure and effects to non-target plants.

5.2.2 Host-Specificity Testing

The most straightforward way to assess direct risks of biological control agents to non-target organisms is to conduct host-specificity trials. Host (or prey) specificity trials have two related goals. The first is to determine whether particular non-target species are at risk of attack from the proposed agent. For example, host-specificity testing for weed biological control agents typically includes crop plants that are conceivable hosts. Second, host-specificity trials provide information on the potential agent's general feeding pattern – whether it is monophagous, polyphagous or somewhere in between. The phylogenetic pattern of host use is of great interest in this context, as we discuss later.

The science of host-specificity testing is most advanced for arthropod biological control agents of weeds. Also, the methodologies, analyses and interpretations of data are more highly developed and standardized for weed biological control agents than they are for biological control agents of arthropods. The procedures for host-specificity testing of parasitoids and predators of arthropod pests are not as well developed, but efforts are being made to close this gap (Van Driesche & Reardon 2004; Bigler et al. 2006). And finally, host-specificity testing of pathogens for arthropods (Simoes & Rosa 1996; Hajek & Butler 2000; Cory 2003) and of plant pathogens for weed control (Barton 2004; Morin et al. 2006) have been conducted as well. Here again, though, standardized methodologies aimed specifically at risk assessment for biological control in these areas tend to lag behind those established for weed biological control.

Before embarking upon a discussion of host-specificity testing, it is worth noting that *target* selection is an important component of the ERA process in biological control as well. Indeed, target selection can be seen as a kind of preemptive risk management, since some targets are inherently more risky than are other targets. As we discuss later, non-target feeding is much more common among herbivores introduced to control weeds that have close relatives in the introduced range than those introduced to control weeds with no close relatives in the introduced range (Pemberton 2000; Louda et al. 2003). Consider, for example, two invasive weeds of New Zealand, the mist flower, *Ageratina riparia*, and hawkweeds, *Hieracium* spp. Both of these weeds cause significant environmental damage, and neither has closely related native species in New Zealand (Barratt & Moeed 2005). Mist flower is the only member of its tribe in New Zealand, while the invasive hawkweeds are the only *Hieracium* species in New Zealand. Finding agents for these weeds that did not feed on native plants was not difficult due to the relative taxonomic isolation of the New Zealand fauna, and permits for the release of herbivores were granted as risks to native species were deemed low. On the whole, plants introduced from Eurasia into North America, or vice versa, are likely to have more close relatives in their introduced range, making it more difficult to avoid the potential for non-target impacts.

5.2.2.1 Choosing Which Non-target Species to Test: Wapshere's Method and Beyond

Once a target species and prospective agent species is chosen, a list of non-target test species for testing against the agent needs to be developed. Wapshere's (1974) 'centrifugal phylogenetic testing' method is a starting point for the development of such a list. Wapshere's method was designed for use with arthropod agents of weeds, but has been adopted for fungal pathogens of weeds (Barton 2004) and arthropod agents of arthropod pests as well (Kuhlmann et al. 2006b). This method consists of exposing non-target species to the prospective agent in sequence, with the most closely related strains and species tested first, and species that have successively lower phylogenetic relationships to the target tested subsequently. In addition to centrifugal phylogenetic testing, Wapshere added a number of additional criteria that allowed for the inclusion of cultivated plants that are more distantly related to the target than those in the centrifugal testing sequence. These include threatened and endangered species that are in the same family as the target weed and species that occur in the same habitat as the target weed. The number of host plants screened in host-specificity trials for herbivorous arthropods in recent decades has ranged from about 40 to more than 100 (Kuhlmann et al. 2006b).

Extensions of Wapshere's testing method for weed biological control have paralleled the growth in the understanding of phylogenetic relationships among plants (Briese 2005). It is now possible to choose plants not only based on *taxonomic* affiliation, but also on actual *phylogenetic* separation, which should lead to a more sensitive estimation of host specificity of the proposed biological control agent. This approach can also increase the efficiency of the testing procedure. For example, consider the following example of a test plant list development for the flea beetle, *Longitarsus* sp., a proposed biological control of blue heliotrope, *Heliotropium amplexicaule*. Table 5.2 provides a list of test plant

Table 5.2. **Taxonomic organization of potential test plants with respect to the target weed, blue heliotrope, *Heliotropium amplexicaule*, by the flea beetle, *Longitarsus* sp. Taxonomic relationships are used to infer the degree of relatedness of the test species to the target, but variation in relatedness within genera or subfamilies or among subfamilies or families is unclear. Modified from Briese (2005) and Briese and colleagues (2005).**

Test Species	Subfamily	Family
Heliotropium nicotianaefolium		
Heliotropium indicum		
Heliotropium curassavicum		
Heliotropium foertheri		
Heliotropium asperrimum	Heliotropioideae	
Heliotropium europaeum		
Heliotropium supinum		
Heliotropium arborescens		Boraginaceae
Euploca ovalifolium		
Euploca brachygyne		
Cordia dichotoma	Cordioideae	
Ehretia acuminata	Ehretioideae	
Myosotis australis	Boraginoideae	
Myosotis discolor		
Phacelia tanacetifolia	Hydrophylloideae	
Lycopersicon esculentum	Solanoideae	Solanaceae
Convolvulus sabatius	Convolvuloideae	Convolvulaceae
Verbena citriodora	(none)	Verbenaceae
Mentha spica	Nepedoideae	Lamiaceae

species organized taxonomically. Under Wapshere's method, these plants would either all be tested or representative taxa within groupings would be tested. With increased knowledge of the phylogenetic relationship among these plants, however, species can be grouped based on phylogenetic separation (Figure 5.3), allowing for representative species with particular levels of phylogenetic separation to be assayed (Briese et al. 2002a; Kelch & McClay 2004). The use of plants in host-specificity testing assays could therefore be correlated to the degree of phylogenetic separation for a more accurate estimation of host specificity (Briese et al. 2005).

Using a phylogenetic approach for choosing test plant species appears to be a robust way to assess risk to non-target species, at least for arthropod biological control agents of weeds. In a retrospective analysis of 112 insects used for weed biological

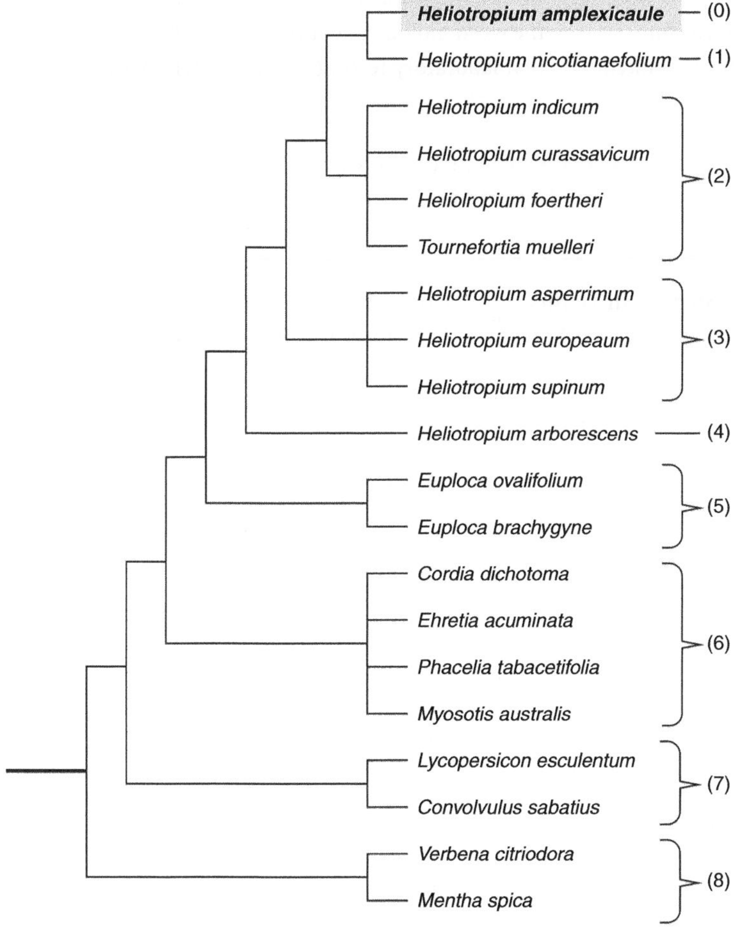

Figure 5.3. Diagram of the phylogenetic relationship between the weed blue heliotrope, *Heliotropium amplexicaule*, and potential non-target species for testing against the flea beetle, *Longitarsus* sp. Numbers in parentheses indicate the degree of phylogenetic separation from the target weed (highlighted in gray). Modified from Briese (2005) and Briese and colleagues (2005).

control between 1902 and 1998, Pemberton (2000) found that more than 95% of the non-target plants used by weed biological control agents were closely related to the weed species. Only one species of herbivore was found that attacked a non-target native plant not closely related to the target host. This was the Mexican lace bug, *Teleonemia scrupulosa*, which was released against Lantana (family Verbenaceae) in Hawaii in 1902 and fed upon an endemic shrub, Naio, which is not in the Verbenaceae. Cases in which non-target native plants that were closely related to target weeds were attacked include *Rhinocyllus conicus* weevils on native *Cirsium* thistles in various areas of the United States and *Cactoblastis cactorum* moths on *Opuntia* cacti in the southeastern United States (Table 5.3). In both of these cases, pre-release and post-release host-specificity testing as well as observations of feeding in the native range were used to estimate host specificity (Dodd 1940; Zwölfer & Harris 1984; Johnson & Stiling 1996), and the results of these studies are broadly consistent with the level of non-target feeding that has been seen. Neither of these agents would likely have been released under the more risk-averse regulatory climate of the early 21st century (Barratt et al. 2006). Other reviews of the weed biological control literature have shown that laboratory testing provides a relatively good

Table 5.3. **Number of native congeneric plant species influences the number of non-target weeds fed upon by insect biological control agents in the United States (including Hawaii). Modified from Pemberton (2000).**

Genus	Number of Native Species in Genus Co-occurring with Target Weed Species	Number of Non-target Native Species Fed Upon by Biological Control Agents
Alternanthera	5	0
Centaurea	3	0
Cirsium	90	22
Convolvulus	2	0
Cuscuta	38	0
Cyperus	10	1
Euphorbia	42	0
Hypericum	46	1
Linaria	2	0
Lythrum	4	0
Opuntia	61	5
Rubus	2	2
Salvia	43	0
Senecio	63	3

indication of the scope of non-target attacks (Van Klinken & Edwards 2002; Paynter et al. 2015).

Host-specificity tests for parasitoids: playing catch-up. During the 1990s, the arthropod biological control community was debating *whether or not* to conduct host-specificity testing for arthropod parasitoids and predators – let alone how best to conduct such tests (Van Driesche & Hoddle 1997). Before this time, host-specificity tests of biological control agents of arthropods as part of pre-release studies were only rarely conducted. Instead of host-specificity testing, information from rearing or biological studies was typically used to estimate the host range of potential agents. Parasitoids are particularly amenable to these kinds of assessments because information on the host species from which they were reared is often available, although errors and omissions in the published literature can seriously compromise host range estimates (Godfray 1994; Shaw 1994). More recent arthropod biological control projects have incorporated host-specificity testing, however, and by the time Kuhlmann and colleagues (2006b) wrote their review, the average number of potential non-target species used in these tests exceeded 10.

Choosing non-target test species for parasitoids and predators of arthropod pests poses a number of challenges beyond those encountered for herbivores (Kuhlmann et al. 2006b). First, arthropods are more speciose than plants are, so that the potential number of non-target species to test will often be greater for entomophagous than for herbivorous biological control agents. Second, the phylogenies of arthropods are not as well understood as those of plants for many groups, making it more difficult to use phylogenetic separation to delimit host range. It has also been suggested that there is a weaker phylogenetic signal for host range in parasitoids than for herbivores (Messing 2001; Hoddle 2004b; Haye et al. 2005), and indeed, some parasitoids show greater habitat than host taxon specificity (Hoffmeister 1992; Godfray 1994; Shaw 1994; Messing 2001). For example, some parasitoids specialize on leafminer hosts, attacking and developing in leafmining flies, beetles, moths or sawflies (Askew 1994). Laboratory studies on *Trichogramma brassicae*, a generalist parasitoid of lepidopteran eggs used in augmentative biological control, also showed very little concordance between host phylogeny and acceptability or suitability of host species (Babendreier et al. 2003). And while laboratory studies of the plant bug parasitoid *Peristenus relictus* show a clear signal of host phylogeny, host species that are both distantly related to the target host and only reluctantly accepted in the laboratory are nevertheless attacked in the field (Haye et al. 2006).

The prevalence of ecologically based host specificity rather than phylogenetically based host specificity is particularly important, since if host ranges of arthropod parasitoids and predators tend to be phylogenetically disjunct, there is little point in using a centrifugal phylogenetic testing method for these agents. The consensus, however, seems to be that despite the caveat of host-habitat specialization in some parasitoid species, host phylogeny is a major determinant of host range for many, if not most, parasitoid species (Godfray 1994; Kuhlmann et al. 2006b). Specialist parasitoids in particular tend to attack closely related host species rather than phylogenetically disjunct host species (e.g., Rosen & DeBach 1979; Carton & Kitano 1981; Dijkerman 1990; Brodeur et al. 1996; Coombs 2004; De Nardo & Hopper 2004; Fuester et al. 2004; Desneux et al. 2012). For various scale insects, the acceptability of hosts by certain parasitoid species has even been proposed as a taxonomic character to use in the differentiation of otherwise cryptic scale insect species (Rosen & DeBach 1977)! An example of a parasitoid that exhibits phylogenetic signal in its host range is shown in Figure 5.4.

From a risk-assessment standpoint, the goal of host-specificity testing is precisely to identify those parasitoid species with a narrow host range that is not phylogenetically disjunct. This consideration alone provides justification for a phylogenetic

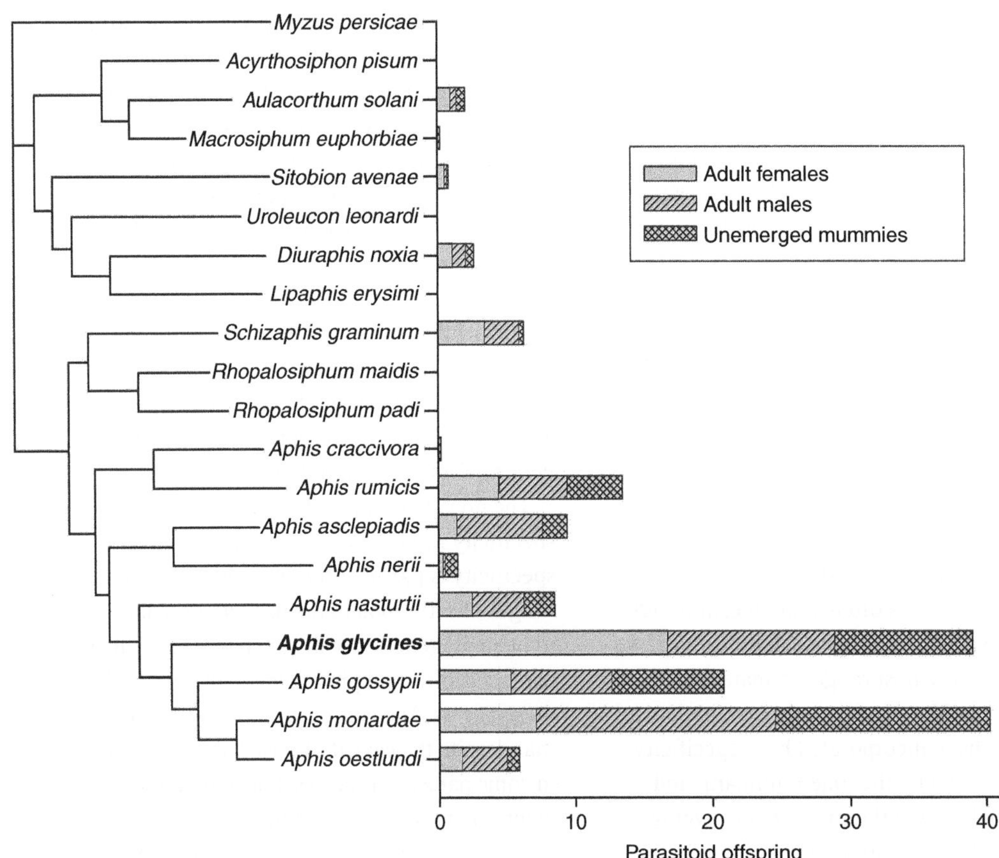

Figure 5.4. Reproduction by the parasitoid *Binodoxys communis* when exposed to 20 aphid species in laboratory assays, including the target host *Aphis glycines* (in bold). The bars show the total number of parasitized aphids ('mummies') produced and whether these emerged as adult males, females or not at all. The pattern of reproduction is overlaid onto a phylogeny of the aphid species, demonstrating a strong signal of host phylogeny on reproduction of the parasitoid. From Desneux and colleagues (2012).

centrifugal approach for host-specificity testing in parasitoids (Van Lenteren et al. 2003). Researchers should exercise caution, however, and be aware that host-habitat specificity or other ecological factors could influence the host range of any parasitoid species. Thus, as is the case for biological control agents of weeds, a phylogenetic testing protocol should use a list of test species with additional species included for ecological or other reasons.

Is it possible to predict parasitoid host specificity from taxonomy or general life history? To some extent, the answer is yes, but only in a general sense. For instance, idiobiont ectoparasitoids (those that paralyze their hosts prior to laying their eggs on them) tend to have a broader host range than koinobiont endoparasitoids, which allow the host to continue growing after laying eggs inside of them (e.g., Askew & Shaw 1986; Belshaw 1994; Shaw 1994; Althoff 2003; but see Mills 1992). Thus, it may be tempting to conclude that idiobionts should be avoided for biological control because of a greater risk to non-target arthropods. However, important exceptions can be found. For instance, *Aphytis* spp. are idiobiont ectoparasitoids, but have relatively narrow host ranges and include some of the most effective biological control agents ever used (Rosen & DeBach 1979; Rosen 1994). Another general trend is that the host ranges of parasitic Hymenoptera as a whole tend to be narrower than those of parasitic flies in the family Tachinidae (Belshaw 1994). And as we saw in Chapter 4, the release of one generalist tachinid, *Compsilura concinnata*, has resulted in serious non-target effects (Boettner et al. 2000). However, not all tachinids are generalists, as is exemplified by the gypsy moth parasitoid *Aphantorhaphopsis samarensis*, which is quite specialized (Fuester et al. 2004), and *Trichopoda giacomellii*, an oligophagous tachinid released to control the southern green stink bug in Australia (Coombs 2004). Thus, despite some general trends based on life history and taxonomy, the host ranges of parasitoids are idiosyncratic enough that every species will require individualized host-range testing.

An added complication with biological control agents of arthropods is that they may attack weed biological control agents, as we have discussed already in Chapters 2 and 4. Thus, host-range testing for arthropod biological control agents increasingly includes weed biological control agents (Kuhlmann et al. 2006a).

5.2.2.2 Fundamental versus Ecological Host Range

G. E. Hutchinson defined the fundamental niche of an organism as its tolerance range for a number of dimensions, including abiotic conditions for survival, diet and other factors. In contrast, the realized niche is an assemblage of these conditions as observed in the field (Futuyma & Moreno 1988). Parallel distinctions can be made for host ranges of insect herbivores, parasitoids and pathogens and the prey range of predators (for simplicity, we lump prey ranges in with host ranges). Thus, the *fundamental host range* is defined as the set of species that can support development of a given consumer species (Onstad & McManus 1996), and the *realized* (or *ecological*) *host range* denotes those species actually used in the field. We follow convention in biological control by using *ecological* rather than *realized host range* for this book, but we consider the terms synonymous.

The term *physiological host range* has also been used to denote the fundamental host range, but this is easier to determine for herbivorous arthropods and predators than for most parasitoids. The reason is that neonate larvae of many herbivore species can easily be placed on plant material to determine whether the plant will support development of the herbivore, making it relatively easy to determine experimentally whether a plant is physiologically suitable for consumption by a given herbivore species. This is a form of force-feeding, which is almost impossible to accomplish with endoparasitoids, which for the most part must deposit eggs within the hosts themselves. Some clever techniques have been devised to force-feed endoparasitoid larvae, however, and these have produced interesting results. For example, the ant parasitoid *Apocephalus paraponerae* only attacks one species of ant in the field, but is able to complete development on eight other co-occurring ant species when its eggs are microinjected into these other species after being dissected out of the preferred species (Morehead & Feener 2000). Using a similar technique, Fuester and colleagues (2001, 2004) showed that nine non-target host species of the gypsy moth parasitoid *Aphantorhaphopsis samarensis* were neither attacked nor suitable for development. In some cases, parasitoids can be 'tricked' into laying eggs into objects that they would otherwise not accept (Strand & Vinson 1983; Godfray 1994), but we are not aware of such techniques being used to estimate the physiological host range of any parasitoid species. The physiological host range of ectoparasitoids can be evaluated in a manner similar to herbivores, since the eggs can rather easily be moved onto hosts that are not accepted for oviposition (e.g., Heimpel et al. 1997).

By necessity, the fundamental host range must be determined under controlled conditions (typically in the laboratory) and in no-choice tests. The ecological host range, on the other hand, is defined as the set of species that a consumer actually feeds upon in the field. Ecological host ranges cannot be completely determined in the laboratory, although laboratory studies can be used to test specific hypotheses related to why the ecological host range may differ from the fundamental host range. For example, olfactometer studies can be used to ask whether a potential biological control agent is attracted to non-target host species (Wyckhuys & Heimpel 2007; Yong et al. 2007). Other methods used to estimate ecological host range include detailed literature analyses (De Nardo & Hopper 2004; Sands & Van Driesche 2004) and field collections in the native and/or introduced ranges (e.g., Van Lenteren et al. 2003; Barratt 2004; Haye et al. 2005; Louda et al. 2005b). Manipulative approaches are possible too, as is illustrated by a study done by Dudley and Kazmer (2005). They outplanted non-target host plants of the saltcedar biological control agent *Diorhabda elongata* near field releases as a test of a 'worst-case scenario' situation and showed that feeding on non-target species even under these conditions was minimal.

The ecological host range is narrower. The most common pattern by far is for the ecological host range to be narrower than the fundamental host range. Indeed, Wapshere (1989) noted that weed biological control agents known to be specialists under field settings often showed a broad host range in laboratory testing, especially when the herbivore's eggs or larvae were placed onto plant species that the females would not be attracted to or accept as oviposition sites. Insect pathogens are also notorious for infecting far fewer host species in the field than laboratory studies would suggest (Hajek & Butler 2000). In perhaps the best-studied case of an entomopathogen used in importation biological control, Hajek and colleagues (1995) found that approximately 25 species of non-target Lepidoptera were infected by the gypsy moth pathogen *Entomophaga maimaiga* in laboratory trials. Field studies, however, revealed an extremely low incidence of non-target attack, with only two individuals (from two species) infected out of more than 1,500 examined (Hajek et al. 1996). In other cases, epizootics are observed on one host species while very closely related host species in the same area remain untouched. Some insect pathogens are known for being comprised of numerous host-specific strains, or 'pathotypes' that can in some cases only be distinguished using molecular methods, and this is a likely explanation for the extreme host specificity seen in these and other cases (Hajek & Butler 2000).

There are a number of reasons that ecological host ranges may be narrower than physiological host ranges. For arthropod agents of weeds or arthropod pests, behavior is perhaps the most important factor. As noted earlier, species accepted under confined, no-choice settings may not be accepted in the field. For pathogens and nematodes, optimal conditions for infection (such as moisture regimes) may be created in laboratory tests, but these conditions may be rarely encountered in the field in association with certain non-target species (Barbercheck & Millar 2000; Hajek & Butler 2000; Barton 2004). Pathogens may also become highly specialized under field conditions through the epizootic process. One or a few strains of pathogens may emerge to dominate from epizootics, leading ultimately to numerous highly specialized strains. This model was proposed by Hajek and Butler (2000) on the basis of unpublished data suggesting that the genetic variability of *Entomophaga aulicae*, a fungal pathogen of the spruce budworm, was reduced greatly during an epizootic, and is supported by the observation that bacterial pathogens of humans can become genetically restricted during epidemics (Maynard Smith et al. 1993).

The fundamental host range becomes restricted through a series of ecological filters to produce the ecological host range. Ecological filters begin with factors that keep the agents from coming into contact with potential host species, such as climate and phenology. For example, the fungal pathogen *Uromyces heliotropii* was considered for biological control of common heliotrope, *Heliotropium europaeum*, in Australia. However, laboratory studies showed that the non-target plant *H. crispatum*, which is native to Australia, was mildly susceptible to the agent, bringing into question the safety of the proposed release. Climate-matching and phenological studies showed that the non-target plant was very unlikely to be successfully attacked based on moisture requirements and phenological separation. In addition, the geographic range of the target and non-target species in Australia did not overlap and prevailing winds would not favor transport of fungal spores the 600+ kilometers from the range of the target to the non-target weed (Hasan & Delfosse 1995). The risk to *H. crispatum* was considered negligible based largely on these considerations and a permit for release was granted.

Wyckhuys and colleagues (2007b) investigated two additional ecological filters with the potential to protect non-target aphids from attack by an introduced parasitoid of the Asian soybean aphid, *Aphis glycines*. *Binodoxys communis* is a Chinese parasitoid of the soybean aphid that has been imported into the United States for release against the soybean aphid. Host-specificity screening prior to release showed that a particular native aphid, *Aphis monardae*, was accepted at a relatively high rate when grown on the leaves of its host plant, *Monarda fistulosa*. Field observations of *A. monardae*, however, showed that it tended to cluster in the flower head of its host plant, and also that it was heavily ant tended (Wyckhuys et al. 2007b, 2009). Additional laboratory studies under quarantine conditions then showed that feeding in the flower head provided protection from parasitism by *B. communis* as did ant tending. The implication was that feeding in flower heads and ant tending would likely serve as ecological filters restricting potential non-target impacts on non-target aphids (Wyckhuys et al. 2007a, b; 2009).

The general trend for a narrower ecological than fundamental host range can in some cases give a false impression of low ecological risk. This is because ecological conditions may favor a non-target host in the field, in some cases even over the target pest or weed. Two examples of this scenario come from herbivores of invasive thistles, both of which were alluded to in the previous chapter. In the first, the flowerhead weevil, *Rhinocyllus conicus*, had larger-than-expected negative impacts on native thistles than were predicted from laboratory studies (Louda et al. 2005b). These impacts were due mainly to the fact that the target and non-target thistles do not co-occur so that a preference for the target weeds could not be expressed in sites where *Rhinocyllus* was attacking native thistles. Also, the phenology of *Rhinocyllus* was well matched with that of the non-target thistle species. The second example involves *Larinus planus*, another weevil. This species was accidentally introduced, but its target weed was ostensibly the Canada thistle, *Cirsium arvense*. However, damage to the native wavyleaf thistle, *C. undulatum*, greatly exceeded that to the Canada thistle (Louda & O'Brien 2002). In this case, the phenology of the weevil is better matched to the non-target than the target plant. These and other examples serve as a reminder that ecological filters can favor as well as disfavor non-target impacts and also that a low rank in a laboratory host-specificity trial does not guarantee low risk.

Broader ecological host range? We have been stressing the general observation that the ecological host range tends to be narrower than the fundamental host range. But can this be reversed? Indeed it can. Many parasitoids, for example, attack hosts in the field on which their larvae cannot develop or develop only poorly (Hoogendoorn & Heimpel 2002; Heimpel et al. 2003). These attacks may or may not kill the host (often they do not), so while they may lead to egg or time wastage on the part of the parasitoid, they may not cause serious non-target or other risk concerns. A documented case in which laboratory studies underestimated the ecological host range involves the Eurasian/North African parasitoid *Microctonus aethiopoides*, which was released against the forage weevil, *Sitona discoideus*, in New Zealand, and the alfalfa weevil, *Hypera postica*, in North America. Initial laboratory studies showed no attack of *M. aethiopoides* on the flowerhead weevil, *Rhinocyllus conicus*, which had been previously introduced to control musk thistle there (Barratt 2004). However, field studies showed that this weevil was indeed attacked by *M. aethiopoides*, albeit at low levels. Subsequent laboratory studies confirmed that *M. aethiopoides* attacked *R. conicus* after all (Barratt et al. 1997; Barratt & Johnstone 2001), and Barratt (2004) hypothesized that the reason for the mismatch between the original laboratory studies and subsequent laboratory and field studies was that *R. conicus* was relatively inactive in the early laboratory studies. This is an important factor as *M. aethiopoides* is a parasitoid of adult weevils and will only attack actively moving hosts. Barratt suggested that if initial quarantine studies had demonstrated a risk to *R. conicus*, permission to release *M. aethiopoides* would likely not have been granted. As we will discuss later, laboratory conditions used for fundamental host-range testing can lead to false negatives – these include odor masking in choice studies, and inappropriate motivational status in no-choice tests.

Finally, the ecological host range can *seem* broader than the fundamental host range when a species is composed of multiple host specialized strains, or pathotypes. In this case, field records of infection or attack are recorded from multiple host species while laboratory studies done on just one strain show host specialization. This was the case for the fungal pathogen *Metarhizium anisopliae*, which can be isolated from many insect species in the field while laboratory studies often show high degrees of specialization to host species from which the pathogen was isolated (Goettel et al. 1990; Hajek & Butler 2000). Host ranges can vary between strains of weed biological control agents, and the failure to test all strains released can lead to non-target impacts (Paynter et al. 2008a).

5.2.2.3 Methodological Concerns

Numerous methodological issues confront researchers setting out to estimate host specificity of potential biological control agents. These include whether choice or no-choice tests should be used, what kinds of enclosures should be used, what levels of hunger or host deprivation the agent should be tested at, to what extent field testing should be done in the native range, and what statistical analyses should be used to analyze host-specificity data. Many of these topics have been well reviewed by other authors (e.g., Van Driesche & Hoddle 1997; Barton Browne & Withers 2002; Van Driesche & Reardon 2004; Sheppard et al. 2005; Bigler et al. 2006; Hoffmeister et al. 2006; Withers et al. 2013), and we do not attempt a comprehensive review here. We do, however, provide some comments on key topics.

Choice versus no-choice tests. In no-choice tests, the agents are evaluated in the presence of a single non-target species at a time, while in choice tests, more than one species is offered to the agent – often a non-target species along with the target species. One common pattern is that agents that accept a non-target species in a no-choice setting will not do so in a choice setting. Van Driesche and Murray (2004) illustrated this point by noting that children will tend to eat pizza over broccoli when given a choice, but that broccoli will be accepted when offered alone – eventually. This logic illustrates how no-choice tests tend to provide more conservative results than choice tests do from the standpoint of risk assessment. On the other hand, the close proximity and confinement of target and non-target plants in a choice test may lead to oviposition cues emanating from the target plant, leading to an unnaturally *high* level of oviposition on the non-target plant. For these and other reasons, it is generally accepted that the no-choice test is a more valuable tool for risk assessment (Mansfield & Mills 2004; Van Driesche & Murray 2004; Briese 2005; Van Lenteren et al. 2006b; Murray et al. 2010), although choice tests can clearly provide interesting supplementary information on host preference and related topics. Van Lenteren and colleagues (2006b) suggested that using large cages to do choice tests alleviates some of the problems just mentioned, and while this may be true in some or even most cases, it would need to be confirmed for each natural enemy being tested. It is also true that biological control agents may actually find themselves in no-choice situations in the field under spillover or other conditions (Blossey et al. 2001; Holt & Hochberg 2001; Rand & Louda 2004, 2006), and this is another justification for no-choice tests. We describe later an innovative experimental design in which choice and no-choice tests were combined in a field setting that simulates the transition from a choice to a no-choice setting that can occur in the field.

Hunger and other physiological states. Herbivores that are starved prior to feeding tests will be more likely to accept low-quality non-target plants than satiated ones are. It has therefore been standard practice to starve herbivores for feeding trials. Parasitoids are subject to an analogous situation: individuals that have a low life expectancy are expected to be more likely to accept low-quality host species than well-fed parasitoids are, which can expect to have more time to find higher-quality hosts (Withers & Browne 2004). The number of mature eggs that a parasitoid carries ('egg load') should affect host choice as well, with acceptance of low-quality hosts expected at high and not low egg loads (Minkenberg et al. 1992). More broadly, any conditions that cause an insect's fitness to be limited by the time available to search for hosts should lead to a broad host range, and conditions that cause a parasitoid's fitness to be limited by the number of eggs available should lead to a narrow host range. A number of studies have shown that parasitoids with high egg loads and short life expectancy accept a broader range of size and quality classes within a single preferred host species (e.g., Roitberg et al. 1992, 1993; Collier et al. 1994; Fletcher et al. 1994; Heimpel et al. 1996; Heimpel & Rosenheim 1998). Despite this, investigations of the influence of parasitoid state on the acceptance of different host species have not revealed any effects for four parasitoid species (Dieckhoff 2011; Hopper et al. 2013; Jenner et al. 2014). As this is a limited number of studies, however, we feel that it is prudent to retain the recommendation to test parasitoids in a time-limited state, so that they will express the maximum host range (Withers & Browne 2004). Whether the condition of time limitation corresponds to the parasitoid's state in the field is another question, and this depends on a number of factors. While many parasitoid species do appear to be time-limited in the field (Heimpel et al. 1996; Heimpel & Casas 2008), others cycle between periods of time and egg limitation (Heimpel et al. 1998). Egg and time limitation can also vary with host density (Segoli & Rosenheim 2013; Dieckhoff et al. 2014), the availability of sugar resources (Lee & Heimpel 2008) and the time of year (Irvin et al. 2014).

Field tests. Field tests in the native range of the biological control agent are desirable because of the insight they give into the ecological host range, which is difficult to assess under quarantine conditions. Conducting the entire host-specificity testing protocol in the field, however, is impractical for a number of reasons, and non-target organisms native to the proposed introduced range may not be available for testing in the native range of the biological control agent.

Directed field studies aimed at addressing particular hypotheses can be used in the native range. A particularly innovative approach is to simulate spillover situations in the field by allowing agents to establish on target plants or hosts in the presence of non-targets and then destroying the targets without killing the agents. Briese and colleagues (2002b) conducted just such a study in the context of biological control of blue heliotrope, *Heliotropium amplexicaule*, a South American native that is an invasive weed in Australia. The studies were done in the open field in Argentina, and data were obtained for three agents: the leaf beetle

Deuterocampa quadrijuga, the thrips *Haplothrips heliotropica* and the flea beetle *Longitarsus* sp. Blue heliotrope was planted alongside six non-target plant species, ranging from closely to distantly related to the target weed. In the first phase of the experiment, the insects were allowed to settle in the plots, and were found to feed only on the target weed and the most closely related non-target plant, *Heliotropium nicotianaefolium*. In the second phase of the experiment, the two host plants used by the insects in the first phase were cut, simulating a no-choice setting for non-targets, and also creating appropriate conditions for a 'spillover effect'. What happened to the insects when the preferred plants were removed? The thrips and the leaf beetle disappeared from the plots entirely, but the flea beetle moved onto *Heliotropium arborescens*. The results thus showed a higher likelihood of attacking less preferred plants for the flea beetle than for the other insects (Figure 5.5). As a result of these field

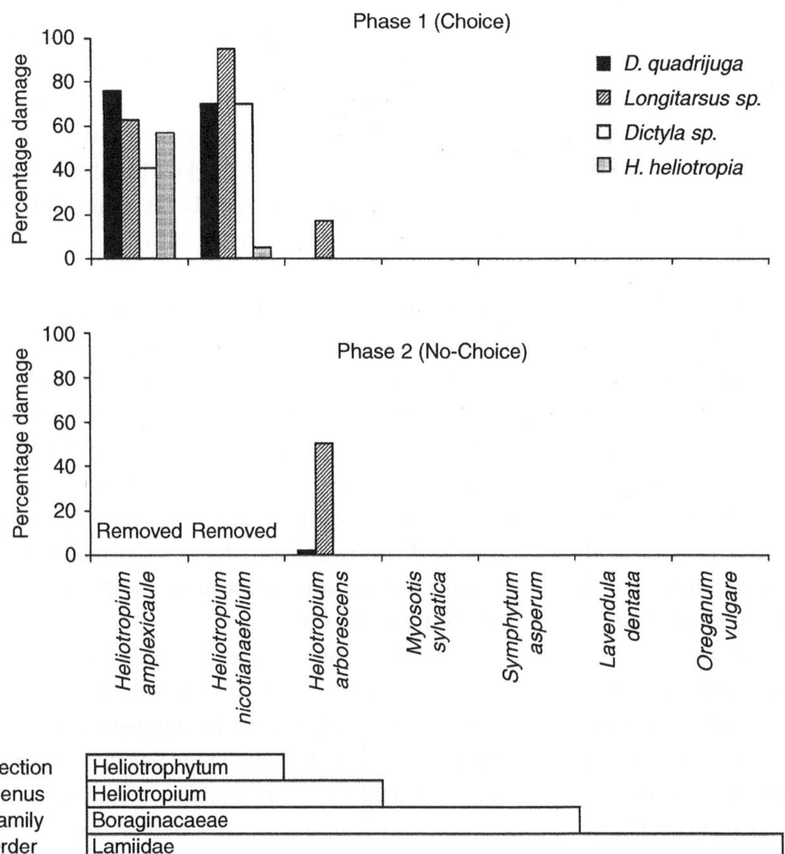

Figure 5.5. Summary of the results of a field choice/no-choice study of four biological control agents of the weed blue heliotrope, *Heliotropium amplexicaule*, which is native to South America and invasive in Australia. The field studies were conducted in South America with an initial choice phase followed by a no-choice phase in which plants experiencing insect attack were cut, placing insects in a no-choice situation with less preferred plant species. The lacebug, *Dictlya* sp., is not discussed in the text because its population crashed before the initiation of phase 2 of the experiment. From Briese and colleagues (2002).

studies, the leaf beetle *D. quadrijuga* went on to the next step in host-range testing required for release in Australia while the flea beetle *Longitarsus* sp. was not considered further (Briese et al. 2002b; Briese & Walker 2002).

5.2.3 Studies Other than Host-Specificity Trials

While host-specificity trials are the cornerstone of risk assessment in biological control, they do not tell the whole story. Other important components of risk include the potential for population growth of the agent, habitat specificity, dispersal capabilities and indirect effects. To this must be added the potential for evolutionary change, including the evolution of host-range expansion or host shifts (Van Klinken et al. 2002). This last topic is covered in Chapter 7.

5.2.3.1 Demographic Modeling and Risk Assessment

While it can be difficult to study ecological processes under quarantine laboratory situations, some interactions are amenable to laboratory studies. For example, as alluded to previously, Wyckhuys and colleagues (2007b) brought an ant colony into a quarantine laboratory to estimate effects of ant tending on the risk of a non-target aphid to attack by a parasitoid proposed for release against the soybean aphid in North America. Another approach is to use laboratory studies to parameterize mathematical models in an attempt to anticipate the risk of a proposed biological control agent to non-target organisms. This approach was adopted by Raghu and colleagues (2007), who predicted the risk of the tortoise beetle, *Charidotis auroguttata*, a potential biological control of cat's claw creeper, *Dolichandra* (= *Macfadyena*) *unguis-cati*, in Australia. Laboratory studies had suggested that this beetle, although relatively host specific, could feed on the native Australian plant *Myoporum boninense australe* (Dhileepan et al. 2005). Raghu and colleagues (2007) used a demographic approach to define risk by considering damage to the non-target plant as 'acceptable' if herbivory did not prevent the plant from replacing its damaged leaves during a 250-day simulation period. Factors investigated in the model included beetle fecundity and survival, duration of immature and adult stages on target and non-target plants, herbivory parameters by the tortoise beetle on target and non-target plants, plant growth rates and the spatial overlap of target and non-target plants. The simulations showed that unacceptable levels of risk to the non-target plant were only likely when the target and non-target plants co-occurred spatially. In particular, risk was considered greatest in stands where the target plant dominates but the non-target plant is present nevertheless, thus creating conditions ideal for spillover effects (Figure 5.6). Areas conforming to this pattern included coastal regions where the non-target plant is grown as an ornamental, since direct effects of cat's claw creeper on the non-target plants would rival such

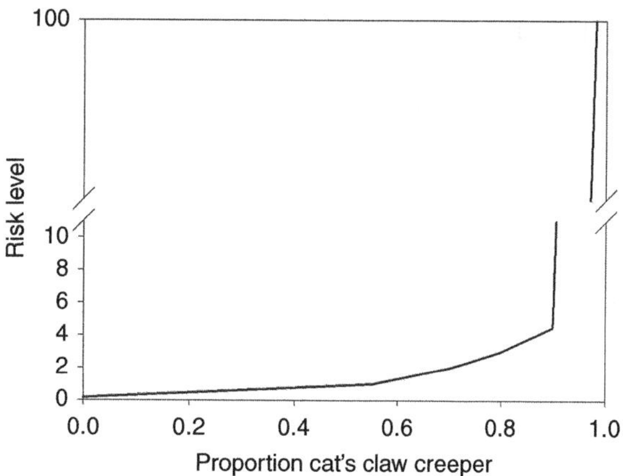

Figure 5.6. The level of risk to the non-target plant *Myoporum boninense australe* as a function of the proportion of the plant patch that contained the target weed cat's claw creeper, *Dolichandra unguis-cati*. A risk level above 1 indicates that the non-target plant is not able to compensate for herbivory by the tortoise beetle *Charidotis auroguttata*. Redrawn from Raghu and colleagues (2007).

spillover effects in these areas. Raghu and colleagues (2007) therefore suggested that consultation with stakeholders such as landowners and the native plant nursery industry could be fruitful in negotiating the risks of releasing biological control agents or not.

It is also possible to anticipate particular non-target impacts for agents that are already introduced and use what is already known about the biology of the agent to estimate risk. This was done for potential effects of the flowerhead weevil, *Rhinocyllus conicus*, on the endangered Pitcher's thistle, *Cirsium pitcheri*, in North America (Louda et al. 2005a). Oviposition and development studies as well as demographic modeling of phenological considerations showed that *R. conicus* indeed posed a serious risk to this rare native plant.

5.2.3.2 Risk Assessment and Dispersal of Biological Control Agents

As we mentioned in the previous chapter, natural enemies used in importation and augmentative biological control have an intended and an unintended geographic/habitat range. A major risk of biological control is that agents disperse out of their intended range into areas where the risk of negative consequences is higher. Two broad issues contribute to increased risk of biological control agents dispersing beyond their intended range. The first is the intrinsic dispersal ability of the natural enemy, and the second is the permissiveness of the environment in which dispersal would take place (Van Lenteren et al. 2006a).

Considering first the intrinsic dispersal characteristics of biological control agents, it would appear to be clear what characters are desirable from the standpoint of safety to non-target organisms: those enemies with low dispersal capabilities are associated with the least risk. An analysis by Kimberling (2004) seems to belie this intuition, however. Kimberling used proportions tests, correlation tests and multiple regression to evaluate the effect of natural enemy dispersal ability (among other variables) on the likelihood of non-target effects from 24 importation biological control projects involving insect agents of insect pests. Dispersal ability was coded as 'weak, medium, or strong' and non-target effects were classified as either absent or affecting native hosts or prey through trophic interactions or competition. The proportion tests suggested that the risks of non-target effects were greatest for 'weak' and 'strong' dispersers, and lowest for 'medium' dispersers. There is some question as to how robust this result is because the effect disappeared when multiple regression was used, and also because dispersal was correlated with searching ability. Additionally, Kimberling found no effect of dispersal ability on the success of biological control projects. We are not sure how to interpret this result, both because of the statistical ambiguity and the difficulty in intuitively understanding the pattern. While it seems clear enough that strong dispersers would have more severe non-target effects, it is less clear why this would also be the case for weak dispersers (and not medium dispersers). The significance of dispersal for either the success or risks of biological control using insects is in need of further study (Heimpel & Asplen 2011).

The risk of biological control agents escaping from their intended to their unintended range has been more carefully studied in the area of augmentative biological control where the intended range is well defined (Mills et al. 2006). To illustrate this, we discuss risk assessment approaches for the 'escape' of two biological control agents that have emerged as model systems for the investigation of this issue. The first is the fungal plant pathogen *Sclerotinia sclerotiorum* against Canada thistle, *Cirsium arvense*, in New Zealand, and the second involves *Trichogramma* parasitoids released against the European corn borer, *Ostrinia nubilalis*, in the United States.

Sclerotinia sclerotiorum. This fungal plant pathogen can be used effectively as a mycoherbicide against

Canada thistle and bindweed, among other weeds. However, *S. sclerotiorum* has a broad host range that includes many crop plants. Additionally, it produces air-borne spores, thereby providing a clear route for non-target impacts. Despite this seemingly high potential for risk, *S. sclerotiorum* is considered a promising weed control agent in both New Zealand and the United States (Quimby et al. 2004). Indeed, a formulation of this pathogen has been used in New Zealand dairy pastures against the giant buttercup, *Ranunculus acris* (Cornwallis et al. 1999; Harvey & Bourdot 2001).

How have non-target risks been assessed and navigated in this case? We discuss here deliberations in New Zealand, where these questions were addressed in some detail. It is important to recognize first that *S. sclerotiorum* is a holarctic pathogen that was already present in New Zealand, so it could be argued that negative effects were occurring already and that use of the fungus as a weed control agent in pastures was not likely to greatly increase negative effects. Still, the process of producing a mycoherbicide for use in the field involved screening various isolates of the pathogen and preparing formulations aimed at effective weed suppression. It is therefore fair to ask what the implications of this process were for risk to non-target plants such as crops. Scientists in New Zealand and elsewhere realized that the addition of spores to the background levels of *S. sclerotiorum* could have non-trivial impacts on crop plants surrounding release sites (de Jong et al. 1999; Bourdot et al. 2000, 2001, 2006), and based on studies of overwintering survival of the resting stage of the fungus (the sclerotia) and the dispersal range of the ascospores, they defined a 'safety zone' as:

> *That distance for a pasture undergoing biological weed control by S. sclerotiorum at which the concentration of dispersing ascospores has declined to that occurring naturally in the air above susceptible market garden crops.* (Bourdot et al. 2006)

Initial work suggested a safety zone of 8 m (Bourdot et al. 2001), but further studies showed that a reasonable safety zone can range between 0 and 300 m depending on climate, ambient spore levels and the type of pasture in which *S. sclerotiorum* is released (De Jong et al. 2002; Bourdot et al. 2006).

Building upon the safety zone work, De Jong and colleagues (2002) used a detailed mathematical model to explore the parameters contributing to the risk of spores dispersing beyond the target pasture. One of the important conclusions was that escape of spores was more likely from more heavily grazed pastures than from those that had a taller grass canopy and a greater leaf area index. This led to the recommendation that grazing by sheep should be limited during the time that ascospores are being formed. This strategy leads not only to decreasing escape of ascospores, but to increased deposition of ascospores onto weed leaves within the pasture, and therefore promotes both safety and increased efficacy (Figure 5.7).

Trichogramma parasitoids. As is discussed in Chapter 8, species of the egg parasitoid genus *Trichogramma*

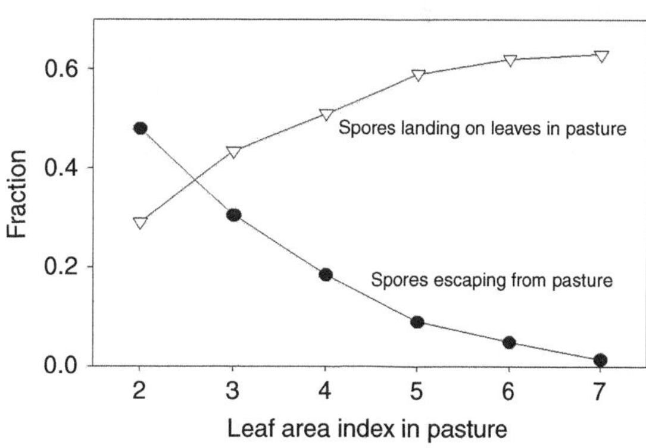

Figure 5.7. Model results showing the fraction of ascospores of the plant-pathogenic fungus *Sclerotinia sclerotiorum* either 'escaping' from a pasture in which the fungus is being used for biological control or landing on the leaves within the pasture as a function of leaf area index within the pasture. Since high leaf area indices are achieved by restricting grazing, it is recommended that grazing be limited during the times that ascospores are being liberated. Modified from De Jong and colleagues (2002).

are among the most important parasitoids used in augmentative biological control. Members of this genus have also figured prominently in the development of risk assessment methodologies for augmentative biological control. In particular, three species, *T. nubilale, T. ostriniae* and *T. brassicae*, have been examined, the former two in releases against the European corn borer, *Ostrinia nubilalis*, in North America, and the latter in releases against the same pest in North America and Europe (Andow et al. 1995; Orr et al. 2000; Babendreier et al. 2003; Kuske et al. 2003, 2004; Wright et al. 2005; Yong & Hoffmann 2006; Yong et al. 2007).

The potential risk of augmentative releases of the native *T. nubilale* was studied by Andow and colleagues (1995) in the north-central United States. These researchers focused on a series of potential host species that were considered 'threatened', 'endangered' or 'of special concern' in the state of Minnesota. In particular, they concentrated on the Karner Blue butterfly, *Lycaeides melissa samuelis*, which was designated an endangered species and suspected of being a suitable host for *T. nubilale*. Andow and colleagues conducted field releases to characterize the dispersal capacity of the parasitoid and used this information to estimate the likelihood that *T. nubilale* could come into contact with Karner Blue eggs at the edges of cornfields based on realistic release rates for augmentative biological control. Although their calculations did not take into account searching in non-corn habitats, their worst-case scenario suggested that there was a 'modest but significant potential risk involved in the release of *T. nubilale* near habitats with rare Lepidoptera'.

Similar studies were done on *T. ostriniae*, an exotic species also released against the European corn borer, but in the northeastern United States (Wright et al. 2005; Yong & Hoffmann 2006). This species was shown in side-by-side comparisons to be superior to *T. nubilale* as a control agent for the European corn borer (Wang et al. 1999). However, *T. ostriniae* has a broad fundamental host range, raising the possibility of negative non-target effects (Wright et al. 2005). Release in the northeastern United States was considered likely to pose little risk to non-target species since it is not able to overwinter there. Wright and colleagues saw this as an opportunity to do intensive field studies of the risk to non-target species from releases of this parasitoid during the field season.

The approach taken by Wright and colleagues (2005) was similar to that taken by Andow and colleagues (1995). They used field studies to estimate the probabilities of *T. ostriniae* (i) dispersing out of cornfields that they were released into, and (ii) attacking suitable non-target species. They then used the joint probabilities of these occurrences to construct 'precision trees', which summarize the risk of both of these hazards. Similar to Andow and colleagues (1995), uncertainty in estimates was based on probability distributions, and allowed an estimate of 'worst-case' scenarios for the precision trees. Dispersal studies showed that *T. ostriniae* released into the center of cornfields rapidly dispersed to field edges, but did not disperse beyond the field edges into adjacent forest (Figure 5.8). However, the

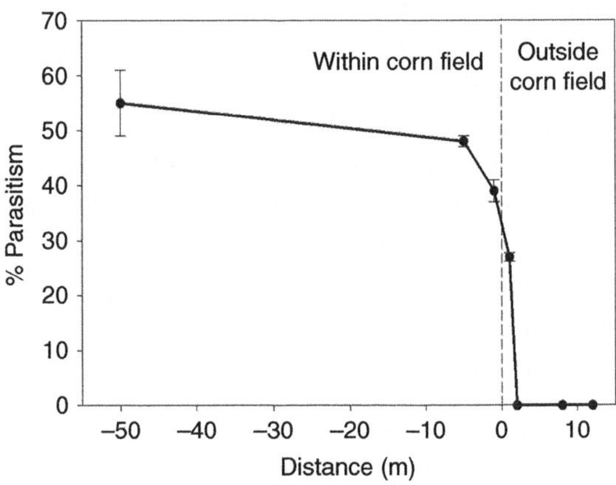

Figure 5.8. Dispersal of *Trichogramma ostriniae* egg parasitoids from cornfields into which they were released into edge habitat. Parasitism was measured by using sentinel host egg masses both within and outside of the fields. From Wright and colleagues (2005).

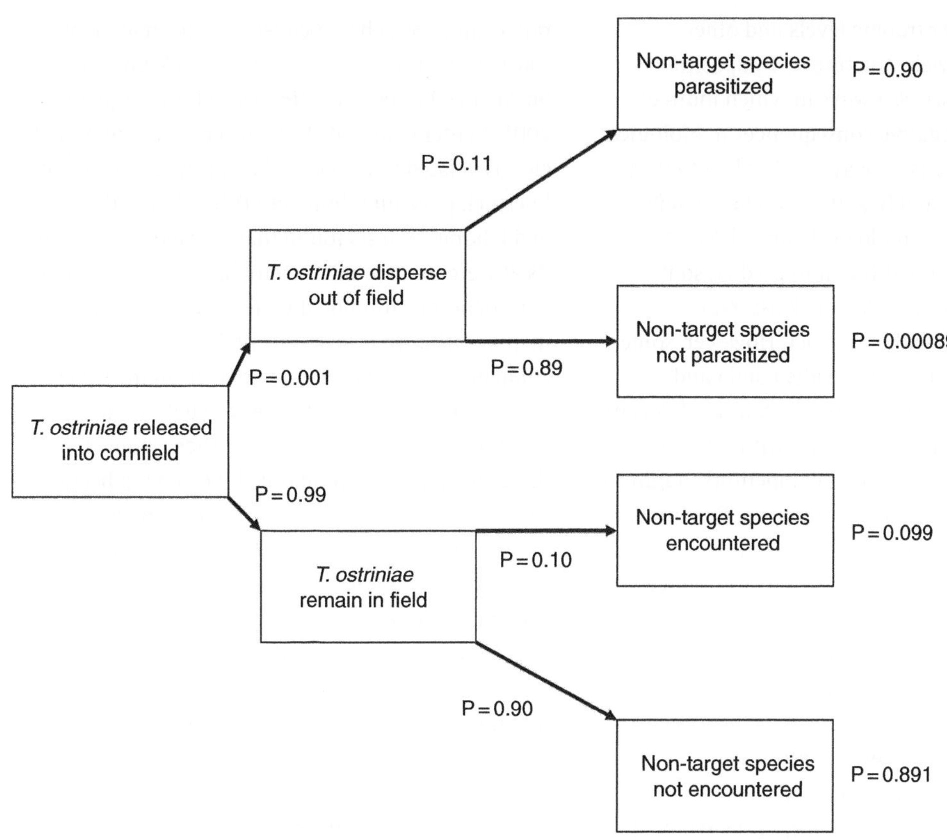

Figure 5.9. Precision tree for non-target risk assessment of releases of the European corn borer parasitoid *Trichogramma ostriniae* into cornfields in New York. From Wright and colleagues (2005).

parasitoids were more willing to enter forest habitats if released between cornfields and forests (Yong & Hoffmann 2006). Parasitoids released directly into forest settings parasitized outplanted European corn borer eggs at a rate less than 1/100 of that found in cornfields. Wright and colleagues (2005) used the likelihood of attacking European corn borer egg masses as a proxy for the likelihood of attacking non-target hosts in constructing their precision tree, and their conclusion was that non-target species occurring in forest habitats adjacent to cornfields are at very low risk of being impacted by *T. ostriniae* (Figure 5.9). Similar studies done with *T. brassicae* foraging in meadows versus cornfields in the United States and Switzerland gave mixed results, with some showing very little movement into non-crop habitat (Orr et al. 2000; Babendreier et al. 2003), and some showing substantial movement beyond crop borders in addition to overwintering 25 m into non-crop habitat on outplanted (laboratory-reared) host eggs (Kuske et al. 2004).

5.2.3.3 Indirect Effects

Negative indirect effects of biological control agents are any effects that do not involve direct interactions between the agent and a non-target organism. They can include competition with non-target species, enrichment at higher trophic levels and other alterations of food web structure or composition. In Chapter 4, we discussed ways in which indirect effects can cause negative consequences in biological control. Here, we focus on ways that indirect effects may be anticipated, and how they can be included in risk assessments for biological control. Some indirect effects are very difficult to predict, so it is inherently difficult to design risk assessment protocols to anticipate their impact. However, some recommendations have been made. Louda and colleagues (2005a) suggested that potential effects on species that the agent may interact with other than the target pest or weed, such as competitors, ought to be considered for laboratory evaluation. Consumers of the agent could be considered as well. For instance, hyperparasitoids often attack introduced parasitoid species and this could lead to increased attack of other, potentially native parasitoids in the area via apparent competition. Testing for the susceptibility of prospective parasitoids to known hyperparasitoids in the area of a proposed release would be feasible in quarantine and could provide valuable information. For instance, of the two major groups of aphid parasitoids – aphidiine braconids and *Aphelinus* spp. – the latter are as a rule less susceptible to hyperparasitism (Brodeur 2000). This decreased susceptibility should decrease the risk of hyperparasitoid enrichment and potential negative indirect effects on non-target parasitoids (Heimpel et al. 2004a). Mills (2006b) has argued that apparent competition through hyperparasitoid enrichment is likely to be important in a relatively restricted number of taxa (such as aphid parasitoids) that are particularly vulnerable to hyperparasitism.

Some potentially effective natural enemies cannot be used as biological control agents because they produce toxins that could endanger animals such as livestock and pets that accidentally ingest the agents or their remains. The sawfly, *Lophyrotoma zonalis*, was found by Burrows and Balciunas (1997) to be an effective defoliator of the Australian paper bark tree, *Melaleuca quinquenervia*, in its native range, and it was being considered for release in southern Florida, where the tree (also known as 'melaleuca') is a devastating weed (Buckingham 2001; Center et al. 2012). However, larvae of a sawfly closely related to *L. zonalis* had been implicated in livestock poisoning (Burrows & Balciunas 1997), and laboratory tests found that extracts of *L. zonalis* itself were toxic to mice (Oelrichs et al. 2001). These considerations prompted vertebrate toxicity tests in quarantine after host-specificity tests had been completed. Other potential agents from the same sawfly family (Pergidae) were rejected on the same grounds (Center et al. 2012). Another example of this general approach involved the vedalia beetle, *Rodolia cardinalis*, which was under consideration for introduction into the Galapagos Islands to control the cottony cushion scale, *Icerya purchasi.* Since the vedalia beetle contains toxic alkaloids, Lincango and colleagues (2011) evaluated the possibility that they could be harmful to Darwin's finches in the event that they fed on the beetles. The results suggested low palatability and toxicity, and this information was included in the overall risk assessment of the project (Causton 2009). While such toxic effects are not indirect effects in the sense that we defined them earlier, they represent a potential negative effect of biological control that goes beyond feeding on non-target plants or arthropods.

Is there a general framework that can be used to predict which classes of indirect effects are (i) most likely and (ii) most detrimental in the context of biological control? As we have mentioned in Chapter 4, a key is likely to be the extent to which abundance of the biological control agents remains elevated after establishment. The likelihood and magnitude of negative indirect effects should be proportional to agent abundance, primarily because of the multitude of indirect effects that can result from biological enrichment (Holt & Hochberg 2001; Pearson & Callaway 2005; Messing et al. 2006). As we noted in Chapter 3, the gold standard in importation biological control involves suppression of the pest or weed along with greatly diminished abundance of the natural enemy as pest or weed densities fall. The likelihood of unintended direct and indirect effects is greatly diminished by such tracking of pest densities by their biological control agent. A key to limiting negative indirect effects of biological control agents therefore lies in establishing effective control that limits the abundance of the introduced enemy. An important component of predicting indirect effects therefore lies in forecasting efficacy, a topic that is beyond the scope of this chapter but is discussed for importation and augmentative biological control in Chapters 3, 6, 7 and 8.

5.2.4 Risk Assessment for Transgenic Biological Control Agents

An additional layer of risk analysis accompanies releases of genetically modified biological control agents, whether these agents are used in the context of importation or augmentative biological control. To date, environmental releases have been made of genetically modified baculoviruses in China and the United Kingdom (Harrison & Bonning 2000; Cory 2007), and entomopathogenic nematodes and predatory mites in the United States (Gaugler et al. 1997b; Hoy 2003). In addition, however, numerous microbial antagonists of soil-borne plant pathogens have been genetically transformed in the laboratory to make them more effective in suppressing pathogens (Spadaro & Gullino 2005). See Chapter 8 for more details on this tactic. Many of these latter cases, however, do not include true *transgenesis* in which genes of one species are introduced into the genome of another, but involve manipulation of genes that already exist in a given species. This approach has also been taken with the fungal entomopathogen *Metarhizium anisopliae*, a modified strain of which was released in field trials in the United States in 2002 (Hu & St. Leger 2002; St. Leger 2007), although more recent work with this agent does involve true interspecific transgenesis (Wang & St. Leger 2007).

Transgenesis is sometimes used to generate neutral markers that can be used to track and detect biological control agents in the field that may carry other transformations or not. For example, Hoy (2003) described the use of a *lacZ* marker gene in a predatory mite to determine the effects of the transgenesis process on fitness and other traits. Green fluorescent protein genes have been incorporated into improved strains of the entomopathogenic fungus *M. anisopliae* to track their movement and persistence in the soil (Hu & St. Leger 2002; Cao et al. 2007). Markers such as these are becoming commonplace in environmental microbiology and show promise for applications in biological control risk assessment (Jansson 2003).

Transgenic baculoviruses, bacteria and entomopathogenic nematodes have emerged as important model systems for investigating the potential risks of releasing transgenic biological control agents. We discuss examples of these approaches next.

Transgenic baculoviruses. Fieldwork on transgenic baculoviruses was initiated with an experimental field cage release of a strain of nucleopolyhedrovirus (a type of baculovirus) of the alfalfa looper, *Autographa californica*, that was genetically engineered to express an insect-selective scorpion toxin to decrease the time to kill (Cory et al. 1994). Questions posed with respect to ecological risk accompanying these releases were articulated by Godfray (1995), who began his treatment of the subject with a light-hearted jab at the media coverage accompanying the first such release in the United Kingdom:

> *Evil Oxford scientists, bent on world domination, release virus containing scorpion-poison gene into English countryside. Well, not quite; but readers of the British press last summer might be forgiven for thinking that a latter-day Central Anarchist Council had taken root among the dreaming spires.*

Despite the alarmist tone of the press reports alluded to in this quote, Godfray found that reasonable ecological points were raised by both the press and the scientific community and he discussed a number of these. Some of the issues, such as the fact that the virus was not native to Britain and that it had a relatively broad host range, were not specifically related to the fact that it was transgenic (Williamson 1991). But what role could transgenesis *per se* play in allowing the virus to infiltrate natural or non-target managed ecosystems? Godfray pointed out that, if anything, transgenesis should *decrease* the likelihood of this happening. The reason is that a more rapid kill puts the virus at a disadvantage with respect to potential competing viruses because fewer infective occlusion bodies are produced by hosts infected by transgenic virus (Cory et al. 1994; Hoover et al. 1995; Dushoff & Dwyer 2001; Bonsall et al. 2005). This is an example of the classical trade-off between virulence and transmissibility in which more virulent pathogens compromise their own transmission rates by killing their hosts too rapidly (Bull 1994). We will return to this trade-off in Chapter 7. In this case, the trade-off leads to an attenuation of risk to non-target species. In fact, Hammock (1992) had argued previously that these factors relegated such transgenic viruses to inoculative augmentative biological control agents or insecticides that had little if any chance of persisting in the environment. Hails and colleagues (2002) used a field experiment to test the hypothesis that the transgenic virus would have lower non-target impacts than a wild-type strain because of these traits. Their study compared infection levels of a non-target host, the cabbage moth, *Mamestra brassicae*, and a target host the cabbage looper moth, *Trichoplusia ni*, by both wild-type and recombinant viruses over a one-week period. The results showed lower infection rates on both hosts by the recombinant than the wild-type virus. This was attributed primarily to an alteration in virus-induced host behavior: hosts infected with wild-type viruses seek elevated locations before dying and disseminate viral occlusion bodies from these elevated locations. This modification of host behavior has been lost in the recombinants, however, and instead of seeking elevated locations on the plant, hosts infected with the transgenic virus become paralyzed and fall off the plant (Cory et al. 1994; Hoover et al. 1995). The result is that transmission rates from target to non-target hosts should be lower for recombinant than wild-type baculoviruses (Hails et al. 2002).

A second important question was whether the transgene could be transferred to another virus. Godfray (1995) concluded that this risk would be remote given that it requires co-infection of the same host nucleus with two baculovirus strains, which is unlikely to occur under field conditions. Also, the very act of rapid host kill would reduce the likelihood of such transfers.

Transgenic nematodes. The first release of a non-microbial transgenic biological control agent occurred in 1996 in the United States, and it involved the entomopathogenic nematode *Heterorhabtitis bacteriophora* engineered to express a heat-shock protein from the nematode *Caenorhabditis elegans* (Gaugler et al. 1997b). Heat-shock protein genes were targeted because the high temperatures that entomopathogenic nematodes sometimes encounter during storage and transport were identified as a critical mortality factor. Once successful transformation was achieved Wilson and colleagues (1999) confirmed that host range was not altered by the genetic transformation process. Field persistence of the transgenic nematode was no greater than a wild-type strain, confirming expectations that the heat-shock protein would not confer an advantage in the temperature-buffered soil environment (Gaugler et al. 1997b).

Transgenic bacteria. There is much interest in genetically modifying bacterial antagonists of plant pathogens to increase their virulence, and Spadaro and Gullino (2005) pointed to the danger that such genes could be transferred to other bacteria by plasmid transfer or other mechanisms, making it very important that transgenes are stable once inserted. Genetic engineering itself can generate such stability, however, and thus lessen the risk of gene transfer. For example, Jones and colleagues (1988) designed a strain of *Agrobacterium radiobacter* that is effective against crown gall disease and in which the transfer region of the plasmid is disabled. Other molecular 'failsafe' methods to improve the stability and safety of transgenes in fungi used for biological control of plant pathogens are discussed by Gressel (2001). These kinds of approaches are more appropriately discussed in the context of risk management than risk assessment, however, as we do in more detail in the next section.

5.3 Risk Management

As we mentioned earlier, choosing less risky targets can be seen as a form of preemptive risk management. The most common way to do this is to choose targets that have no close relatives in the introduced range. For example, there are no native aphids on the Hawaiian islands, making biological control of aphids in Hawaii with strict aphidophages less risky than biological control of, say, stink bugs, which have native representatives in Hawaii (Follett et al. 2000). We have discussed similar cases involving weeds in New Zealand previously.

Beyond this, 'partial quarantine' measures are sometimes used, where releases are allowed within field cages only or in field plots with a specified amount of surrounding bare ground while estimates of efficacy and/or field safety can be made (Sheppard et al. 2003). This approach has been used for a few arthropod biological control agents of weeds without incident, but its usefulness has been questioned for pathogens after the escape of a rust fungus being evaluated for musk thistle control in Virginia, in the United States, in 1987 (Baudoin et al. 1993), and the escape of rabbit hemorrhagic disease virus (RHDV) from a containment area in Australia in the 1990's. In this latter case, RHDV was moved from an indoor quarantine facility to an outdoor quarantine compound to do efficacy tests once specificity had been established. The compound was on Wardang Island, 4 km off the Australian coast. Despite precautions, however, the virus was found in rabbits outside the compound on Wardang Island and then, within a year on the Australian mainland, where it spread rapidly and with a strong effect on rabbit populations (See Chapter 6; Kovaliski 1998; Story et al. 2004).

5.3.1 Reversible Biological Control?

Importation biological control is often called irreversible because of the ability of biological control agents to disperse, reproduce and evolve in the field (e.g., Howarth 1991). The effects of augmentative biological control have the potential to be irreversible as well if the agents 'escape' and reproduce in the field, but on the whole any negative effects of augmentation can in principle be halted by ceasing to make releases (Goettel & Hajek 2001). A few caveats can be raised with respect to the irreversibility argument for importation biological control, however. These include the deliberate use of sterile biological control agents, and control of unwanted biological control agents.

5.3.1.2 Sterile Biological Control Agents

Interest in introducing the herbivorous grass carp, *Ctenopharyngodon idella*, from southeast Asia into the United States dates to the 1950s and concern over the use of herbicides in freshwater bodies. The first introductions were made in 1963 for research purposes, and intentional stocking into open waterways, primarily in the southeastern United States, began in the early 1970s (Mitchell & Kelly 2006). During this time, the use of grass carp (known also by the common name 'white amur') was promoted by federal and state agencies, as illustrated in this 1972 quote reprinted by Mitchell and Kelly (2006):

> *The Arkansas Game and Fish Commission has decided to use the white amur for its intended purpose, and has stocked them in many lakes in all areas of the*

state. While the grass carp controversy rages, and thousands of tons of unnecessary chemicals are dumped into our waters, the white amur is slowly doing its job in the public waters of Arkansas. Nature has provided a valuable 'tool' to solve another problem if only 'technical man' is able to recognize it.

The controversy alluded to involved potential long-term ecological damage to native aquatic flora and fauna, and it continues to this day (Kapuscinski & Petronski 2005). One response to these concerns was the development of non-breeding grass carp. Early efforts focused on producing monosex carp through mechanisms such as irradiation, interspecific crossing and use of sex reversal hormones (Stanley 1976; Jensen et al. 1983) and culminated in cold-shock methods to produce triploid grass carp females that gave rise to sterile offspring (Cassani & Caton 1985; Allen et al. 1986; Van Eenennaam et al. 1990). Sterile triploid grass carp were made available throughout much of the United States but are only mandated in some states (other states either allow diploid grass carp to be released or prohibit both diploid and triploid grass carp).

The rationale for developing sterile biological control agents was that population growth and spread are limited, leading to a reduced potential for negative effects. In essence, importation biological control is transformed into augmentative biological control. Sterilization has been advocated as a strategy to limit reproduction and spread of transgenic organisms in the environment already (Snow et al. 2005), and such an approach could be adopted for transgenic biological control agents as well. We saw earlier that the reduced reproduction of baculoviruses transformed to express scorpion toxin was seen as a risk-reducing characteristic (Godfray 1995; Hails et al. 2002).

5.3.1.3 Control of Unwanted Biological Control Agents

Of course, a variety of control options are available if biological control agents become unwanted. In the words of Elton (1958), the 'counterpest' has become the pest in this case, and it may therefore be subjected to management strategies as would any other pest. For example, when myxoma virus was illegally introduced into France in 1952 and spread throughout most of Europe to the consternation of rabbit breeders and hunters, vaccines were developed and used in France and Italy with mixed success (Fenner & Fantini 1999). And when *Cactoblastis cactorum* moths island-hopped from the Caribbean to Florida in the 1980s, putting native *Opuntia* cacti at great risk, scientists initiated research into biological control of *C. cactorum*. A particularly tidy example of suppressing biological control agents comes from a case in which biological control was seen from the outset as temporary and geographically restricted. A small group of dingoes was introduced to an island off of the Australian coast to control a population of feral goats. The dingoes reduced the goat population from 3,000 to 21, at which point the remaining goats were shot and the dingoes poisoned (Hickling 2000).

A somewhat ironic response to unwanted biological control agents is further biological control. Some of these cases come from the early phases of biological control when generalist vertebrate predators were introduced with disastrous results (see Chapter 4). In one such case, the lizard *Anolis grahami* was introduced into Bermuda in 1905 from Jamaica as a control agent of the recently established Mediterranean fruit fly, *Ceratitis capitata* (Cock 1985). *A. grahami* became widely established in Bermuda despite having little effect on medfly populations, but apparently fed upon beneficial coccinellid beetles and thus came to be considered a pest (Simmonds 1958). The response was importation of the (largely insectivorous) kiskadee bird from Trinidad in the late 1950s, which became widely established without controlling the lizard (Cock 1985). In a similar case, the monitor lizard, *Varanus indicus*, was imported to some Micronesian islands from nearby islands to control rats (Schreiner 1989). Apparently, though, they had little effect on the rats but became a pest largely because they preyed on domestic chickens. In an attempt to control the monitor lizards, the cane toad, *Bufo marinus*, was introduced. The hope was that the lizards would attempt to eat the toads and become poisoned. This apparently led to declines of monitors on Guam (Strecker et al. 1962), but caused problems of its own, as we have discussed more generally in Chapter 4.

The cane toad in Australia has become a target for biological control research (Shanmuganathan et al. 2010). An early proposal for introducing a dung beetle that would burrow out of the toad's alimentary tract upon being eaten (Waterhouse 1974) was not carried out, but a search for parasites within the toad's native range in South America was mounted in the early 1990s. This search uncovered seven viruses and a number of bacterial species, but all of these exhibited too broad a host range to justify release (Shanmuganathan et al. 2010). Attempts were also made to genetically engineer a virus that would compromise the immune system of cane toads, but without success. Conservation biological control strategies, in which the impact of resident (Australian) enemies of cane toads would be improved, have been considered as well. These include the use of native frogs as 'typhoid Marys' to infect invasive cane toads with parasitic lungworms (Pizzatto & Shine 2012) and attracting predatory ants to areas where young cane toads aggregate (Ward-Fear et al. 2010). In this latter strategy, cat food bait is placed at the margin of water bodies adjacent to aggregations of recently metamorphosed cane toads. The bait attracts meat ants, *Iridomyrmex reburrus*, leading to a high rate of toad mortality via ant attacks.

As alluded to earlier, *Cactoblastis cactorum* has become a target of biological control research since it island-hopped to the southeastern United States, where it is putting native prickly pear cacti, *Opuntia* species, at risk (see Section 4.3.2 for a general treatment of this case) (Pemberton & Cordo 2001). Strategies under consideration include sterile male releases (Hight et al. 2005; Tate et al. 2007) and augmentative biological control using egg parasitoids native to Florida (Paraiso et al. 2013a). Host-specificity studies revealed, however, that these parasitoids had a strong preference for the eggs of native moth species over those of *C. cactorum*, and Paraiso and colleagues (2013a) recommended against release. Jezorek and colleagues (2011) found that ant predation on *C. cactorum* is increased when host plants grow adjacent to plants bearing extrafloral nectaries that attract ants and suggested that conservation biological control tactics could be developed.

5.4 Risk Communication

Formal public consultation is sometimes used prior to making a decision on allowing a release (Sheppard et al. 2003; Warner 2012). At least eight countries have formalized risk communications protocols for biological control – Australia, Canada, India, Mexico, New Zealand, Switzerland, the United Kingdom and the United States (Paraiso et al. 2013b). Public notification and participation processes for the subset of those countries that require these steps are listed in Table 5.4, which show that Australia and New Zealand involve the public at the highest levels. In Australia, the process of pre-release public consultation regarding biological control of the invasive weed Patterson's curse, *Echium plantagineum*, led to that country's 1984 Biological Control Act. The nectar from Patterson's curse produces high-quality honey and can be used by livestock when other forage is not available in dry areas and was thus considered a *beneficial* plant to beekeepers and some ranchers. Members of these groups (who refer to the plant as *Salvation Jane* as opposed to *Patterson's curse*) raised concerns about biological control that led to the drafting of the 1984 Act, which incorporates a requirement to publicly advertise the contemplation of biological control (Cullen & Delfosse 1985; Delfosse 1985). After an initial injunction barring biological control releases against *E. plantagineum*, releases of at least six biological control agents (all insects) have been allowed, presumably because consensus had been reached that the plant was better

Table 5.4. **Processes of public notification and participation for biological control releases in six countries. AUS – Australia, CAN – Canada, IND – India, UK – United Kingdom, NZ – New Zealand, USA – United States. Compiled from Paraiso and colleagues (2013b).**

Process	AUS	CAN	IND	NZ	UK	USA
Public notification						
Notification of proposed release	X	X		X		X
Public has access to release applications					X	
Risk assessment published in local newspaper				X		
Risk assessment posted by government	X			X		X
Local community informed about safety issues	X		X			
Public participation						
Solicit public comments prior to importation	X					
Solicit public comments prior to release	X			X		X
Procedures for hearings during decision process				X		
Approval process includes public comment period	X			X		X

characterized as a 'curse' than as a 'salvation' for Australia (Nordblom et al. 2002).

Paraiso and colleagues (2013b) conducted a survey on the topic of risk communication among biological control scientists in the United States and found that most participants felt that risk communication was an important activity; this survey also highlighted how risk communication can be integrated into an overall risk analysis program and a number of ways that the risk communication process could be improved. One of the main outcomes of the survey was a desire to more effectively include stakeholder input. A few case studies show how important effective communication with stakeholders can be in implementing biological control releases. A prime example involves releases of genetically modified mosquitoes for control of human diseases (see Section 1.2.4) in which gaps can exist between the assumptions held by local communities and scientists despite extensive efforts at risk communication (Ochanda 2010; Willoquet 2010; Subramaniam et al. 2012).

The importance of effective risk communication is highlighted in the case of biological control of the invasive strawberry guava tree, *Psidium cattleianum*, by the scale insect *Tectococcus ovatus* in Hawaii. In this case, communication of the pros and cons of a potential release to the local public was made difficult by suspicions of conservation scientists (Warner & Denslow 2012). The result was resistance by local activists, delaying a release despite federal and state permits.

5.5 Risk-Benefit Analyses

Until this point in the chapter, we have discussed only the risks of biological control. But what of the benefits? As we saw in Chapter 3, biological control is capable of delivering very substantial environmental and economic benefits. How should these benefits be weighed against the risks outlined in this and the previous chapter? And what of the monetary costs of biological control versus the savings that may be achieved? It is widely recognized that a complete risk-benefit analysis for biological control is a daunting, complicated task (Delfosse 2005; Bigler & Kölliker-Ott 2006; Thomas & Reid 2007; Morin et al. 2009; De Clercq et al. 2011), and McEvoy and colleagues (2012) have advocated adding evolutionary potential of the agent into such analyses, adding another dimension. Even without this addition, however, some of the difficulties include placing the risks and benefits in the same quantitative terms; uncertainties inherent in assessing risks and benefits of biological control; the different time scales that risks and benefits may operate on; and the different (and often changing) perceptions of risks and benefits held by diverse stakeholders. Cost-benefit analyses are particularly difficult when the benefits accrue to one set of stakeholders (e.g., farmers) and the risks to another (e.g., conservationists). While it is relatively straightforward to calculate economic costs and benefits (Naranjo et al. 2015), it is difficult to compare these to the risks that non-target species of conservation value are subjected to. Estimates of the role biological control can play in reducing greenhouse gas emissions (Heimpel et al. 2013) captures one category of environmental benefits but still cannot easily address non-target effects. Biological control of invasive species of conservation concern (Van Driesche et al. 2010) presents a more tractable comparison between risks and benefits since the risks and benefits both accrue in natural ecosystems.

Regardless of the setting, the analysis that is most relevant to our discussion here involves a framework known as Environmental Risk-Benefit Analysis (ERBA), although terminology is variable, including, for example, 'risk-benefit-cost assessment', 'benefit-risk analysis' and other expressions (Sheppard et al. 2003; Morin et al. 2009). ERBA is an extension of ERA in which risks and benefits are evaluated against a baseline scenario, which, for the case of biological control, typically comprises the current pest management methods (Bigler & Kölliker-Ott 2006). Thus, the environmental benefits to be captured by biological control will be lower in the face of environmentally benign management practices such as host-plant resistance than in the face of environmentally damaging methods such as frequent use of broad-spectrum pesticides. Stripped down, such an analysis has four components: (1) the probability that adverse effects will occur, (2) the magnitude of these adverse effects, (3) the probability that benefits will accrue and (4) the magnitude of these benefits. Clearly, accurate quantification of these four elements of the analysis will be difficult for most, if not all biological control systems. Thus, a qualitative approach can be adopted, in which probabilities and magnitudes of risks or benefits may be deemed 'insignificant', 'low', 'medium' or 'high' (Bigler & Kölliker-Ott 2006; Moeen et al. 2006). Generally speaking, the benefits and risks trade off such that a greater degree of risk can be accepted with greater anticipated benefits. Thus, any enhancement in the ability to ensure efficacy of biological control agents improves the benefit-to-risk ratio. Clearly, a biological control agent that establishes but is not effective results in risk without the attendant benefits (Pearson & Callaway 2005).

Not many biological control projects have explicitly used the risk-benefit approach. Cases include importation biological control of forage weevils and mealybugs in New Zealand by parasitoids (Moeed et al. 2006), and augmentative biological control of the European corn borer in central Europe by the egg parasitoid *Trichogramma brassicae*. In this latter case, the risks and benefits of *T. brassicae* releases were compared to the baseline scenario of applying the broad-spectrum pyrethroid insecticide deltamethrin. The risks associated with these releases was deemed low while the risks that deltamethrin posed to non-target arthropods were high (Bigler & Kölliker-Ott 2006).

Less formal comparisons have been made for particular projects (De Clercq et al. 2011) and one of the most fascinating of these involves *Compsilura concinnata*, the tachinid fly released against the gypsy moth that was discussed in Chapter 4. As we mentioned there, the negative effects of *C. concinnata* on various species of native giant silk moths in North America greatly outweighed the meager benefit in terms of biological control of gypsy moth. But what of the documented control of another invader, the browntail moth, *Euproctis chrysorrhoea*, in the northeastern United States by *C. concinnata*? A study by Elkinton and colleagues (2006) used historical data and experimentation to build a very strong case that *C. concinnata* prevented this European moth from spreading throughout the forests of eastern North America and possibly causing high levels of defoliation in forests. By a number of accounts, browntail moth had the potential to be a more devastating pest than the gypsy moth due to its ability to defoliate many tree species and also through its urticating hairs, which can cause a severe skin rash in humans (Elkinton & Boettner 2012). So what was the net effect of the *C. concinnata* introduction? On the one hand, *C. concinnata* caused devastating negative effects on native forest Lepidoptera, but, on the other hand, it seems to have prevented the spread of a potentially destructive and dangerous forest pest. There is no answer to this question since it is difficult to know how native natural enemies or more specialized biological control agents would have interacted with the browntail moth in the absence of *C. concinnata*. Elkinton and colleagues caution that this example should not be taken to support the introduction of generalist biological control agents, but they also recognize the irony that *C. concinnata* is held up as a prime example of a biological control agent that should not have been released.

5.5.1 Trade-Offs between Efficacy and Safety

In many cases, safety and efficacy are positively correlated. As discussed in Chapter 3, Kimberling (2004) found that high specificity was one of the best predictors of both success *and* safety in biological control agents. Also, as has been stressed throughout this chapter, highly successful biological control agents control their pest or weed species at low equilibrium densities, thereby keeping their own densities low in the process and minimizing the possibilities of spillover effects (Beddington et al. 1978; Holt & Hochberg 2001; Pearson & Callaway 2005).

However, circumstances do exist when the most effective hosts may not be the safest. In arthropod biological control, effective parasitoids or predators often use alternative hosts or prey when the pest is either present only in a non-vulnerable life stage or at low densities. Indeed, in earlier days of biological control, the introduction of an agent that could use alternative hosts was seen as a desirable trait. For example, Murdoch and Bence (1987) noted that *Gambusia* fish are effective control agents of mosquitoes precisely because they are generalist feeders and can therefore persist during times when mosquitoes are scarce (a point made more generally by Murdoch et al. 1985). As we noted in Chapter 4, however, this generalist feeding habit has made *Gambusia* spp. some of the most notorious biological control agents from the standpoint of non-target impacts. The lack of suitable alternative hosts has been cited as a reason for the failure of some biological control projects (Stiling 1993), and Mackauer (1976) summed up the possible trade-offs between safety and establishment by hypothesizing that

> *[B]iological control agents are also required to be host specific, a condition that appears to limit their ecological versatility and hence their chances for establishment in a new environment.*

Kaser and Heimpel (2015) explored interactions between safety and efficacy in importation biological control with a one-parasitoid, two-host mathematical model. The results confirmed that the presence of alternative hosts can increase the efficacy of biological control via apparent competition (see also Section 2.2.4.2). The models also suggested that the highest level of target suppression (i.e., efficacy) is achieved when the suppression of non-target species is at intermediate levels. Indeed, intermediate levels of suppression of the non-target host are needed to produce the additional parasitoids that lead to the highest levels of target host suppression (Figure 5.10). Together, these analyses suggest that higher efficacy of biological control can be achieved at a cost to non-target species. However, strong non-target population declines are possible in these models only when the non-target host is highly suitable to the parasitoid, a situation that is unlikely for specialized biological control agents.

In augmentative biological control, a broad host range is often seen as a positive characteristic, especially when multiple pests are present. This has been noted in particular for entomopathogenic nematodes with the clear understanding that this increased potential efficacy inevitably comes with increased environmental risk (Barbercheck & Millar

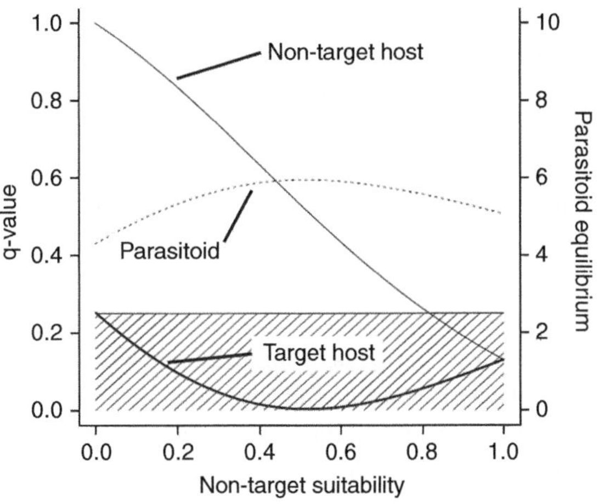

Figure 5.10. Results of a population model of two host species (one a biological control target and the other a non-target) that are attacked by one parasitoid species. q-values (proportional reduction of equilibrium below the carrying capacity) are shown for both the target and the non-target hosts as a function of the suitability of the non-target host. The hatched area indicates target q-values below that obtained without the non-target host. The dotted line indicates the parasitoid equilibrium density as shown on the right axis. From Kaser and Heimpel (2015).

2000). What should be done when such trade-offs are suspected? Certainly in weed biological control, agents deemed less effective but more safe than others have been chosen for release (Wapshere 1989). Ideally, such cases should be evaluated within a risk-benefit framework.

5.5.2 Guidelines for Best Practices

Codes of conduct designed to navigate the potential costs and benefits of biological control are expressed in national legislation regulating biological control (Sheppard et al. 2003) and also in guidelines for best practices. One such guideline has been put forward by the UN Food and Agriculture Organization (FAO) through the International Plant Protection Convention (IPPC), which is meant as a guide for individual countries in drafting their own legislation (IPPC 2006). The code outlines suggested responsibilities of biological control agent importers, exporters and legal authorities. While some of the principles of ecological risk analysis are alluded to implicitly in this code, no formal risk analysis or risk-benefit analysis framework is advocated.

A general theme in this chapter has been that the weed biological control community has led the way in risk analysis in biological control, and codes of best practice are no exception. In the United States, a so-called Technical Advisory Group (TAG) has published a manual for use by scientists embarking upon the process of acquiring a permit to release an exotic weed biological control agent. This manual discusses topics such as the formation of a test plant list, host-specificity trials and potential environmental impacts (USDA 2003). As an extension of the ideas discussed in the TAG guidelines, Balciunas (2000) developed a 'Code of Best Practices for Biological Control of Weeds' that included considerations related to redistribution and post-release testing. This code was distilled into 12 guiding principles that are reprinted in Table 5.5 and has been viewed favorably by the international weed biological control community (Balciunas 2004b).

Table 5.5. **Guiding principles of the International Code of Best Practices for importation Biological Control of Weeds. From Balciunas (2000).**

1	Ensure target weed's impact justifies release of exotic agents
2	Seek multi-agency approval for target
3	Select agents with potential to control target
4	Release safe and approved agents
5	Ensure only the intended agent is released
6	Use appropriate protocols for release and documentation
7	Monitor impact on target
8	Stop releases of ineffective agents, or when control is achieved
9	Monitor impacts on potential non-target organisms
10	Encourage assessment of changes in plant and animal communities
11	Monitor interaction among species of agents
12	Communicate results to the public

5.6 Conclusions

A number of recommendations of how to assess the risks of biological control have been introduced since the turn of the 21st century (e.g., Van Lenteren et al. 2003; 2006; Van Driesche & Reardon 2004; Sheppard et al. 2005; Kuhlmann et al. 2006b; Van Lenteren & Loomans 2006). These recommendations reflect a growing interest in risk assessment in biological control over this time period that can be clearly traced to the identification of non-target impacts covered in Chapter 4. Common elements include an acknowledgment that host ranges have a phylogenetic basis and that various behavioral and ecological filters may restrict the ecological host range with respect to the fundamental host range. Conversely though, it is recognized that host ecology may lead to phylogenetically disjunct species being attacked, and this needs to be taken into account. And finally, there is recognition that certain 'safeguard' species will have to be tested outside of the ecological or phylogenetic frameworks due to conservation or other concerns. How different are these considerations from those made by an earlier generation of biological control scientists (e.g., Harris & Zwölfer 1968; Wapshere 1974; 1989)? Perhaps not so very different. What is important, though, is that these recommendations are being made specifically with environmental safety in mind, and also that they are being directed for the first time toward arthropod biological control agents. Considerations in need of further development in the area of risk assessment include target selection, and estimation of ecological host range, including modeling of combined exposure and effects of biological control agents. And last, the role of legal and regulatory processes still varies greatly across the globe and standardization along these lines would be useful (Sheppard et al. 2003). For instance, while the process is clearly defined and rigorous in Australia and New Zealand (Barratt & Moeed 2005; Moeed et al. 2006), it is uneven and poorly defined in the United States, particularly for arthropod biological control agents of arthropod pests, and focused primarily on augmentation in Europe (Messing 2000; Van Lenteren et al. 2003; Messing & Wright 2006).

Also, a number of researchers have called for efficacy studies of potential biological control agents in their native range as a component of risk analysis (Balciunas 2004a; McClay & Balciunas 2005; Pearson & Callaway 2005). This reflects a goal of limiting the introduction of ineffective agents that nevertheless build to high densities on the target pest or weed, a scenario Holt and Hochberg (2001) identified as particularly risky (see also Veldtman et al. 2011).

II
Dynamics of Biological Control

6 Population Dynamics in Biological Control

When one considers the historical records of huge fluctuations in abundance of forest insect pests in Germany, and the dramatic success of the introduction of the vedalia beetle from Australia that saved the citrus industry in California from destruction by the cottony cushion scale, it is easy to see why ecologists were stimulated to develop the field of population ecology. Why forest insect populations can fluctuate in abundance by a factor of 100,000, and how an introduced natural enemy can lead to a reduction in scale abundance of a similar magnitude were clearly phenomena in need of explanation, and ecologists were eager to rise to the challenge (Turchin 1995). While pests and biological control have contributed tremendously to the development of population ecology, it remains questionable to what extent population ecology has contributed to the development and improvement of the success of biological control (Kareiva 1990, 1996). In this chapter, we consider the ecological theory that forms the basis for both importation and augmentative biological control and the theoretical models applied to different functional groups of natural enemies. These models provide the framework for a mechanistic understanding of the dynamics of biological control systems.

6.1 Introduction and History

Importation biological control has attracted far greater attention than augmentation from theoretical ecologists as it is concerned with the longer-term dynamics of a system of interacting populations. There have been two different approaches to the use of models in biological control: one, a more heuristic or theoretical approach that seeks a unified conceptual framework for understanding the underlying mechanisms that lead to different types of dynamics in consumer–resource interactions; and the other, a more applied or empirical approach to evaluate (retrospective) or guide (prospective) the use of specific natural enemies in biological control applications. The former approach seeks broad generalities while the latter seeks an explanation for observed patterns of abundance among interacting populations under field conditions. Among the different functional groups of natural enemies, microbial antagonists have only once been used for importation biological control of plant pathogens, but microbial pathogens have been used in both augmentation and importation programs for arthropod pests and weeds. However, it is the predators, parasitoids and herbivores that have been the dominant groups used in the importation biological control of pests and weeds. Models of importation biological control share the same basic framework of a consumer-resource relationship in which a consumer population makes gains from a resource population through trophic interactions that include herbivory, predation and parasitism. Some notable differences do occur among importation biological control systems, however, such as the extent to which the interaction kills resource individuals, the occurrence of multiple attacks on the same resource individuals and the generation times of the consumer relative to its resource.

In contrast, for augmentative biological control, natural enemies are not intended to have a long-term impact on pest or weed populations, and consequently the interaction is necessarily transient. This reaches an extreme in the case of inundative biological control where the action of the natural enemy is confined to a single generation. Under such circumstances, the dynamics of the natural enemy population is no longer of interest as the outcome of inundation is more simply determined by the number of natural enemies released, their per capita 'killing power' and their speed of kill. Thus models for augmentative biological control have focused more on questions of the timing of natural enemy release (e.g., Morales Ramos et al. 1996), the potential for secondary cycling of the natural enemy population (e.g., Knipling 1992), the killing power or speed of kill of the natural enemy (e.g., Sun et al. 2006) and the potential for refuge effects resulting from the dose-response curve of numbers released versus level of suppression achieved (e.g., Johnson 1999, 2010). Further details of the use of models in augmentation biological control can be found in Chapter 8.

As a historical prelude, Verhulst was among the first to address the question of population dynamics, developing the famous logistic equation for population growth (Verhulst 1838). Subsequently, both Lotka (1923) and Volterra (1926) independently developed a set of first-order, nonlinear, differential equations to describe the coupled interaction between predator and prey populations, classically known as the Lotka-Volterra (L-V) model. The continuous time formulation of the L-V model is particularly well suited to interactions in which the resource population shows continuous reproduction or replacement, or to those in which the generation times or replacement rates of the two populations are very different. The L-V model forms the basis of an extensive theoretical literature on the dynamical consequences of consumer-resource interactions, with application to host-pathogen (Briggs et al. 1995), predator–prey (Gutierrez 1996), host-parasitoid (Murdoch et al. 2003) and plant-herbivore (Crawley 1997) interactions. An alternative framework, proposed by Nicholson and Bailey (1935), used a set of discrete time equations to describe the interaction of a host population with non-overlapping generations and that of a specialist parasitoid population with synchronized generations. The N-B model has subsequently been developed extensively to explore the dynamics of host-parasitoid (Hassell 2000) interactions, but has also been used to explore plant-herbivore (Buckley et al. 2005) and host-entomopathogenic nematode (Stuart et al. 2006) interactions.

One of the central themes of importation biological control is the equilibrium concept, based on the notion that success is not only characterized by the suppression of the pest, but also by the stability of the new equilibrium to which the pest population has been driven by the action of the natural enemy (Huffaker et al. 1976; Beddington et al. 1978). This led to what Briggs (2009) described as the quest for persistence and stability, a period in which decades of research were devoted to the quest of identifying all possible mechanisms that could regulate and stabilize consumer-resource interactions. In this context, it is interesting that Murdoch and colleagues (1985) questioned the equilibrium concept, indicating that there was little or no evidence for stable local populations in six (winter moth, larch sawfly, olive scale, cottony cushion scale, walnut aphid and California red scale in Australia) out of seven (California red scale in California being the only exception) successful biological control projects with sufficient documentation to address the question. This was an important step as it not only revitalized a comment made earlier by Nicholson (1933), but also led to the now generally accepted view that importation biological control can be characterized by local instability or extinction, with the system persisting globally through asynchronous dynamics among the coupled subpopulations of a metapopulation (Murdoch et al. 2003). It is also debatable whether pest and natural enemy populations ever reach a state of equilibrium due to disturbances caused by seasonality and/or management practices. In this context, transient rather than equilibrium dynamics may be of more practical significance for biological control (Hastings & Higgins 1994; Kidd & Amarasekare 2012). We will return to these and other issues in greater detail at several points throughout this chapter.

6.2 Framework for Consumer-Resource Models

Consumer-resource models are applicable to a variety of biological control systems in which the resource population can be a plant, invertebrate or vertebrate and the consumer population is a predator, a herbivore, an insect parasitoid, a nematode parasite or a microbial pathogen. In using a theoretical framework to explore and understand the dynamical properties of consumer-resource interactions, it is important to consider whether consumer and resource populations have synchronous generations, and whether their generations are discrete in time or overlapping. When generations of the resource population overlap or generations of the consumer population are not synchronized with those of the resource, a differential equation framework provides the best representation of a coupled consumer-resource system in continuous time:

$$dR/dt = g(R)R - h(R,C)C$$
$$dC/dt = \gamma h(R,C)C - \delta C$$

where R and C are the population abundance of resource and consumer, respectively, $g(R)$ is the per capita net rate of increase of the resource population representing the balance of births and deaths due to other factors, $h(R,C)$ is the per capita functional response of the consumer or the rate of resource use, γ is the conversion efficiency of resource to consumer, and δ is the per capita death rate of the consumer population. This is the L-V model that updates both consumer and resource populations continuously and so includes within-generation dynamics as well as between-generation dynamics (Lotka 1923; Murdoch 1990; Murdoch et al. 2003).

The simplest form of the L-V model includes exponential growth of the resource population $g(R) = r$, where r is the per capita rate of increase or productivity of the resource population, and a linear functional response for the consumer population $h(R,C) = aR$, where a is an instantaneous attack rate. This simple model is neutrally stable and generates regular cycles in the abundance of both resource and consumer populations.

In contrast, if the consumer population has generations that are synchronized with those of the resource population, and the resource population has discrete (non-overlapping) generations, as is common among phytophagous insects in temperate regions, then a difference equation framework is particularly effective in representing the coupled consumer-resource relationship. The basic difference equation model is of the form:

$$R_{t+1} = d(R_t)R_t f(R_t, C_t)$$
$$C_{t+1} = cR_t\{1 - f(R_t, C_t)\}$$

with R_t and R_{t+1} and C_t and C_{t+1} the resource and consumer population abundance in generations t and $t+1$, respectively, $d(R_t)$ the per capita net rate of increase of the pest population, $f(R_t, C_t)$ the proportion of resource individuals that escape attack by the consumer, and c embodies the numerical response or the average number of consumers produced per resource individual consumed. This is the N-B model, and it differs from the L-V model in updating each population just once each generation (Nicholson & Bailey 1935; Hassell 1978, 2000). It thus represents between-generation dynamics and is unable to capture any dynamics that occur within generations.

The simplest version of the N-B model includes a constant rate of increase for the resource population $d(R_t) = \lambda$, and a linear escape response such that encounters by consumers are randomly distributed among resource individuals $f(R_t, C_t) = \exp(-aC_t)$, where a is now a search efficiency representing the proportion of resource individuals that an individual consumer could encounter in its lifetime. This simple model has an implicit lag of one generation between the timing of attack by consumers and the resultant effect on the resource population, and as a consequence is dynamically unstable, such that any perturbation from the locally stable equilibrium generates cycles of increasing amplitude that eventually lead to the extinction of the consumer population.

Both the continuous and discrete time models assume that the populations are local and closed, with no immigration or emigration, and neither of the two models includes any form of age or stage structure within the populations. Nonetheless, for both models, it is often convenient to assume that the consumer population represents the age or stage most active in acquiring resources, and that the resource population represents the age or stage most vulnerable to attack. The key to the difference in the stability properties of the two model frameworks lies in the destabilizing influence of a one-generation time lag in the N-B model, a point we return to later.

Neither the L-V model nor the N-B model really captures the timing of events and seasonality of consumer–resource interactions for populations with discrete generations. While reproduction by the resource population is often a seasonal pulse event, resource acquisition by the consumer population is more continuous over time and resource abundance declines continuously over the course of the interaction. Consequently, discrete generation interactions may actually be better represented by semi-discrete time models (Singh & Nisbet 2007). There has been a brief history of use of semi-discrete models to address specific issues in both host-parasitoid (Munster-Swendsen & Nachman 1978; Godfray et al. 1994; Bonsall & Hassell 1999; Dugaw et al. 2004) and host-pathogen (Briggs & Godfray 1996) models. However, Singh and Nisbet (2007) and Pachepsky and colleagues (2008) provide a persuasive argument for the need to replace the classical discrete N-B model with an equivalent semi-discrete framework for the future.

6.3 Functional and Escape Responses

Both the L-V and N-B models, as originally conceived, assumed that the interaction between consumer and resource populations was a function of the product of their densities (e.g. aRC). Implicit in this assumption is that resource acquisition is a linear function of resource density for an individual consumer. However, due to the difference in time frames of the two models, an important distinction appears between the corresponding functions that couple the two populations in these models. For the L-V model, the coupling function $h(R,C) = aR$ is the functional response that describes the number of resource individuals attacked in relation to both resource and consumer densities. In contrast, for the N-B model, the same coupling function must be integrated over a full generation to give $f(R_t,C_t) = \exp(-aC_t)$. This function can be derived either as the zero term of a Poisson distribution, assuming that consumer encounters are randomly distributed among individuals in the resource population (Hassell 1978), or as an integral form of the L-V model (Murdoch et al. 2003). To distinguish the two coupling functions, the latter has been described as an escape response rather than a functional response (Mills & Getz 1996), as it describes the fraction of the resource population that escapes from attack by the consumer population rather than the instantaneous number of resource individuals attacked.

The assumption that consumer-resource interactions are proportional to the product of their densities has persisted in the literature due to its simplicity, and the fact that it often allows elegant analytical solutions for equilibrium densities and stability boundaries to be derived for such models (Hassell 2000; Murdoch et al. 2003). However, such simplicity fails to incorporate the wealth of information that exists on the behavior of consumers, which can have important consequences for the outcome of the consumer-resource interactions. Here we will briefly explore the many different behavioral components included in such models, focusing primarily on the instantaneous form of the interaction or functional response, with the understanding that the corresponding escape response must be represented as an integrated form of the functional response.

6.3.1 The Influence of Resource Density on the Functional Response

Holling (1959) was the first to explore how the functional response varies in relation to resource density, and he recognized three different forms of response that he termed types I–III (Figure 6.1). He concluded that the functional response is seldom linear (type I, Figure 6.1a), although it can be representative of consumers that are sedentary filter feeders. A linear response may also occur for some invertebrates that search at random, detect the resource only at close proximity and have a very brief handling time (e.g., Mills & Lacan 2004). A type I response is also particularly well suited for microbial consumers, as they show little or no behavior in their interaction with their resources and encounters are governed by the law of mass action. For most invertebrates, however, Holling (1959) considered that the functional response more frequently takes an asymptotic form (type II, Figure 6.1b) due to the influence of handling time, the time interval from first encountering a resource individual to resuming search. Thus handling time reduces the time available for search by a consumer, with the consequence that as resource density increases, more and more of the available time is spent handling rather than searching for resource individuals. As a result, attacks by a consumer will rise at a decreasing rate to an upper asymptote or the maximum attack rate. This led to the so-called disc equation for which $h(R,C) = aR/(1 + aT_hR)$, with a as attack rate and T_h as handling time. Similar asymptotic (type II) functional responses can be generated through the effects of consumer satiation (Ivlev 1961; Mills 1982; Jeschke et al., 2002) or parasitoid egg limitation (Getz & Mills 1996; Heimpel 2000; Heimpel et al. 2003; Kaser & Heimpel 2015). Although type II responses are typical of invertebrates, Holling (1959) considered that a sigmoid (type III, Figure 6.1c) response is more frequent among vertebrate consumers and can also occur among invertebrates, particularly when

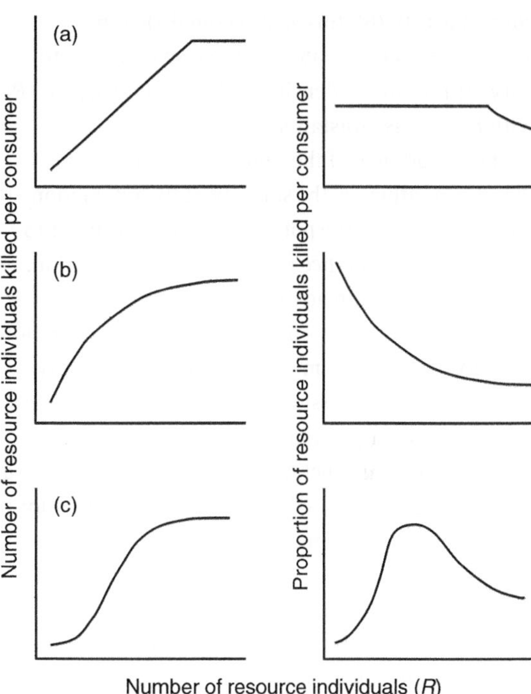

Figure 6.1. Holling (1959) defined three types of functional response relating the number of resource individuals attacked per consumer to resource density (R). A type I response (a) is linear, generating a constant mortality rate over a broad range of resource densities (neutral); a type II response (b) decelerates to a plateau, causing the rate of mortality to decline with resource density (destabilizing); and a type III response (c) is sigmoid such that the rate of mortality initially increases with resource density (stabilizing).

learning is involved. For a type III response, there is an initial acceleration in resource acquisition at lower resource densities due to consumers becoming more efficient at finding their resource, but subsequently at higher resource densities the same asymptotic effect occurs due to handling time, satiation or egg limitation. Murdoch and Oaten (1975) and Hassell and colleagues (1976) have both incorporated type III responses into consumer-resource models, and in the case of Hassell and colleagues (1976), this was accomplished by allowing the attack rate a of the disc equation to vary with resource density such that $a = bR/(1 + cR)$ with b and c as constants.

The importance of the functional response is that it can influence the stability of the interaction and consequently the potential for the consumer to regulate the abundance of the resource population. As we have seen already from the basic L-V and N-B models, a type I functional response does not promote stability, but at the same time it does not pose any constraint on suppression of the resource population by the consumer population (Figure 6.1a). In contrast, a type II functional response is destabilizing as the mortality imposed on the resource population is inversely density dependent (Figure 6.1b). The equilibrium abundance of the consumer population increases with handling time (T_h) in the L-V model, and as the equilibrium is unstable the dynamics are represented by cycles of increasing amplitude (Murdoch et al. 2003). Similarly, handling time causes increased fluctuations and even less persistence before extinction for the N-B model (Hassell 1978). The influence of a type III response is notably different, however, as the initial acceleration in the response imposes density-dependent mortality on the resource population at lower resource densities (Figure 6.1c). This density dependence has the potential to stabilize the L-V model depending on parameter values (Oaten & Murdoch 1975; Murdoch et al. 2003), but Hassell and Comins (1978) showed that it was not sufficient to stabilize the N-B model. The apparent contradiction between the two models regarding the influence of a type III response has been resolved by Singh and Nisbet (2007), who show that the response is also stabilizing in a semi-discrete version of the N-B model. Thus the apparent contradiction is a result of the difference in time frames of the two models, such that the escape response of the N-B model is based on initial resource density, which does not allow for exploitation of the resource over the course of the interaction within a generation.

For both type II and type III responses, the integral form for the N-B model differs between consumers that can re-encounter previously attacked resources, as is the case for parasitoids, and those that completely remove the resource, such as predators (Rogers 1972). However, both types of consumers can be represented by inclusion of an additional term in the model (Arditi 1983). When this additional term is zero, the consumer is a predator or herbivore, and when it is greater than zero, the consumer is a parasitoid and it represents the time spent handling a previously parasitized host.

6.3.2 The Influence of Consumer Density on the Functional Response

Far less attention has been paid to the fact that the functional response is influenced as much by consumer density as by resource density. Although addressed earlier by Watt (1959), it was the notion of mutual interference between consumers as they search for resources that first highlighted the importance of consumer density (Hassell & Varley 1969). Mutual interference is based on the idea that search time is wasted through direct encounters between consumers, as the individuals concerned may either move away from the resource altogether or at least stop searching for a period of time in response to such encounters. This simple notion of mutual interference was defined as $f(R_t, C_t) = \exp(-QC_t^{1-m})$with Q the attack rate when $C_t = 1$ and m the interference constant that reduces per capita efficiency as consumer density increases. The importance of interference is that it imposes density dependence on the population growth of consumers and generates a more effective source of stability for consumer-resource models than can be achieved through density dependence in the functional response to resource density (Hassell 1978; Murdoch et al. 2003). The consequence of this stability for both the L-V and N-B models is that the equilibrium densities of both consumers and resources are increased. A number of other forms of consumer interference have been suggested, and in all cases they reduce the per capita efficiency of consumers (Rogers & Hassell 1974; Beddington 1975; DeAngelis et al. 1975; Ruxton et al. 1992; Skalski & Gilliam 2001), and have the same stabilizing influence (Arditi et al. 2004). In addition, the stability generated by density dependence in attack rate is sufficiently strong to allow multiple consumers to coexist on the same resource population, at least under certain conditions (May & Hassell 1981; Murdoch et al. 2003).

6.3.3 The Combined Influence of Resource and Consumer Densities on the Functional Response

In some cases, the functional response of a consumer-resource model includes the combined effects of both consumer and resource densities, such as the disc equation with mutual interference (Crowley & Martin 1989; Arditi & Akcakaya 1990). An alternative approach to combining the influences of resource and consumer densities, proposed by Getz (1984), is to consider the functional response in terms of the ratio of resource/consumer densities. Arditi and Ginzburg (1989, 2012) coined the terms *prey-dependent* for a functional response that is dependent on resource density alone, and *predator-dependent* for a response that is dependent on consumer density alone. They further argue that while prey dependence captures the sharing of resources in a homogenous environment, ratio dependence $h(R,C) = h(R/C)$ is more appropriate for a heterogeneous environment or for consumers that show variation in spatial behavior. Although intuitively appealing, ratio dependence has found little favor as a simplification of the response of consumers to independently varying consumer and resource densities (Murdoch & Briggs, 1996; Abrams & Ginzburg 2000; Abrams 2015). Few direct measurements of functional responses have been made in the field, and while some empirical support for ratio dependence exists (Arditi & Ginzburg 2012), it continues to be a topic for debate (Abrams 2015).

6.4 Other Sources of Stability in Consumer-Resource Dynamics

6.4.1 Resource Density Dependence

While the functional response can generate density dependence under specific conditions, the long-term persistence of consumer-resource interactions led many to search for other sources of stabilizing influences (Briggs 2009). Of course, one of the most obvious points is that the resource population may experience density-dependent per capita population growth. It has often been argued that in successful examples of biological control, resource populations are driven to very low levels where the action of density dependence can be ignored. While this argument has some appeal, it assumes that we are only interested in the equilibrium rather than transient dynamics, and in success rather than failure of biological control. In addition, there has been much debate on the topic of whether local populations experience density dependence, as spatial population processes are sufficient to account for persistence (Turchin 1995). Nonetheless, in the case of local consumer-resource models, in which spatial processes play no role, we feel that there is a need to include density dependence in the per capita growth of the resource population.

One of the most commonly used forms of resource density dependence is the logistic model giving $g(R) = r(1-R/K)$ for the L-V model and $d(R_t) = \exp[r(1-R_t/K)]$ for the N-B model, where r is the net rate of increase (the product of per capita fecundity, female sex ratio and survivorship from density independent mortalities other than from the consumer population) and K is the carrying capacity or equilibrium density of the resource population. Many other forms of density dependence have also been suggested (Getz 1996; Hassell 2000). The effect of density dependence is to add stability to consumer-resource models, making the equilibrium of the L-V model locally stable (Murdoch et al. 2003), and generating considerable parameter space for stability in the N-B model (Beddington et al. 1975; May & Hassell 1981). It is interesting to note, that in the latter model, as the consumer becomes more effective in suppressing the resource population, the degree of stability is reduced due to the decreasing role of density dependence in the system.

6.4.2 Consumer Density Dependence

We have already explored how a predator-dependent functional response can add stability to consumer–resource interactions, but in the case of the L-V model it is also possible to add density dependence to the per capita death rate of the consumer population (δ). Murdoch and colleagues (2003) point out that while this adds stability to the model at the inevitable cost of an increase in resource equilibrium density, unlike the basic L-V model, the equilibrium now scales with the productivity (r) of the resource population. Density-dependent death rates are also sufficient to allow multiple consumers to coexist on the same resource population (Murdoch et al. 2003). As the N-B model does not include a consumer death rate, no specific consumer density-dependent term can be added to the model. However, the semi-discrete version of the N-B model developed by Singh and Nisbet (2007) does allow inclusion of consumer density dependence, and shows a similar effect as in the L-V model, including the scaling of the resource equilibrium with productivity.

6.4.3 Refuges from Consumer Attack

There are several reasons why variation may exist among individuals in their susceptibility to consumer attack, which leads to a refuge for part of the resource population. Spatial refuges are probably the most common, and a good example of such a refuge comes from the detailed analysis of biological control of California red scale by *Aphytis melinus* in California, where scale on the bark of the main trunk and scaffold branches are less susceptible to parasitism than those on twigs in the exterior canopy of citrus trees (Murdoch et al. 1989). Temporal refuges from a mismatch in the synchronization of consumer activity and the susceptible stages of the resource may also be relatively common (Barlow et al. 1994), while refuges that arise from genetic or behavioral variation among resource or consumer individuals may also occur, but are less well documented. In the context of consumer-resource models, refuge effects can either apply to a fixed number of individuals or to a fixed proportion of individuals in the resource population. A fixed-number refuge is always stabilizing for the L-V model (Murdoch et al. 2003), and is also stabilizing over a broad range of parameter values for the N-B model (Holt & Hassell 1993). In contrast, however, a proportional refuge adds no stability to the L-V model (Murdoch et al. 2003) and is stabilizing for the N-B model only under a restricted range of intermediate refuge sizes (Hassell 1978; Hochberg & Holt 1995). The reason for the greater influence of a fixed number refuge is that the proportion of the resource population exposed to consumer attack varies with resource density, clearly a density-dependent effect.

Refuges provide another mechanism by which multiple consumers can coexist on the same resource (Hochberg & Hawkins 1992, 1993), and have been shown to correlate with the success of biological control introductions (Hawkins et al. 1993, Hawkins & Cornell 1994). Using maximum observed rates of parasitism to estimate refuge size, Hawkins and Cornell (1994) showed that the likelihood of success for biological control introductions against insect pests increased inversely with the size of the host refuge. As for other stabilizing influences, the presence of a refuge from consumer attack also resulted in an increase in resource equilibrium density.

6.4.4 Immigration of Resources or Consumers

While the classical framework of consumer-resource models assumes closed populations, a fixed number of resource or consumer individuals as recruits from the surrounding landscape is another important stabilizing influence. In the case of resource immigration, it is easy to see that a constant number of recruits would function in exactly the same way as a fixed number refuge from consumer attack. Thus even if small in magnitude, an outside source of recruitment generates density dependence in the resource population. It is less clear, however, in the case of consumer immigration, to see how a constant number of recruits would generate stability. Nonetheless, provided that the consumer recruitment is not sufficient to drive the resource population to extinction, the recruitment is both stabilizing and leads to a reduction in resource equilibrium density (Nisbet et al. 1997; Gutierrez et al. 1999). This is perhaps the only exception to the more general rule that increased stability compromises the ability of a consumer to suppress a resource population.

6.5 Stage-Structured Populations

For many different types of consumer-resource interactions, resource individuals are susceptible to attack for only a part of the life cycle, and consumers may only attack during a particular phase of their life cycle. As an example, parasitoids frequently attack only certain instars of their hosts and only adult female parasitoids attack hosts. Similarly, herbivores that are flower and seedhead feeders attack plants only during the reproductive phase, and often only adult herbivores are involved in the attacks. Even in the case of pathogen–host interactions, infection by microbes is often confined to juvenile stages of the host. Thus the ecology of consumer-resource interactions is inherently stage structured, which introduces both time lags and refuges into the basic dynamics of the L-V and N-B models.

Some early examples of age-structured consumer-resource models were developed by Auslander and colleagues (1974) and Smith and Mead (1974), and a number of detailed age- and stage-structured models for specific systems were subsequently developed (Royama 1984; Shaw 1984; Gutierrez et al. 1988a, b, c). However, it was the lumped age-class or stage-structured approach to population models developed by Gurney and colleagues (1983) that, once applied to the L-V framework, provided a tractable stage-structured consumer-resource model (Murdoch et al. 1987; Murdoch et al. 2003). This approach allows ages that experience the same demographic processes to be grouped into the same stage, and allows individuals to move between stages according to development schedules. The stage-structured model Murdoch and colleagues developed is a modified L-V model with all of its basic assumptions of closed populations consisting of females only, and attacks that are proportional to the product of resource and consumer densities. The difference is that the resource population consists of a vulnerable immature stage and an invulnerable adult stage, and that both immature and adult stages are also represented in the consumer population (Figure 6.2a). All individuals are assumed equal within a stage, development times are fixed for each stage, and once attacked, vulnerable immature resource individuals immediately become immature consumers. The invulnerable adult stage acts much like a refuge from attack for the resource population, and thus the basic stage-structured model is stable when the duration of the invulnerable adult stage is greater than that of the vulnerable immature stage, and the duration of the immature consumer stage is considerably less than that of the vulnerable immature resource stage (Figure 6.2b). Other forms of the functional response and additional sources of density dependence can also be included in the basic stage-structured L-V model, and have been explored in detail by Murdoch and colleagues (2003). In essence, these additional components have the same influence on the dynamics of the stage-structured L-V model as they do for the basic L-V model.

Stage structure has only rarely been explored in a N-B model, although both Beddington (1974) and Wang and Gutierrez (1980) have shown that it can under some circumstances also lead to greater stability. A less explicit form of stage structure

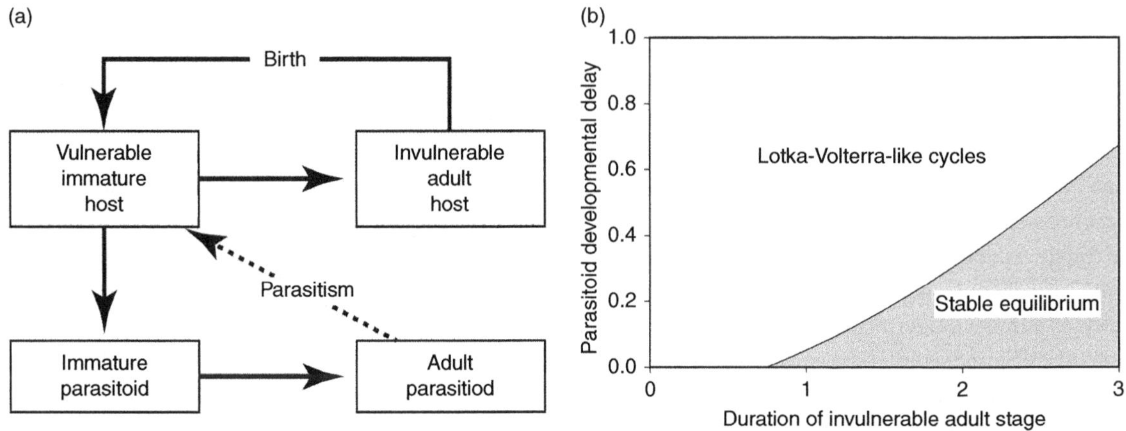

Figure 6.2. The Lotka-Volterra, parasitoid-host model of Murdoch and colleagues (1987). A schematic representation of the stage structure of the interaction (a) showing the coupling of host and parasitoid populations (from Briggs et al. 1999), and the stability boundary of the model (b) illustrating that a longer invulnerable adult stage adds stability to the model (from Briggs et al. 1999).

has also been explored in an N-B framework by altering the order of demographic processes in the model (Wang & Gutierrez 1980; May et al. 1981; Lane et al. 2006). One interesting result is that when resource density dependence acts after consumer attack, rather than before consumer attack, the resource equilibrium density can be greater than in the absence of the consumer population (May et al. 1981). Another is that the introduction of two different parasitoids for biological control of the same pest can lead to very different levels of equilibrium pest suppression depending on the sequence of attack, relative fecundity and relative efficiency of the two parasitoids (Lane et al. 2006).

6.6 Heterogeneity in the Risk of Consumer Attack

6.6.1 Variation between Hosts in the Risk of Parasitism

One of the central themes in population ecology over the past few decades has been the role of spatial processes, and the dynamics of populations that are spatially structured in a heterogeneous environment (Hanski 1999; Walde & Nachman 1999; Cronin & Reeve 2005, 2014; Hoopes et al. 2005). While the exploration of spatial heterogeneity in consumer-resource systems was initiated by Hassell and May (1973), Hilborn (1975) and Murdoch and Oaten (1975), it was May's (1978) negative binomial model that really captured attention. In this model, the standard derivation of the N-B model was altered to replace the assumption of a random distribution of attacks with one of an aggregated distribution of attacks. The escape response was thus based on the zero term of the negative binomial distribution $f(R_t, C_t) = (1 + aC_t/k)^{-k}$ with k a parameter that inversely describes the level of aggregation. The interest in this model was that, for the first time, a variant of the N-B model could provide both stability (for $k < 1$) and a sufficient level of resource suppression to match that observed in the successful biological control of insect pests (a reduction in resource density by a factor of 0.03 or more). An additional feature of the negative binomial model was that, similarly to consumer interference in the functional response, aggregated attacks are sufficiently stabilizing to allow the coexistence of multiple consumers on the same resource population (May & Hassell 1981; Hassell 2000).

Here we will return briefly to a point raised earlier, namely that the inherent one-generation time lag in the N-B models is the source of the instability in this model in contrast to the neutral stability of the L-V model. Hassell (2000) very nicely demonstrates this using an L-V model with a negative binomial functional response and an explicit time lag of up to one generation for recruitment to both consumer and resource populations (Figure 6.3). In the absence of a time lag ($\tau = 0$), the model simplifies to the standard L-V model in which there is neutral stability in the absence of aggregation ($k \rightarrow \infty$). As the time lag increases, greater levels of aggregation are needed to stabilize the interaction, until the lag reaches one generation ($\tau = 1$), at which point the model simplifies to the discrete negative binomial model of May (1978), in which aggregation must be less than unity ($k < 1$) to generate stability.

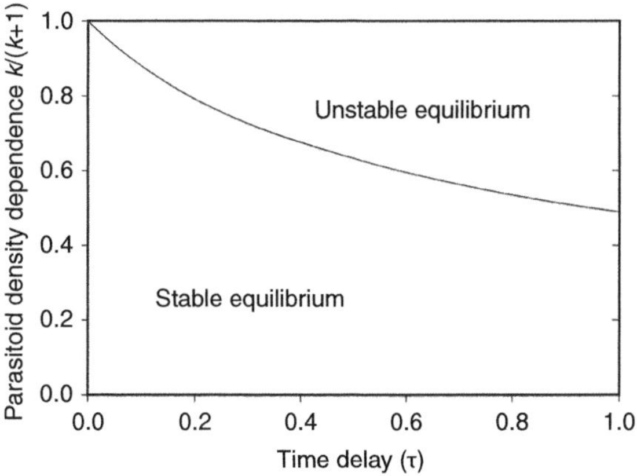

Figure 6.3. The influence of time lags on the stability of the Lotka-Volterra and Nicholson-Bailey models of consumer–resource interactions (from Hassell 2000). The neutral stability of the continuous time L-V model is represented at the top left of the graph where $\tau = 0$ and $k \rightarrow \infty$. In contrast, stability in the discrete-time N-B model, with a time lag of one generation, can be achieved only through the addition of sufficient parasitoid density dependence. This is represented on the right-hand side of the graph by the negative binomial version of the N-B model of aggregated attacks for which k must be less than 1 to achieve stability (May 1978).

As Chesson and Murdoch (1986) pointed out, the importance of spatial heterogeneity in consumer-resource models is that it leads to variation in the risk of attack among individuals in the resource population. Heterogeneity in the relative risk of attack among resource individuals can arise through dispersal or non-random search by the consumer, or from the differential susceptibility or non-uniform distribution of resource individuals. Thus heterogeneity leads to an aggregated distribution of attacks that can take one of two forms, resource density dependent or resource density independent (Pacala et al. 1990), and these are often referred to in the literature as host density dependent (HDD) or host density independent (HDI). May's (1978) negative binomial model was initially developed as a phenomenological model for aggregated attacks, but it has become clear that it represents the specific scenario of HDI aggregation (Chesson & Murdoch 1986). For the N-B framework, we have already seen from May (1978) that HDI is strongly stabilizing, and Hassell and May (1973) and Hassell (1984b) provide good evidence that HDD (both direct and inverse) is also stabilizing if the degree of aggregation is sufficient. In addition, in the case of HDD, the greatest suppression of resource equilibrium density occurs when the distribution of consumer attacks exactly matches the distribution of resource densities.

In contrast to the N-B framework, HDD does not lead to stability in the L-V model and can even be destabilizing (Murdoch & Stewart-Oaten 1989). This contradiction between the two model frameworks became a hotly debated issue that was finally clarified by Rohani and colleagues (1994), who used a spatially explicit N-B model in which parasitoids were redistributed between subpopulations of a host at different rates within a host generation. In the absence of within-generation redistribution of parasitoids, the model showed that density-dependent aggregation was stabilizing. However, even with moderate rates of within-generation redistribution, the stabilizing influence of density dependent aggregation was lost, matching the stability properties of a corresponding L-V model (Murdoch & Stewart-Oaten 1989).

Similarly, in contrast to the strongly stabilizing influence of HDI in the N-B model, Murdoch and Stewart-Oaten (1989) found that HDI had no influence on stability in an L-V framework. In this case, it appears that the difference between the two model frameworks is not due to the presence or absence of within-generation redistribution of consumers, but to a difference in the relative persistence of the variation in relative risk of attack. Thus the model of Rohani and colleagues (1994) showed that HDI was stabilizing even when parasitoids were allowed to redistribute among patches within a generation. Murdoch and colleagues (2003) point out, however, that the fraction of consumers visiting a patch remained constant throughout the generation in Rohani and colleagues' (1994) model, and thus that HDI was stabilizing due to a persistent distribution of the relative risk of attack. Using a more flexible model they showed that if the distribution of risk of attack is allowed to vary within a generation, the more transient the risk to any given resource individual, the less stabilizing the influence of HDI.

Thus in the context of biological control, one can conclude that while aggregation of consumers in direct response to local resource density is likely to lead to greater success in the suppression of the resource populations, it is unlikely to be a stabilizing influence. In contrast, aggregation of consumers that is independent of local resource density will not help suppress resource populations, but may confer some stability if they spend sufficiently long in local patches to provide persistent aggregation in the relative risk of attack among resource patches.

6.6.2 Spatially Explicit Consumer-Resource Models

The potential stabilizing influence of aggregated attacks has also led to a proliferation of spatially explicit models that focus on the role of dispersal in stabilizing consumer–resource interactions between distinct patches or subpopulations (Briggs & Hoopes 2004). Dispersal itself does not appear to be a stabilizing influence in the patch dynamics of consumer-resource interactions. However, aside from the 'statistical stabilization' that results from an apparent constancy in the global densities of consumers and resources that occur in a large enough set of asynchronous patches (de Roos et al. 1991), limited dispersal between asynchronous patches can increase persistence through one of two mechanisms. First, random dispersal decoupled from local patch densities can lead to increased persistence for both N-B (Reeve 1988) and L-V models (Weisser et al. 1997) through a variety of possible mechanisms (Briggs & Hoopes 2004). Second, increased persistence can occur if consumers have suboptimal foraging strategies, spending more time in patches of lower resource density, coupled with a nonlinear response to host density such as a type II functional response (de Roos et al. 1998).

That explicit spatial structure can contribute to greater persistence has been well illustrated experimentally for the interaction between a parasitoid *Anisopteromalus calandrae* and its bruchid beetle host, *Callosobruchus chinensis*, in a simple laboratory arena (Bonsall et al. 2002). When the arena consisted of a single cell the host was able to persist alone for 75 weeks (20 overlapping generations, Figure 6.4a), but in combination with the parasitoid, extinction occurred on average 5.4 weeks after introduction of the parasitoid (Figure 6.4b). As the number of linked cells was increased, the persistence time of the interaction increased, provided that the extent of dispersal between cell was restricted (Figure 6.4c, d).

Of greater interest, however, in the context of biological control, is the degree of pest suppression that can be achieved through importation biological control. In this regard, the trade-off between stability and suppression in consumer-resource models seen in local models also applies to spatially explicit models. Of particular note is the study of Kean and

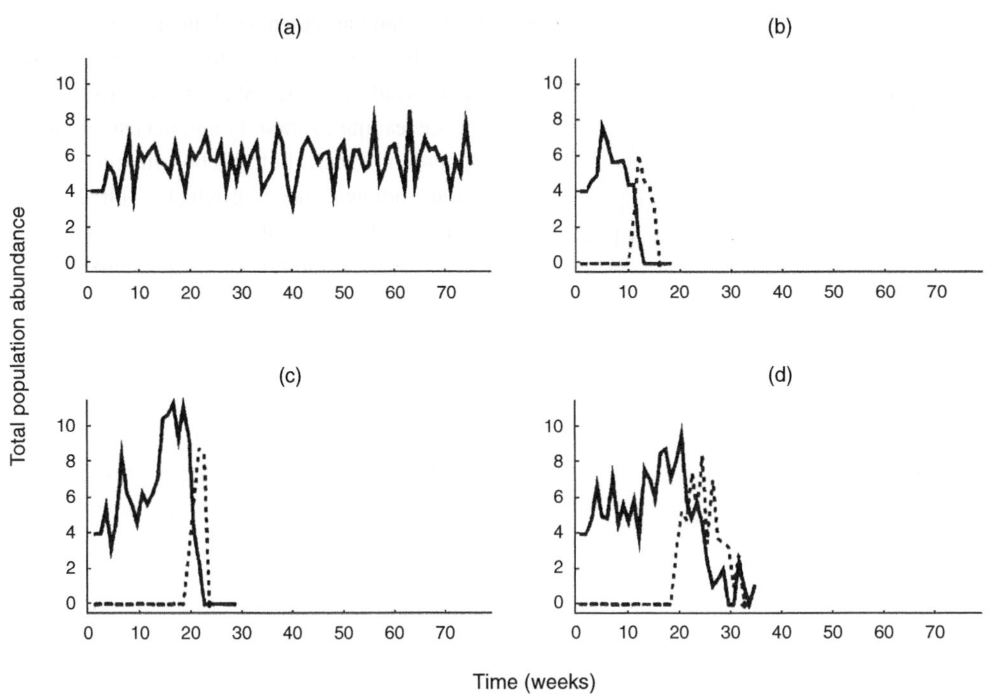

Figure 6.4. Metapopulation persistence of a parasitoid, *Anisopteromalus calandrae* (broken line), and its bruchid beetle host, *Callosobruchus chinensis* (solid line), in a simple laboratory arena (from Bonsall et al. 2002). The weevil persists in a single cell (a) in the absence of the parasitoid, but in the presence of a parasitoid with limited dispersal capabilities, persistence time is dependent on the number of interconnecting cells in the arena; (b) 1 cell, (c) 4 cells and (D) 49 cells.

Barlow (2000a), which used a lattice model to more specifically address the influence of metapopulation structure on the success of biological control of insect pests by introduced parasitoids. Although the global population projections of the lattice model were the same as for a single patch model, a higher level of global pest population suppression by a parasitoid was possible in the lattice model than in a single patch model.

6.7 Density Dependence and Biological Control

One of the central concepts of population dynamics through the early part of the 20th century was that density dependence is required for a population to be regulated around an equilibrium level of abundance (Turchin 1995). Consequently, it was believed that the sustained host suppression observed following the establishment of a successful biological control agent must require the involvement of density dependence generated through some aspect of the consumer-resource interaction. It was a surprise, therefore, when an early set of observations on the biological control of greenhouse whitefly, *Trialeurodes vaporariorum*, by the parasitoid *Encarsia formosa* showed no evidence of stability (Burnett 1958; Figure 6.5). The dynamics of the interaction, observed from small cages in a glasshouse setting, was one of cycles of abundance followed by extinction of the parasitoid when the host population reached a sufficiently low level of abundance. While the observed changes in

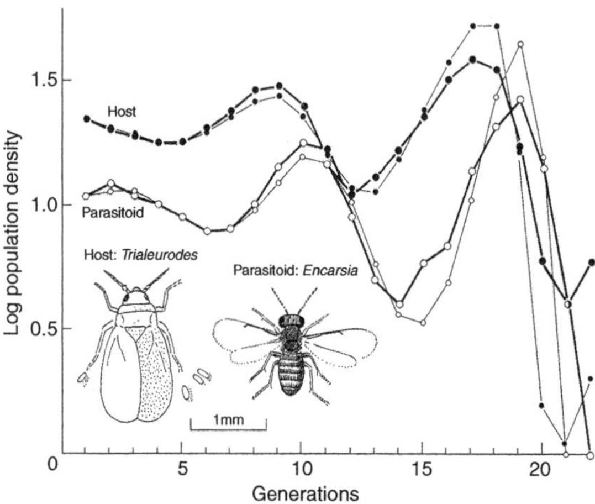

Figure 6.5. The dynamics of the interaction between the parasitoid *Encarsia formosa* and its host, *Trialeurodes vapaporiarum* (after Varley et al. 1973). The heavy lines are observed data from Burnett (1958) and the fine lines are from the simplest Nicholson-Bailey model with $\lambda = 2$ and $a = 0.068$.

both whitefly and parasitoid populations could be matched very closely by an unmodified N-B model (Varley et al. 1973; Figure 6.5), this observation seemed at odds with the fact that successful biological control by natural enemies was known to have persisted over considerable lengths of time, as in the case of the introduction of the vedalia beetle to control cottony cushion scale in California in 1889.

However, it was Murdoch and colleagues (1985) who questioned the conventional wisdom that successful biological control is achieved through the establishment of a locally stable consumer-resource equilibrium. In analyzing six examples of biological control that had earlier been heralded as successes by Beddington and colleagues (1978), it became clear that only one (California red scale in California) out of the six examples (winter moth in Nova Scotia, olive scale in California, larch sawfly in Manitoba, walnut aphid in California and California red scale in Australia) provided evidence of stability at a local scale. That extinction may be more characteristic of biological control than stability at a local scale was not generally accepted until the late 1980s, and this realization switched attention from density dependence leading to local stability to density dependence leading to stability and persistence at a metapopulation scale. Both Nicholson and Bailey (1935) and Andrewartha and Birch (1954) had earlier suggested that populations can persist at a large spatial scale, even though independent smaller groups of individuals experience local extinctions. Levins (1969) had formalized this concept in a cell-occupancy model that represented biological control systems as a set of distinct patches of interacting populations, characterized by the fraction of patches occupied. However, as noted earlier, interest in the spatial dynamics of consumer-resource models rapidly gained momentum (Walde & Nachman 1999), switching the focus on density dependence in biological control from local to global stability.

Natural and managed environments are becoming increasingly fragmented and thus the persistence of higher trophic levels may well be driven by metapopulation processes, although we know that this is not the case for two systems, California red scale (Murdoch et al. 2003) and arrowhead scale (Matsumoto et al. 2003), which provide evidence of stability at a local scale. More broadly, it is also surprising that there continue to be few documented examples of metapopulation structures under field conditions (Harrison & Taylor 1997; Cronin & Reeve 2005). Hanski and Simberloff (1997) define several specific types of metapopulation structures, while Thomas and Kunin (1999) consider them to reflect more of a continuum. Classic metapopulations, in which patch dynamics occur at a faster rate than metapopulation processes, leading to a significant probability of local extinction and a sufficient number of patches acting as a refuge from consumer attack, appear infrequent. Consequently, mainland-island metapopulation dynamics, in which small, extinction-prone patches are colonized by resource migrants from larger, more stable patches, may be more representative of biological control systems (Kruess & Tscharntke 1994; Cronin & Reeve 2005,

2014; Tscharntke et al. 2007). While consumer-resource populations are almost certainly fragmented in nature, it remains largely unknown whether spatial processes affect consumer and resource populations at the same spatial scale, and there is very limited knowledge of the relative rates of dispersal of consumer and resource populations (Coll et al. 1994). In addition, it remains unknown whether the coupling of demographic and evolutionary processes in metapopulation models could facilitate the persistence of introduced biological control agents (Fauverge et al. 2012).

Thus, from our current state of knowledge, the answer to the question of whether density dependence is a requirement for successful long-term biological control is a qualified yes. While density dependence leading to stability at a local level may indeed occur among a subset of examples of biological control, it is not a necessary characteristic for success and probably does not occur for many examples of biological control. However, for an interaction that is unstable at a local level to persist at a global scale requires asynchronous dynamics with limited dispersal between patches that can function as a form of density dependence at a metapopulation level. However, the extent to which spatial processes affect demographic and evolutionary change in biological control systems has yet to be fully explored, and will be a fruitful area for further research.

6.8 Application of Models to Biological Control

Various adaptations of the basic L-V and N-B models have been applied to specific examples of biological control, and a number of these have been discussed by Barlow (1999). Applications have been made in the context of parasitoids for control of insect pests, herbivores for control of weeds, pathogens for control of insect pests, weeds and vertebrates and microbial antagonists for control of plant pathogens. Specific applications often require modifications of the basic L-V and N-B models, and a range of different systems will be considered in greater depth through the rest of this chapter.

6.8.1 Parasitoid-Host Models

Parasitoid-host interactions are often considered the simplest of consumer-resource interactions for two reasons – first, that as only adult females search for hosts, this greatly reduces the need for stage structure, and second, as only a single host is needed to support the development of parasitoid offspring, this simplifies the coupling of host and parasitoid models. Even with these simplifications, however, an absence of stage structure in the basic L-V and N-B frameworks can obscure important within-generation effects that are generated by an invulnerable host stage (Murdoch et al. 1987), the relative sequence of mortalities (Wang & Gutierrez 1980; May & Hassell 1981) and the pulse nature of host reproduction (Singh & Nisbet 2007). Various types of models have been applied to insect biological control, including discrete models for pests with a single generation a year (e.g., Hassell 1980; Barlow et al. 1996), continuous models for pests with overlapping generations (e.g., Godfray & Waage 1991; Gutierrez et al. 1993; Murdoch et al. 2005), spatially explicit (lattice) simulation models (Gutierrez et al. 1999; Kean & Barlow 2001) and stage-structured matrix models (Mills 2005, 2008). In some cases, the models have been used retrospectively to enhance basic understanding (e.g., Hassell 1980; Gutierrez et al. 1993; Kean & Barlow 2001), while in other cases, they have been used prospectively for guidance (e.g., Godfray & Waage 1991; Barlow et al. 1994; Mills 2008).

One of the best-studied cases of importation biological control concerns the California red scale, *Aonidiella aurantii*. This armored scale originated in southeast Asia, was accidentally introduced to California around 1870, and has since become a worldwide pest of citrus (Murdoch et al. 2005, 2006b).

The scale feeds on a broad range of woody plants, and is a particularly important pest of citrus, feeding on all above ground parts of the plant, including bark, foliage and fruit. It has two generations per year, and in the absence of control can reach population densities of millions per tree, causing defoliation, dieback of branches and even death of the trees. Immature scales take approximately 40 days to develop through to adult, and female scales live for about six weeks and lay from 10–35 eggs (Murdoch et al. 1996b). The biological control program against California red scale in California was unusual in that it holds the record for the greatest number of exotic natural enemy species introduced against a single pest at 28 coccinellid and 2 nitidulid predators, and 17 aphelinid and 2 encyrtid parasitoids. From these introductions, a total of eight natural enemies became established, at least temporarily, and three introduced parasitoids are found consistently in different citrus production regions in California. However, the successful control of this pest is mostly attributed to a single aphelinid parasitoid, *Aphytis melinus*, which has a shorter development time of approximately 12 days and an assumed longevity in the field of about seven days (Murdoch et al. 1996b, 2006b).

California red scale also provides one of the best-documented examples of competitive exclusion among parasitoids used in a biological control program (Luck & Podoler 1985; Murdoch et al. 1996a; Reitz & Trumble 2002). In 1947, the parasitoid *Aphytis lingnanensis* was introduced to California from India, and this led to satisfactory control of the scale along the coast, eliminating the earlier accidentally introduced *A. chrysomphali*. However, *A. lingnanensis* failed to provide sufficient control of California red scale in the inland valleys, and so *A. melinus* was imported from China in 1957. In coastal citrus groves where scale had been adequately suppressed, *A. lingnanensis* coexisted with *A. melinus*, but for groves in the inland valleys, *A. lingnanensis* had been completely excluded by 1972 and scale densities were more effectively controlled. The displacement of *A. lingnanensis* by *A. melinus* occurred surprisingly rapidly, taking from one to three years or two to nine scale generations, and occurred over an extensive region. It is interesting to note that, from laboratory studies, *A. lingnanensis* appeared to be the better competitor, being intrinsically superior in larval competition and extrinsically superior in search rate (Rosen & DeBach 1979). Although both species were subsequently found to avoid multiple parasitism, *A. melinus* was also able to produce female progeny from younger stages of scale than was the case for *A. lingnanensis*, and was able to produce two progeny from larger-sized scale in comparison to one for *A. lingnanensis* (Luck & Podoler 1985).

Using a series of increasingly sophisticated stage-structured L-V models (Murdoch et al. 1987, 1992, 2005), and experimental field tests of these models (Murdoch et al. 2003, 2005), Murdoch and colleagues contributed greatly to our understanding of this system. A simplified version of the model (Figure 6.6) consists of three stages of scale development, young juveniles that are perceived by A. *melinus* as poor-quality hosts and used for host feeding old juveniles that are perceived as good-quality hosts and receive either a single female egg or both a male and female egg (gregarious) and an invulnerable adult stage that produces new young juveniles. The parasitoid is represented by two stages, immatures that are produced from parasitized scales, and adult females that attack the vulnerable stages of the scale population. Some of the key points that have arisen from this system are:

- The California red scale-*Aphytis melinus* interaction is unusual in that it shows stability at a local scale of individual trees (Murdoch et al. 1995).
- The long-lived invulnerable adult stage of California red scale not only contributes to the local stability of the interaction, but also to the control of the scale by *A. melinus* (Murdoch et al. 1987; Murdoch et al. 2006b).
- Spatial processes do not appear to play a role in the dynamics of the California red scale-*A. melinus* interaction, either in terms of aggregation

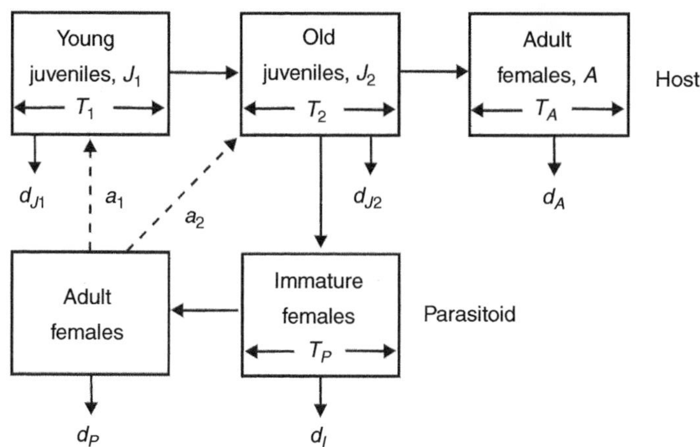

Figure 6.6. A schematic representation of the interaction between the parasitoid *Aphytis melinus* and its host, the California red scale (from Murdoch et al. 2006b), in which the juvenile stage of the host in Figure 6.2a is separated into two stages. Younger, immature scale (J_1) are killed by *A. melinus* at rate (a_1) by host feeding, and have developmental duration (T_1) and background mortality (d_{J1}). Older, immature scale (J_2) are parasitized by *A. melinus* at rate (a_2), leading to developmental duration (T_P) and background mortality (d_I), while unparasitized individuals have duration (T_2) and background mortality (d_{J2}). Adult scale (A) are invulnerable to parasitism and have longevity (T_A) and background mortality (d_A).

of risk or metapopulation asynchrony (Murdoch et al. 1996b).

- Several key life history attributes of *A. melinus* (encounter rate, clutch size, sex allocation, host feeding and survival) increase with scale size, creating an accelerating gain to the future female parasitoid population with scale age. This effect induces an indirect delayed density dependence that generates local stability in the presence of a long-lived invulnerable adult stage (Murdoch et al. 1992).
- While the extent of host suppression is dependent on the per-capita search rate of the parasitoid, it is also dependent on the parasitoid having a shorter development time than that of the host (Murdoch et al. 2006b).

While it is not possible to manipulate the life history attributes of the California red scale and *A. melinus* to test their effect in direct experiments, Murdoch and colleagues (2005, 2006b) have used a more complete version of the model to predict the response of the system to an artificially created outbreak of the scale. They used individually caged trees in a lemon grove in California to create an outbreak through the addition of scale crawlers to the trees over a period of three months, and then monitored the dynamics of the ensuing host and parasitoid populations. The match of the output from the model to the observed trend in both scale and parasitoid abundance was remarkable (Figure 6.7). Following the artificially created increase in scale abundance from less than 1 per 100 cm^2 to more than 40 per 100 cm^2, the parasitoid population responded within a single generation of the scale to reduce the experimental populations to their typical background level of abundance. That the model could predict these responses with such accuracy gives us confidence that the mechanisms included in the model capture the dynamical properties of the system. In addition, a sensitivity analysis of the model showed that the prediction was robust to a doubling of the parameter estimates, and that the two most important parameters were the duration of the

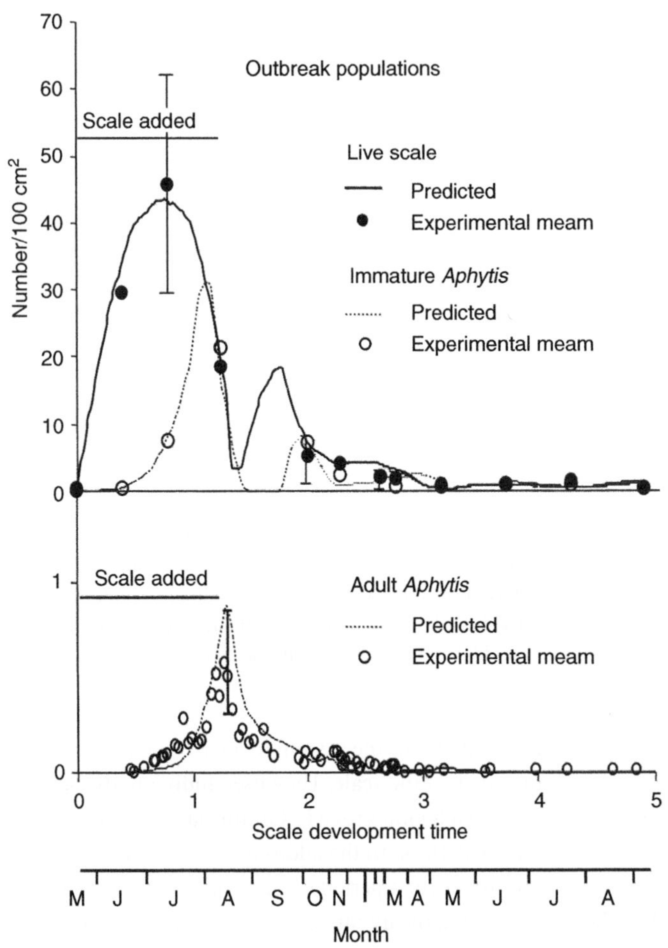

Figure 6.7. Mean density of California red scale and *Aphytis melinus* (both immature and adult) predicted by the model (curves) and observed in outbreak trees (data points). In the top panel, the range is indicated for three dates on which the observed mean densities of live scale is furthest from the predicted value; in the bottom panel, the range is given for the date closest to the predicted peak density of adult *Aphytis* on which counts were made from outbreak trees. The parameter values for the model were estimated independently of the experiment. From Murdoch and colleagues (2006b).

invulnerable adult stage of the scale and the shorter generation time of the parasitoid in relation to that of its host (Murdoch et al. 2005).

This case study illustrates very effectively the value of theoretical models in helping us to understand the mechanistic basis for biological control. However, as the California red scale system is perhaps unusual in exhibiting stability at a local scale, the generality of this model in representing mechanisms of biological control for other insect pests may be limited. Fortunately, it is not the only application of host-parasitoid models in biological control systems, and two other detailed analyses are the cassava mealybug, *Phenacoccus manihoti* (Gutierrez et al., 1988a, b, c, 1993, 1999), and the alfalfa weevil, *Sitona discoideus* (Barlow & Goldson 1993, Kean & Barlow 2000c, 2001). For the cassava mealybug, a unique elaboration of a stage-structured

L-V model was used to represent a plant-pest-parasitoid interaction in which a common set of metabolic functions governs the ecological processes at each trophic level. The model includes a greater degree of biological realism than the California red scale model, and each of the metabolic functions can also be influenced by climate. The model was used very successfully to confirm, based on field monitoring, that control of mealybug populations was due to parasitism by *Anagyrus* (= *Epidinocarsis*) *lopezi* in the dry season, and by rainfall acting either directly or indirectly through fungal disease during the wet season. In contrast, Barlow and Goldson (1993) developed a modified N-B model with parasitoid interference to explore the suppression of the alfalfa weevil in New Zealand by the introduction of the parasitoid *Microctonus aethiopoides*. The model was shown to provide an accurate prediction of the subsequent patterns of pest abundance, although it slightly overestimated the levels of parasitism (Kean & Barlow 2000b, Figure 6.8). With further modification, the model could also account for the lack of success of this parasitoid under Australian climatic conditions relative to its success in New Zealand (Kean & Barlow 2001).

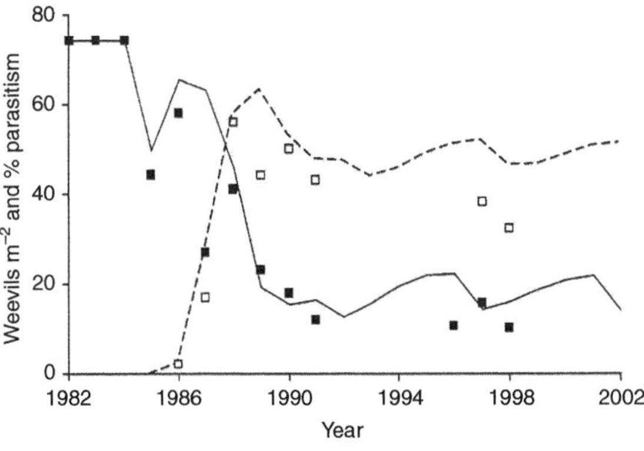

Figure 6.8. Observed (symbols) and predicted (lines) autumn densities of alfalfa weevils (solid) and levels of parasitism by *Microctonus aethiopoides* (broken) in grass paddocks in New Zealand from the model by Kean and Barlow (2000b).

Barlow (1999) reviewed a number of earlier, less well-explored models of insect biological control, and more recently, additional models have addressed various aspects of the biological control of whiteflies (Mills & Gutierrez 1996, 1999), coffee berry borer, *Hypothenemus hampei* (Gutierrez et al. 1998), olive scale, *Parlatoria oleae* (Rochat & Gutierrez 2001), spotted alfalfa aphid (Gutierrez & Ponti 2013), green stinkbug, *Nezara viridula* (Liljesthroom & Rabinovich 2004), codling moth, *Cydia pomonella* (Mills 2005), vine mealybug, *Planococcus ficus* (Gutierrez et al. 2008), and light brown apple moth, *Epiphyas postvittana* (Mills 2008).

Simple models can also be used to improve our understanding of non-target impacts in biological control. While several studies have documented mortality from introduced parasitoids among non-targets (see Chapter 4), the significance of the mortality in terms of suppression of non-target abundance can be assessed only at a population level. Barlow and colleagues (2004) outlined a technique for using simple models to assess non-target impacts of introduced parasitoids, and the approach has been applied to both *Microctonus hyperodae* (Barlow et al. 2004) and *Pteromalus puparum* (Barron 2007). While in both cases, the level of non-target suppression was less than predicted by the level of mortality alone, Barlow and colleagues (2004) provide an example of where the reverse could also be the case.

6.8.2 Predator–Prey Models

The key distinction between predator–prey models and parasitoid-host models is that the search for, and conversion of, resource to consumer occurs at a series of different life stages with differing efficiencies. Sometimes referred to as a developmental response, these behavioral and energetic complications cry out for a stage-structured modeling approach. Although initial steps were taken to build such relationships into an N-B framework (Beddington et al. 1976), the continuous nature of the L-V framework is clearly more suitable for the differing time scales inherent to the within-generation dynamics of predation.

The complexities of stage-specific search, energy acquisition and conversion have frequently led to the development of large, multi-parameter simulation models of specific predator and prey species (e.g., Barlow & Dixon 1980; Carter et al. 1982, Gutierrez & Baumgärtner 1984; Holst & Ruggle 1997; Skirvin et al. 1997; Pekar & Zd'Arkova 2004). The energetic complications of predation can be simplified to some extent by changing the currency of the model from individuals to relative nutritional value (Nachman 1987) or biomass (Gutierrez et al. 1994), but in general predator–prey models have remained more cumbersome than the simpler host-parasitoid models and there has been far less effort directed toward the development of models for biological control by predators. A further complication that has received greater attention in recent years is that many arthropod predators are omnivorous, consuming both prey and other predators, introducing a certain amount of indeterminacy into the predation function (Rosenheim & Corbett 2003).

Predators have not been used as frequently in importation biological control and consequently few models have been developed for such purposes (but see Dixon 2000; Hanna et al. 2005). More typically predator–prey models have been applied to augmentative biological control and in particular to inoculative biological control. An interesting example of such an application is that of Nachman (1987). This spatially explicit simulation model was developed to provide mechanistic insights for inoculative biological control in acarine predator–prey systems, with particular emphasis on control of two-spotted spider mites, *Tetranychus urticae*, in glasshouse crops using the phytoseiid predator *Phytoseiulus persimilis*. The model included both within-plant dynamics and inter-plant dispersal and was based on plant condition, prey density and predator density. The model included several simplifying assumptions, such as invariant age distributions and sex ratios and linear functional response relationships with the exception of a type II response for the predatory mite, but also included greater realism through demographic stochasticity and variation in the probability of successful dispersal. The model has subsequently been compared to experimental data from two different acarine predator–prey systems, glasshouse and orchard (Walde & Nachman 1999). The key points arising from this comparison are:

- The dynamics of *T. urticae-P. persimilis* on individual plants in a patchy environment are characterized by a sequence of local colonizations and extinctions – no local stability.
- The mean dynamics in a patchy environment are characterized by persistent predator–prey oscillations of constant period, but variable amplitude.
- *P. persimilis* is a specialist predator with a short development time, but high reproductive rate, search efficiency, dispersal and tendency to eliminate local prey colonies.
- Increased dispersal between patches (by reducing the distance between plants or use of bridges) reduces prey density and plant injury in the case of a specialist predator, but the reverse may occur for a generalist predator in the presence of alternative food.
- Plant size, plant susceptibility to injury, predator specialization and dispersal behavior are likely to influence the outcome of biological control in acarine systems.

Nachman and Zemek (2003) and Hardman and colleagues (2013) have subsequently developed more detailed simulation models to explore the relative roles of plant condition versus predation in driving the dynamics of tetranychid mite populations. Additional models that address the importance of plant type on augmentative biological control of tetranychid mites by phytoseiid predators include Saito and colleagues (1996) and Skirvin and colleagues (2002).

In an interesting attempt to simplify the complexity of arthropod predator–prey models, Kindlmann and Dixon (1999, 2001) focused on an important distinction between juvenile and adult predators. Juveniles are frequently confined to a single patch where their survivorship is dependent on prey availability in the patch chosen for them by the parent female. Thus survivorship can be low if the availability of prey declines during the course of juvenile development, and cannibalism is advantageous in systems where prey are more likely to be ephemeral. In contrast, adults move freely between patches where their oviposition strategy (or numerical response) determines their overall fitness. Adults must assess a patch of prey individuals not only in terms of its current quality, but also in terms of its potential to support the future development of their offspring. If prey patches are more ephemeral, as is the case for predators that have a longer generation time than their prey, then it is advantageous for females to respond to the age of a patch rather than to the number of prey present, with the consequence that the impact of predation on the global prey population will be low. However, for predators with a similar generation time to that of their prey, patch quality for offspring survival is more predictable, and they can distribute their eggs between patches in relation to prey density with the consequence that predation has a much greater impact on the global prey population. Using a simulation model, Kindlmann and Dixon (1999) show that the impact of predation on prey population density is inversely related to predator development time, suggesting that the effectiveness of biological control will always be low when the ratio of predator to prey generation times is large.

An additional aspect of predation that has received very little attention in the context of predator–prey models is the indirect effect of non-consumptive predation (Preisser et al. 2005; and see Chapter 1). The fear factor induced by the presence of predators can lead to sub-lethal effects on prey populations through reductions in growth, reproduction and survivorship from factors other than direct predation. Non-consumptive effects of predation can be influenced by relative body size (Krenek & Rudolf 2014), and often lead to the movement of prey to safer, but lower-quality refuge habitats (Orrock et al. 2013). The relative importance of consumptive versus non-consumptive effects of predation is likely to vary considerably among communities, but could have a significant influence on predator–prey dynamics and the outcome of multiple predator interactions.

Due to the complexity of predation and the limited use of predators in importation biological control, few specific models have been developed either to predict or evaluate specific pest systems other than tetranychid mites and their phytoseiid predators. Exceptions include models to assess the impact of augmentative releases of the pentatomid predator *Podisus maculiventris* on the Mexican bean beetle, *Epilachna varvivestis*, in soybean (O'Neil et al. 1996), the influence of non-crop aphids on conservation biological control by the coccinellid *Coccinella septemepunctata* (Bianchi & Van der Werf 2004), control of the cassava green mite, *Mononychellus tanajoa*, by the introduced predatory mite *Typhlodromalus aripo* (Hanna et al. 2005) and the importance of the coccinellid predator *Cryptolaemus montrouzieri* in the biological control of the vine mealybug, *Planococcus ficus* (Gutierrez et al. 2008).

6.8.3 Herbivore-Weed Models

In the case of herbivore-weed models, added complications include the ability of modular plants to compensate for herbivory (Strauss & Agrawal 1999) and the potential for the impact of herbivory to be manifest as a reversal in the competitive relationships between plant species (Crawley 1989a). One of the first applications to weed biological control was that of Caughley and Lawton (1981), who used the standard L-V model to examine the biological control of prickly pear by the cactus moth *Cactoblastis cactorum*. Subsequently, however, almost all models of weed biological control systems have focused solely on the plant population, uncoupling the herbivore and relegating its effect to a reduction in one or more vital rates in the life cycle of the plant. The framework of the weed models has varied from discrete models for annual weeds (Lonsdale et al. 1995), to either spatially explicit (coupled lattice) simulation models (Rees & Paynter 1997; Buckley et al. 2004) or stage-structured matrix models (Shea & Kelly 1998; McEvoy & Coombs 1999; Davis et al. 2006) for perennial weeds. In each case, the models have been designed more for management recommendations for specific weeds than for mechanistic understanding, and have focused on identification of vulnerabilities in the life cycle, relative contributions of herbivory and plant competition to population growth rate and selection of effective control agents.

An interesting example of a weed biological control model is that used to predict the impact of a defoliating chrysomelid beetle, *Calligrapha pantherina*, on the tropical annual weed *Sida acuta* in northern Australia (Lonsdale et al. 1995). The model was based on the simple discrete time model for density dependent plant population growth of Watkinson (1980):

$$N_{t+1} = \frac{\lambda N_t}{(1+aN_t)^b + m\lambda N_t}$$

where N_{t+1} and N_t are the densities of weeds in generations t+1 and t, λ is the per capita population growth rate (the balance of fecundity and density independent mortality), a and b define the dependence of fecundity on plant density, and m defines the intensity of seedling mortality due to self-thinning. Lonsdale and colleagues (1995) estimated the parameters of the model through field experiments, in both the presence and absence of herbivory, and found that the key impact of defoliation was a reduction in seed production in relation to plant biomass (Figure 6.9a). An additional unexplained loss factor was needed for the model to generate weed densities equivalent to those observed in the field in the absence of the defoliator. A series of 1,000 runs of the model over a period of nine generations predicted a steady state population in the absence of herbivory, and a steady decline by 97% in the presence of a constant level of herbivory (Figure 6.9b). Ten years after the introduction of the chrysomelid beetle into northern Australia, populations of *S. acuta* had indeed declined by 84–99% (Flanagan et al. 2000), providing a good match to the prediction of the model.

The trend of uncoupling herbivore dynamics from that of the plant has recently been reversed in a discrete time model for the annual weed *Echium plantagineum* and the root-crown weevil, *Mogulones larvatus* (Buckley et al. 2005). The model has the form:

$$S_{t+1} = \rho S_t + s_p S_t f(s_p S_t) g(aW_t)$$
$$W_{t+1} = s_p S_t a W_t h(W_t)$$

where S_{t+1} and S_t are the density of seeds in the seed bank in generations t+1 and t, W_{t+1} and W_t are the weevil densities in generations t+1 and t, ρ is the combined probability of seed not germinating and surviving in the seedbank, s_p is the probability of a seed from the seedbank germinating and surviving to become a reproductive plant, $f(s_pS_t)$ is the per capita plant fecundity as a density dependent function of reproductive plant density, $g(aW_t)$ describes the proportional reduction in plant fecundity caused by weevil herbivory, a is the attack rate (eggs/weevil/plant) of the weevil, and $h(W_t)$ is a function representing density-dependent survivorship of the weevil larvae.

Data from a variety of sources were used to parameterize the model, and simulations were used to explore the dynamics of the model. In the absence of herbivory, the plant model was stable and provided a good quantitative description of field densities of the weed before introduction of the weevil. The strength and form of the density-dependent competition among weevil larvae $h(W_t)$ was found

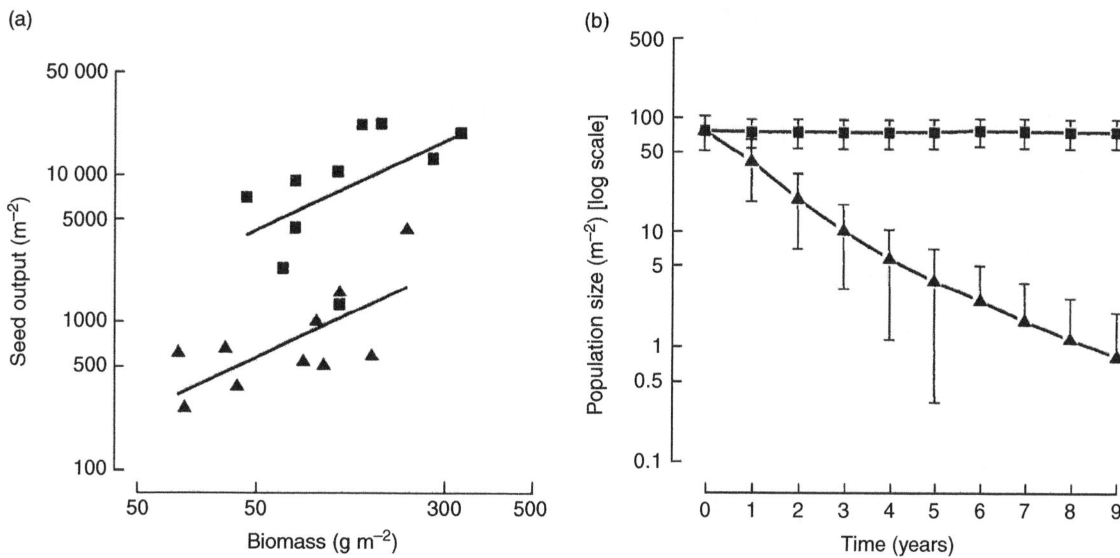

Figure 6.9. The relationship (a) between seed output and plant biomass for an insecticide-treated population (squares) of the weed *Sida acuta* and a population defoliated by the leaf beetle *Calligrapha pantherina* (triangles). The output of the model for *Sida acuta* (b) for control (squares) and defoliated populations (triangles), showing the potential for weed suppression over a period of nine years given the constant rate of reduction in seed output observed in (a). From Lonsdale and colleagues (1995).

to be instrumental for the stability of the coupled interaction. Assuming a conservative estimate of weevil attack rate (0.1 eggs/plant/weevil) simulations predicted a 67% reduction of the seed bank irrespective of the form of weevil competition (Figure 6.10a, c), and an 87% reduction in plant biomass. A greater attack rate (0.5 eggs/plant/weevil) led to greater reductions of the seedbank, and while these reductions were stable for contest competition among weevil larvae (Figure 6.10b), there was transient instability for scramble competition among weevil larvae (Figure 6.10d). The instability induced by a high attack rate combined with scramble competition may help account for the total destruction of plants seen at one field site (Yanco) in 1998 and 2001 where weevils had been established for at least five years (Sheppard et al. 2002).

A number of earlier models for evaluation of weed biological control were reviewed by Barlow (1999), and more recent models have been developed for Scotch broom, *Cytisus scoparius* (Rees & Paynter 1997); gorse, *Ulex europeas* (Rees & Hill 2001); scentless chamomile, *Tripleurospermum perforatum* (Buckley et al. 2001); *Melaleuca quinquenervia* (Pratt et al. 2004); *Mimosa pigra* (Buckley et al. 2004); yellow starthistle, *Centaurea solstitialis* (Gutierrez et al. 2005); and *Buddleia davidii* (Watt et al. 2007).

6.8.4 Pathogen–Host Models

Pathogen–host interactions are fundamentally different from predator–prey interactions in two important respects. First, as microparasites, pathogens typically have a much shorter generation time and a greater rate of replication than that of their hosts. Then second, infection may not necessarily result in the death of the host, and if not,

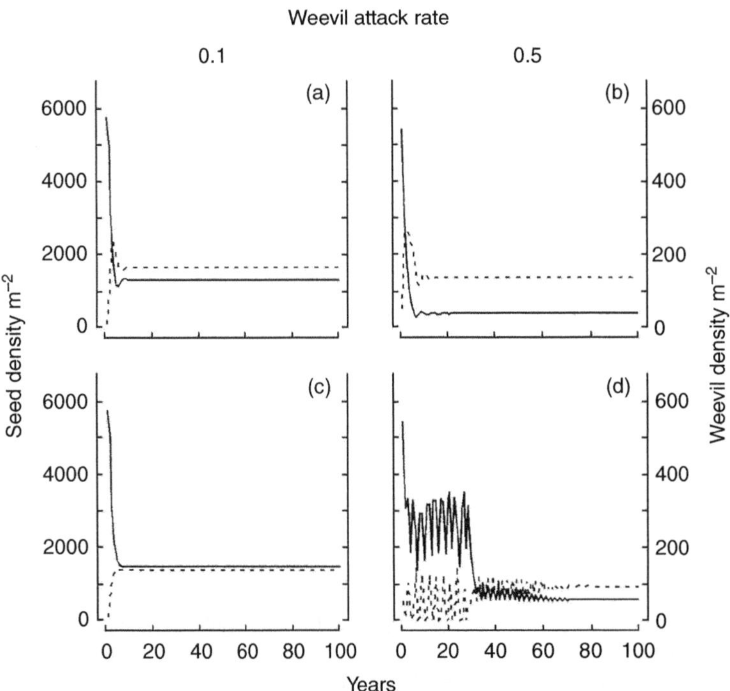

Figure 6.10. The predicted density of the weevil *Mogulones larvatus* (broken line) and its impact on the seed bank density (solid line) of the weed *Echium plantagineum* at two different attack rates, for contest competition among weevil larvae in (a) and (b) and for scramble competition in (c) and (d). From Buckley and colleagues (2005).

this leads to recovery from infection and in some cases resistance to further infection. This not only requires the use of a differential equation model to capture the different time scales of the generation cycles of host and pathogen, but also requires the host population to be partitioned into different classes of individuals that are either Susceptible, Infected or Recovered. Kermack and McKendrick (1927) developed the well-known SIR model that describes the epidemiology of an infectious disease based on the simplifying assumptions that disease transmission and recovery occur at a constant rate, that transmission results from direct contacts between infected and susceptible individuals, and that the host population is of constant size. This epidemiology model was later modified and popularized by Anderson and May (1978, 1981), to produce a more general population model in which the host population has both a birth and natural mortality rate, and the process of infection occurs through contact between susceptible hosts and free-living infectious pathogen particles, rather than between susceptible and infectious hosts. This generated a standard set of four coupled differential equations of the form:

$$dS/dt = aN - bS - \beta SW$$
$$dI/dt = \beta SW - (b + \alpha + \gamma)I$$
$$dR/dt = \gamma I - bR$$
$$dW/dt = \delta I - dW - \beta SW$$

where S, I and R are the abundance of susceptible, infected and recovered hosts with $N = S + I + R$, W is the abundance of free-living infective pathogen particles, α the rate of disease-induced mortality of infected hosts, β is the probability or rate of disease transmission, γ is the rate of recovery of infected hosts, δ is the number of infective pathogen particles produced per infected host, a is the per capita birth rate of the host, b is the per capita natural mortality rate of hosts, and d is the per capita natural rate of loss of infective pathogen particles in the environment. The first of the four equations indicates that in the absence of the pathogen, the host population exhibits exponential growth and that in the presence of the pathogen, disease transmission is governed by 'mass action' or 'proportional mixing'. In the second of the four equations, infected individuals are recruited through disease transmission, but are lost from the population by natural and disease-induced mortality or through recovery. The third of the four equations represents recovered hosts recruited from infected individuals through a constant recovery rate, but lost from the population through natural mortality. The final equation describes the rate of production of infective particles and their loss due to natural mortality or infection of susceptible hosts. This series of coupled equations is directly applicable to vertebrate hosts for which recovery is the norm, leading to lasting immunity and a resistant class of the host population. However, the coupled equations require some simplification for other groups of hosts, such as plants and invertebrates. For some plant hosts, there may be no recovery from infection ($\gamma = 0$), and if so the third of the four coupled equations becomes redundant. In the case of invertebrate hosts, such as insects, infection can lead to recovery ($\gamma > 0$), but recovered individuals become immediately susceptible to re-infection. Thus again the third of the four coupled equations becomes redundant, but the first equation requires the addition of γI representing recruitment of susceptible individuals through recovery from infection. Models for insect hosts frequently also require a stage-structured approach, as insects are often only susceptible to infection for a part of the life cycle (Briggs & Godfray 1995a).

SIR models for pathogen infection of plant hosts, with or without the recovery (R) component, have also been extended to include the action of microbial antagonists and consequently to provide a theoretical framework for the biological control of plant diseases (Gilligan 2002; Jeger et al. 2009; Xu et al. 2010; Cunniffe & Gilligan 2011). These models can be modified to represent the different modes of action of plant antagonists, and those for soil-borne pathogens (Cunniffe & Gilligan 2011) are slightly more complex than those for foliar pathogens (Xu et al. 2010) due to the inclusion of secondary as well as primary rates of pathogen infection and plant growth. Some of the most important parameters affecting the outcome of the interactions in these SIR models are the epidemiology of the pathogen and the rates of colonization of healthy and infected plant tissue by the antagonist. In terms of modes of action of the antagonist, the foliar pathogen model (Xu et al. 2010) suggests that competition and induced resistance may be more effective than antibiosis or mycoparasitism in reducing disease, but that this efficiency can be greatly compromised by a delay in the timing of application of the antagonist relative to the inoculation of the pathogen. In addition, a single antagonist with a mode of action that combines competition with either antibiosis or mycoparasitism provides the most effective suppression of disease.

Although pathogen–host models have been applied to the importation biological control of insect pests (Hochberg & Waage 1991; Briggs & Godfray 1995b; Godfray et al. 1999) and weeds (Smith et al. 1997), a particularly effective example of the use of an SIR model in the context of biological control is that developed to explore the outcome of the introduction of a calicivirus to New Zealand that induces hemorrhagic disease in rabbits, *Oryctolagus cuniculus* (Barlow & Kean 1998; Barlow et al. 2002). Barlow and Kean (1998) firstly used simple epidemic SIR models to estimate the parameters of the disease process from epidemics that had occurred in both Spain and Australia. They then incorporated the epidemic model into a full population model with host reproduction and density dependent mortality, using data from rabbit populations in a semiarid region of New Zealand, to examine the persistence of the virus and its likely impact on suppression of rabbit populations. Local persistence of the disease was found to be dependent on the unknown natural rate of loss of virus particles in the environment, but was considered unlikely given the rapidity and severity of the disease epidemics observed in Australia. When the rapidity of epidemics was relaxed to match those observed in Spain, the model predicted disease persistence and the typical two-yearly cycle of abundance of rabbits observed in that region. Including the recovery and immunity of juvenile rabbits during the first few weeks of life into the model, as seen in Australia, enhanced the potential for disease persistence and predicted yearly rather than biennial epidemics of disease. Under circumstances where the disease could persist locally, the model predicted a 75% reduction in the density of rabbit populations and in the absence of disease persistence, the model suggested that a 60% reduction in rabbit densities could be achieved if the virus was reintroduced every three years.

Following the accidental introduction of the calicivirus into the South Island in 1997, outbreaks of rabbit hemorrhagic disease occurred throughout New Zealand, and rabbit populations in some areas were reduced by as much as 95% (Parkes et al. 2002). The virus has shown local persistence, and it is now known that it is likely able to persist in dead rabbit carcasses in New Zealand for up to three months providing a reservoir for future infections (Henning et al. 2005). Barlow and colleagues (2002) found that the original full model of Barlow and Kean (1998) gave a surprisingly good fit to the data collected over a three-year period in New Zealand (Figure 6.11a), but tended to predict a slightly lower level of suppression than observed with distinct peaks of recruitment to the rabbit population in summer that were not actually observed. The predicted rabbit densities were found to be sensitive to the transmission rate of the disease, and a doubling of the transmission rate provided an improved fit to the observed rabbit densities and a damping of the summer recruitment that was not observed in the New Zealand field sites (Figure 6.11b). The greater rate of disease

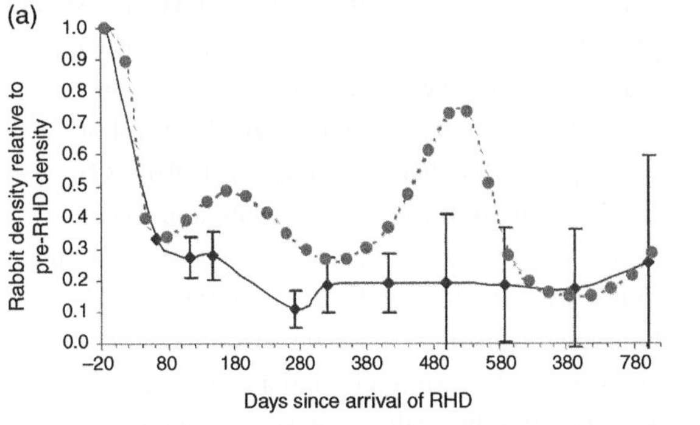

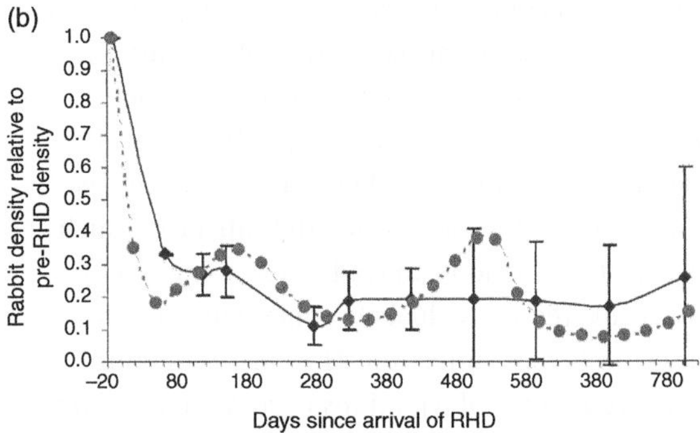

Figure 6.11. The observed relative rabbit densities in New Zealand (diamond and solid line) following introduction of the rabbit calicivirus, which causes rabbit hemorrhagic disease (RHD), in comparison with those predicted from the SIR model (circles and dotted line) of Barlow and colleagues (2002). The model in (a) corresponds to the original model of Barlow and Kean (1998), and that in (b) is modified to include a transmission rate twice that of the original model.

transmission than expected suggests the involvement of unknown modes of transmission such as could occur via direct rabbit-to-rabbit encounters (Barlow 1998) or avirulent chronic infections in immune rabbits (White et al. 2002). This study represents one of the very few attempts to use models to predict the potential impact of biological control before introduction of a natural enemy, and to test the validity of the model using observed field data following the introduction. The model proved correct in predicting annual epidemics of disease, and following adjustment to a higher rate of disease transmission than originally envisaged, also captured the very rapid suppression of rabbit densities by 75% in the field sites where observations were made. Whether the longer-term dynamics of rabbits at these field sites in New Zealand will continue to be matched by this simple model remains to be seen, but this study again shows the value of models in biological control.

6.9 Conclusions

As we have seen throughout this chapter, models can play a valuable role in helping us to understand the population dynamics of biological control and to provide a framework for thinking about the mechanisms that form the basis of biological control. From a historical perspective that natural enemies must drive local stability as well as suppressing pest population abundance, we can now appreciate that suppression alone is at the heart of successful biological control, that equilibrium is a state that may seldom be attained in real systems, and that asynchronous dynamics among metapopulation patches provides a sufficient basis for the long-term persistence of consumer-resource interactions. Nonetheless, while general models can be invaluable in helping us to think about the population-level consequences of the different formulations of simple consumer-resource relationships, we must not forget that the idiosyncrasies of real systems can also tailor the dynamics of specific cases in biological control. In the application of basic modeling frameworks to specific cases, however, a balance must be struck between simplicity for clarity of understanding and complexity for sufficiency of realism. For example, application to real case studies may require models to be extended to a multitrophic perspective to capture trophic influences both from below and above those directly coupled in the interaction of interest, and may need to include the influence of temperature, rainfall or other aspects of seasonality as external drivers of the players at all trophic levels. On the other hand, models that incorporate too much complexity rapidly become opaque and intractable, and thus such influences must be captured as simply as possible to be of value in furthering our understanding of biological control and our ability to predict the likely outcome of using natural enemies for the suppression of pests, weeds and plant diseases.

It must also be stated that the value of population models in helping us to understand the population dynamics of biological control systems is only advanced when the predictions of such models can be tested through post-release monitoring or rigorous experimental manipulation. The post-release monitoring of an introduced parasitoid of the alfalfa weevil, *Sitona discoideus* (Kean & Barlow 2000b), an introduced leaf beetle of the weed *Sida acuta* (Flanagan et al. 2000) and the accidentally introduced calicivirus of rabbits *Oryctolagus cuniculus* (Barlow et al. 2002) provide excellent examples in this regard, as does the very effective experimental manipulation of the California red scale system (Murdoch et al. 2005). Without such testing, population models cannot inform the practice of future biological control programs, and conversely biological control introductions that are not properly monitored cannot inform our understanding of ecological theory. One of the greatest barriers to improved cross-fertilization of biological control and ecological theory is the availability of sufficient funding to complete a biological control program from the point of initiation and prediction to completion of monitoring and experimental verification. Funding agencies support biological control programs through to the point of natural enemy introduction, but fail to appreciate the importance of continued support for monitoring and evaluation. With renewed interest in the environment and the critical importance of sustaining ecosystem services in the face of global change, we can hope for improved funding in the future, and biological control practitioners would do well to appreciate the value of population models as a tool for increasing scientific rigor and the predictability of future success.

7 Biological Control and Evolution

Biological control interactions take place over time scales of years and decades, spans traditionally considered within 'ecological', but not 'evolutionary' time frames. The view has emerged, however, that evolutionary processes can take place over these time frames, and that this is especially likely under non-equilibrium circumstances such as may occur during invasions and under strong selection pressure (Thompson 1998; Reznick & Galambor 2001; Lee 2002; Lambrinos 2004). These are precisely the conditions within which biological control interactions typically occur, so it is not unreasonable to ask whether evolutionary change plays a role in biological control interactions. Furthermore, evolutionary processes at these time scales have the potential to interact with ecological factors such as demographic parameters and density-dependent competition or attack rates among species. These interactions, sometimes labeled 'community genetics' effects (Antonovics 1992; Neuhauser et al. 2003; Whitham et al. 2003), are particularly likely to be relevant in cases of biological control, in which multiple species interact under non-equilibrium conditions while exerting strong selection on one another.

Explicit investigations of evolutionary dynamics in biological control interactions are not common. Despite this, a number of reviews have valiantly gleaned the evolutionary insights available from the biological control and related literatures (Simmonds 1963; Wilson 1965; Force 1967; Mackauer 1976; Messenger et al. 1976b; Myers & Sabbath 1981; Roush 1990; Hopper et al. 1993; Holt & Hochberg 1997; Roderick & Navajas 2003; Ehler et al. 2004; Hufbauer & Roderick 2005), and a Special Feature in the journal *Evolutionary Applications* in 2012 has added considerable insight (Roderick et al. 2012). And of course, some cases of biological control *have* been analyzed from an evolutionary standpoint in some detail, and much can be learned from these. We begin this chapter by reviewing three case studies that we find particularly instructive and that also represent a wide breadth of biological control interactions. These cases are (i) control of rabbits using myxoma viruses; (ii) control of rush skeleton weed using a *Puccinia* rust fungus; and (iii) control of the pea aphid using the parasitoid *Aphidius ervi*. After covering these cases in some detail, we go on to discuss known and suspected evolutionary dynamics during the process of biological control in general, focusing primarily on importation biological control.

7.1 Case Studies

The case studies described in this chapter all contribute to our understanding of the importance of evolutionary dynamics during the process of importation biological control. Issues addressed include the evolution of resistance to biological control agents, the effects of restricted genetic variability on the efficacy of biological control agents and the potential for local adaptation by biological control agents once established.

7.1.1 Case Study 1: Rabbits and Myxoma Virus

The European rabbit, *Oryctolagus cuniculus*, was first introduced into Australia in 1788 as a game animal. By the 1860s, it was starting to assume pest status in various parts of Australia, with a member of the Victorian parliament proclaiming: 'the rabbit nuisance in this colony promises to be as great as that of the locusts in the land of Egypt' (Fenner &

Fantini 1999). These words proved prophetic, and the explosive spread of rabbits in the 1870s ushered in a long period during which rabbits were reviled as severe rangeland and environmental pests. This attitude did not begin to subside until the 1950s, when control by myxoma virus was implemented. In the meantime, various control tactics were attempted, including the passage of laws mandating the destruction of rabbits by landowners, the payment of government 'scalp bonuses' for rabbit skins and the construction of some of the longest fences in the world. Before the use of myxoma, a number of other biological control solutions were attempted, including introductions of mongooses, weasels, stoats and cats. Also, when it was discovered that large predatory monitor lizards were capable of killing and consuming rabbits, these predators were afforded legal protection – an attempt at conservation biological control of sorts. None of these biological control measures had an appreciable impact on rabbit spread or population growth. Fenner and Ratcliffe (1965) hypothesized that these measures were doomed to failure because the predators were all generalists that tended to attack immature rather than mature rabbits.

A disease outbreak in a laboratory colony of European rabbits in Uruguay led to the discovery of the myxoma virus by Giuseppe Sanarelli in 1896. Myxoma was actually one of the first viruses ever discovered, and it has been placed in the family poxviridae, which includes smallpox. Myxoma is transmitted by mosquitoes, fleas and other biting insects, and its major host in South America is the tropical forest rabbit (or tapeti), *Sylvilagus brasiliensis* (Fenner & Fantini 1999). The virus produces small, benign skin lesions (fibroma) in its natural host and few disease symptoms. In the European rabbit, however, the disease that early researchers observed was severe and lethal. Fenner and Fantini (1999) described the progression of myxomatosis as follows:

> *A skin lesion appeared 4–5 days after the bite of an infective mosquito and enlarged to become a hard, hemispherical tumour about 3 cm in diameter by the ninth or tenth day. The eyelids became thickened on the sixth day and the eyes were usually completely closed by the ninth day and there was a semipurulent ocular discharge. Secondary lesions were widely distributed over the body from the sixth or seventh day and there was an oedematous swelling of the head, base of the ears and genitalia. Death was almost invariable, 8–15 days after infection.*

Brazilian scientist H. B. R. Aragão proposed that myxoma could be used to control European rabbits in Australia as early as 1918, but the first experimental releases were not made until 1950. The first inoculated rabbits were released in the Murray-Darling river basins in New South Wales and Victoria, where it was hoped that native mosquitoes would aid establishment and spread of the virus. The initial trials appeared to have failed, and by the end of 1950, enthusiasm for myxoma releases was low (Fenner & Fantini 1999). However, the disease had apparently not completely died out at one of the experimental sites, where large numbers of rabbits were reported to be sick and dead in December 1950. This was the beginning of one of the most spectacular epizootics ever observed in the history of infectious diseases. By May 1951, myxomatosis had spread throughout the river valleys of southeastern Australia, an area exceeding 1 million square kilometers. The spread was accompanied by mortality rates of approximately 90% in many areas and over the next three years, myxomatosis had spread to encompass the entire range of rabbits in Australia. This was accompanied by a complete collapse (but not extinction) of the rabbit population. An example of the population trends between 1951 and 1953 seen at Lake Urana, New South Wales, a site chosen to study epidemiology of myxomatosis, is shown in Figure 7.1.

It was not long after these spectacular population crashes that trends began to be seen that have made this project an important case study for host-parasite evolutionary dynamics. In early

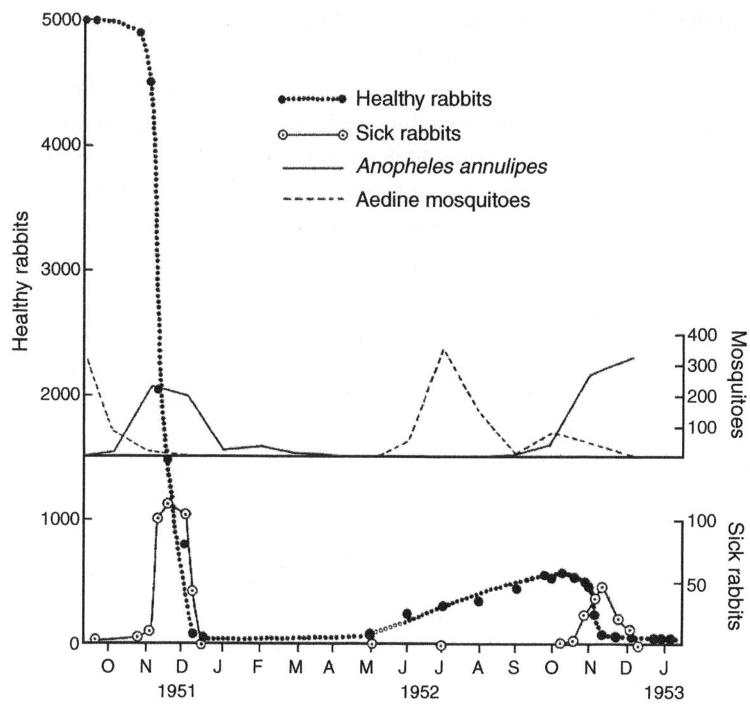

Figure 7.1. Population levels of healthy and sick rabbits, as well as *Anopheles annulipes* and *Aedes* spp. mosquitoes in one of two study areas at Lake Urano in Australia. The outbreak was initiated in October 1951 by capturing hundreds of rabbits, inoculating them with myxoma virus and releasing then on site. From Myers and colleagues (1954).

1952, a rabbit recovering from myxomatosis was found and viruses isolated from this rabbit were experimentally inoculated into other rabbits. None of these rabbits died, and it was concluded that this and other strains of myxoma were exhibiting decreased virulence (Fenner 1953). At the Lake Urana site, the mortality rate of infected rabbits was 99.8% during the generation of rabbits following inoculation, and this rate dropped to 90% by the second generation.

7.1.1.2 Evolution of Intermediate Virulence in Myxoma Virus

For myxomatosis, the survival time of inoculated rabbits was used as a measure of virulence, and 'virulence grades' were established in Australian studies. Characteristics of these virulence grades, in terms of mortality rate and mean survival times for rabbits that succumb, are as follows (from Fenner & Fantini 1999):

Virulence Grade:	Extreme	Very High	Moderate	Low	Very Low
Mortality rate	> 0.99	0.95–0.99	0.70–0.95	0.50–0.70	< 0.50
Mean days survival	< 13	13–16	17–28	29–50	–

Between 1951 and 1993, hundreds of strains of myxoma were tested for virulence by isolating viral material from wild infected rabbits and testing each isolate on rabbits with no history of the disease in the laboratory. Multiple infections (i.e., rabbits infected with more than one virulence strain) were rarely encountered in the field (Williams et al. 1973; Dwyer et al. 1990). Results of these studies are shown in Figure 7.2. A striking outcome is the demise of the highly virulent strains through the 1980s and the increasing dominance over this time period of a strain exhibiting intermediate virulence. Equally impressive is the reemergence of the highly virulent strain in the 1990s. Interested readers are encouraged to consult Kerr and colleagues (2012, 2013) for insights into the genetic mechanisms accompanying these strain changes.

How can we explain the dominance of viral strains with intermediate levels of virulence through the 1980s? The generally accepted reason is that rabbits infected with strains of intermediate virulence lived longer, and were thus able to transmit virus to other rabbits (via insect vectors) for a longer period of time. This was especially important in the winter, when mosquito densities are relatively low and high-virulence strains are at risk of dying out along with their hosts (Fenner & Fantini 1999). The low-virulence strains were not favored because they did not produce high enough concentrations of virus to be transmitted by mosquitoes with high efficiency. More generally, the dominance of the strains with intermediate virulence is consistent with the outcome of the importation trade-off between virulence and transmissibility in host-parasite systems (Bull 1994). High virulence reduces transmissibility by killing the host too quickly, low virulence reduces transmissibility by allowing the host to recover, and an equilibrium is reached at an intermediate level of virulence. Models incorporating parameters from the rabbit-myxoma system into this trade-off indeed predicted that virus strains with intermediate levels of virulence would prevail (Levin & Pimentel 1981; Anderson & May 1982; Massad 1987; Dwyer et al. 1990).

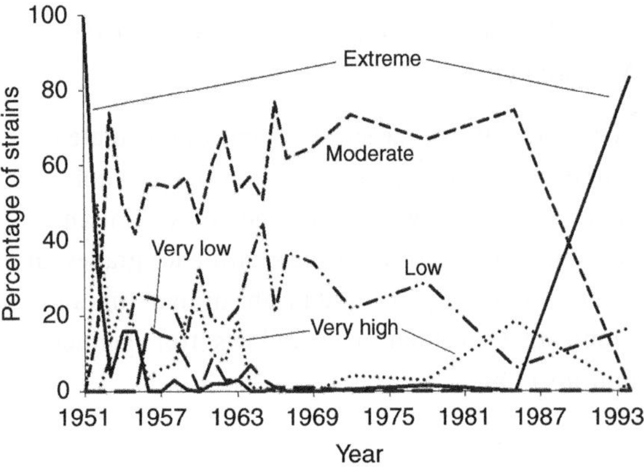

Figure 7.2. Relative percentages of five myxoma virulence grades, ranging from 'extreme' to 'very low', found in Victoria and New South Wales, Australia, between 1951 and 1994. Characteristics of the virulence grades are described in the text. Data taken from Fenner and Fantini (1999).

Density-dependent processes can make the trade-off between virulence and transmissibility even more stark. In horizontally transmitted diseases like myxoma, transmission rates tend to drop as host density decreases simply because the rate of encounter among hosts declines (Bull 1994). Thus, the precipitous drop in rabbit densities prior to the 1980s would have favored the strains with lower virulence in order to keep rabbits alive long enough to encounter other rabbits, which were becoming increasingly scarce.

But what of the sharp jump in virulence between the 1980s and 1990s (Figure 7.2)? It appears that the 'equilibrium' at intermediate virulence was transient. To address this, we need to consider the evolution of resistance on the part of the rabbits.

7.1.1.3 Evolution of Resistance in Rabbits

While virulence of myxoma was dropping, evidence was mounting that rabbits were developing resistance to the myxoma virus as well. Studies from the 1950s indicated that rabbits from populations

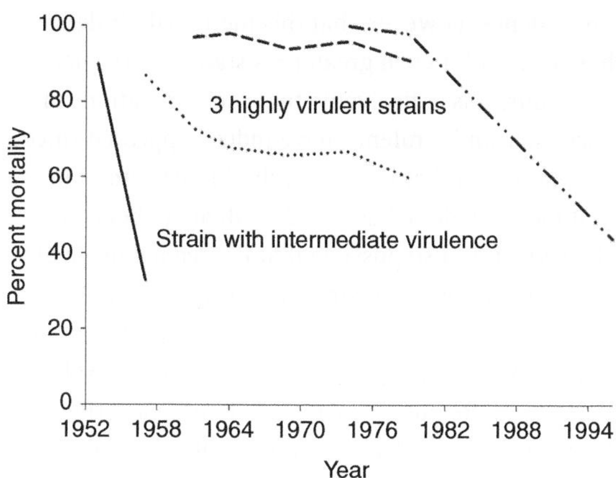

Figure 7.3. Mortality rates for rabbits field-collected in areas with annual (or at least frequent) myxomatosis epizootics between 1952 and 1996 and inoculated with virus strains that exhibited either intermediate or high virulence in control rabbits with no history of myxomatosis. Data from Fenner and Fantini (1999).

that had been exposed to numerous epizootics of myxoma were better able to survive inoculation with a standard intermediate-virulence strain of myxoma (Marshall & Fenner 1958; Fenner & Fantini 1999). This trend was also seen with respect to the highly virulent strains through the 1990s (Figure 7.3). Selection studies showed that it was possible to breed for resistance to myxoma, with selected rabbit strains exhibiting greatly reduced mortality rates. By using survival time as an index of resistance, Sobey (1969) arrived at a heritability estimate of 35–40%. However, subsequent work by Sobey and Conelly (1986) suggests that some of this apparent heritability was in fact non-genetic, and due in part to unknown factors passed within the semen of recovered male rabbits to susceptible females. In either case, it would appear that the heightened resistance is what selected for the highly virulent viral strains detected in the 1990s. Given resistance on the part of rabbits, high virulence was needed to produce the same infection and transmission rates that it had previously only taken intermediate virulence to achieve (Fenner & Fantini 1999). Later laboratory studies also confirmed that a virulent strain of myxoma produced much more severe symptoms in laboratory rabbits with no history of myxomatosis than in field-collected rabbits, and that an attenuated strain had essentially no effect on field rabbits and produced only mild symptoms in laboratory-bred rabbits (Best & Kerr 2000; Best et al. 2000).

7.1.1.4 Coevolution

This brings us quite naturally to the topic of coevolution. Resistance on the part of the rabbits should select for higher virulence on the part of the virus, and so on in a reciprocal, ever-escalating, coevolutionary 'arms race' (Dawkins & Krebs 1979; Dwyer et al. 1990). Indeed, the myxoma-rabbit case would seem to be a likely candidate for Van Valen's (1973) 'red queen hypothesis', which is taken from Lewis Carroll's *Alice Through the Looking Glass.* In this children's book, the red queen remarks to Alice that 'it takes all the running you can do to stay in place.' And just as the red queen had to keep running just to stay in place, it would appear that the myxoma virus had to increase its virulence just to keep up with the evolving resistance of the rabbits.

The battleground for such an interaction appears to be the rabbit's immune system, as the myxoma virus is able to subvert immune function in susceptible rabbits while using immune cells to replicate and spread throughout the host. Myxoma replicates within dendritic cells that produce major histocompatability complex II proteins, and within T lymphocytes. The virus then spreads throughout the body within these cells and concentrates in the lymph nodes, where it replicates to high levels and causes the symptoms that eventually lead to death in susceptible rabbits. High viral replication in the lymph nodes also induces the loss of lymphocytes from susceptible rabbits, and this further limits their capability to mount an immune response (Best et al. 2000). The genome of the myxoma virus has been sequenced and some of the virulence genes encoding these capabilities have been identified (Kerr & McFadden 2002; Zuniga 2002; Stanford et al. 2007). Knockout studies have also shown that disruption of a single one of these genes is sufficient to interfere with pathology (Macen et al. 1996; Mossman et al. 1996). However, not much research has been done to identify differences in virulence genes between viruses exhibiting high and intermediate or low virulence. In a genetic study of 37 field isolates of myxoma from Australia, Saint and colleagues (2001) found that one of the attenuated strains harbored five mutations in the region of the viral genome that contains a high density of virulence genes, and Kerr and colleagues (2012) discovered additional mutations in a broader study. Thus, the molecular basis of differences in virulence is being characterized.

The mechanisms of resistance on the part of rabbits are less well studied than the mechanisms of virulence on the part of the virus, although Abrantes and colleagues (2011) have suggested that genes encoding for cytokine receptors are likely candidates. While virus concentrations in skin lesions at the site of inoculation are similar in resistant and susceptible rabbits (Best & Kerr 2000), differences occur in the lymph nodes, where viral loads are one to two orders of magnitude higher in susceptible rabbits. Best and colleagues (2000) speculated that resistant rabbits can mount an appropriate antiviral response in lymph nodes and other parts of the body that are not the immediate inoculation site. Does the myxoma-rabbit system constitute an example of a gene-for-gene reciprocal coevolution? Perhaps, but the actual genes that are involved on the part of the rabbit have not been identified.

Coevolution or density-dependent cycling? Pairwise, reciprocal coevolution between the virus and the rabbit should in principle result in ever-escalating levels of virulence and resistance (Dwyer et al. 1990). Thus far, the pattern that most fits this general model is the increase in viral virulence in the 1990s following decades of resistance evolution. There is no evidence, however, that this increased virulence has selected for even greater resistance on the part of rabbits. It is conceivable that such gradations of resistance and virulence have indeed appeared since the 1980s and that the relatively limited sampling that has been done has failed to identify them all. However, it is also possible that the reemergence of a highly virulent viral strain in the 1990s is not a response to increased resistance on the part of rabbits but to relatively simple density-dependent effects that are analogous to the effects described earlier for the evolution of viral strains with intermediate virulence.

How would such a density-dependent effect work? For a horizontally transmitted disease such as myxoma, transmission rates tend to increase with host density since proximity of hosts facilitates transmission (Bull 1994). Thus, under high host densities, the costs of high virulence (decreased transmission by killing the host too fast) are diminished. Highly virulent strains can therefore be favored at high host densities when opportunities for horizontal transmission are high. As host populations crash, high virulence becomes a liability because of the decreased transmission opportunity (Bull 1994). Once strains of intermediate virulence predominate, and rabbits evolve resistance, however, rabbit densities are in a position to recover. And indeed, rabbit populations in Australia made a steady comeback over the 40 years following the initial crash in the early 1950s (Fenner & Fantini 1999). These higher host densities can then allow the reemergence of highly virulent strains because of the greater transmission opportunity that comes with high host densities.

In principle, it would seem that these processes could lead to indefinite cycling of myxoma virulence and rabbit densities in the absence of escalating resistance evolution. The finding of resistance in rabbits, however, has two implications – first, it could lead to a stronger density-dependent effect since rabbits could build to higher densities. Second, it could select directly for increased virulence, as others have suggested (Dwyer et al. 1990; Fenner & Fantini 1999). Thus, the coevolutionary and density-dependence hypotheses explaining increased virulence in the 1990s are not mutually exclusive. The density-dependence hypothesis, however, predicts a future drop in virulence of myxoma, while the coevolution hypothesis predicts ever-increasing virulence.

In summary, myxoma virus released against European rabbits in Australia was wildly successful in the early years, but the initial success was tempered by the rise to prominence of viral strains of intermediate virulence. Intermediate virulence likely evolved due to a trade-off between transmissibility and virulence, although interactions between host demographics and optimal virulence levels also may have played a role. Resistance to myxoma was documented in rabbits from the beginning, and this likely contributed to the development of a second round of virulent strains in the 1990s. The hypothesis of genetic coevolution seems likely based on evolutionary and demographic patterns, but specific resistance genes have not been identified and alternative hypotheses (such as an important role of demographics) cannot be excluded.

7.1.2 Case Study 2: Skeleton Weed and *Puccinia* Rust

Our next case study is centered on the topic of founder effects and genetic bottlenecks during invasion. This case illustrates how restricted genetic variability can make introduced natural enemies less effective biological control agents. The case involves rush skeleton weed, *Chondrilla juncea*, which is native to Eurasia and an invasive weed in Australia and North America, and the rust fungus *Puccinia chondrillina*, a pathogen of *C. juncea* that has been used as an importation biological control agent. In Australia, skeleton weed is primarily of concern in wheat fields, where it decreases grain yield and interferes with harvesting practices. A number of arthropod herbivores, including a midge, a mite and a moth, were released to control skeleton weed in Australia beginning in the 1970s, but the most effective agent against the dominant and widespread 'narrow-leaf' form of the weed proved to be the rust *P. chondrillina* (Burdon et al. 1981). Like many other rust fungi, *P. chondrillina* is a highly specific plant pathogen that attacks leaf tissue. Severe infestations can lead to significant defoliation and because spores are wind-borne, dispersal can be quite rapid under epidemic conditions (Evans & Gomez 2004).

Puccinia chondrillina was introduced to wheat fields in southeastern Australia from Italy in 1971, and control of skeleton weed was spectacular over a wide area within just one year and the introduction was hailed as a great success (Cullen et al. 1973). This was the first-ever introduction of a plant pathogen as an importation biological control agent, and remains one of the few cases of importation biological control by any microbial agent attempted for a weed of an annual crop.

7.1.2.1 Self-Defeating Biological Control?

However, three distinct forms of the weed are known in Australia, two of which exhibit resistance to the isolate of *P. chondrillina* that was released. These resistant forms are the intermediate-leaf and broad-leaf forms of the weed (Figure 7.4), and they replaced the susceptible form upon release of the rust fungus, with little or no net long-term suppression of the weed (Burdon & Thrall 2004). While the narrow- and broad-leaf forms of skeleton weed were geographically restricted prior to the control of the narrow-leaf form, the release from competition with the narrow-leaf form that was achieved through biological control allowed the other forms to spread throughout southeastern Australia (Burdon et al. 1981, 1984). An additional strain of the rust specific to the intermediate-leaf form of skeleton weed was introduced in 1980, and indications are that the broad-leaf form increased in response to control of the intermediate-leaf form (Thrall & Burdon 2004). This general process, in which a biological control

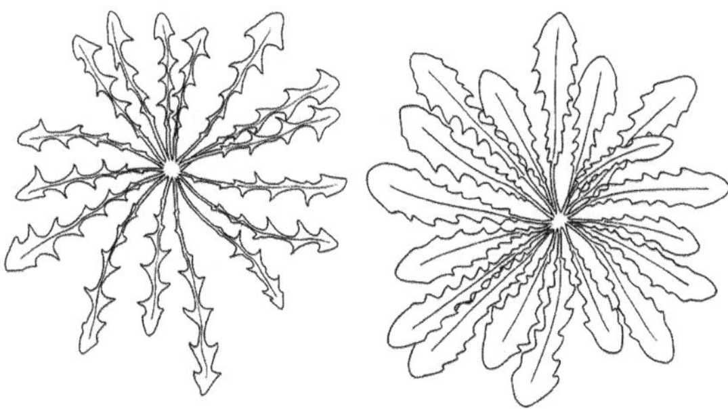

Figure 7.4. Two forms of skeleton weed present in Australia: narrow-leaf and broad-leaf (from left to right). Only the narrow-leaf form is susceptible to the genotype of *Puccinia chondrillina* that was originally released in Australia. Drawing by Mary Marek-Spartz.

agent suppresses susceptible genotypes of a pest, only to allow resistant genotypes to dominate, has been called 'self-defeating biological control' by Garcia-Rossi and colleagues (2003), who warned of such an effect in the control of invasive clones of cord grass in western US estuaries by highly specialized herbivorous insects.

Later work on the skeleton weed-rust system has dissected the factors responsible for the failure of the rust to control all of the forms of skeleton weed in Australia. Collections of *C. juncea* in Turkey (within the native range) uncovered eight distinct resistance phenotypes against *P. chondrillina*, and seven distinct 'pathotypes' of the rust, each one of which could attack a different combination of weed resistance phenotypes (Espiau et al. 1998). The original release in Australia was represented by only a single pathotype – one that was capable only of attacking the narrow-leaf variety of skeleton weed (Cullen et al. 1973; Burdon et al. 1981, 1984).

At least two strains of *P. chondrillina* have also been introduced into the United States to control skeleton weed there (Emge et al. 1981; Hasan et al. 1995; Piper et al. 2004). The particular strains introduced were arrived at using an innovative approach developed by Hasan and colleagues (1995). They planted genotypes of skeleton weed that had invaded the western United States in areas of Turkey thought to be the center of origin of *C. juncea* in an attempt to capture pathotypes matched to the skeleton weed forms that had invaded the United States. While two suitable strains were identified using this technique, some skeleton weed genotypes escaped control in this case as well. Early- and late-flowering strains in parts of Oregon, Washington and Idaho are not impacted by the rust strains that have been released, but skeleton weed forms in California appear to be better-controlled (Piper et al. 2004). The difference between the Australian and North American cases is that in Australia, the resistant weed strains replaced the susceptible strain, nullifying the effects of biological control, while in North America, control by the rust occurs, but is patchy, occurring only where susceptible strains of skeleton weed are established.

7.1.2.2 The Importance of Genotype Matching

The high level of intraspecific host specificity by *P. chondrillina* is consistent with a gene-for-gene coevolutionary interaction between resistance and pathogenicity (Espiau et al. 1998), similar to some other plant-rust systems (Thompson & Burdon 1992; Ellison et al. 2004). Matching weed and fungal genotypes for maximum suitability is very important in these cases (Thrall & Burdon 2004; Morin et al. 2006). The control of one genotype but not the other(s) can lead to increases in the previously rare weed form. This kind of displacement can reflect frequency-dependent selection in which virulence is highest against the most common plant genotypes (Chaboudez & Burdon 1995; Clay & Kover 1996). While the advantage of rare plant genotypes should erode as they become more common, this has apparently not happened yet in the skeleton weed-*Puccinia* system, where the broad-leaf form is still an important weed in Australia, and early- and late-flowering varieties have not been controlled in the western United States at the time of this writing.

7.1.3 Case Study 3: Pea Aphids and the Parasitoid *Aphidius ervi*

The pea aphid, *Acyrthosiphon pisum*, is a pest of alfalfa, peas, red clover and other herbaceous legumes. Like many other aphid species, it is parthenogenetic, and produces more than 8–10 clonal generations of females per summer in temperate areas. The pea aphid was accidentally introduced from Europe to North America during the 19th century, where it became a major pest of alfalfa. It has been subjected to a number of biological control releases, including at least six parasitoid species, all in the braconid genus *Aphidius*. The species that emerged as the most important is *A. ervi*, which was introduced into the United States from France in 1959, but may have also been present at low densities in North America before this time (Stary 1974; Angalet & Fuester 1977; Marsh 1977). *Aphidius ervi* has been considered a highly successful biological control agent of the pea aphid, and has spread rapidly throughout much of North America, apparently displacing previously released *Aphidius* species during its establishment (Gonzalez et al. 1995).

Of all host-parasitoid combinations that form an importation biological control interaction, this system has probably received the most attention from an explicit evolutionary standpoint. This work has included considerations of the genetic basis of host range (Henry et al. 2008, 2010), the importance of endosymbionts in mediating host-parasitoid interactions (e.g., Oliver et al. 2003) and post-release evolution, which we focus on in this section. The outcome of these studies has provided insights into genetic and evolutionary changes that can take place in biological control interactions while also raising important questions that are not yet answered. In describing this body of work, it is useful to start with a rather enigmatic outcome of a local adaptation study performed by Hufbauer (2002b). She found that *A. ervi*

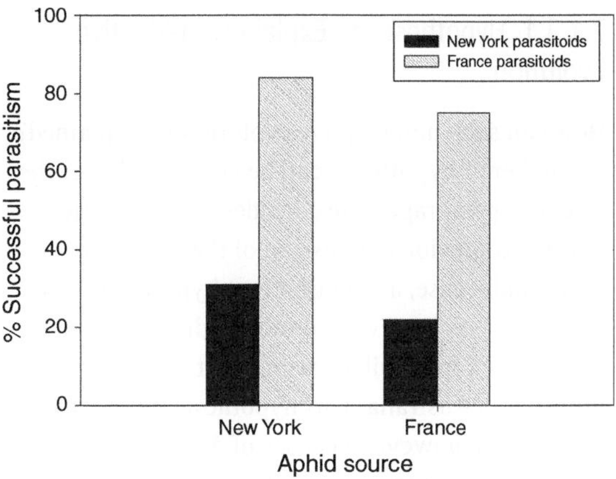

Figure 7.5. Percent successful parasitism of pea aphids from New York or France exposed to *Aphidius ervi* females from New York or France. Data from Hufbauer (2002b).

collected in New York in 1997 (i.e., presumably the offspring of parasitoids introduced from France 34 years earlier) performed *more poorly* on pea aphids grown on alfalfa in New York than did their French counterparts. More specifically, the New York parasitoids had far lower virulence (here defined as the ability to overcome host defenses) on New York aphids than the French parasitoids did on either New York or French aphids. The differences were striking, as is shown in Figure 7.5. This was unexpected under a hypothesis that the parasitoids would exhibit local adaptation to pea aphids in their introduced environment. Clearly they had not. The studies were done as reciprocal transplant experiments so that the influence of environmental factors can be excluded. Assuming that the French strain performed similarly in Hufbauer's experiments as it would have in 1959 (when *A. ervi* was moved from France to the United States), these experiments suggest genetic change of the introduced strain. Since performance of the introduced strain was inferior to that of the French strain, Hufbauer called the genetic change 'nonadaptive evolution'.

7.1.3.1 Hypotheses to Explain Nonadaptive Evolution

How can such nonadaptive evolution be explained? A number of hypotheses can be posed, and over the next few paragraphs, we consider some of these. Given our previous discussion of the myxoma virus-rabbit case, a tempting first hypothesis may be that *A. ervi* has evolved intermediate virulence in New York for similar reasons that the myxoma virus did in Australia. This hypothesis can be rejected out of hand, however, because of differences in the biology of parasitoids and horizontally transmitted pathogens such as myxoma. While intermediate virulence can evolve in the myxoma-rabbit system because of a trade-off between virulence and transmission, no such trade-off operates in parasitoids. Low virulence in the sense that Hufbauer used it (limited ability to overcome host resistance) is synonymous with low fitness for the parasitoid since allowing a host to survive parasitism almost invariably results in death of the developing parasitoid.

A related hypothesis, that New York aphids evolved resistance to the introduced parasitoids, seems more plausible. Indeed, earlier research on this same system had shown that pea aphids from New York and other locations harbor substantial variability for resistance to *A. ervi*, thus making the evolution of resistance not only possible, but expected (Henter & Via 1995; Ferrari et al. 2001; Li et al. 2002). However, if low virulence is explained by the evolution of increased resistance in New York aphids since the introduction of *A. ervi*, how is it that the French aphids also show high levels of resistance to New York parasitoids (Figure 7.5)? This is one of the reasons that Hufbauer's reciprocal transplant design is valuable – it allows us to exclude this hypothesis. Another caveat is that the resistance may not be genetic after all, but instead conferred by mutualistic bacterial endosymbionts and associated phages (Oliver et al. 2003, 2005, 2006; Ferrari et al. 2004). Different strains of these symbionts confer different levels of resistance to *A. ervi* in pea aphids (Oliver et al. 2005), making it conceivable that aphids from one location may be more resistant to *A. ervi* than from another. In Hufbauer's (2002b) observation, however, both New York and French aphids were more resistant to *A. ervi* from New York than from France. For differences in symbiont effectiveness to be responsible for this pattern, the different populations of aphids would need to harbor different symbiont strains *and* parasitoids would need to vary both between population and symbiont strain in their susceptibility to symbiont-conferred resistance. While variability in susceptibility to defensive symbionts is known in other aphid parasitoid systems (e.g., Rouchet & Vorburger 2012), it seems unlikely that just such a combination of symbiont and parasitoid strains is responsible for the pattern seen in Figure 7.5. In fairness though, this hypothesis has not been directly tested.

Perhaps the most appealing hypothesis is that the New York *A. ervi* were subject to a strong genetic bottleneck during introduction, and that this led to decreased virulence. As we explain in more detail later, the process of importation biological control provides ample opportunities for genetic bottlenecks and founder effects. Henter (1995) had previously demonstrated variance in virulence for *A. ervi* attacking pea aphid in New York, so that it is indeed conceivable that a restriction of genetic variability could have led to the introduction of a subset of the French population with low virulence. In addition, the variation in virulence of parasitoids collected from various fields in New York was much lower than that of parasitoids collected from various fields in France (Hufbauer 2002b). This suggests genetic variance for virulence is lower in New York than in it is in France. If indeed *A. ervi* strains vary in the susceptibility to symbiont-induced resistance, one could also hypothesize that the imported parasitoids constituted a highly susceptible subset of parasitoids.

As plausible as the scenario for a genetic bottleneck seems, however, the evidence for it is not very strong. First, approximately 1,000 *A. ervi* were originally collected in France in 1959 (most directly from the field), and this population grew rapidly in quarantine before being released less than three months (at most six parasitoid generations) after collection (Hufbauer et al. 2004). This initial release was also followed by other releases in the eastern United States in the late 1960s (Angalet & Fuester 1977). These introduction characteristics should, in principle, result in only minimal loss of alleles or heterozygosity in the founding population (Nei et al. 1975), and, indeed, Unruh and colleagues (1983) found no appreciable loss of allozyme heterozygosity in even very small populations of *A. ervi* over six generations. Also, although explicit genetic comparisons between North American and French populations of *A. ervi* did reveal a slightly more restricted gene pool in the North American than the French populations, the bottleneck was rather mild (Hufbauer et al. 2004). It is doubtful that this mild bottleneck could be responsible for the strong reduction in virulence of the North American *A. ervi*.

Another possibility is laboratory selection. If the parasitoids became adapted to aphids in the laboratory that had low levels of resistance, then virulence may have deteriorated during laboratory adaptation. As noted earlier, however, the parasitoids were kept in the laboratory for at most three months between being collected in France and released in North America in 1959. It is during this time that laboratory adaptation would have taken place. This explanation seemed unlikely to Hufbauer (2002b) given the apparent lack of evolution of increased virulence in the years since introduction and the substantial genetic variability in virulence for *A. ervi* attacking pea aphid (Henter 1995).

Three other hypotheses were posed by Hufbauer (2002b) as well, but little or no data are available with which to evaluate them. The first is a trade-off between virulence and some other trait important for colonization and/or survival in New York (in which case the evolution would not be nonadaptive), the second is the fixation of a deleterious mutation by genetic drift during times of low population size, and the third is the loss of epistatic interactions that enabled high virulence. The latter two hypotheses can result from a founder effect, but as we noted previously, there is little evidence for a strong founder effect.

In summary, while sufficient genetic variability in both host resistance and parasitoid virulence were documented (Henter 1995; Henter & Via 1995), there was no indication of local adaptation of this parasitoid to its host in the introduced range. In addition to the dynamics involving pea aphid on alfalfa in New York and France discussed earlier, a lack of local adaptation was found when pea aphids growing on alfalfa versus red clover were compared as hosts for *A. ervi* (Hufbauer & Via 1999; Hufbauer 2001). Despite this lack of local adaptation, *A. ervi* is considered an effective biological control agent of the pea aphid in alfalfa in both eastern and western North America (Angalet & Fuester 1977; Gonzalez et al. 1995; Hufbauer 2002a), showing that successful biological control is possible even in the face of apparently nonadaptive evolution.

7.1.4 Summary of Case Studies

The pattern both for the rabbit-virus and the skeleton weed-rust cases was that spectacular biological control success was followed by erosion of biological control efficacy due to evolutionary or genetic reasons. Also, ecological forces seemed to play a role in both cases – for myxoma, demographics (relationships between host density and parasite virulence) could have played a role in virulence evolution, and in the rust case, the release from competition by a susceptible plant genotype leading to spread of the resistant (but competitively inferior) plant genotypes. The aphid-parasitoid case is different in the sense that no evolutionary erosion of biological control was seen over time. Instead, biological control seemed robust in the face of apparently nonadaptive changes in parasitism for as-yet-unknown evolutionary reasons.

7.2 Evolution during the Process of Importation Biological Control – A Chronological View

We now turn to a more general treatment of evolution during the process of importation biological control. Since the cases discussed were idiosyncratic, we have not yet presented a systematic treatment of evolutionary dynamics in importation biological control. We will attempt this here by discussing the process of importation biological control chronologically, and considering the known or suspected role of evolutionary change at each step. The steps that we consider are (i) collection of natural enemies; (ii) quarantine rearing and testing; (iii) release; and (iv) post-release establishment and control evaluation. Many of these topics discussed in this chapter have been covered in Chapter 3, but here we take a more explicitly evolutionary perspective.

7.2.1 Collection of Natural Enemies for Importation Biological Control

An important decision to make in the collection of biological control agents is whether they should be collected from a wide range of localities in order to maximize genetic variability, or from a restricted locality to provide an adaptive match to the known or presumed origin of the pest (Simmonds 1963; Messenger et al. 1976b). We will call the first of these strategies the 'genetic variability' strategy, and the second one the 'local adaptation' strategy.

7.2.1.1 The Genetic Variability Strategy

The genetic variability strategy has been supported by biological control scientists for two reasons. The first is that by collecting potential agents from a variety of geographic locations and sites, one increases the chance that a suitable genotype is released or that the genetic material will be available for the production of a genotype that will be well adapted to the introduced habitat. This line of reasoning follows from a long history of documentation of significant intraspecific variability in life history traits of biological control agents collected in different localities. Variation in a wide variety of traits has been found, including host range, fecundity, virulence, longevity, development time, search rate, insecticide resistance, sex ratio and climatic tolerances, among others (e.g., Messenger et al. 1976b; Hopper et al. 1993; Wajnberg 2004; Benvenuto et al. 2012; Cory & Franklin 2012). These observations provide circumstantial evidence for the hypothesis that genetic variability for traits of interest to biological control exists in the native range of biological control agents. By collecting widely, one increases the chances of either collecting a genotype that is fortuitously pre-adapted to the introduced habitat and pest or of collecting a wide enough range of traits so that these traits can be combined or modified in adaptive ways in the introduced range. Most of the early recommendations take this approach (Messenger et al. 1976b and references therein), and Remington (1968) and Lucas (1969) refined this general recommendation to suggest that the goal of maximizing genetic variability of natural enemies is best attained by sampling from the center of a larger population rather than from small, marginal populations. Further research, however, has cast doubt on the hypothesis that individuals from central portions of population ranges are more genetically variable in the first place (Myers & Sabbath 1981).

The second rationale for the genetic variability strategy is the hypothesis that matings among members of divergent populations will produce higher-quality or fitter offspring through heterosis (hybrid vigor). Indirect support for this general hypothesis comes from a number of studies across various taxa that have demonstrated that intraspecific hybridization (admixture) can facilitate accidental invasions (e.g., Kolbe et al. 2004; Lavergne & Molofsky 2007). Admixture also appears to be partly responsible for the worldwide invasion of the ladybeetle *Harmonia axyridis*, which was originally released as a biological control agent but was subsequently considered a damaging invasive species (Lombaert et al. 2014). Legner (1972) was the first to explore admixture in the context of importation biological control empirically, and he demonstrated heterosis under laboratory conditions between different populations of two species of parasitoids of muscoid flies (negative heterosis, or outbreeding depression, was shown in a third species) and other laboratory studies of biological control agents have produced similar results since (Mathenge et al. 2010, Szűcs et al. 2012a). A particular case of hybrid vigor involving the sex determination mechanism of some parasitoid wasps with strong implications for biological control is discussed in detail later (Section 7.2.3.1). Legner (1972) suggested that previous cases in which the supplementation of additional strains of natural enemies improved biological control may have been due to the emergence of a new hybrid strain, rather than replacement of one strain by another as is typically assumed. This interesting hypothesis remains untested as far as we are aware. And while not within the realm of importation biological control, targeted hybridization to improve the efficacy of biological control agents has been achieved for entomopathogenic nematodes in augmentative biological control and attempted for predatory mites (Hoy 1985; Shapiro et al. 1997; Shapiro-Ilan et al. 2005; Nimkingrat et al. 2013). This topic is discussed in Chapter 8 as well (Section 8.5.3). Despite these indications of the advantages of outcrossing for biological control, there are strong arguments *against* facilitating outbreeding, as we discuss in the next paragraph.

Force (1967) was the first to point out that seeking heterosis in biological control agents could backfire because of outbreeding depression (also termed *hybrid breakdown*). Reduced fitness in F_2 progeny stemming from crosses of various *Drosophila* spp. populations collected in different locations provided some of the first observations consistent with the hypothesis that crosses between geographically separated populations can disrupt co-adapted gene complexes and lead to reductions in fitness (Vetukhiv 1956; 1957). Since then, outbreeding depression has been found in a number of insects (Alstad & Edmunds 1983; Blows 1993; Lynch 1991; Armbruster et al. 1997; Aspi 2000), including at least five parasitoid wasp species used for biological control (Messing & AliNiazee 1988; Goldson et al. 2003; Wu et al. 2004; Rincon et al. 2006). One of these latter cases involves a European strain of *Microctonus aethiopoides*, which was a candidate for introduction into New Zealand to control the forage weevil *Sitona lepidus*. A Moroccan strain of this parasitoid had already been introduced, and was providing control of a different weevil, *S. discoideus*. Mating studies of the Moroccan and European strains under quarantine conditions indicated that hybrids of these two strains had reduced fitness, leading to the potential of reduced biological control efficacy of *S. discoideus* if both strains were present in New Zealand (Goldson et al. 2003). This led to the decision not to release the European strain (Barratt et al. 2006). Rincon and colleagues (2006) also found various levels of outbreeding depression and incompatibility among five strains of the diamondback moth parasitoid, *Cotesia plutellae* (= *vestalis*), and suggested that this may have been at the root of some failures to establish this parasitoid in various localities. Last, combining strains of host-specific agents may produce more generalist F_1 offspring (Hoffmann et al. 2002). While this may improve biological control under some circumstances, it can also increase the risk to non-target species.

An extreme form of outbreeding depression occurs when two strains of one insect species are reproductively incompatible due to infection of one of the strains by cytoplasmic-incompatability (CI) inducing endosymbionts such as *Wolbachia*. This situation was modeled by Mochiah and colleagues (2002) for the parasitoid *Cotesia sesamiae*, which contains strains that are infected with *Wolbachia* and strains that are not. The daughters of uninfected females that mate with infected males die, and Mochiah and colleagues (2002) therefore cautioned that releasing both infected and uninfected strains could result in reduced overall population growth. Numerous parasitoids used in importation and augmentative biological harbor CI *Wolbachia*, making this a possible widespread problem (e.g., Perlman et al. 2006; Vasquez et al. 2011).

Finally, cryptic and other sibling species may engage in reproductive behavior in the final stages of the speciation process, and this can lead to the production of offspring with low fitness or to the effective sterilization of females, with negative consequences for both the efficacy and the safety of biological control (Stouthamer et al. 2000; Hopper et al. 2006).

7.2.1.2 The Local Adaptation Strategy

The basis of the local adaptation strategy for natural enemy introductions is the hypothesis that establishment and pest suppression is most likely when a precise match is attained between the climate and habitat in the native and introduced ranges of the introduced agent. A related hypothesis holds that the most promising natural enemies are collected as closely as possible to the site from which the pest immigrated, a hypothesis that was confirmed for a gall-forming phytophagous mite of an invasive fern in Florida in the United States (Goolsby et al. 2006a). The former hypothesis, that matching the climate of the native and introduced range improves biological control, also has some support. As discussed in Chapter 3, collecting natural enemies in geographic regions that match the climate where the agent is to be introduced does tend to improve the chances of successful biological control (e.g., Stiling 1990; Goolsby et al. 2005), although accurate climate matching by no means guarantees successful biological control (Van Klinken et al. 2003; Goolsby et al. 2006b), and Mason and Hopper (1997) failed to find evidence for local adaptation by the aphid parasitoid *Aphelinus asychis* to natal climate as reflected in walking speed. Despite these considerations, it is likely that biological control agents that are well matched climatically are more effective than agents that are poorly matched climatically.

Whether more subtle forms of adaptation, such as adaptation to local genotypes of a host or prey species is common in biological control cases, is even less well understood. Hufbauer and Roderick (2005) summarized the general considerations used to determine whether local adaptation of this kind is likely. They focused on three factors that increase the likelihood of local adaptation of parasites to their hosts: (1) a narrow host range, (2) a short enemy life cycle in comparison to the host and (3) restricted host migration with respect to the enemy. By these criteria, biological control agents with particularly high potential for local adaptation would appear to be plant pathogens attacking perennial plants and diseases of vertebrates with low migration rates.

In the event that natural enemies do indeed become locally adapted to host or prey genotypes – would these locally adapted strains necessarily perform better as biological control agents? Perhaps, but not if this means that the host or prey have evolved resistance or tolerance to the enemy. Local adaptation can cut both ways in the context of an evolutionary 'arms race' depending on who is 'winning' (Hufbauer & Roderick 2005). A related counterargument to the hypothesis that the best biological control agents are those that are well adapted to the pest in its natal range comes from the 'new associations' literature, in which cases have been uncovered where natural enemies that have no historical association with the pest are more effective than ones that have such an association (Hokkanen & Pimentel 1984, 1989). In these cases, it is precisely this lack of potential for the pest or weed to evolve resistance to the natural enemy that leads to high biological control efficacy. The rabbit-myxoma system discussed earlier exemplifies this case, in which a virus collected from South American rabbits (where it produced only mild disease symptoms) was used to devastating effect on European rabbits, with which it had had no previous association. We discuss the post-establishment effects of local adaptation in biological control in more detail later (Section 7.2.4.1). First, however, we turn to the potential for evolutionary dynamics during quarantine rearing.

7.2.2 Quarantine Rearing

After a natural enemy species is collected, it must be reared in quarantine before being released in a new geographical region. This rearing process can last from less than five to hundreds of generations, depending on the goals of the quarantine phase. During this time, the population is subject both to genetic bottlenecks and to adaptation to laboratory conditions, either of which can lead to a change in the genetic makeup of the population between collection and release (i.e., evolution).

Significant genetic bottlenecks can occur if the founding population brought into quarantine is small, or if the numbers drop during quarantine rearing. Potential detriments of founder effects such as these include restricted genetic variability through the loss of rare alleles, and an increased susceptibility to genetic drift, which can further erode genetic variability. Inbreeding is also more likely in small populations. Hopper and colleagues (1993) have suggested that the loss of rare alleles may not constitute a detriment to biological control since rare alleles often do not confer great fitness benefits. Regardless, some of the impacts of founder effects can be alleviated by increasing the population size in quarantine shortly after importation (Nei et al. 1975; Roderick 2004).

Selection during quarantine rearing (or laboratory rearing in general) can lead to adaptation to laboratory conditions (Roush 1990). It has generally been assumed that this form of local adaptation is detrimental to natural enemy establishment potential and/or efficacy in the field, but it can be difficult to disentangle actual effects of laboratory adaptation from detrimental effects of suboptimal rearing conditions (Mackauer 1976; Hopper et al. 1993). Examples of laboratory-selected traits detrimental to biological control include loss of diapause and cold-temperature tolerance in the parasitic fly *Pseudosarcophaga affinis* (House 1967). In this case, the loss of diapause was inadvertently selected for since non-diapausing flies were disproportionately used in mass-rearing programs because of their convenience in maintaining large colonies. The deterioration of traits such as virulence, host-seeking, fecundity, sex ratio and stress tolerance during mass rearing seems to be quite widespread in entomopathogenic nematodes (Stuart & Gaugler 1996; Wang & Grewal 2002; Bai et al. 2005; Bilgrami et al. 2006; Chaston et al. 2011). These effects are not likely due to inbreeding or genetic drift due to the large population sizes at which nematodes are cultured, and neither is it likely that they are due to disease or suboptimal rearing conditions because control lines were maintained in the experiments cited. The most plausible explanation for these effects is thus adaptation to artificial laboratory rearing conditions, likely linked to the need for cost-effective, high-volume rearing operations. Gaugler (1993) has suggested that the limited success of nematodes in field application in general may be largely due to the prevalence of just this kind of laboratory adaptation. One practical solution to the problem of laboratory selection is the maintenance of highly inbred homozygous lines that harbor insufficient genetic variability for laboratory adaptation (Roush & Hopper 1995; Bai et al. 2005). These can then be crossed prior to release to recover useful traits (Chaston et al. 2011).

7.2.3 Release and Establishment of Biological Control Agents

By the time biological control agents are released, they have undergone the filters of collecting and quarantine rearing, and they may represent a restricted subsample of the genetic variability present in the native range with possible additional changes introduced by selection to laboratory conditions. And the very act of releasing the agents can introduce further bottlenecks. For example, parasitoids are subject to Allee effects when released as small populations (Hopper & Roush 1993; Fauvergue et al. 2012), possibly due to the difficulty of finding mates upon release. If this is so, the effective population size declines disproportionately with decreasing size of the released population. Genetic bottlenecks could therefore be exacerbated by making many small releases as opposed to few large releases. On the other hand (and in the absence of Allee effects), multiple smaller releases may be advisable if they facilitate local adaptation, a point to which we return later.

In sum, the process of importation biological control presents a high risk of genetic bottlenecks. Evidence for such bottlenecks come from various systems (Baker et al. 2003; Hufbauer et al. 2004; Lloyd et al. 2005), but confirmation that genetic bottlenecks have impeded the performance of biological control agents in the field is difficult to come by. We have already discussed the case of pea aphids and the parasitoid *Aphidius ervi*. In this case, introduced populations of *A. ervi* did perform more poorly than native populations, but the bottleneck was deemed mild, and not likely a cause of the poor performance (Hufbauer et al. 2004). A population of another aphid parasitoid, *Diaeretiella rapae*, which was introduced to western Australia, had a stronger genetic bottleneck than did *A. ervi*, but it is unclear whether this bottleneck has compromised *D. rapae* as a biological control agent in this area (Baker et al. 2003). Myers and Sabbath (1981) made the point that there is little evidence for bottlenecks being an important impediment in biological control, and, indeed, introductions of fewer than 20 individuals have in some cases provided excellent biological control. In the extreme, a release of only two individuals of the lace bug *Teleonemia scrupulosa* was sufficient to lead to establishment on the invasive weed Lantana in Zanzibar (Cock 1986). More recently, Fauvergue and colleagues (2007) showed that the establishment rate from introductions of just a single mated female of a dryinid parasitoid against an introduced planthopper species in southern France exceeded 85%.

7.2.3.1 Genetic Bottlenecks and Inbreeding Depression

Despite these cases, it is premature to state that genetic bottlenecks are irrelevant to biological control. In particular, genetic bottlenecks are expected to have the strongest negative effect on the success of biological control for agents that are especially prone to inbreeding depression. The parasitoid Hymenoptera deserve special treatment in this regard because of their mode of sex determination and the relationship of some species with endosymbionts that manipulate reproduction. Depending on the parasitoid species, these considerations can either decrease or increase their susceptibility to inbreeding depression. We discuss these scenarios in turn.

Why some parasitoids have *weak* inbreeding depression. Parasitoid wasps, like other hymenopterans, exhibit arrhenotokous sex determination – males develop as haploids from unfertilized eggs and females develop as diploids from fertilized eggs (Heimpel & de Boer 2008). One outcome of this system is that deleterious recessive alleles are exposed to selection within haploid males, facilitating the purging of genetic load. This has been cited as an explanation for the generally lower genetic diversity of Hymenoptera than other insects (Berkelhamer 1983; Graur 1985), and it should also make the Hymenoptera less prone to inbreeding depression. Perhaps in part due to this restricted genetic variability, many species of parasitoid Hymenoptera routinely engage in inbreeding (Hamilton 1967; Godfray 1994), a behavior that should serve to further purge deleterious recessive alleles. Another factor associated with haplo-diploidy that can influence genetic diversity is endosymbiont-induced thelytoky, which is the production of daughters by unmated females. Thelytoky can be genetically determined as well (Heimpel & de Boer 2008; Sandrock & Vorburger 2011), but some endosymbiotic bacteria in the genus *Wolbachia*, when present within male parasitoids, duplicate the genome of their hosts, effectively turning male individuals into females that are homozygous at every locus (Stouthamer 2004). While this homozygosity may be deleterious, it is not expected to impede fixation of the endosymbiont-mediated thelytoky (Stouthamer et al. 2010) and it certainly should lead to purging of deleterious recessive alleles that affect female function. Endosymbiotic bacteria in other genera – *Cardinium* and *Rickettsia* – can also cause thelytoky in some parasitoids (Zchori-Fein et al. 2004; Adachi-Hagimori et al. 2008; Giorgini et al. 2009), but in these cases the role of genome duplication is either unknown or does not lead to whole-genome homozygosity.

Why some parasitoids have *strong* inbreeding depression. Despite arrhenotoky and endosymbiont-induced parthenogenesis, inbreeding depression can still be strong in some hymenopteran parasitoids as well as in sawflies, which can be used as biological control agents of weeds. Inbreeding depression is particularly strong in species in which sex is determined by a variant of arrhenotoky known as complementary sex determination (CSD) (Heimpel & de Boer 2008). Under CSD, sex is determined by one or more sex loci. In *single-locus* CSD, fertilized eggs that are heterozygous at a single sex locus develop as females, while fertilized eggs that are homozygous at this locus develop as diploid males. Under *multiple-locus* CSD, homozygosity at all of two or more loci is needed to produce diploid males. Unfertilized eggs develop as haploid males, as in standard arrhenotoky under both forms of CSD. The reason that CSD causes inbreeding depression is that diploid males typically have low fitness. In some species, they die early in development, and in others they develop as adult males that are sterile. Females that mate with diploid males may be constrained to producing only haploid males if they are monogamous or if the diploid male sperm somehow interfere with the sperm of haploid males.

Stouthamer and colleagues (1992) modeled the effect of single-locus CSD on the sex ratio and population-level reproductive rate of parasitoid species. They found that three factors strongly influenced the extent to which CSD negatively affected these parameters. The first was sex allele diversity. When sex allele diversity is greater than six, CSD did not greatly affect population growth in their model, but below this number, strong increases in male bias and strong reductions in population growth can be expected. In wild parasitoid populations, sex allele diversity often greatly exceeds six, but this may be much reduced during genetic bottlenecks such as those that can occur during biological control (Heimpel et al. 1999; Butcher et al. 2000; Heimpel & Lundgren 2000). The second factor was diploid male survivorship. If diploid male survivorship is high, then sex ratios become more male-biased under CSD than when diploid male survivorship is low. However, diploid male survivorship has no influence on the population-level reproductive rate unless diploid males participate in mating. In Stouthamer and colleagues' (1992) models, females that mated with diploid males were constrained to only producing haploid males, which is for the most part a reasonable assumption (Holloway et al. 1999; de Boer et al. 2007; Harpur et al. 2012). In this case, high diploid male survivorship has a strong negative effect on reproductive rate, with relative population growth rates below half of those of species without CSD.

Further modeling of the effect of CSD on hymenopteran population dynamics has shown that CSD can greatly increase the risk of extinction, making hymenopterans with CSD more extinction-prone than both hymenopterans without CSD and organisms with diploid ('regular') sex determination (Zayed & Packer 2005; see also Hein et al. 2009). Zayed and Packer's models culminated in a 'diploid male extinction vortex', in which extrinsic factors such as founder effects lead to a loss of sex alleles, which itself leads

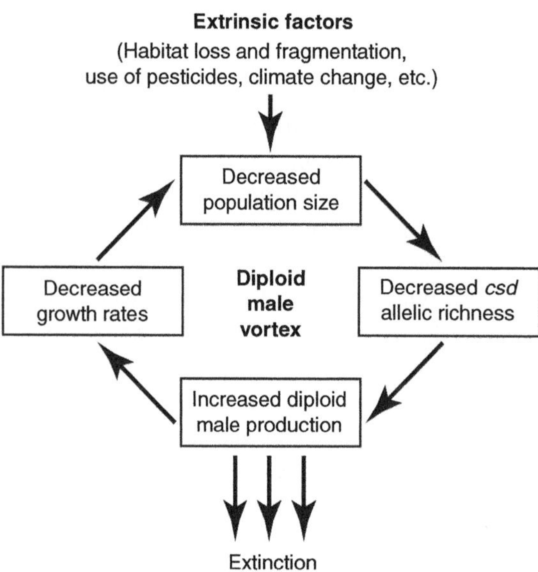

Figure 7.6. Elements of the 'diploid male extinction vortex' that can affect parasitoid wasps (or other hymenopterans) with complementary sex determination. From Zayed and Packer (2005).

to decreased population growth via diploid male production, exacerbating and intensifying population declines to the point where the extinction probability rises to nontrivial levels (Figure 7.6).

At least three factors have the potential to greatly diminish the extent to which CSD can harm biological control: reproductive viability of diploid males, multiple-locus CSD and inbreeding avoidance. Reproductive viability of diploid males has been found in one species of non-parasitoid Hymenoptera. In the inbreeding hunting wasp, *Euodynerus foraminatus*, diploid males are reproductively functional (Cowan & Stahlhut 2004). Within parasitoids, *Cotesia glomerata* – a biological control agent of the imported cabbageworm, *Pieris brassicae* – produces diploid males that are viable, but the fecundity of females that mate with these males is slightly lower than that of females that mate with haploid males (Elias et al. 2009). The cost of CSD is clearly attenuated in these cases, but the ability of females to produce 'optimal' sex ratios would still be compromised. Multiple-locus CSD has now been documented in three species of parasitoids – all of them biological control agents (de Boer et al. 2008, 2012; Carabajal Paladino et al. 2015). The costs of CSD are greatly diminished by multiple-locus CSD under moderate inbreeding, but approach those of single-locus CSD under strong inbreeding (Heimpel & de Boer 2008; de Boer et al. 2015). Last, the cost of CSD can be attenuated by inbreeding avoidance, as has been shown for an important biological control agent of stored-grain infesting moths, *Habrobracon* (= *Bracon*) *hebetor*. A number of pre-mating behaviors such as brood-mate avoidance, lekking and sex-specific dispersal reduce the probability of inbreeding, and therefore, the probability of diploid male production (Antolin & Strand 1992; Ode et al. 1995; Ode et al. 1998).

It is important to realize that not all species of hymenopteran parasitoids have CSD. Among parasitoid taxa, CSD has only been discovered in the superfamily Ichneumonoidea (families Ichneumonidae and Braconidae) (Asplen et al. 2009). In addition, it is known from tenthredinoid sawflies (superfamily Symphyta), some of which are used as weed biological control agents. It is not clear how widespread CSD is within these two superfamilies because not many species have been studied, but even within the Ichneumonoidea, a number of species have been found *not* to have CSD (Asplen et al. 2009). Still, it appears that this is the only parasitoid superfamily that has CSD, and Stouthamer and colleagues (1992) hypothesized that the relatively poor performance of ichneumonoids as biological control agents compared to the chalcidoids may be due in large part to CSD. They especially cited the high likelihood of genetic bottlenecks during the process of importation biological control as making ichneumonoids (and not chalcidoids) vulnerable to inbreeding depression. And indeed, ichneumonoids are more commonly associated with male-biased sex ratios in laboratory cultures than are parasitoids in other superfamilies (Heimpel & Lundgren 2000).

In Table 7.1 we summarize information on CSD for parasitoids in the superfamily Ichneumonoidea (families Braconidae and Ichneumonidae).

For the parasitoid Hymenoptera, an obvious hypothesis is that species with CSD are more prone to negative effects of genetic bottlenecks than are species without CSD, simply because CSD imposes inbreeding depression above and beyond that already experienced by parasitoids (Antolin 1999; Henter 2003). This hypothesis has led to

Table 7.1. **Ichneumonoid biological control agents for which complementary sex determination (CSD) status has been determined or is suspected. CSD has been studied in many other hymenopterans that we do not consider biological control agents and is absent in many important hymenopteran biological control agents (van Wilgenberg et al. 2006; Asplen et al. 2009).**

Species	Pest	CSD Status	References
Ichneumonoidea: Braconidae			
Aphidius rhopalosiphi	Various aphids	*sl* or *ml*-CSD	Salin et al. 2004
Asobara spp.	Various drosophilid flies	no CSD	Ma et al. 2013
Bracon (Habrobracon) brevicornis	Various Lepidoptera	*sl* or *ml*-CSD	Thiel & Weeda 2014
Bracon (Habrobracon) hebetor	Stored-grain infesting moths	*sl*-CSD	Heimpel et al. 1999
Bracon (Habrobracon) sp.	Stored-grain infesting moths	*sl*-CSD	Holloway et al. 1999
Cotesia flavipes	Stem-boring moths	no CSD	Niyibigira et al. 2004a, b
Cotesia glomerata	*Pieris* cabbage butterflies	*sl*-CSD	Zhou et al. 2006
Cotesia rubecula	Small cabbage white butterfly	*ml*-CSD	de Boer et al. 2012
Cotesia plutellae (= vestalis)	Diamondback moth	*ml*-CSD	de Boer et al. 2008
Diachasmimorpha longicaudata	True fruit flies (Tephritidae)	*ml*-CSD	Carabajal Paladino et al. 2015
Heterospilus prosopidis	Store-legume-infesting bruchids	no CSD	Wu et al. 2005
Microplitis croceipes	Noctuid moths	*sl* or *ml*-CSD	Steiner & Teig 1989
Ichneumonoidea: Ichneumonidae			
Bathyplectes curculionis	Alfalfa weevil	*sl* or *ml*-CSD	Stouthamer et al. 1992
Diadegma spp.	Plutellid and yponomeutid moths	*sl*-CSD	Butcher et al. 2000
Diadromus pulchellus	Diamondback moth	*sl*-CSD	Periquet et al. 1993
Venturia canescens	Stored-grain-infesting moths	*sl*-CSD	Beukeboom 2001

specific recommendations for how parasitoids should be collected and reared for the purposes of importation and augmentative biological control, with the express purpose of conserving sex alleles. One method involves keeping parasitoids in a few large colonies to preserve sex allele diversity and minimize inbreeding (Stouthamer et al. 1992). In another method, multiple isofemale lines that harbor as few as two alleles each are kept separately and then recombined prior to release (Cook 1993). This latter method is analogous to the recommendations for avoiding adaptation to laboratory conditions discussed earlier (Section 7.2.2).

7.2.4 Post-establishment Evolution

Two evolutionary processes that are of great interest once biological control agents have established are (i) adaptation to the new habitat, including the evolution of host specialization; and (ii) the evolution of resistance to the agent by the pest (i.e., local adaptation by the pest). We consider these dynamics in turn.

7.2.4.1 Local Adaptation by Biological Control Agents

If climate and habitat matching between the native and introduced ranges of the natural enemy are strong, and the agent is released against the same pest or weed that it was collected on, local adaptation in the introduced range may not occur (Hopper et al. 1993). Alternatively though, it has been suggested that introduced biological control agents are able to adapt to local climates, resources, enemies or competitors in ways that make them more effective natural enemies (e.g., Myers & Sabbath 1981). For instance, Harris and colleagues (1975) noted that establishment of the cinnabar moth against tansy ragwort in areas of Canada where previous releases had been made was greater when 'local stock' was used as opposed to imported populations. In this and other cases, though, alternative factors, such as disease in imported insects, could not be ruled out. Hufbauer and Roderick (2005) have pointed out that the evolution of insecticide resistance by importation biological control agents is perhaps the strongest proof of post-release local adaptation available.

As we have noted earlier, local adaptation to introduced host genotypes apparently did not occur in the case of *Aphidius ervi* and the pea aphid. However, data consistent with local adaptation of parasitoids to host species has been obtained in a number of other biological control projects, although definitive proof of local adaptation in some of these cases is lacking. Here we discuss four cases that are suggestive of, or demonstrations of, local adaptation in importation biological control agents for parasitoids and herbivores.

Local adaptation in parasitoids released as importation biological control agents. The first case involves the ichneumonid parasitoid *Bathyplectes curculionis*, which was introduced into the United States from Europe to control the alfalfa weevil, *Hypera postica*, but also attacked a related pest species, the Egyptian alfalfa weevil, *H. bruneipennis*. Initially, resistance of *H. bruneipennis* to *B. curculionis* was high. The host encapsulated approximately half of the parasitoid eggs in the mid-1950s, about 20 years after the establishment of both the weevil and the parasitoid in southern California. This encapsulation rate dropped to less than 15% by 1966, however (Van den Bosch & Dietrick 1959; Salt & Van den Bosch 1967). In support of the hypothesis that this difference was due to local adaptation on the part of the parasitoid, these low levels of encapsulation were only seen with parasitoids collected from the geographic range of *H. bruneipennis*, not that of *H. postica*. Berberet and colleagues (2003) also documented a significant decrease in encapsulation of *B. curculionis* by alfalfa weevil between 1975 and 2000 in Oklahoma in the United States, but in this case it was not clear whether the decrease was due to local adaptation or mixing of alfalfa weevil strains with differing encapsulation abilities.

The second case involves the braconid parasitoid *Cotesia glomerata*, which was introduced into North America in the late 19th century from the United Kingdom to control the imported small cabbage white butterfly, *Pieris rapae*. Although *C. glomerata* was capable of attacking larvae of the small cabbage white, it was known in Europe primarily as a parasitoid of the large cabbage white, *P. brassicae*, which has not, as of this writing, invaded North America. Le Masurier and Waage (1993) tested the hypothesis that *C. glomerata* from the United States had evolved to better exploit the small cabbage white – not an unreasonable hypothesis since *C. glomerata* had been present in the United States in excess of 300 generations at the time of the test. They found that, indeed, attack rates of American *C. glomerata* on the small cabbage white were significantly higher than the attack rate of British *C. glomerata* on the small cabbage white, a conclusion that was confirmed and expanded upon by later studies (Vos & Hemerik 2003; Vos & Vet 2004). Also consistent with local adaptation was the finding that the number of mature eggs carried by females (the 'egg load') was lower in American than in British *C. glomerata* (Le Masurier & Waage 1993). The significance of this difference is that the small cabbage white lays eggs singly and the large cabbage white lays eggs in batches, so that members of American populations of *C. glomerata* are likely not to need as many eggs as members of European populations (Heimpel 2000).

This situation was somewhat reversed in Japan, where a subspecies of the small cabbage white, *P. rapae crucivora*, is native, and the large white invaded from Siberia around 1995. Here, *C. glomerata* has a long historical association with the Japanese small cabbage white, and was very reluctant to attack the large cabbage white upon its introduction (Sato & Ohsaki 2004). This host was readily accepted in the field within 20 generations of its invasion, however, and the number of eggs laid in the novel hosts increased gradually over these 20 years as well, suggesting rapid adaptation to the new host species (Tanaka et al. 2007).

Research by Antolin and colleagues (2006) has come tantalizingly close to demonstrating the evolution of increased specialization in a parasitoid post-release. They studied the aphid parasitoid, *Diaeretiella rapae*, which is native to Eurasia and was introduced to North America, where it attacks a number of aphid species, including the cabbage aphid and the Russian wheat aphid. Working in the United States, Antolin and colleagues (2006) showed that *D. rapae* that were collected on cabbage aphid performed better on cabbage aphid than on Russian wheat aphid, and vice versa, a result consistent with local adaptation to a host species. Two caveats accompany these findings, however. The first is that phylogeography studies failed to find host-associated lineages of this species (Baer et al. 2004). And second, the results are open to an alternative explanation, namely that the different lineages collected in the United States stem from two introductions: an earlier one against cabbage aphid and a later one against Russian wheat aphid. Similar results have been reported for *Aphidius ervi* and two aphid host species in western Canada (Henry et al. 2010), but the history of this project also cannot rule out separate introductions.

Last, Phillips and colleagues (2008) compared the establishment trajectories of two geographic strains of the thelytokous parasitoid *Microtonus hyperodae* that were introduced from South America to New Zealand to control the forage weevil *Listronotus bonarensis*. They found that one strain – collected at various sites east of the Andes – came to dominate a strain collected west of the Andes within a few years of the introduction of both at most New Zealand sites. The authors were able to eliminate genetic drift as an explanation and therefore concluded that clonal selection of the eastern over the western strain had occurred.

Local adaptation in herbivores released as importation biological control agents. The leaf beetle *Diorhabda carinulata* was introduced from China to the western United States beginning in 2001 to control saltcedar shrubs (tamarisks), which are highly invasive weeds in this area and cause widespread ecological damage. Establishment and saltcedar defoliation were noted at northern sites where the latitude was similar to collection sites in the region of origin, but the beetles went into early diapause farther south, leading to lower population growth and levels of feeding (Bean et al. 2007). At the most southern sites, the longest day lengths were shorter than the critical day length below which the beetles entered diapause, as we have already noted in Chapter 3. This appeared to impede establishment from initial releases in these areas entirely. However, within four years of release, researchers noted that *D. carinulata* were entering diapause later than initially observed at some of the southern sites (Bean et al. 2012). Further study showed that the critical day length had undergone evolutionary change; over a period of five years (= ca. 10 generations) it had shifted down by almost 30 minutes – enough to allow establishment and feeding later into the season at these sites.

In addition to this case, two insect biological control agents of the rangeland weed tansy ragwort – the flea beetle *Longitarsus jacobaeae* and the moth *Tyria jacobaeae* – have exhibited adaptive evolution since their introduction into the northwestern United States (McEvoy et al. 2012; Szűcs et al. 2012b). In both cases, careful studies and numerous lines of evidence were brought to bear to demonstrate the evolution of shorter larval development times at higher elevations and other appropriate environments, allowing improved establishment and better synchronization to weed populations. As with the previous example involving the evolution of critical day length in *Diorhabda*, both of these cases are characterized by rapid evolution in a biological control context. These and similar cases illustrate just how rapid evolution can be in the context of invasions (Reznick & Ghalambor 2001). As we noted in Chapter 5, McEvoy and colleagues (2012) used the results of their study to advocate for the inclusion of evolutionary potential in pre-release risk-benefit assessments of biological control agents.

Most of the cases of local adaptation resulted in improved biological control (Berberet et al. 2003; Vos & Vet 2004; Tanaka et al. 2007; Phillips et al. 2008; Bean et al. 2012; McEvoy et al. 2012; Szűcs et al. 2012b). This is expected because fitness optimization should in most cases lead to increased population suppression, although these two factors can be decoupled (Mills & Wajnberg 2008; Wajnberg et al. 2016). We turn next to a particularly important form of local adaptation for biological control, the evolution of host specialization. While increased host specialization upon introduction could lead to improved biological control, the incorporation of new hosts into the host range of a parasitoid or pathogen as part of an evolutionary broadening of the host range could lead to non-target effects, as mentioned in Chapters 4 and 5.

7.2.4.2 Evolution of Host Specialization

In importation biological control, specialist natural enemies are taken from their home range and released against what is often their most suitable host in a new geographic range that by definition contains fewer species with which the enemy has an historical association. Does this scenario favor the evolution of increased host or diet specialization? In this section, we review studies on the genetic basis of host specificity in biological control agents, and discuss the general selective regimes that can lead to host specialization, asking whether the process of importation biological control leads to the evolution of host specialization. No evidence for evolutionary changes in host specialization has been found for herbivores introduced as biological control agents (Van Klinken & Edwards 2002) or for pathogens introduced against insect pests (Cory & Franklin 2012), but genetic variation in traits mediating host range have been identified in these and other groups of biological control agents. In the following paragraphs, we consider the potential for the post-release evolution of host specialization in importation biological control agents.

Host range genes in pathogens. Host range genes have been discovered in pathogens of insects, mammals and weeds. For instance, specificity of the bacterium *Bacillus thuringiensis* (*Bt*) to different insect groups is famously conferred by recognition of host midgut binding sites by protein crystal toxins (Federici & Maddox 1996). Host specificity is thus conferred by the genes that encode these toxins. These genes reside on plasmids and different host-specific strains or subspecies of *Bt* may contain different plasmid complements. Among the baculoviruses used in biological control, a number of genes with different modes of action have been identified that mediate host range (Table 7.2). These findings suggest that there is ample scope for genetic variation within *Bt* and baculoviruses. However, neither of these entomopathogens are typically used in importation biological control, so the question of variation in host-specificity traits and post-release local adaptation based on such variation has not been addressed. Genetic variability for host specificity has also been found in pathogens of vertebrates and plants. We have already discussed a host specificity gene in the myxoma virus, and similar host range genes have been found in other poxviruses (McFadden 2005). These findings, however, have not been extended to studies of genetic variation in host specificity and the potential for host adaptation in biological control settings. And as we have noted for the case of skeleton weed and *Puccinia* rust, the trend is to introduce strains that are already highly adapted to particular weed strains, leaving little additional scope for local adaptation in the area of host specificity.

Genetic basis for host specialization in insect herbivores and parasitoids. Ground-breaking work on the genetics of host specialization in herbivorous

Table 7.2. **Genes mediating host range in baculoviruses used in biological control. Based, in part, on Thiem (1997).**

Virus	Gene	Mechanism	References
Autographa californica nucleopolyhedrovirus (AcMNPV)	*p35*	Inhibits apoptosis in host cells (a host defense mechanism)	Clem et al. 1991; Clem & Miller 1993
AcMNPV	*p143* helicase	Facilitates expression of viral genes	Maeda et al. 1993; Croizier et al. 1994; Argaud et al. 1998
AcMNPV	*hcf-1*	Facilitates expression of viral genes	Lu & Miller 1995
AcMNPV	*lef-7*	Facilitates expression of viral genes	Chen & Thiem 1997
Cydia pomonella granulovirus	Cp-*iap*	Inhibits apoptosis in host cells	Crook et al. 1993
Orgyia pseudotsugata nucleopolyhedrovirus	Op-*iap*	Inhibits apoptosis in host cells	Birnbaum et al. 1994
Lymantria dispar nucleopolyhedrovirus	*hrf-1*	Relieves host cell's ability to block translation of viral genes	Thiem et al. 1996; Ikeda et al. 2005

Drosophila and some agricultural pests is starting to dissect the genetics of host attraction and host use in these insects (e.g., R'Kha et al. 1991; Dambroski et al. 2005; Jones 2005; Oppenheim et al. 2012), and work with the pea aphid has shown that nuclear genetic factors and bacterial endosymbionts can both mediate host specificity (e.g., Simon et al. 2003; Leonardo 2004; Tsuchida et al. 2004). Regrettably, similar studies have not yet been performed on arthropod biological control agents of weeds. Still, the case of *Drosophila seychellia* is instructive as it involves a natural history that makes it in some ways similar to a biological control agent. *D. seychellia* is endemic to the Seychelles in the Indian Ocean and specializes in feeding on fruits of the shrub *Morinda citrifolia*, which is toxic to other species of *Drosophila*. *D. seychellia* likely migrated from the eastern coast of Africa to the Seychelles after the arrival of *Morinda*. The presumed ancestor of *D. seychellia* was likely a generalist feeder based on information from related *Drosophila* species on the African mainland and Madagascar, and in all probability had a pre-adaptation allowing it to feed on *Morinda* fruits. While the selective forces leading to specialization are not known (although hypotheses abound; see Jones 2005), the genetic changes accompanying this specialization have been elucidated thanks to its close relationship with *D. simulans* and *D. melanogaster*, for which a plethora of genetic tools are available. These studies have revealed the genetic basis of a number of traits that mediate host use, such as resistance to compounds that are toxic to other *Drosophila* species, oviposition site preference, egg maturation in the presence of the host and larval morphology facilitating host use (Jones 2005; Matsuo et al. 2007; McBride 2007). What is the relevance of these findings for biological control? They suggest that adaptation to a single host plant in the introduced range of an herbivore can lead to increased host specialization (the implication is that this could occur for biological control agents as well). Also, as the introduction of *D. seychellia* presumably occurred passively via wind currents from Africa or Madagascar (Jones 2005), it seems likely that the increased host specialization occurred in the context of a founder effect.

And what of the genetic basis of host specialization in parasitoids? The success or failure of parasitoid development within a host (endoparasitism) is clearly determined in large part by interactions between host resistance genes and parasitoid virulence genes, although host resistance genes may be carried by endosymbiotic bacteria (Hsiao 1996; Oliver et al. 2005), and parasitoid virulence genes may be carried by mutualistic viruses such as polydnaviruses (Pennachio & Strand 2006). Research beginning in the 1990s showed variability in both host resistance and virulence in the pea aphid-*Aphidius ervi* system (Henter 1995; Henter & Via 1995), as we mentioned earlier. This was an important result because it showed that there was the scope for coevolutionary interactions between hosts (and/or defensive symbionts) and parasitoids. In addition, studies comparing *A. ervi* collected from, and developing on, two host species showed a genetic basis for performance-related traits on the different host species (Henry et al. 2010). Evidence for within-species genetic variation in virulence factors has also been found in two species of *Drosophila* parasitoids (Kraaijeveld & Van Alphen 1995; Dubuffet et al. 2007), and in one of these, *Leptopilina boulardi*, results of directed crosses suggest that two unlinked genes are involved – one overcoming resistance by *D. melanogaster* larvae, and one overcoming resistance by *D. yakuba* larvae (Dupas & Carton 1999).

A recent study by K. R. Hopper and colleagues (unpublished) is beginning to reveal the genetic architecture of host specificity in a host-parasitoid biological control system: aphid parasitoids in the genus *Aphelinus* attacking various aphid species. They found that one species, *A. certus*, rarely attacked the Russian wheat aphid, while a closely related species, *A. atriplicis*, readily attacked this host. Taking advantage of the fact that these species are not completely reproductively isolated, hybrids and backcrosses were produced to study the genetic basis of Russian wheat aphid parasitism. These researchers found eight quantitative trait loci (QTL) that were significantly correlated with parasitism of the Russian wheat aphid. Separate studies had shown that differences in parasitism were due to host acceptance rather than physiological host suitability in these species, so these QTL may indicate the presence of genes that mediate host recognition.

7.2.4.3 Evolution of Resistance to Biological Control Agents

For this last topic of the chapter, we recall one of the original case studies – myxoma virus and rabbits – as the prime example of evolution of resistance of a pest to a biological control agent. The same can be said for the second case study, where the frequency of resistant genotypes of skeleton weed increased upon the control of susceptible strains. But how general is this phenomenon? How does it differ between taxa used in biological control and how does resistance evolution differ between chemical and biological control? We deal with these questions in this section, borrowing heavily from an important paper by Holt and Hochberg (1997) titled: 'When is biological control evolutionarily stable (or is it)?'

Holt and Hochberg began their review by making the point that while resistance to chemical pesticides is commonly documented in arthropod pests, plant pathogens and weeds, the same cannot be said for resistance to parasitoids, predators and herbivores used in biological control. We review the evidence for or against resistance evolution in targets of pathogens and then parasitoids in the following paragraphs (we could find no discussion of this topic for herbivores or predators used for biological control). We end by considering Holt and Hochberg's hypotheses to explain evolutionary stability in biological control.

Resistance to pathogens used in biological control. Resistance to pathogens used for biological control is not particularly rare. Aside from the myxoma and *Puccinia* cases discussed previously, it has been documented for bacterial pathogens of pest nematodes (Channer & Gowan 1992) and other bacteria, baculoviruses, fungi and protozoa used in biological control (Briese 1986b,c; Keller et al. 1999; Ferrari et al. 2001). Notable among these are entomopathogenic fungi, which can encounter resistance that is strong enough to make them completely non-infectious in some host strains (Hughes & Bryce 1984). In the case of fungi attacking the pea aphid, resistance may be linked to bacterial endosymbionts (Scarborough et al. 2005). Resistance to *Bt* sprays has been documented as well (Tabashnik 1994), but as we noted in Chapter 1, *Bt* sprays do not constitute clear cases of biological control when *Bt* toxin, rather than living bacteria, constitute the bulk of the formulation.

While it is clear that pathogens of various kinds used for biological control can encounter host resistance, cases in which the effectiveness of biological control has actually been eroded in the face of resistance to pathogens in a manner analogous to what is seen in pesticide resistance are hard to find beyond the rabbit-myxoma case. The next-best example involves the initially successful control of the rhinoceros beetle, *Oryctes rhinoceros* (a scarab that feeds on economically important palm trees) by an unclassified virus (Wang et al. 2007) in the islands of the South Pacific. Biological control of rhinoceros beetle has eroded on some islands, along with a marked reduction in virulence, suggesting the evolution of resistance to the virus (Zelazny et al. 1989; Jackson et al. 2005). Another compelling case study that has similarities to a true biological control introduction involves the European cockchafer, *Melolontha melolontha* (another scarab beetle – in this case, one whose larvae feed on the roots of young trees) and the fungal pathogen *Beauveria brongniartii* (Keller et al. 1999). Various isolates of this pathogen were much more virulent in cockchafer populations from Italy than from Switzerland and interestingly, the cockchafer-*Beauveria* association is older in the Swiss than in the Italian population. Keller and colleagues hypothesized that the longer association in Switzerland allowed adaptation of the host to the pathogen, resulting in lower virulence.

Two counterexamples in which the virulence of biological control pathogens was expected not to erode in the face of documented variation in host resistance or tolerance include a hypovirus of the chestnut blight fungus, and a fungal pathogen of the gypsy moth. Chestnut blight fungus, *Cryphonectria parasitica*, decimated mature American chestnut trees throughout eastern North America during the first half of the 20th century. Sprouts continue to grow from the root systems of killed trees only to be infected by chestnut blight and eventually be killed themselves. A number of attenuated, or 'hypovirulent' strains of the fungus were found in Italy, France and the United States, which have proven effective in controlling chestnut blight in some parts of Europe and in Michigan in the United States (Anagnostakis 1982; Milgroom & Cortesi 2004). The hypovirulence factor turned out to be an RNA virus, and field trials in Italy showed variation in both host tolerance and viral virulence traits, suggesting a scope for resistance evolution (Peever et al. 2000). However, Peever and colleagues concluded that the net effect would not likely be a reduction of biological control, in part because of a relatively low magnitude of tolerance, and the potential for coevolutionary dynamics. Nielsen and colleagues (2005) similarly compared tolerance to and virulence by, various strains of the gypsy moth fungal pathogen *Entomophaga maimaiga*, and concluded that erosion of biological control was unlikely despite variation both in resistance and virulence traits. Both of these cases are consistent with the general hypothesis that coevolutionary interactions between resistance and virulence traits impede erosion of biological control by pathogens (Briese 1986b), which is one of the hypotheses considered by Holt and Hochberg (1997) to explain the apparent evolutionary stability of biological control.

Resistance to parasitoids used in biological control. Even before the review by Holt and Hochberg, a number of authors had marveled at the apparent lack of erosion of biological control by parasitoids (Waage & Greathead 1988; Henter & Via 1995). The only cited example that we know of remains the ichneumonid *Mesoleius tenthredinis*, which was first imported from England to North America to control the larch sawfly, *Pristiphora erichsonii*, in 1910. While control was impressive at first, outbreaks of the sawfly were observed in the 1940s in Canada, along with evidence of encapsulation of parasitoid eggs (Muldrew 1953). This is certainly consistent with resistance evolution, but Ives and Muldrew (1981) proposed an alternative hypothesis: the repeated introduction of larch sawfly strains from Europe with greater encapsulation abilities. In particular, they noted that parasitoids were released as pupae and not as adults, such that groups of parasitized larch sawfly cocoons were released in the field. These releases were thought to contain genotypes of the larch sawfly that had the ability to encapsulate *M. tenthredinis*. It is not clear whether the erosion of biological control of the larch sawfly was due to the evolution of resistance, the inadvertent introduction of resistant strains of sawflies from Europe, or both of these factors. Perhaps most important, though, is the observation that this is the only case of an imported parasitoid in which such an erosion of biological control associated with host resistance has been documented (to our knowledge).

Explaining evolutionary stability. Holt and Hochberg (1997) offered five hypotheses (which we collapse into four here) to explain more rapid resistance evolution to chemical pesticides than to biological control agents:

(1) Lack of genetic variation for resistance traits against biological control agents.
(2) Constraints on the evolution of resistance such as costs of resistance.
(3) Weak or variable selection pressure, as can be produced by spatial or temporal refuges from natural enemy attack.
(4) Balanced coevolutionary dynamics in which biological control agents evolve counter-adaptations to pest resistance traits (the red queen hypothesis).

Of course these hypotheses are not mutually exclusive, but it is fair to say that hypothesis (1) – the lack of genetic variability in resistance traits – is likely not a strong contender for explaining the pattern of little evolutionary decay in biological control (see references cited earlier). Hypothesis (2) has been supported to some extent by the documentation of costs of standing resistance in *Drosophila* parasitoids (Fellowes & Godfray 2000; Kraaijeveld et al. 2002) and pea aphids, where the costs appear to be in part mediated by endosymbiont presence (Gwynn et al. 2005). Note that in these cases, only the costs of *standing* resistance (i.e., being prepared to mount an attack even in the absence of parasitism) is relevant since hosts that are successfully parasitized inevitably die, implying that any cost of resistance would be adaptive in the face of an actual attack. Such costs of resistance could be important in maintaining susceptible genotypes during times (or places) of low natural enemy pressure. The importance of such refuges is further emphasized in hypothesis (3), which posits that ecological aspects of the pest-enemy interaction, acting primarily through spatial and/or temporal heterogeneity can lead to weak selection. A particularly compelling aspect of this hypothesis is the possibility of weak selection stemming from disproportionate local extinction of pest patches that are colonized by natural enemies within a metapopulation structure. This hypothesis was extended in a model by Holt and colleagues (1999) that showed that unstable pest-enemy interactions at the population level can retard the evolution of resistance because it interferes with the intimate interactions between species that favor resistance evolution. Hypothesis (4), reciprocal coevolution, was favored by Holt and Hochberg in conjunction with hypothesis (3), and the recent studies demonstrating variability in virulence in pathogens and parasitoids (discussed earlier) surely strengthen this view.

7.3 Conclusions

Biological control interactions are clearly subject to evolutionary dynamics. This is particularly true in importation biological control because relatively long time frames are involved. Indeed, the case of the release of myxoma virus to control rabbits stands out as a seminal case that illustrates a number of important principles in evolutionary biology (Fenner & Fantini 1999; Di Giallarondo & Holmes 2015b). Beyond this case, though, a number of other biological control interactions are emerging as instructive models for the study of evolution. Many of these have been discussed in this chapter.

A unifying theme that we have detected in the context of evolutionary dynamics within biological control interactions is stability, with questions including the following. First, are host ranges of introduced agents evolutionarily stable enough to protect non-target species? This appears to be the case for some taxa (Van Klinken & Edwards 2002; Cory & Franklin 2012), but may not be the case for other taxa, and also we still have a lot to learn about the dynamics of host specificity in released biological control agents (Brodeur 2012). Second, do the various steps of importation biological control allow for the maintenance of traits that confer efficacy, or do processes such as genetic bottlenecks, drift and laboratory selection erode agent efficacy? This question becomes more critical as the permitting processes require more and more tests, which increases the time agents spend under quarantine conditions (Messing & Wright 2006). And last, is importation biological control robust to resistance on the part of the target pests? We have seen that this can vary greatly depending on taxa with resistance to microbial agents apparently more likely than to insects.

III

Local Facilitation of Biological Control

8 Augmentation: Orchestrating Local Invasions

Augmentation becomes an important approach to biological control when either the indigenous natural enemies of native pests or the introduced natural enemies of invasive pests fail to provide the level of suppression desired even after conservation measures have been considered and implemented. In some cases, natural enemies are unable to persist year-round or to increase in numbers quickly enough to suppress pest damage, and it is under these circumstances that augmentative biological control, involving the periodic release of mass-produced natural enemies, can be effective. This has led to the development of the only commercial application of biological control through a highly specialized industry based on the mass production of natural enemies. Augmentation, of necessity, is focused on natural enemies that are amenable to mass production, which are primarily fungal and bacterial pathogens and antagonists, but also includes insect parasitoids and arthropod predators. As an approach to biological control, augmentation has been widely used for the suppression of arthropod pests in agricultural crops and forest plantations, for the suppression of noxious weeds in crops, rangelands and forests and for the suppression of soil-borne and aerial pathogens in agricultural and forest environments.

8.1 History and Scope of Augmentation

8.1.1 History of Augmentation

Augmentation has a surprisingly long history of operational use that can be traced back to the third century AD in China when nests of the weaver ant, *Oecophylla smaragdina*, were sold in Canton for the control of citrus pests such as the stink bug, *Tesseratoma papillosa* (Konishi & Ito 1973). It was not until the late 1800s, however, that the full range of possibilities for augmentation of natural enemies and the potential for microbial pathogens in this endeavor was realized. Louis Pasteur (1874) was among the first to envisage the use of fungal pathogens for insect control, and it was in the Ukraine in 1884–1888 that the first known application of a mycoinsecticide took place. In this case, spores from cultures of the fungal pathogen *Metarhizium anisopliae* were successfully field tested by Krassilstschick against both the grain chafer, *Anisoplia austriaca*, and the sugar beet weevil, *Cleonus punctiventris* (Steinhaus 1956). Although anticipated as early as 1893, the use of fungal pathogens for the suppression of noxious weeds lay dormant until the 1940s, perhaps due to fear of potential spread to crop plants. The first known successful application of a mycoherbicide appears to have been the use of *Fusarium oxysporum* for suppression of the white form of the prickly pear, *Opuntia megacantha*, in Hawaii (Wilson 1969). The potential for mass production and augmentation of insect parasitoids was inspired by the work of Speyer (1927) on the whitefly parasitoid, *Encarsia formosa*, and of Flanders (1930) on the egg parasitoid *Trichogramma minutum*, although much of the subsequent development of *Trichogramma* for field augmentation took place in the former USSR and China in the 1940s. Although the complexity of microbial interactions in the soil had been recognized in the early 1900s, inoculation of antagonists for the biological control of plant diseases began in the 1920s with studies of damping-off of forest nursery seedlings caused by the fungal pathogen *Pythium debaryanum* (Garrett 1965; Baker 1987). The culture and field release of entomopathogenic nematodes then began in the 1930s, following the discovery of *Steinernema glaseri* as a parasite of Japanese beetle grubs in New Jersey (Glaser & Farrell 1935).

As a result of these early studies, there are currently approximately 219 species of insect parasitoids and arthropod predators commercially available worldwide (Van Lenteren 2012), 226 commercial formulations of microbial products registered for use against pests and diseases in Europe and North America (Kabaluk & Gazdik 2007), but only 8 commercial microbial products registered worldwide for use against weeds (Charudattan 2001). Most of these products have been developed over the past 40 years in response to escalating interest in alternatives to chemical pesticides. However, the availability of such products is dynamic and subject to frequent change, and the extent of use is probably marginal for many, but consistent and significant for perhaps 10–15% of them.

8.1.2 Scope of Augmentation

Flanders (1930, 1951) was perhaps the first to distinguish between inundative and accretive methods of releasing mass-reared natural enemies, the former based on the activity of the released individuals alone, and the latter based primarily on the activity of subsequent generations. DeBach and Hagen (1964) replaced the term *accretive* with *inoculative* and introduced both augmentation and periodic colonization to encompass both methods of release. More recently, augmentation has been defined very simply as the release of large numbers of natural enemies with the goal of augmenting natural enemy populations or inundating pest populations with natural enemies (Collier & Van Steenwyk 2004).

DeBach and Hagen (1964) also considered both augmentation and conservation different aspects of a more general facilitation of natural enemies. While this concept has not gained broad acceptance and the two components of facilitation are more generally considered distinct approaches to biological control, Hoy (2008) included environmental manipulation to enhance the effectiveness of naturally occurring natural enemies as an aspect of augmentation. Here, however, we treat environmental manipulation as a component of conservation biological control (see Chapter 9). Despite some early successes, augmentation was not given serious consideration for the biological control of plant diseases until the 1970s when application of antagonists as seed or soil treatments, or to vegetative structures were typically referred to as *inoculation* (Baker & Cook 1974). Increased use of microbial control agents for suppression of insects, weeds and plant diseases has also led to widespread use of the term *biopesticide* (Crump et al. 1999; Glare et al. 2012). However, as the term *biopesticide* is frequently used for biological products other than natural enemies, following Crump and colleagues (1999), we encourage the use of more specific terms such as *mycoherbicide* and *viropesticide*. In addition, as the action of microbial control agents is not always to directly kill the target organism, particularly in the case of plant diseases and sometimes for weeds also, in such cases, the use of '-icide' may not be appropriate. In some cases, applications of microbial control agents have been considered distinct from other forms of augmentative biological control (Van Driesche et al. 2008), but we consider them equivalent here.

From a multidisciplinary perspective, Eilenberg and colleagues (2001) have suggested dropping the term *augmentation*, as it does not clearly describe the nature of the processes involved. Instead, these authors recommend using either *inoculation* or *inundation*, two terms that have long been used to distinguish between the two extremes of a continuum of natural enemy release strategies. Inoculation is often based on the use of smaller numbers of natural enemies and is used to ensure successful colonization at a critical period during the season or during an outbreak phase of a target organism. This approach aims to establish a sufficient number of natural enemies in a particular locality to allow their populations to grow, and over a series of generations or replication cycles, to suppress the abundance of their hosts or plant disease activity. The most important elements of the approach are the timing of the initial inoculation, and the potential for reproduction or replication between generations. Early-season inoculation can be an effective approach for augmenting the parasitoids and predators of arthropod pests in protected cropping, and the best-known example of this is the management of greenhouse whitefly, *Trialeurodes vaporariorum*, on tomato by the parasitoid *Encarsia formosa* (Hoddle et al. 1998). It can also be an effective approach to use with parasites and microbial pathogens that have an effective means of spreading through a host population. Examples of this are the inoculation of nucleopolyhedroviruses into forest plantations or stands at the start of an outbreak of Douglas-fir tussock moth (Shepherd et al. 1984) and the early induction of rust fungus epidemics on nut sedge through inoculation with *Puccinia caniculata* (Phatak et al. 1987). Similarly, the use of antagonists for control of plant diseases is frequently a case of inoculation, although sometimes mistakenly referred to as inundation (Hajek 2004; Glare et al. 2012). Despite the use of very large numbers for the inoculation of microbial antagonists, their populations turn over very rapidly following the colonization of plant surfaces, and thus it is subsequent generations that provide the effective disease suppression. One of the earliest examples of this is the application of *Phlebiopsis gigantea* to tree stumps after thinning of pine plantations for control of root and butt rots, *Heterobasidion* spp., in Europe (Rishbeth 1963).

Inundation differs from inoculation in that it is frequently based on the use of very large numbers of natural enemies employed to blanket an area for immediate effect at any time of the season or stage of an outbreak cycle. In contrast to inoculation, inundation relies on the capacity of the released natural enemies to provide immediate suppression and depends on the reliability and quality of mass production. Although insect parasitoids and predators can also be used, inundation is a particularly effective approach for microbial pathogens and some antagonists (Glare et al. 2012; Lacey et al. 2015). Good examples of the use of microbial control agents include various nucleopolyhedrovirus formulations for control of *Helicoverpa* species in a range of field crops in Australia (Buerger et al. 2007), the use of *Colletotrichum gloeosporioides* spp. *aeschynomene* for control of northern jointvetch, a weed of rice and soybeans in the southern United States (TeBeest & Templeton 1985), and the use of *Bacillus subtilis* strain QST-713 for control of a variety of fungi, including powdery mildews, white molds, *Sclerotinia* spp., and anthracnose, *Colletotrichum* spp., in many different crops (Marrone 2002).

One of the difficulties in distinguishing between inoculation and inundation as release strategies in augmentative biological control relates to differences in the biological traits of the natural enemies used. In this context, microbes differ from other groups of natural enemies in two important ways – their limited powers of dispersal necessitate the release of a very large number of individuals, and their very rapid generation times or replication cycles ensure that it is the progeny of released individuals that contribute to the process of suppression. For example, application rates of 10^{12} spores/ha are typical for mycoherbicides used in weed control (e.g., Kadir et al. 2000), and for mycoinsecticides used in insect control (e.g., Hunter et al. 1999), and these rates are very similar to the rates of 10^{13} colony-forming units (CFU)/ha for microbial antagonists used for suppression of plant diseases (e.g., Collins & Jacobsen 2003). Similarly, generation times of microbial pathogens (latent periods) are typically a few days while those of microbial antagonists are a matter of minutes or hours. Thus by focusing on application rates alone it might be concluded that all augmentative releases of microbial control agents would be examples of inundation. By returning to the original definition of Flanders (1930), however, it becomes clear that inundation should be restricted to releases where immediate effects are expected from the released individuals themselves rather than their progeny. Here we retain *augmentation* as a more general term to define the overall approach of periodic releases or applications of natural enemies, and use the terms *inoculation* and *inundation* to more specifically refer to the two extremes of augmentation, realizing that many applications of augmentation are likely to fall along the continuum between these two extremes.

8.2 Defining Success in Augmentative Biological Control

The success of augmentative biological control can be judged on the basis of five key criteria: technical effectiveness, public good, ease of use, commercial viability and safety (Gelernter & Lomer 2000; Van Lenteren 2000, 2003; Ravensberg 2011; Alabouvette et al. 2012; Jaronski 2012). *Technical effectiveness* refers to the ability of the released natural enemy to provide a predictable level of suppression under natural field conditions for the desired period of time, which may be the duration of a cropping cycle for an agricultural pest or pathogen or the duration of an outbreak cycle for a forest pest or pathogen. It is dependent on the initial selection of a natural enemy that is well matched to the target organism and environmental conditions, and subsequent quality control during mass production. In contrast, *public good* is an externality that refers to the extent to which the use of natural enemies can improve the quality of human life in terms of benefits to the environment and to human health and well-being. Environmental and human safety has been a critical driving force in the development of augmentative biological control agents as alternatives to the use of synthetic pesticides, and thus success could also be measured in terms of reductions in pesticide use. *Ease of use* is an undervalued criterion that can strongly influence the extent to which a product is adopted by end users, as practical difficulty can often detract from the overall success of an otherwise technically effective program. An unpredictable shelf life, poor formulation, ineffective delivery, a need for frequent applications or a requirement for complex instructions or application techniques can all present important barriers to acceptance and widespread adoption. *Commercial viability* has been a limiting factor for the success of many inundative biological control agents due to the economics of commercial production and the cost of registration relative to the size of the market and the price the end user can bear. As many inundative control products are based on natural enemies that are effective against a restricted range of target organisms and have a limited shelf life, commercialization is often characterized by niche markets, specific and limited periods of annual production and sales, and a need to remain competitive with chemical pesticides. Finally, *safety* is needed for success in augmentation and involves a risk assessment to evaluate potential effects on human and environmental health and on non-target organisms (see Chapter 5).

Despite these challenges, a number of augmentation programs have achieved a sufficient level of success to be viable operational programs with increasing usage on a worldwide scale.

8.3 Ecological Basis for Augmentation

8.3.1 Release Rate, Intrinsic Rate of Increase, and Persistence in Inoculative Biological Control

Inoculation as an approach to biological control is characterized by the release of natural enemies at critical times during the season or outbreak cycle. The ecological basis for effective inoculation has seldom been considered from a theoretical perspective, but at least in the case of trophic interactions involves the same types of consumer–resource interactions as considered for importation biological control (see Chapter 6). Using simplistic models to represent the seasonal inoculation of parasitoids for suppression of insect pests, Knipling (1992) estimated that successful control of a holometabolous insect pest would require as much as 80% parasitism by larval parasitoids. He further estimated that this level of parasitism could be achieved for many such pests if inoculative release rates were made at the start of the season using ratios of approximately two parasitoid adults to each pest adult. In addition, simulation models developed for specific pests suggest that the timing of inoculative releases may be even more important than the parasitoid-host ratios (Yano 1989; Flinn & Hagstrum 1995). Given the extensive use of analytical models to provide a framework for understanding potential mechanisms for success in importation biological control, it is perhaps surprising how little attention has been paid to inoculative biological control. Recent interest in the transient dynamics of consumer-resource interactions, however, may provide a better theoretical basis for further exploration of the theory of inoculative biological control (Kidd & Amarasekare 2012).

From an alternative perspective, both Van Lenteren and Woets (1988) and Janssen and Sabelis (1992) suggested that a comparison of the intrinsic rates of increase of natural enemy and pest could be an effective way to estimate the potential for successful inoculative biological control of arthropod pests in protected crops. Hance (1988) was perhaps the first to show that the success of inoculation with the predatory mite *Phytoseiulus persimilis* may stem from a development time that is half that of its prey, *Tetranychus urticae*, at 24°C, which gives it a superior intrinsic rate of increase. Additional observations of superiority in the intrinsic rate of increase of *P. persimilis* have subsequently been made by Krips and colleagues (1999). Similar comparisons have also been made of the intrinsic rates of increase of several parasitoids for inoculative control of the cotton aphid *Aphis gossypii* and of *Orius* species for inoculative control of western flower thrips, *Frankliniella occidentalis* (Yano 2006). Using a simple analytical model, Urano and colleagues (2003) further showed that the outcome of an inoculative release is determined not only by the relative intrinsic rates of increase of pest and natural enemy, but also by the per capita killing capacity of the natural enemy. They also showed that if the intrinsic rate of increase of the natural enemy is greater than that of the pest, any release ratio will lead to successful inoculation, but if the reverse is true, success can be achieved only through early releases and the use of specific release ratios.

The close link between the timing of a release and release rates in determining the potential for success was also apparent to the practitioners of inoculative biological control. This led Huffaker and Kennett (1956) to propose the 'pest-in-first' strategy for early season inoculation of predatory mites for control of the cyclamen mite, *Phytonemus pallidus*, in strawberries, in which phytophagous mites are introduced ahead of predatory mite releases to provide an early food source (see Section 8.4.1 for more on this method). This strategy has subsequently been used for inoculative biological control of both spider mites and whiteflies in protected cropping (Hussey & Scopes 1985). The linkage has also been shown to be of importance for the inoculative biological control of plant pathogens by antagonists (Kessel et al. 2002; Carisse & Rolland 2004). In some cases, however, it seems that release thresholds may be more important than release rates, as observed in the use of fluorescent pseudomonads for suppression of take-all disease in wheat (Figure 8.1). In this case, Raaijmakers and Weller (1998) found that a threshold density of 10^5 CFU/g of root was needed in the rhizosphere, below which disease suppression was poor, but above which there was no benefit of greater application rates. As indicated earlier, one alternative to matching timing and release rate in inoculative biological control is to consider release ratios of natural enemies to target organisms. These have frequently been tested for inoculative control of spider mites (Gillespie & Raworth 2004; Opit et al. 2004), with a ratio as low as 1 predator to 20 prey being sufficient to

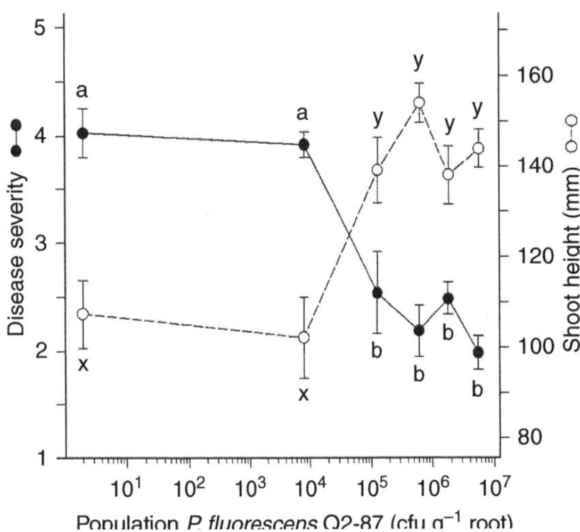

Figure 8.1. Antibiosis by *Pseudomonas fluorescens* provides an example of a threshold inoculum size (population density) needed to achieve successful suppression of take-all disease and improved growth of wheat in inoculative biological control (from Raaijmakers & Weller 1998).

achieve control, but a ratio of 1 predator to 4 prey recommended for rapid suppression.

Few studies have addressed the question of persistence of the natural enemy at lower pest densities in the context of success in inoculative biological control. The so-called killer strategy of the specialist predatory mite *P. persimilis* (Pels & Sabelis 1999) provides evidence that a natural enemy can drive its prey to extinction in a closed, spatially homogeneous environment (Ellner et al. 2001; de Oliveira et al. 2013). Thus at least in this case, persistence at lower prey densities in protected crops may not always occur, and either habitat structure or additional mid-season releases may be needed to ensure effective season-long suppression. Natural enemy persistence is also likely to be dependent on generation time, dispersal ability and diet specificity. Thus specialist natural enemies that have a short generation time and limited powers of dispersal, such as many insect baculoviruses, may have short persistence intervals. Fuller and colleagues (2012) estimated persistence interval for the baculovirus of gypsy moth, *Lymantria dispar*, to be less than three days, indicating that these traits would be better suited to inundative rather than inoculative releases.

From this it should be clear that our current understanding of inoculative biological control is based largely on insights gained from trial and error experiments in the laboratory and field. A more experimental and mechanistic approach to inoculation is clearly needed and would help to resolve which traits of a pest, natural enemy and environment are best suited to the development of successful inoculation programs.

8.3.2 Dose-Response, Attack Rate, and Release Strategies in Inundative Biological Control

Inundative biological control is characterized by the inundation of a local area with very large numbers of natural enemies for immediate suppression of crop damage. In his review of inundative releases, Stinner (1977) noted a lack of modeling studies with direct relevance to inundative biological control. One of the central tenets of inundative biological control was that effectiveness can be increased by releasing a greater number of mass-produced natural enemies at the target site (DeBach 1964; Templeton et al. 1979). However, Crowder (2007) noted that of 31 studies of predator and parasitoid releases, 64% did not show a significant effect of release rate. One of the key reasons for this is that various forms of environmental heterogeneity can create a dynamic refuge for the pest, allowing a proportion of the population to be protected from the natural enemies released, resulting in a steady reduction in per capita return as release rate is increased. Johnson (1994, 1999, 2010) introduced the idea of saturating dose-response models to describe 'the proportion of plant pathogen propagules rendered ineffective' as a result of the introduction of an antagonistic microbial control agent. However, the same model is applicable to all instances of inundative biological control, where the dose is represented by the magnitude or density of the natural enemy application, and the response can be represented either as an increase in the level of damage reduction caused by the pest to an upper asymptote (Figure 8.2 a, b) or as the decline in extent of damage caused by a pest to a lower asymptote (Figure 8.2c, d). The most important aspect of these dose-response models is the asymptote, which seldom corresponds to complete pest suppression and represents the size of the proportional refuge from natural enemy attack for the pest (see also Chapter 6). In the context of inundative biological control, the extent to which the asymptote departs from complete suppression also represents the proportion of the pest population that cannot be controlled by increasing the magnitude of the release. The only way to overcome the limitation of a pest refuge is to use more than one natural enemy species or strain such that the

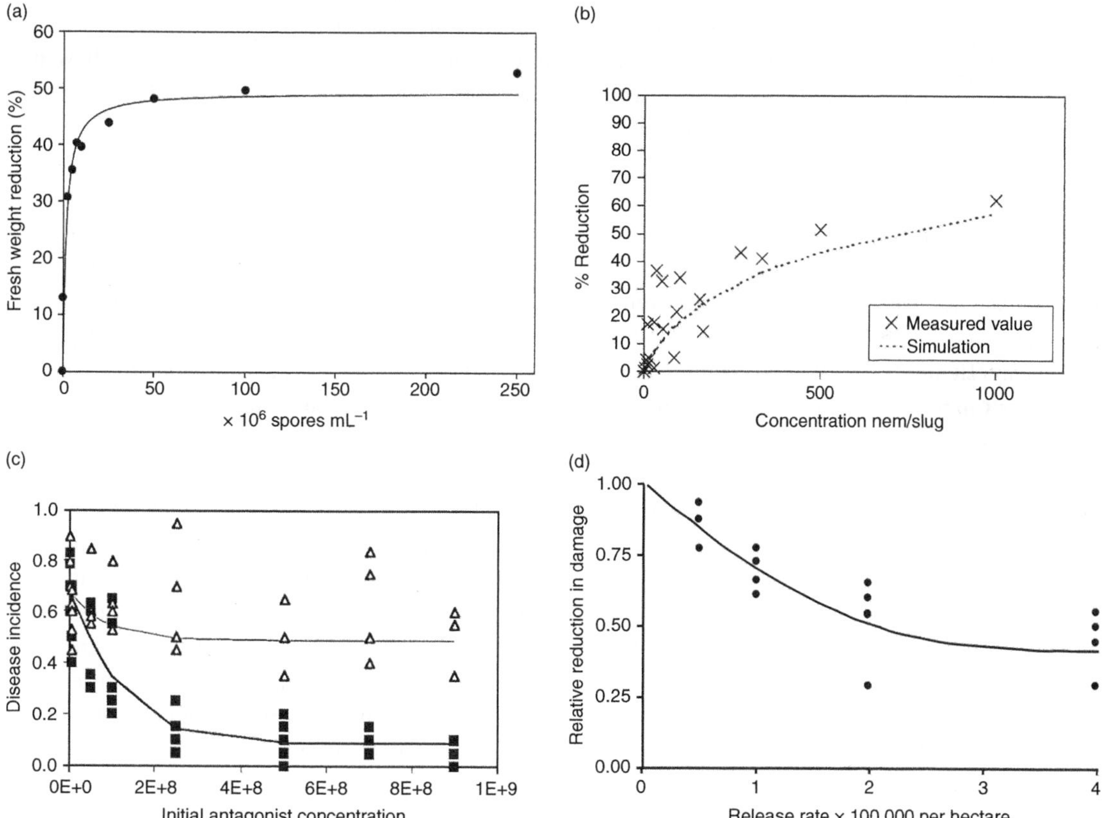

Figure 8.2. Examples of dose-response curves from inundative biological control: (a) fresh weight reduction of scentless chamomile, *Matricaria perforata*, in relation to spore concentration of the fungal pathogen *Colletotrichum truncatum* (from Graham et al. 2006); (b) reduction in feeding by slugs *Deroceras reticulatum*, in relation to concentration of the nematode parasite *Phasmarhabditis hermaphrodita* (from Wilson et al. 2004); (c) reduction in disease incidence of tomato plants from infection with *Agrobacterium tumefaciens* in relation to initial concentration of the antagonist *A. radiobacter* (from Johnson & DiLeone 1999); and (d) reduction in fruit and nut damage by codling moth, *Cydia pomonella*, in relation to release rate of the egg parasitoid *Trichogramma platneri* (after Mills 1998).

refuge from one natural enemy is broken by another that occupies a rather different ecological niche (see Pedersen & Mills 2004; Xu et al. 2011).

Why should a pest refuge exist in inundative releases? The factors most likely to contribute to a refuge are spatial heterogeneity, genetic heterogeneity of the pest and efficiency of the natural enemy. In many cases, the spatial activity of a natural enemy and its target pest may not completely overlap, as discussed earlier for California red scale and its parasitoid *Aphytis melinus* on the trunk versus twigs of citrus trees (see Chapter 6). Similarly, in almost all cases, the natural distribution of a pest is aggregated while the release distribution of the natural enemy is uniform, and this inevitably generates heterogeneity in the risk of the pest to the action of the natural enemy (Adams 1990; Cronin & Reeve 2005). In addition, most pest populations will be genetically diverse, unless they are invasive species that have gone through a substantial genetic bottleneck, and even so may show variation in their susceptibility to the natural enemy (Henter & Via 1995; Mazzola et al. 1995; and see Chapter 7). Finally, efficiency can vary extensively among populations or strains of natural enemies and can result in distinct differences in their ability to impact the pest (Smith et al. 1997; Kraaijeveld 2004).

So far, dose-response models have been used to better understand natural enemy inundation for control of only a few plant diseases (Johnson 1999; Shishido et al. 2005), insects (Mills et al. 2000), slugs (Wilson et al. 2004) and weeds (Graham et al. 2006). However, there is still a need for greater recognition that refuges can set upper limits to the level of pest suppression that can be achieved through inundative biological control. Dose-response models provide an effective way to estimate refuge size and the diminishing returns associated with greater release rates, and should be used more extensively in the development of inundative biological control programs.

Host-parasitoid models have also been used to investigate the minimum release rates needed for pest suppression. Using a wide variety of analytical host-parasitoid models, Barclay and colleagues (1985) were able to show that the minimum per generation release rate required to reduce a host population to zero is given by the ratio of the intrinsic rate of increase of the host to the attack rate of the parasitoid. Suppression of the host population to zero can take a large number of generations at the minimum release rate, but a release rate of double the minimum can do so in five to six generations. Using a simple individual-based model of a host-parasitoid interaction, Grasman and colleagues (2001) showed that if the growth rate of the parasitoid is greater than that of the host, host suppression will be achieved for all initial release rates (i.e., an inoculative release is sufficient). If the reverse is true, however, then the minimum parasitoid release rate needed to achieve host suppression must satisfy the inequality that the initial number of hosts per parasitoid should not exceed the ratio of the maximum per capita attack rate of the parasitoid and the minimum growth rate of the host population. Thus both the inoculation model of Urano and colleagues (2003) and the inundation model of Grasman and colleagues (2001) agree that inoculation is a sufficient strategy when the growth rate of the enemy is greater than that of the pest. Similarly, the inundation models of Barclay and colleagues (1985) and Grasman and colleagues (2001) agree that minimum release rates need to be greater for parasitoids with lower attack rates, but differ with respect to whether release rates depend on current levels of host abundance.

Simulation models have also been developed to address specific aspects of inundative releases for particular pests. For example, Smith and You (1990) explored the frequency of releases of the egg parasitoid *Trichogramma minutum* for suppression of spruce budworm, *Choristoneura fumiferana*, a univoltine forest pest. While a double release of parasitized eggs four days apart provided greater suppression than a single release, a single release of parasitized eggs with parasitoid emergence staggered over a five-day period provided an equivalent level of suppression. Similarly, Heinz and colleagues (1993) addressed timing and release rates for *Diglyphus begini*, a parasitoid of leafminers *Liriomyza trifolii* on chrysanthemums in greenhouses. They found that in order to keep larval leafminer densities below 1 per 1,000 chrysanthemum leaves within 40 days of planting, releases must be initiated within the first 14 days, and that while influenced by current larval density, effective release rates did not conform to a simple ratio of parasitoids to hosts. A detailed simulation model is also currently under development for inundative releases of a pupal parasitoid *Spalangia cameroni* for control of stable flies *Stomoxys calcitrans* in livestock facilities (Skovgard & Nachman 2015a, b). While simulation models can be of considerable value in guiding the development of parasitoid release programs for specific pests, they provide few generalizations that can be applied more broadly for inundative biological control.

8.4 Spider Mites, Locusts and Take-All: Case Studies of Augmentation

While it is not our intention to provide an exhaustive review of the various augmentative biological control projects implemented around the world, a selection of three case studies will be helpful to illustrate both the ecological basis and practical development of augmentation programs. These case studies have been selected to represent some of the different groups of pests and natural enemies that have been used and to highlight some of the more interesting successes in the context of both inoculation and inundation. The three case studies include an arthropod predator of an arthropod pest, a fungal pathogen of an insect pest and a bacterial antagonist of a fungal plant pathogen.

8.4.1 *Phytoseiulus persimilis* for Inoculative Biological Control of Spider Mites

One of the better known applications of inoculative biological control concerns the use of the predatory mite *P. persimilis* for control of the two-spotted spider mite, *Tetranychus urticae*, and the carmine spider mite, *T. cinnabarinus*, in glasshouse vegetable crops and field-grown strawberries (McMurtry 1991; Gillespie & Raworth 2004). *P. persimilis* was initially collected in Chile in 1959 and has been mass reared since 1968, with its extent of use rising steadily through the 1970s and 1980s to its current level of at least 6,000 ha of glasshouse vegetable crops and 20,000 ha of field-grown strawberries (Van Lenteren & Woets 1988; Van Lenteren 2000). The lifestyle of *P. persimilis* has been categorized as type I, specialist predators that are well adapted to feeding on web-spinning *Tetranychus* mites, and particularly suitable for use in inoculative biological control (McMurtry & Croft 1997; McMurtry et al. 2013). Such predators have stronger functional and numerical responses to their prey, and thus lower ratios of predators to prey are needed to achieve effective control. The predation rate of *P. persimilis* has been estimated to be as high as 45 spider mite eggs/day (Skirvin & Fenlon 2003), its reproductive capacity peaks at 5.6 eggs/female/day over a 22-day reproductive period (Drukker et al. 1997; Gillespie & Raworth 2004), and its sex ratio can increase from 50% to 80% female at high prey densities (Toyoshima & Amano 1998). Its dispersal strategy has also been characterized as 'killer', indicating that it drives local prey populations to extinction before the final individuals disperse from a patch (Pels & Sabelis 1999). The latter is an unusual trait among arthropod natural enemies, as they more typically leave patches well before completely exploiting their hosts (Heimpel & Casas 2008; Mills & Wajnberg 2008). *P. persimilis* is attracted to plants with spider mites via volatile cues induced by feeding damage (Janssen 1999), and can learn to respond to particular volatile components, such as methyl salicylate, which appears to play an important role in the foraging behavior of this predatory mite (de Boer & Dicke 2005).

P. persimilis is mass reared by commercial insectaries and sold worldwide (Van Lenteren 2012). It is generally formulated with vermiculite as a carrier in plastic bottles from which they can be sprinkled onto the surface of leaves after gentle mixing. An alternative formulation includes either spider mite eggs or spider mites on bean leaves to supply the predatory mites with a food source during transit. For control of spider mites in protected crops, the predatory mites are applied using one of two strategies: pest-in-first or in response to detection of infestation (Hussey & Scopes 1985). For the pest-in-first strategy on cucumbers, 10 female spider mites are released per plant, followed 10 days later by 1 predatory mite on every fifth plant. While this approach works well, most growers prefer to use the release on detection of infestation strategy, for which an application rate must be determined. Application rates can either be fixed, commonly at 5–15 per plant (approximately 50,000–200,000 per ha), or can be based on predator:prey ratios, commonly 1:10 or 1:20 (Gillespie & Raworth 2004). An example using predator:prey ratios of 1:20 and 1:4 for control of two-spotted mite, *Tetranychus urticae*, on ornamentals is shown in Figure 8.3. The

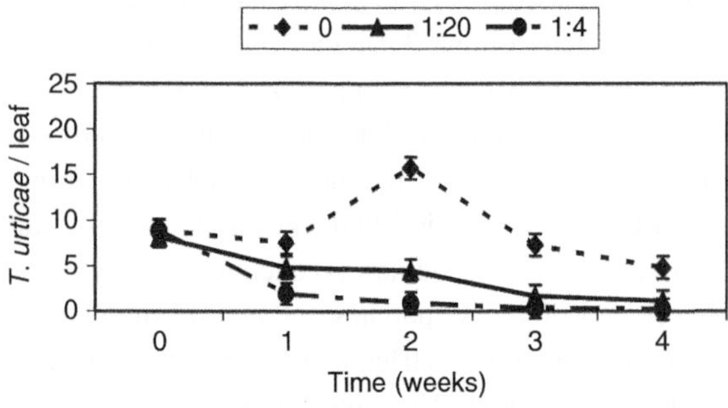

Figure 8.3. Inoculative release of *Phytoseiulus persimilis* for control of two-spotted mite, *Tetranychus urticae*, on ivy geraniums in a glasshouse, showing the effect of different predator:prey release ratios (from Opit et al. 2004).

fixed-rate applications are less precise as the spider mite population size is unknown and the time of detection of infestation is dependent on grower monitoring practices, while predator:prey ratios are theoretically more efficient, but suffer from the need for more detailed monitoring and the knowledge that spider mite populations could increase 10-fold in the 7–10 days between monitoring and application. *P. persimilis* is less effective on crops, such as tomatoes, that have numerous trichomes, and their use on ornamentals is limited by the near-zero tolerance for spider mite damage. In addition, in warm or dry environments *P. persimilis* may not be sufficiently effective as its population growth is unable to match that of its spider mite prey at temperatures above 30°C or relative humidities below 60% (Gillespie & Raworth 2004).

Inoculative applications of *P. persimilis* have also been effective in field-grown strawberries in southern California (McMurtry 1991), southern Florida (Decou 1994) and southeastern Queensland (Waite 1988). However, as *P. persimilis* is more tolerant of some miticides (e.g., abamectin and acramite) than *T. urticae* is, combination IPM treatments have attracted attention and provide the best economic return in some cases (Trumble & Morse 1993). Perhaps due to the greater environmental variation and levels of mite colonization that are inherent in field crops in comparison to protected crops, integrated mite control has become more typical of field-grown strawberries (Cross et al. 1996; Rhodes & Liburd 2006).

8.4.2 *Metarhizium acridum* for Inundative Biological Control of Locusts

One of the most promising recent developments in the use of inundative biological control began with the LUBILOSA project in Africa. This project resulted from the desert locust outbreak in the 1980s, and was initiated and driven by the need to find a biological control alternative to the organophosphates that were used extensively at the time (Lomer et al. 2001;

Hunter 2005; Langewald & Kooyman 2007). In the search for suitable natural enemies, an important breakthrough came with the discovery that fungal pathogens have much greater effectiveness if formulated in oil rather than water (Prior et al. 1988). Oil formulation lends itself to ultra-low-volume (ULV) application, a distinct advantage in arid regions where non-water-based formulations are needed for locust control. By focusing on the collection of diseased individuals from field cages, *Metarhizium* was found to be the most frequent fungal pathogen of locusts and grasshoppers in West Africa, although its prevalence in natural host populations was typically low at 2–6% (Shah et al. 1998; Lomer et al. 2001). A series of isolates was obtained and screened for virulence before IMI330189, an isolate found to be virulent against a range of Sahelian grasshopper and locust species (family Acrididae), was selected for development as a mycoinsecticide (Bateman et al. 1996). Initially known as *M. flavoviride*, this isolate has since been designated as *M. anisopliae* var. *acridum* (Driver et al. 2000), and now as *M. acridum* (Bischoff et al. 2009). A particularly interesting aspect of the interaction of *M. acridum* with grasshoppers and locusts is that the latter can actively thermoregulate by altering their orientation to the wind and sun. As a result, there is evidence that in some cases, hosts can overcome infection through a behavioral fever response to elevate their normal body temperature to a level that exceeds the upper tolerance of the fungus (Blanford et al. 1998; Ouedraoga et al. 2004). Similarly, in its gregarious phase, the desert locust has been found to be more resistant to infection, but in this case due to elevated antimicrobial activity rather than behavioral fever (Wilson et al. 2002). Both of these dynamics have important implications for application of *M. acridum* against locusts in warm climates.

As a deuteromycete fungus, *M. acridum* has a simple life cycle with conidia produced as the only durable infective spore stage, while unstable blastospores are formed after infection of a host by budding of the mycelium into the hemocoel (Lomer et al. 2001). However, mass production as well as formulation proved a key issue for the development of *M. acridum* as a mycoinsecticide, as aerial conidia produced through solid-state fermentation with a lower moisture content both had a greater shelf life than submerged conidia produced through liquid fermentation (Moore et al. 1996; Watanabe et al. 2006), and proved more effective in field trials (Kassa et al. 2004). In addition, while blastospores from submerged culture are more pathogenic than either submerged or aerial conidia, they have less stability, and are intolerant of drying (Leland et al. 2005). Solid-state fermentation is ideal for small-scale production (Cherry et al. 1999), but scaling up production of *M. acridum* for use in Africa proved difficult and the commercial product Green Muscle™ has yet to achieve full operational use in Africa. This is due both to the technical problem of temperature control in solid-state fermentation for consistent yield and quality, and the sporadic nature of the outbreaks of the target pest, which does not provide a consistent market for production (Lomer et al. 1999; Langewald & Kooyman 2007). The initial success of the LUBILOSA project also led to testing of local isolates of *M. acridum* in Australia, Brazil, Madagascar and Mexico against different species of locust and grasshopper (Hunter 2005). While recent advances may offer new opportunities for large-scale solid-state fermentation (Ye et al. 2006), only Australia has found a sufficiently consistent market to produce the local isolate FI-985 for operational use under the name Green Guard™. This latter isolate not only has a higher yield than IMI330189, but also has greater UV tolerance, which could make the secondary pick up of spores from residues on vegetation a more important route of infection given a greater duration of field activity (Scanlan et al. 2001).

Initial field trials with Green Muscle™ were made against bands of red locust, *Nomadacris septemfasciata*, nymphs in an outbreak zone in Mozambique using knapsack sprayers and

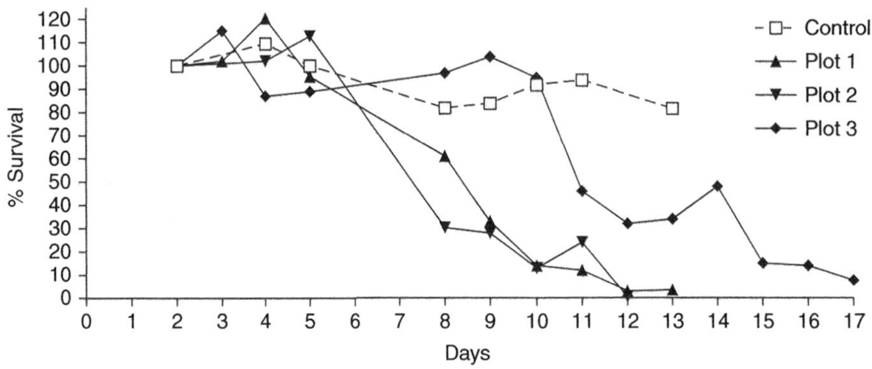

Figure 8.4. Survival of the Australian plague locust, *Chortoicetes terminifera*, in three treatment plots treated with a low dose application (12.5 g/ha) of a fungal pathogen, *Metarhizium acridum* (Green Guard™), relative to that in a control plot with no application (from Milner & Hunter 2001).

application rates of 5×10^{12} conidia/ha (Price et al. 1999). From samples caged after application, these trials resulted in 98–100% kill of third instars in 8–9 days (median time to kill 3.8 days) and > 90% kill of fourth instars after 21 days (median time to kill 6.0 days). Early field trials of *M. acridum* against the Australian plague locust, *Chortoicetes terminifera*, were conducted using a high dose of 100 g conidia/ha (Hunter et al. 1999), but as this locust is a more susceptible species, lower doses of 25 g/ha and even 12.5 g/ha have given 90% control of nymphs in 14 days (Milner & Hunter 2001; Figure 8.4). Green Guard™ has been used operationally for control of the Australian plague locust since 2000, and has subsequently been field tested for registration and use against the oriental migratory locust, *Locusta migratoria manilensis*, in China (Hunter 2005).

8.4.3 *Pseudomonas fluorescens* for Inoculative Biological Control of Take-All Disease

Plants commonly have resistance genes for foliar pathogens, but not for soil-borne diseases, and so rely on antagonistic microorganisms in the rhizosphere as an alternative means of defense (Weller et al. 2007). Plant roots release relatively large amounts of organic material into the soil in the form of exudates and secretions, which leads to the enrichment of microbial populations, including antagonists (Whipps 2001). Although many microbial antagonists appear effective during trials at a laboratory or a greenhouse scale, few have yet to provide consistent activity under field conditions, suggesting that we still have much to learn about the effectiveness of individual strains within the soil microbial community and how biological control activity is influenced by environmental conditions. In this context, take-all disease caused by *Gaeumannomyces graminis* var. *tritici* is not only one of the most important root diseases of wheat worldwide, but also provides one of the best-studied examples of the role of suppressive soils in reducing the prevalence of disease (Cook 2007; Weller et al. 2007; Kwak & Weller 2013). The well-known phenomenon of take-all decline, in which soil in continuous monocultures of wheat becomes suppressive to take-all after one or more outbreaks of disease and is both transferable to conducive soils and eliminable by sterilization, is suggestive of the action of living microbial antagonists (see Chapter 9). Suppressive soils have both general activity against take-all from the standpoint of total microbial biomass, and specific activity, which in the Pacific Northwest of the United States has been shown to be due to the action of *Pseudomonas fluorescens*, which produce the phenazine antibiotic 2,4-diacetylphloroglucinol (DAPG) (Raaijmakers & Weller 1998).

Pseudomonas species are aerobic gram-negative bacteria that are ubiquitous in soils and have high rhizosphere competence, meaning that they are particularly well adapted to colonizing and growing on the roots of plants (McSpadden Gardener 2007; Mercado-Blanco 2015). The mode of action of the plant-growth-promoting pseudomonads include production of a variety of antibiotics, production of siderophores that mediate competition for iron, induced systemic resistance and production of lytic enzymes (Haas & Défago 2005; Bakker et al. 2007). The key role of the phenazine antibiotics produced by *P. fluorescens* in biological control of take-all disease was very elegantly demonstrated by Thomashow and Weller (1988) in comparative experiments using bacterial colonies that were either parent, had a single mutation to delete the biosynthetic gene or had genetic complementation to restore the target gene. In addition, early experimental and commercial scale trials of strain 2–79, isolated from take-all suppressive soils in the Pacific Northwest, reduced disease incidence by 17% and 11%, respectively (Weller 1988). Suppression of take-all disease continues to be focused on strains with the *phlD* gene that produces DAPG, but it is now known that there are many different genotypes that are detectable and quantifiable using allele specific primers (De La Fuente et al. 2006; Weller et al. 2007).

As noted earlier, suppression of take-all disease requires a threshold population density of 10^5 CFU/gm of the DAPG-producing isolate Q2–87 (Raaijmakers & Weller 1998; Figure 8.1). These densities can be found naturally in suppressive soils and can be achieved by adding suppressive soil to a conducive soil in a ratio of 1:9, by inoculation of conducive soils with strain Q8r1–96 using a dose of 10^2 – 10^4 CFU/g of soil, or by using seed treatments with strain 2–79 (Weller & Cook 1983; Raaijmakers & Weller 1998). The competitive status of the different genotypes of *P. fluorescens* can be influenced by the crop plant and crop variety, and isolate Q8r1–96 (now known as *P. brassicacearum*) has a particularly strong affinity for the roots of wheat, with populations responding to wheat root exudates more rapidly and maintaining higher population densities in the rhizosphere than other *phlD+* genotypes (Figure 8.5).

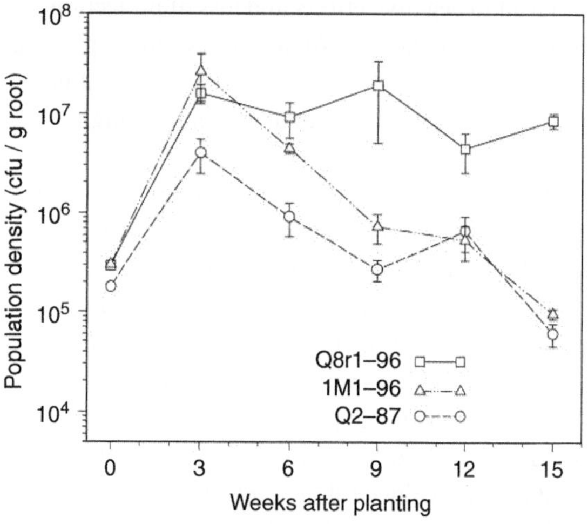

Figure 8.5. The rhizosphere competence of different *phlD+* genotypes of fluorescent pseudomonads showing the superior performance of strain Q8r1–96 that maintained population densities consistently greater than the threshold of 10^5 CFU/g of root, with other strains being 100 to 1000 fold lower in density after eight successive growth cycles (from Weller et al. 2002).

In some cases, the effectiveness of fluorescent pseudomonads is more limited or inconsistent, indicating the equivalent of a refuge for the soil-borne pathogen from antagonist activity. In such cases, it has been suggested that the combination of two or more antagonists with different modes of action can break the refuge and provide better suppression of disease (Pierson & Weller 1994; Duffy et al. 1996; Guetsky et al. 2002). However, Becker and colleagues (2012) caution that too rich a combination of fluorescent pseudomonad genotypes can lead to negative effects and loss of protection from fungal pathogens. An alternative for improving the success of pseudomonad inoculation might be to combine multiple modes of action in a single recombinant isolate. An example is the insertion of the *phzABCDEFG* operon from isolate 2–79 that produces the antibiotic phenazine-1-carboxylic acid (PCA) into isolate Q8r1–96. This recombinant strain is not only suppressive of take-all through production of DAPG, but is also suppressive of rhizoctonia root rot through production of PCA (Huang et al. 2004). In principle, it would take only a one-liter culture of this transgenic at 10^{10} CFU/ml to treat all the seed needed to plant 4,000 ha of wheat and to provide suppression of both diseases.

8.5 Ecological Aspects of the Selection of Natural Enemies for Augmentation

One point that all disciplines emphasize is that a thorough understanding of the ecology of a natural enemy and its interaction with its host is vital, and that an absence of knowledge is almost certain to lead to failure in the development of augmentative biological control programs. In this context, there are a number of important questions that need to be asked before selecting the most effective natural enemies to use in an augmentation biological control program.

8.5.1 Is It Necessary to Match the Natural Enemy to the Pest?

Many mistakes can be made in the selection of natural enemies for augmentation programs. For both inoculation and inundation, the natural enemy must be amenable to production if it is to be widely available for implementation. For a production facility, the broader the host range of a natural enemy, the easier the production, the lower the cost and the greater the market size for the product. From a technical viewpoint, however, the specificity of the natural enemy can be an essential component of its effectiveness, which sets up a potential conflict between commercial viability and technical effectiveness. For most groups of natural enemies, host specificity seems key to success in augmentative biological control (Jackson et al. 2000; Van Lenteren 2003; Charudattan 2005; Glare et al. 2012). There are some examples, however, of microbial control agents that are either insect pathogens or antagonists of plant diseases where a broad host range appears compatible with effectiveness (Lacey et al. 2001; McSpadden Gardener 2004; Harman 2006).

The well-known bacterial pathogen of insect pests, *Bacillus thuringiensis*, presents no trade-off between ease of production and technical effectiveness and has a broad spectrum of activity. As a result, it has achieved much greater market penetration than any other microbial product, and this has led to its prominence as the most successful and best-known microbial control agent (Federici 2005; Lacey et al. 2015). Similarly, the multiple modes of action of *Bacillus* (Marrone 2002; McSpadden Gardener 2004) and *Trichoderma* (Harman 2006), which can include induction of host resistance and promotion of plant nutrition and growth in addition to antagonism of plant pathogens, offer more extensive opportunities for marketing in the control of plant diseases. In contrast, the widespread commercialization of *Trichogramma* egg parasitoids as inundative control agents for lepidopteran pests in a wide range of agricultural crops and forests has been driven almost entirely by ease of production and breadth of application rather than effectiveness in suppressing pest abundance (Smith 1996), although large-scale implementation continues to be restricted to regions where costs of production are lower (Mills 2010). The importance of technical effectiveness versus commercial viability is far better illustrated by the meager representation of mycoherbicides for weed suppression among the full spectrum of microbial control agents, where specificity of the natural enemy has been paramount for success (Charudattan 2005). Thus, for augmentative biological control to play an effective role in pest and disease management in the future, technical effectiveness is the first priority, and this may often mean that natural enemies have to be selected very carefully for match to the target pest. This can often necessitate the screening of a large number of isolates for microbial natural enemies, particularly those isolated from the target host itself (e.g., Pilz et al. 2007). However, Gressel (2002) and Hallett (2005) argue that a broad host range should not be excluded from consideration in the development of mycoherbicides, as this would alleviate the conflict between specificity and marketability, if effectiveness could be improved with targeted enhancement of virulence through recombinant gene technology. For example, insertion of the virulence gene *NEP1* from *Fusarium oxysporum* into *Colletotrichum coccodes* increased its virulence by a factor of nine against velvetleaf, *Abutilon theophrasti*, and also showed an increase in host range and a reduction in the leaf wetness requirement of the fungus (Amsellem et al. 2002).

8.5.2 Does the Natural Enemy Need to Be Matched to the Environment?

Knowledge of the ecology of a natural enemy is critical for understanding potential limitations imposed by the environmental conditions in which the target pest occurs. Many generalist arthropod natural enemies can be habitat specialists in terms of either vertical stratification or the range of plants or plant parts on which they will search for hosts. Natural enemies from low-growing vegetation are ineffective in trees, and vice versa, and such behavioral adaptations are well known, for example, among *Trichogramma* egg parasitoids (Thorpe 1985; Mills 2003; Romeis et al. 2005), and green lacewing predators (Stelzl & Devetak 1999; Tauber et al. 2000). A similar vertical stratification occurs among entomopathogenic nematodes, with *Steinernema carpocapsae* active near the soil surface and *S. glaseri* and *Heterorhabditis bacteriophora* better adapted to search within the soil profile (Gaugler et al. 1997a). Habitat specialization of generalist natural enemies also extends to plant species and even to plant parts. Plant volatiles and plant surface chemicals can both be important in influencing the search of highly mobile natural enemies. Plant volatiles that are induced by herbivore feeding or oviposition can vary between host plants and are used by parasitoids and predators as long-range foraging cues (Takabayashi & Dicke 1996; Mumm & Dicke 2010). Similarly, some plants may be rejected by natural enemies due to surface chemistry, which can act as a deterrent or toxicant for *Trichogramma* egg parasitoids (Romeis et al. 2005). Finally, trichomes are an important component of the microstructure of some plants and can affect the success of natural enemy foraging. For example, cucumber is a far less suitable glasshouse crop than tomato for successful inoculative release of the parasitoid *Encarsia formosa* due to its higher quality for whitefly population growth and its retentiform venation and large trichomes that impede parasitoid search (Woets & Van Lenteren 1976).

Abiotic factors can also limit the potential of augmentative control with temperature, humidity or soil moisture and UV radiation being among the most important. For example, Bourchier and Smith (1996) found that temperature could account for 75% of the variation in parasitism of spruce budworm from inundative releases of *Trichogramma minutum* in the temperate forests of Canada, while Grewal and colleagues (1994) suggested that entomopathogenic nematode species have distinct thermal niches that may be invariant across populations from different localities. Humidity is the limiting factor for fungal spores applied to aerial plant parts, as they require sufficient free moisture for several hours to achieve germination. Fortunately, formulation in oil can greatly enhance infection of hosts under conditions of low relative humidity and has become more widely used in the development of mycopesticides following the success of the LUBILOSA project discussed previously. UV radiation can rapidly degrade microbial natural enemies, and considerable effort has gone into screening microbial control agents for UV tolerance (Braga et al. 2001; Lacey & Arthurs 2005; Fernandes et al. 2015).

8.5.3 Can the Effectiveness of the Natural Enemy Be Improved?

Enhancing the characteristics of natural enemies through selection, hybridization or recombinant DNA technology has been considered for some time (Miller et al. 1983; Beckendorf & Hoy 1985). In the case of arthropod natural enemies, there have been few developments in this area due to a poor understanding of the genetic basis for virulence, and a lack of consensus about which factors are limiting for this group of natural enemies. One form of improvement for an arthropod natural enemy was the use of selection through traditional breeding to produce a pesticide-resistant strain of the predatory mite *Galendromus occidentalis* for more effective control of spider mites in almonds in California (Hoy 1985). However, technical difficulties associated with insertion of genes into arthropod nuclear genomes and safety concerns have prevented the development of any transgenic parasitoids and predators (Hoy 2000b). Similarly, few enhancements have been achieved with entomopathogenic nematodes, although they are considered good targets for genetic improvement (Segal & Glazer 2000; Grewal et al. 2006), as sensitivity to environmental extremes has been identified as one of the major limitations for effectiveness (Kaya & Gaugler 1993). However, targeted hybridization to improve effectiveness has been achieved in *Steinernema carpocapsae* by successfully combining the characteristics of a strain that had high virulence against the pecan weevil, *Curculio caryae*, but low heat and desiccation tolerance with a different strain with low virulence and high heat and desiccation tolerance (Shapiro et al. 1997; Shapiro-Ilan et al. 2005). Alternatively, improved heat tolerance in *Heterorhabditis bacteriophora* was achieved through transformation of a wild-type strain with the heat shock gene *hsp70A* from the free-living nematode, *Caenorhabditis elegans*. The transgenic strain proved 18 times more tolerant of heat shock than the wild-type strain (Hashmi et al. 1998), with no impairment of its virulence against 11 different invertebrate hosts (Wilson et al. 1999). In contrast to arthropod natural enemies and entomopathogenic nematodes, gene technology has advanced more rapidly in the context of microbial control agents where individual gene products have been shown to enhance characteristics such as virulence (Hallett 2005; Inceoglu et al. 2006; Loper et al. 2007).

Baculoviruses have been extensively used in the control of insect pests, but the time taken to kill an infected host is slow and is considered an important barrier to wider adoption (Lacey et al. 2015). Early attempts at genetic modification led to little improvement until recombinant forms that incorporated insect selective toxins, particularly the *AaIT* gene isolated from the desert scorpion *Androctonus australis*, were developed in the early 1990s (Inceoglu et al. 2006). Speed of kill was improved by 40% (Figure 8.6), and

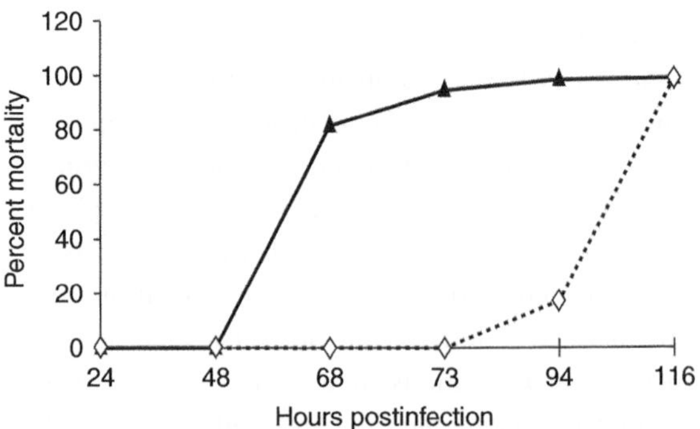

Figure 8.6. The speed of kill of neonate larvae of *Heliothis virescens* infected by two different strains of a baculovirus isolated from *Autographa californica*, the wild-type AcMNPV (diamonds and dotted line) and a recombinant strain that incorporates a scorpion toxin gene AcAaIT (triangles and solid line). The speed of kill by the recombinant strain is about 40% faster (from Inceoglu et al. 2006).

the paralyzing action of the toxin reduced host larval feeding by 60–70% (Hoover et al. 1995). As a result, recombinant forms of two different nucleopolyhedroviruses have been successfully field tested, AcMNPV-AaIT against *Trichoplusia ni* (Cory et al. 1994; Cory 2007), and HaSNPV-AaIT against *Helicoverpa armigera* (Sun et al. 2004). Speed of kill has also been a concern in the development of fungal pathogens of insects as mycopesticides, but in this case the limiting step is the successful penetration of the host cuticle by extracellular chitinases and proteases. Transformants of *M. anisopliae* with overexpression of a protease gene *Pr1A* (St. Leger et al. 1996, St. Leger 2007) and of *B. bassiana* expressing a hybrid chitinase gene *Bbchit1-BmChBD* (Fan et al. 2007) have both been shown to improve effectiveness by reducing time to death by 23–25%.

Recombinant technology has also been used to engineer vertebrate viruses as vectors for immunocontraception. Reducing fertility rather than increasing mortality is an approach that is favored for the biological control of vertebrate pests (Tyndale-Biscoe 1994). While autoimmune infertility or immunocontraception has been shown to be effective for free-ranging wildlife through injection (Kirkpatrick & Frank 2005), species-specific viral vectors offer an opportunity to apply immunocontraception to more widespread wildlife populations. Research in Australia has focused on recombinant viral vectors that express contraceptive antigens derived from the zona pellucida, a protein matrix that surrounds the oocyte (Hardy et al. 2006). Laboratory experiments have shown that high levels of infertility can be achieved in mice infected with recombinant ectromelia virus and in rabbits infected with recombinant myxoma virus. It is believed that sterility in as many as 70–80% of females is needed to suppress population size, and while initial trials have shown that the recombinant viral vectors can persist and spread in the field, there are significant regulatory and safety hurdles to overcome, and it seems unlikely that they will be pursued for management of invasive wildlife populations (Saunders et al. 2010).

Species in the genus *Trichoderma* have been extensively used for the control of aerial and soil-borne plant diseases, and there are many commercially registered products (Kabaluk & Gazdic 2007). A series of cell wall degrading enzymes, chitinases, gluconases and proteases play a crucial role in mycoparasitism by *Trichoderma*, and transformations to overexpress the genes coding for these enzymes have improved their effectiveness in suppressing root diseases (Baek et al. 1999; Pozo et al. 2004; Djonovic et al. 2007). For example, double overexpression transformants of *T. virens* that simultaneously overexpress two gluconase genes, *TvBgn2* and *TvBgn3*, have been effective in providing greater protection of cotton seedlings against *Pythium ultimum, Rhizoctonia solani* and *Rhizopus oryzae* (Djonovic et al. 2007). Similarly, a transgenic strain of *Trichoderma atroviride* that expresses the glucose oxidase gene *goxA* from *Aspergillus niger* was more effective than the wild type in protecting beans from high inoculum levels of *P. ultimum* and *R. solani* (Brunner et al. 2005).

Despite the potential for gene technology to enhance the beneficial traits of microbial control agents, there has been little interest in developing such products for commercial use. Risk assessment and regulation continue to be significant barriers for the commercialization of microbial products (Alabouvette et al. 2012; Jaronski 2012), which together with negative public opinion and strict regulations for genetically modified organisms may have prevented recombinant products from reaching the market (Glare et al. 2012).

8.5.4 Would Multiple Natural Enemies Reduce the Potential for a Pest Refuge?

As discussed earlier, in many situations, individual natural enemy species fail to provide a sufficient level of suppression due to refuge effects that limit natural enemy access to a pest. In such cases, it is worth considering whether a combination of natural enemy species rather than just one would provide a greater level of pest suppression. It must be remembered, however, that when two or more species of natural enemy share a common target pest, they may interact through interference or intraguild predation/antagonism, in which case it is important to be able to predict whether a combination of natural enemies will have a positive, neutral or negative impact on pest suppression. This may be of greatest concern for inoculative releases where interactions over several natural enemy generations are anticipated, and of less concern for inundative releases, where the interaction is confined to a single generation or replication cycle. However, some negative effects have been recorded even for inundative releases. For example, a combination release of the egg parasitoids *Trichogramma nubilale* and *T. ostriniae* proved less effective than a release of *T. ostriniae* alone in suppressing European corn borer, *Ostrinia nubilalis*, in sweet corn (Wang et al. 1999).

From a theoretical perspective, since the dynamics of inoculative biological control are more transient in nature, the outcome of intraguild predation may be difficult to predict (Briggs & Borer 2005). From an empirical perspective, both Rosenheim and Harmon (2006) and Janssen and colleagues (2006) concluded that there is no evidence of a consistent effect of intraguild predation on the success of pest suppression in arthropod predator–prey systems. This does not mean that intraguild effects are always neutral, however, as both significant positive and negative effects have been observed. For example, the presence of intraguild predators (*Zelus renardii* and *Nabis* species) in cotton fields prevented experimental releases of the lacewing predator, *Chrysoperla carnea*, from suppressing populations of cotton aphid (Rosenheim et al. 1993). Interference can also occur among predator species that do not share a common prey species, as is the case when predatory mites are used for control of thrips at the same time as *Aphidoletes aphidimyza* is used for control of aphids on sweet peppers in greenhouses (Messelink et al. 2011). In contrast, however, a combined inoculative release of two generalist predatory bugs resulted in improved control of aphids and thrips in greenhouse-grown sweet peppers than either predator alone, despite *Orius laevigatus* being an intraguild predator of *Macrolophus pygmaeus* (Messelink & Janssen 2014). Thus the outcome of combined releases appears to depend on the particular traits of individual natural enemy species and so need careful consideration and testing before implementation.

The use of pathogens for control of arthropods has primarily involved a single natural enemy species, and there appears to have been less interest in the use of combinations of species (Thomas et al. 2006). However, an interesting example of an effective combination stems from the intractable problem of developing an effective microbial program for inundative control of the Colorado potato beetle. In this example, a combination of the mycopesticide *Beauveria bassiana* and the bactopesticide *Bacillus thuringiensis tenebrionis* provided consistently more effective control in field trials (Wraight & Ramos 2005, Figure 8.7).

In contrast, combination releases of microbial control agents have been used more frequently for the suppression of plant diseases (Siddiqui & Shaukat 2002; Jacobsen et al. 2004; Domenech et al. 2006), often apparently with positive synergistic effects. For example, Domenech and colleagues (2006) showed that biological control of *Fusarium* wilt and *Rhizoctonia* damping-off in tomatoes and peppers could be enhanced by combining the siderophore production of *Pseudomonas fluorescens* CECT 5398 with the plant-growth promoting action of LS213, a product that combines *Bacillus subtilis* GB03 and *B. amyloliquefaciens* IN937a. From a theoretical perspective, using a simulation model for inoculative biological control of a foliar pathogen in a homogenous environment, Xu and colleagues (2010) showed that two different modes of action in

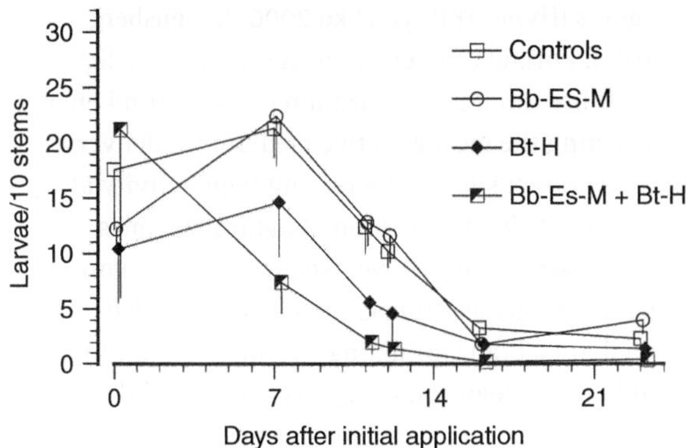

Figure 8.7. The change in larval population densities of the Colorado potato beetle, *Leptinotarsa decemlineata* after application of either a single pathogen species, the fungal pathogen *Beauveria bassiana* (Bb-ES-M) or the bacterial pathogen *Bacillus thuringiensis tenebrionis* (Bt-H), or a combination of the two pathogen species relative to a no application control treatment. The combination application gave both a greater and more rapid reduction in pest larval densities (from Wraight & Ramos 2005).

the same antagonist is generally more effective in reducing disease than two separate antagonists each with a different mode of action. However, with the inclusion of spatial heterogeneity in the model (Xu & Jeger 2013), both additive and synergistic effects of combined antagonists are possible, with synergism more frequent when the combined antagonists have different modes of action. In addition, a recent analysis of 465 published combination treatments using a well-defined statistical model for independent effects (Xu et al. 2011) showed that only 10 provided statistical evidence of synergism (positive effect), 70 provided evidence of additivity (neutral effect), and 41 provided evidence of antagonism (negative effect). Although beneficial effects of combined antagonists (including both synergistic and additive effects) were twice as frequent as detrimental effects were, it is clear that, as for arthropod natural enemy combinations, caution is needed in the choice and implementation of microbial control agent combinations.

8.6 Commercial Production and Application

8.6.1 Mass Production and Quality Control

To be cost-effective, mass production often involves culturing of natural enemies on artificial media or alternative hosts under optimal conditions. Mass production of parasitoids frequently makes use of alternative hosts (Hare & Morgan 1997; Luck & Forster 2003), while that of predators is increasingly reliant on synthetic or semi-synthetic diets (Cohen & Smith 1998; Cohen 2004). Most microbial control agents are produced *in vitro* through either liquid culture (bacteria and entomopathogenic nematodes) or solid-substrate fermentation (fungi), but baculoviruses are still produced *in vivo* (Nagayama et al. 2007; Ravensberg 2011; Jaronski & Jackson 2012). In each case, this requires a cost-efficient and reliable production process that can be scaled up to meet demand and that minimizes contaminants. For example, during development of the culture media, carbon:nitrogen ratios must be fine-tuned to maximize growth and yield (Gao et al. 2007) and virulence (Shah et al. 2005). In each case, the artificial conditions of mass production raise questions of quality control and genetic stability of the natural enemies produced. Issues of quality control include the potential for contamination with competing organisms, and the viability and performance of the natural enemies produced. The consequences of a failure to implement standards for quality control are more evident among arthropod natural enemies than among microbial control agents, by virtue of their size, mobility and behavior (e.g., O'Neil et al. 1998). As a result, the question of quality control began to be addressed for arthropod natural enemies in the 1980s and has been the subject of considerable development since, such that tests are currently in place for at least 26 species (Van Lenteren et al. 2003). These tests are typically based on quantity, sex ratio and fecundity, but may also include identification, survivorship, natural host parasitism or flight activity. Although quality control is equally if not more important for microbial control agents, due to the potential for unknown contaminants, the question of quality control testing has yet to be developed and implemented in a consistent manner (Jenkins & Grzywacz 2000, 2003). Current approaches are typically based on the number of infective propagules, their physical and chemical stability, the presence of microbial contaminants and efficacy in terms of viability or virulence (Ravensberg 2011).

The genetic composition of the natural enemy population presents different concerns for the mass production of microbial versus arthropod natural enemies. As the virulence and other modes of action of microbial enemies is often strain or genotype specific, mass production should always be started from a single DNA-profiled isolate, or a single spore in the case of fungi. This can help to avoid genetic drift during mass production and maximize the predictability of the virulence of the end product (Jenkins & Grzywacz 2000). In contrast, for arthropod natural enemies genetic variability is considered important in maintaining field performance, as we have argued in Chapter 7, since genetic bottlenecks and inbreeding depression due to small population size and adaptation to captive rearing can rapidly reduce effectiveness during mass production (Nunney 2003). Roush and Hopper (1995) suggest that maintaining a mass production as a series of 25–50 isofemale lines that can be hybridized 2–3 generations before release is the best solution for preventing adaptation to captive rearing, although up to 100 lines would be needed to ensure the level of genetic variability seen in field populations (Nunney 2003).

8.6.2 Formulation as a Key to Shelf Life and Duration of Activity

The formulation of mass-produced natural enemies is one of the most neglected components of the development of inundative biological control agents (Hynes & Boyetchko 2006; Ravensberg 2011). Formulation encompasses extending the shelf life through stabilization of the natural enemy, selecting the most effective medium for delivery to the target site and increasing their activity at the target site. For arthropod natural enemies, formulation can involve paper coverings to protect parasitoid pupae from climate and generalist predators, bran (sprinkling), vermiculite (air blown) or liquid suspension (spray) as carriers for delivery of predators, and staggered-release formulations to extend activity (Van Lenteren & Tommasini 2003). For example, 'Trichocaps' have been developed to formulate *Trichogramma brassicae* for use against European corn borer in Europe (Orr & Suh 2000). These paper capsules enclose about 500 diapausing parasitoids that can be stored for up to nine months. The capsules can be applied by hand or by air, diapause termination can be controlled for staggered emergence of parasitoid adults, and the capsule provides protection from predators. Similarly, the thrips predator *Neoseiulus cucumeris* has been formulated in sachets that can be hung from plants containing bran and bran mites to generate an extended slow release of mites to ensure effective timing for control of thrips as they colonize a crop (Jacobson et al. 2001).

For microbial control agents, formulation is essential to stabilize the product and extend its shelf life, to aid application and delivery, to increase persistence through protection from harmful environmental factors, and to enhance activity at the target site (Jones & Burges 1998; Schisler et al. 2004; Grewal & Peters 2005; Ravensberg 2011; Behle & Birthisel 2014). For example, a shelf life of six months has been achieved for the bacterial pathogen *Serratia entomophila* using a formulation of zeolite and biopolymers (Glare et al. 2012) and a shelf life of two years has been demonstrated for some formulations of the granulovirus of codling moth *Cydia pomonella* (Lacey et al. 2008). In addition, oil-based formulations have proved particularly effective in extending the shelf life of mycopesticides (Batta 2004; Vandergheynst et al. 2007), reducing the humidity requirement for germination (Greaves et al. 2001) and increasing effectiveness under field conditions (Shabana 2005).

Perhaps one of the most important aspects of formulation for microbial control agents, particularly baculoviruses, concerns persistence following application in the field, where UV radiation is a particularly destructive environmental factor. A wide variety of sunscreens have been tested, most of which act by absorbing UV radiation (e.g., amino acids, B vitamins, carbon, cosmetic sunscreens, dyes, optical brighteners and proteins), but many could be environmental pollutants, and some may even have toxicity to the microbial agents (Burges & Jones 1998). A few other sunscreens work either by reflecting UV radiation, such as iron and titanium oxide (Farrar et al. 2003; Asano & Miyamoto 2007) or by negating active oxygen radicals, such as carbon, ascorbic acid and peroxidase. In addition to the development of sunscreens for formulation, microbial control agents vary in their natural UV tolerance and strains can be screened for greater tolerance (Braga et al. 2001; Lacey & Arthurs 2005; Fernandes et al. 2015).

In addition to carriers and sunscreens, many other ingredients can be included in the formulation of natural enemy products such as stabilizers, buffers, binders, dispersants, stimulants or food sources (Burges 1998). Stabilizers, buffers, binders and dispersants are generally linked to the type of carrier being used in the formulation, while stimulants and food sources are more specific to the natural enemy. The role of these components in determining the effectiveness of the product is generally less well known, although priming the parasitoid *Aphytis melinus* with a contact kairomone prior to field release has been shown to increase the effectiveness of adults that are mass reared on alternative hosts (Hare et al. 1997; Figure 8.8).

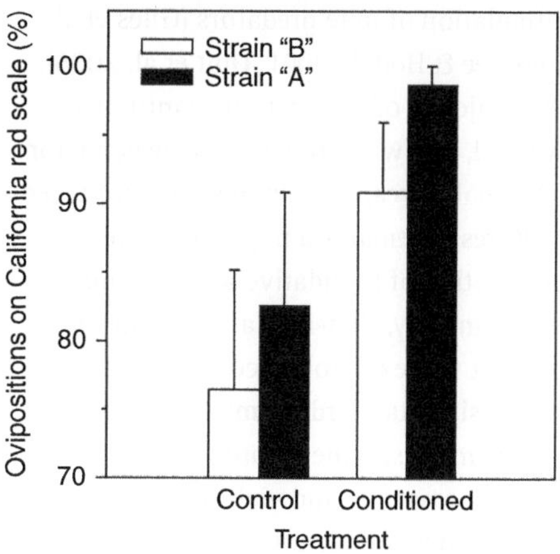

Figure 8.8. The influence of priming (conditioning) for two distinct strains of the parasitoid *Aphytis melinus*, after mass production on oleander scale, *Aspidiotus nerii*. Priming consisted of exposure of adult parasitoids to the contact kairomone *O*-caffeoyltyrosine that is present on the scale covering of California red scale, *Aonidiella aurantii*, the target pest. The percentage of ovipositions on California red scale was recorded for individual parasitoids exposed to three third instars of each armored scale simultaneously (from Hare et al. 1997).

8.6.3 Delivery Systems as a Key to Ease of Use

An aspect of the operational use of inundative releases of natural enemies that often does not attract sufficient attention is the need for cost-effective, broadcast application. In the case of arthropod natural enemies, several attempts have been made to develop and test mechanized delivery systems, including aerial application of dry formulations of *Trichogramma* egg parasitoids (Mills et al. 2000) and mite predators (Drukker et al. 1993), ground application of an aqueous suspension of *Trichogramma* egg parasitoids (Gardner & Giles 1997; Knutson 2003) and lacewing eggs (Gardner & Giles 1997; Giles & Wunderlich 1998) and mechanical blower application of a granular formulation of mite predators (Giles et al. 1995; Takano-Lee & Hoddle 2001; Opit et al. 2005). However, the majority of applications continue to be made by hand, and while this may be feasible for the relatively small acreages of intensively produced protected cultures, it remains a major obstacle to the broader adoption of inundative applications in field crops. Similarly, Gan-Mor and Matthews (2003) argue that while microbial control agents can be applied using standard chemical application equipment, in some cases, the equipment can compromise both coverage and viability. However, more efficient application techniques that minimize the amount of inoculum required to achieve successful control remain an important obstacle for greater commercial development (Lacey et al. 2015). One interesting approach that was developed for application of the fungal antagonist *Clonystachis roseum* to suppress the incidence of gray mold *Botrytis cinerea* in strawberries and raspberries was to use bees as an efficient delivery system (Sutton et al. 1997). By mounting an inoculum dispenser on the front of hives, both bumblebees and honeybees proved more efficient in the delivery of inoculum to raspberry flowers than conventional spray equipment, and provided a level of suppression of gray mold that was equal to that of a fungicide (Yu & Sutton 1997). Pollinators as a targeted delivery system have subsequently been used for other microbial antagonists as well as for delivery of entomopathogens (Kevan et al. 2007). Another interesting development is the use of 'lure and infect' strategies that attract insect pests to a simple device where they become infected with a pathogen and subsequently disseminate the pathogen to other individuals in the population (Vega et al. 2007). This approach has been used to provide greater suppression of Mediterranean fruit flies *Ceratitis capitata* in citrus than was achieved using insecticide applications (Navarro-Llopis et al. 2015). In this case, a combination of male and female lures was used to attract the flies to petri-dishes where they were inoculated with conidia of the fungal pathogen *Metarhizium anisopliae*. Novel and effective application techniques such as these can greatly improve the ease of use and commercial viability of inundative biological controls and deserves greater attention in the future.

8.7 Conclusions

In assessing the success of augmentative releases of natural enemies of arthropod pests in small field plots, Collier and Van Steenwyk (2004, 2006) found that a desired level of suppression was achieved in only 15% of 31 studies deemed sufficiently rigorous in terms of replication and experimental controls. However, this does not take into account the much more effective use of augmentation in closed protected culture, nor the extent of use based on public good (the desire to reduce reliance on synthetic chemical pesticides) as a criterion for success (Van Lenteren 2006). Nonetheless, it does provide an objective evaluation of the low level of adoption and difficulty in use of arthropod natural enemies in open field crops. This contrasts with the steady increase in representation of microbial control agents in the global agrichemical market which was estimated to be 7.7% by 2014 (Glare et al. 2012). In addition, while some of the greatest successes for augmentative biological control with arthropod natural enemies have been achieved in protected culture (Van Lenteren 2000), those for microbial control agents have primarily been implemented in the field (Charrudattan 2005; Lacey et al. 2015). Thus the potential for greater adoption of augmentative biological control in the future will likely arise through further replacement of synthetic pesticides with cost-effective and reliable microbial products.

The critical role of ecology for each step in the development of an augmentative biological control program, from natural enemy selection to mass production and delivery, cannot be overemphasized. However, technical effectiveness remains the single most important criterion for success, and a refuge in the pest system is likely to be the most important ecological factor limiting the effectiveness of a well-matched natural enemy. The limitation of a pest refuge in augmentative biological control can be overcome in one of three ways: improved strains of natural enemies, combinations of natural enemies or integration of natural enemies with other methods of control. Improved strains can either be developed through recombinant technology or identified using high-throughput screening methods. Although the commercialization of recombinant natural enemies appears to be on hold at the current time, this may change in the future if genetically modified organisms become more widely acceptable to the general public. The use of combined natural enemy species offers an alternative to improved recombinant strains, and has proved a viable approach, when carefully tested, for all groups of natural enemies. Finally, the integration of natural enemy augmentation with other approaches to pest, weed and disease management could similarly lead to more effective and reliable control and deserves greater attention in the future. One component of integration that must be treated with considerable caution, however, is the increasing availability of 'reduced risk' pesticides. The categorization of 'reduced risk' applies primarily to human toxicity rather than to natural enemy toxicity, and thus 'reduced risk' products are not necessarily compatible with natural enemies (see Chapter 10 for further discussion).

9 Conservation Biological Control I: Facilitating Natural Control through Habitat Manipulation

We have noted in previous chapters that biological control frequently occurs without intervention. This is called *natural biological control*, and we argued in Chapter 1 that such control may prevent many native species from reaching pest status, and in Chapter 2 that it may limit the extent to which exotic species can establish, spread and attain pest status. In this and the following chapter, we consider ways that human intervention can improve upon natural biological control. Such improvements are termed collectively *conservation biological control* and they consist primarily of two classes of approaches. The first involves habitat manipulation that favors natural enemies at the expense of pests, and this strategy is the focus of the current chapter. The second encompasses strategies that lead to the reduction of pesticide-induced harm to natural enemies, and we turn our attention to this strategy in Chapter 10. Here we discuss conservation biological control through habitat manipulation, which is often attempted through some form of plant diversification in agricultural settings. Conservation biological control does not exist in a vacuum, however, and we end this chapter with a discussion of how it can be combined with augmentative or importation biological control.

The topic of habitat manipulation as a means for improving naturally occurring biological control may be the most extensively reviewed subject in biological control (e.g., Van den Bosch & Telford 1964; Rabb 1971; Altieri & Whitcomb 1979; Altieri & Letourneau 1982; Risch et al. 1983; Powell 1989; Russell 1989; Van Emden 1990; Andow 1991; Trenbath 1993; Bugg & Waddington 1994; Barbosa 1998; Pickett & Bugg 1998; Gurr et al. 2000, 2003, 2004; Landis et al. 2000; Bommarco & Banks 2003; Snyder et al. 2005; Wackers et al. 2005; Tscharntke et al. 2007; Jonsson et al. 2008, 2010; Isaacs et al. 2009; Lundgren 2009; Wyckhuys et al. 2013). Most research on conservation biological control by habitat manipulation has been on arthropod biological control by predators and/or parasitoids, but significant progress has been made in conservation biological control of both weeds and soil-borne plant pathogens using management of naturally occurring microbial antagonists (Windels 1997; Lucas & Sarniguet 1998; Hoitink & Boehm 1999; Kremer & Li 2003; Mazzola 2004; Ghorbani et al. 2005; Timper 2014), although plant pathologists often term these methods 'general' or 'indirect' suppression of disease (Hoitink & Boehm 1999; Boulter et al. 2000). The topic of conservation biological control has been broached for other areas of biological control as well, including arthropod biological control using entomopathogens (Fuxa 1998; Lewis et al. 1998; Meyling & Eilenberg 2007; Meyling et al. 2009), weed biological control using herbivores (Newman et al. 1998) and biological control of foliar and fruit-infesting plant pathogens (Wilson 1998), but these areas remain relatively undeveloped.

We begin this chapter with a consideration of the ecological underpinnings of conservation biological control via habitat manipulation, and go on to outline the main mechanisms by which conservation biological control is hypothesized to work. Our main goal is to assess tests of these hypotheses while highlighting landmark examples. We also discuss ways in which this kind of habitat diversification can backfire by *increasing* pest problems, in some cases by *decreasing* the effectiveness of biological control.

9.1 Ecological Underpinnings

In *On the Origin of Species*, Darwin (1859) remarked on an estate of one of his relatives that contained two distinct habitats. One was 'an extremely barren heath, which had never been touched by the hand of man', and the other 'had been enclosed twenty-five years previously and been planted with Scotch fir'. He went on to describe some of the differences between these habitats:

> *The change in the native vegetation of the planted part of the heath was most remarkable, more than is generally seen in passing from one quite different soil to another: not only the proportional numbers of the heath-plants were wholly changed, but twelve species of plants (not counting grasses and carices) flourished in the plantations, which could not be found in the heath. The effect on the insects must have been still greater, for six insectivorous birds were very common in the plantations, which were not to be seen on the heath.*

Darwin seemed to be suggesting that increased plant diversity begets increased herbivore diversity, which in turn begets increased predator diversity. This hypothesis has been voiced by a number of authors since Darwin (Hutchinson 1959; Murdoch et al. 1972; Hunter & Price 1992), and it also happens to be at the root of conservation biological control by habitat manipulation with the added stipulation that the abundance of one key herbivore species (a pest) declines in the process. The hypothesis can also be seen as a subset of the so-called insurance hypothesis, which holds that higher biodiversity enhances long-term ecosystem stability in the face of environmental change because of the higher likelihood that species able to respond to environmental change are present (Naeem & Li 1997; Yachi & Loreau 1999; Loreau et al. 2003). The insurance hypothesis is itself a subset of broader theories linking biodiversity to ecosystem stability (reviewed by Ives & Carpenter 2007), but it is particularly relevant for conservation biological control because of its emphasis on environmental change. Pest invasions and outbreaks can be linked to environmental change in the form of disturbances due to practices such as cultivation and the use of pesticides and/or by habitat fragmentation (Swift et al. 1996; Wilby & Thomas 2002; Thorbek & Bilde 2004; Moreau et al. 2006; King & Tschinkel 2008). Applied to conservation biological control, the insurance hypothesis would hold that pest outbreaks are less likely in diverse communities that support abundant (including redundant) trophic linkages that are capable of maintaining ecosystem function in the face of environmental change. The literature on conservation biological control that we review later in this chapter can be seen as a test of this hypothesis, and as we shall see, the evidence in support of such a hypothesis is mixed in agricultural landscapes. Before we discuss these cases, however, we will remark on a set of particularly relevant studies done outside the realm of biological control.

The Cedar Creek Ecosystem Science Reserve of the University of Minnesota in the United States was (and remains) the site of an ambitious set of experiments aimed at determining the effects of plant diversity on various components of ecosystem functioning in a prairie setting. Most of the experiments were done in fields containing two sets of plots – 147 smaller plots (3 × 3 m) and 289 larger plots (13 × 13 m) – that were established in 1994 with various combinations of 32 North American native prairie plant species (Tilman et al. 1996, 1997). A subset of the data sets extracted from the various studies done at Cedar Creek addressed the hypothesis that increased plant diversity leads to increases in the diversity of herbivores and consumers of herbivores (coined the 'diversity-trophic structure hypothesis' by Knops et al. 1999). Most of these analyses supported the diversity-trophic structure hypothesis (Siemann et al. 1998, 1999; Knops et al. 1999; Symstad et al. 2000; Haddad et al. 2000, 2001). In one study, increasing plant diversity had a direct positive effect on both herbivore and predator

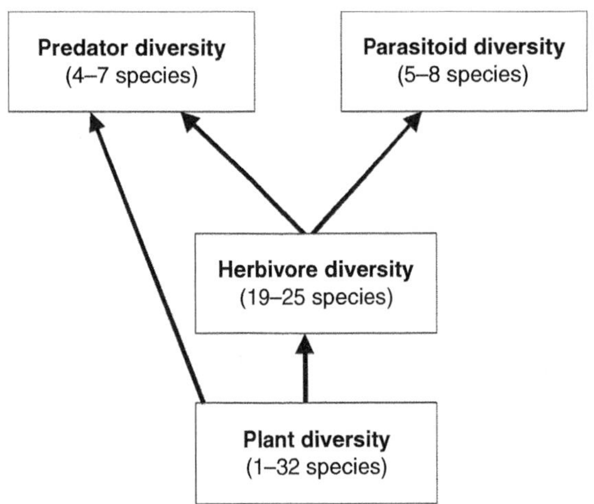

Figure 9.1. Schematic diagram of the relationships between manipulated plant diversity and herbivore, predator and parasitoid diversity (species richness) in a study done in a prairie setting in Minnesota in the United States. Arrows indicate significant positive direct effects. The range of species richness found is shown; for plants, this was the fixed independent variable, and for animal species, the ranges represent average values across all of the plant diversity treatments. From Siemann and colleagues (1998).

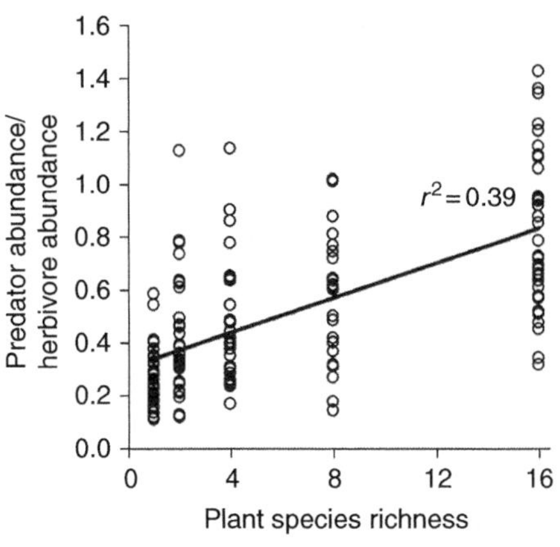

Figure 9.2. The relationship between plant species richness in experimental plots at the Cedar Creek Ecosystem Science Reserve, Minnesota in the United States, and the ratio of arthropod predator to herbivore abundance. From Haddad and colleagues (2009).

diversity, but only an indirect positive effect on parasitoid diversity (through increased herbivore diversity) (Siemann et al. 1998) (Figure 9.1). The direct positive effects of herbivore diversity on predator and parasitoid diversity could reflect either an increase in the number of specialist natural enemy species scaling with the increase in herbivore species or more species of enemies in general responding to the greater variety or abundance of prey types. The direct positive effect of plant diversity on predator diversity could reflect either a greater variety or abundance of resources used by predators (such as nectar or pollen) or a more favorable microclimate for predators at higher plant densities. As we shall see, versions of these hypotheses form the major pillars of explanations for how conservation biological control works.

The most comprehensive of the analyses done on this topic at Cedar Creek incorporated data gathered over an 11-year period and focused on arthropod abundance as well as diversity (Haddad et al. 2009, 2011). While this study showed an increase in both herbivore and predator species richness with plant diversity as had the others, it also revealed a marked shift in the relative abundances of herbivores and predators. While monoculture plots supported only a single predator individual for every four herbivore individuals, this ratio rose almost to 1:1 as plant diversity increased to 16 species per plot (Figure 9.2). Plant productivity (biomass) increased with plant diversity in these plots as well, an effect previously ascribed to niche complementarity (Tilman et al. 2001, 2006). In light of the dramatic shifts in trophic structure, however, Haddad and colleagues (2009) suggested that increased plant biomass may have been due to a trophic feedback mechanism where increased plant diversity led to stronger top-down (predator) control of herbivores, which allowed increased plant productivity. This hypothesis need not supplant niche complementarity; rather it suggests that release from herbivores may be an important dimension of niche complementarity in the Cedar Creek experiments.

The work at Cedar Creek manipulated plant diversity at a relatively narrow spatial scale. Other studies have investigated the diversity-trophic structure hypothesis at the landscape scale, asking whether trophic structure changes as the entire landscape becomes more diverse. Bianchi and colleagues (2006) reviewed the literature on the effects of landscape-level habitat diversity on the diversity and abundance of arthropod predators and parasitoids with a focus on crop versus non-crop habitats. It is clear from their review that a higher diversity of plants, herbivores and natural enemies can be found in habitats with a greater fraction of non-crop natural habitat in farming areas, at least in temperate Europe and North America. The studies also indicate that overall, both natural enemy abundance and the ability of natural enemies to suppress agricultural pests increases with increased habitat diversity. Higher levels of natural enemy abundance or activity were found in 74% of 24 studies that evaluated the effects of landscape-level habitat complexity on components of natural biological control. Pest pressure was reduced in more complex habitats in 45% of the studies that made appropriate comparisons, although a later meta-analysis failed to find a significant effect of landscape diversity on pest abundance (Chaplin-Kramer et al. 2011).

Since the review of Bianchi and colleagues (2006), a number of studies have shown increased diversity, abundance or activity of natural enemies in more diverse landscapes (e.g., Bianchi et al. 2008; Letourneau & Bothwell 2008; Tscharntke et al. 2008; Werling & Gratton 2008; Chaplin-Kramer et al. 2011; Plećaš et al. 2014). Included in these is a study by Gardiner and colleagues (2009b), in which the extent of biological control of an aphid pest of soybeans was quantified in the north-central United States as a function of landscape diversity. In this study, the extent to which predatory insects suppressed populations of the soybean aphid, *Aphis glycines*, were assessed by comparing caged with non-caged plants in 26 soybean fields in four US

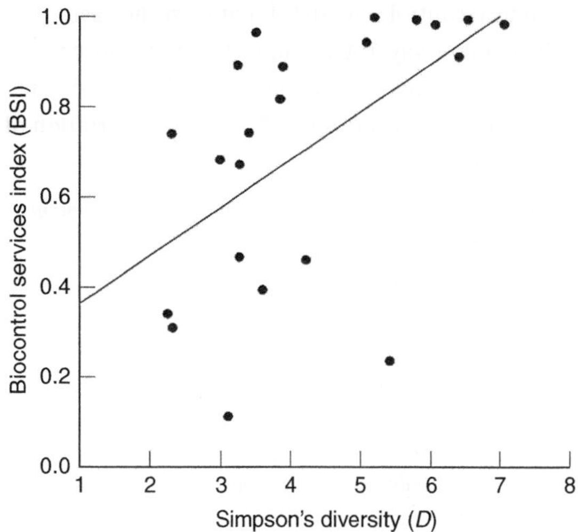

Figure 9.3. The relationship between Simpson's diversity index applied to landscape types over 22 soybean fields in the north-central United States and a biocontrol services index that quantifies the extent to which resident predators suppress the soybean aphid, *Aphis glycines*. From Gardiner and colleagues (2009b).

states over two years spanning a variety of landscape types. The difference between aphid population change on caged and uncaged plants was used to calculate a 'Biocontrol Services Index', which was regressed onto a measure of the per-site landscape diversity. The results (shown in Figure 9.3) show a clear increase in biological control with increased landscape diversity.

9.2 From Ecology to Application

While these results are on the whole consistent with the hypothesis that manipulations that increase plant diversity should improve biological control, they emphatically do not guarantee the success of any particular conservation biological control strategy. In fact, the proposition that increased plant diversity *per se* should lead to improved pest control is misguided and has been held in low esteem by biological control scientists for at least the past half century (Way 1966; Smith & Van den Bosch 1967; Van Emden & Williams 1974). We provide the following quote from Way (1966) as an illustration of this attitude:

> *[I]t is too facile to conclude that since the environment of the crop monoculture is unstable and since pest outbreaks do not occur in the stable 'climax' vegetation of some complex environments, any approach towards complexity in agricultural areas such as the maintenance of hedges and other non-crop habitats tends to decrease pest outbreaks.... Undoubtedly introduction of the right forms of diversity is fundamental to the modern approach to pest control.*

Identifying the 'right forms of diversity' has emerged as the core research program in conservation biological control by habitat diversification, and Way's insight has been echoed by a number of authors calling for 'selective' diversification (Baggen & Gurr 1998; Gurr et al. 2000; Balmer et al. 2014), and 'directed' rather than 'shotgun' approaches (Gurr et al. 2005; Wade et al. 2008a). Approaches based on identifying the specific interactions that link habitat diversity to improved biological control are sought (Van Emden 1990; Landis et al. 2000; Snyder et al. 2005; Straub et al. 2008). Insights into what constitutes the right forms of diversity have come from three classes of investigations. The first are empirical studies in which particular plant taxa or diversification strategies are investigated for their potential ability to conserve natural enemies (e.g., Wilkinson & Landis 2005; Griffiths et al. 2008) or in which particular taxa (or traits) of natural enemies are investigated with the aim of determining which are most amenable to conservation (e.g., Jervis et al. 2004; Wäckers et al. 2008). The second are modeling studies that aim to identify effective manipulation strategies from a number of points of view (Corbett & Plant 1993; Kean et al. 2003; Van Rijn & Sabelis 2005; Banks et al. 2008), and the third are quantitative reviews and meta-analyses that identify patterns leading to the most effective outcomes (Bommarco & Banks 2003; Langellotto & Denno 2004; Bianchi et al. 2006; Chaplin-Kramer et al. 2011; Russell 2015). A comprehensive review of these studies is beyond the scope of this chapter, but we discuss some of the major mechanisms by which conservation biological control can work in the next section.

9.3 Mechanisms of Conservation Biological Control by Habitat Manipulation

Hypotheses linking ecological mechanisms to pest suppression via habitat diversification were crystallized by Root (1973) in probably the most influential contribution to the discipline of conservation biological control by habitat manipulation. Root compared the arthropod communities on collard plants that were either planted within monoculture plots or in single-plant-width rows that were surrounded by diverse vegetation typical of abandoned agricultural fields of the northeastern United States (Figure 9.4). He found that although herbivore *diversity* was higher in the more diverse rows than in the monoculture plots, the herbivore *load* – defined as the biomass of herbivores per unit mass of consumable plant material – was consistently and significantly lower on collards in the diverse rather than the monoculture plots. The ratios of predatory arthropods:herbivores was higher in diverse rows as well, but this was not the case for aphid parasitoids and neither was the overall density or diversity of predators and parasitoids higher in the diverse strips than in the monoculture plots.

Root used his data as impetus to articulate two hypotheses to explain decreased pest pressure in simplified agricultural settings such as monocultures. The first was the 'enemies hypothesis', stated as follows:

> *A greater diversity of prey/host species and microhabitats is available within complex environments, such as most natural, compound communities. As a result, relatively stable populations of generalized predators and parasites can persist in*

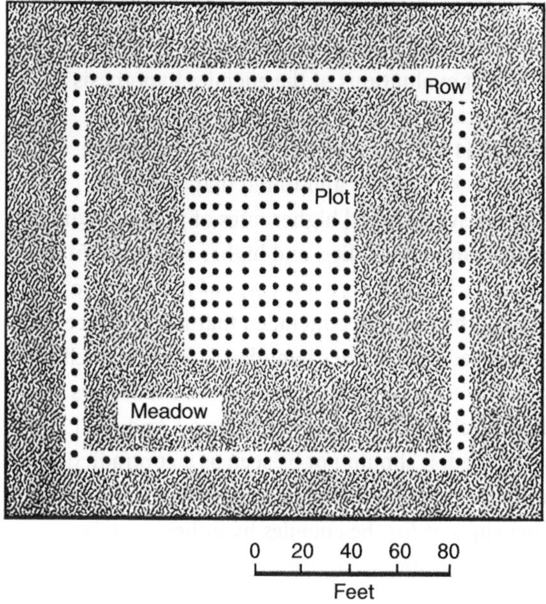

Figure 9.4. Schematic of Root's experimental design (see text). Collard plants are represented by dots and were planted in a monoculture plot and in single-plant rows embedded within meadow vegetation in Ithaca, New York, in the United States. From Root (1973).

> *these habitats because they can exploit the wide variety of herbivores which become available at different times or in different microhabitats.... Similarly, specialized predators and parasites are less likely to fluctuate widely because the refuge provided by a complex environment enables their prey/host species to escape widespread annihilation.... Finally, diverse habitats offer many important requisites for adult parasitoids and predators, such as nectar and pollen sources, that are not available in a monoculture. Thus incipient outbreaks of herbivores are checked early by the functional response of enemies whose numbers have been maintained by the diverse resources available in the complex environments.*

The second hypothesis, coined the 'resource concentration hypothesis', was summarized as follows:

> *Many herbivores, especially those with a narrow host range, are more likely to find hosts that are concentrated (i.e., occur in dense or nearly pure stands). The ease of location, however, is not an essential factor. Species that arrive in a clump of host plants, by whatever means, and find conditions particularly suitable will tend to remain in that area.*

In terms of explaining his own data, Root sided strongly with the resource concentration hypothesis, mainly because of the lack of strong effects of his treatments on predator and parasitoid abundance and diversity. He noted that this was driven in part by the rather idiosyncratic fact that the most important herbivore (in terms of biomass) was the flea beetle, *Phyllotreta cruciferae*, which was nearly entirely free of natural enemy attack in his study. Root by no means excluded the potential importance of the enemies hypothesis in explaining other cases of lower herbivore pressure under conditions of higher plant diversity.

A comment on host quality. The resource concentration hypothesis as stated by Root implied that host-plant quality effects can be included within resource concentration via its effects on herbivore movement patterns. Reduced interspecific competition in monocultures can lead to plants in monocultures providing a higher-quality resource to herbivores than plants in a species mixture (Andow 1991; Costello & Daane 2003; Bukovinszky et al. 2004; Schmidt et al. 2007; Lundgren & Fergen 2010), and this can lead to greater emigration rates from polycultures as predicted by the resource concentration hypothesis. However, those herbivores that remain on lower-quality plants in the polycultures may experience reduced population growth – and this itself can lead to decreased pest pressure. Root did not include this latter mechanism within the resource concentration concept, largely because he saw no evidence for it in his system. However, we include physiological effects of host-plant quality within the resource concentration idea here (as did Trenbath 1993). While this broadens the definition of resource concentration somewhat, we find it preferable to separating physiological effects of host-plant quality from behavioral effects.

9.3.1 Is It Enemies or Resource Concentration?

Once the resource concentration and enemies hypotheses were established, attention turned to which of these hypotheses best explained reductions of herbivore pressure in polycultures versus monocultures. Reviews of more than 200 studies comparing pest densities in monocultures versus polycultures showed that at least half of the studies found lower herbivore (pest) abundance in the complex habitats (Risch et al. 1983; Andow 1991; Tonhasca & Byrne 1994). Whether these cases could be best explained by resource concentration or by enemy conservation was not clear, however, because of insufficient information on the dynamics of pest or natural enemy populations in many of the reviewed studies. A number of lines of reasoning using indirect evidence have been used, however, to support one or the other hypothesis. Risch and colleagues (1983) reasoned that the resource concentration hypothesis would be supported if cases of herbivore suppression by habitat diversification were higher for specialist than generalist herbivores since specialist herbivores were more likely to be affected by resource concentration than were generalist herbivores. This pattern was indeed confirmed, suggesting an important role of resource concentration (Risch et al. 1983; Andow 1991). Risch and colleagues reasoned further that the enemies hypothesis would be a more likely explanation for herbivore suppression by habitat diversification in stable rather than ephemeral habitats (such as orchards versus annual cropping systems) because natural enemies tend to be more effective in stable habitats, as we have noted previously in Chapter 3. There was, however, no indication that habitat diversification increased pest suppression more effectively in stable rather than ephemeral habitats. On the contrary, habitat diversification led to *increased* herbivore abundance more frequently in stable rather than ephemeral agricultural habitats (Risch et al. 1983; Andow 1991). This further bolsters the resource concentration hypothesis.

Direct evaluations of the enemies hypothesis appear to tell a different story, however. Russell (1989) reviewed 16 explicit tests of the enemies hypothesis and found that diversified agroecosystems had higher rates of predation or parasitism in 9 studies, and higher enemy:herbivore ratios in 11, with very few cases of the opposite effect. Thus Russell concluded that the evidence supported the enemies hypothesis. Since Russell's analysis only considered a subset of the studies that Risch and colleagues (1983) and Andow (1991) examined, it should not be seen as a contradiction of those other reviews, but rather as an indication that support for the enemies hypothesis can be found in studies designed specifically to test it. A meta-analysis by Langellotto and Denno (2004) that incorporated later studies also found that arthropod natural enemies were more abundant in more complex habitats. Their analysis also showed a much stronger positive effect on natural enemies in response to additions of amendments such as thatch, leaf litter or mulch than the diversification of living plants.

The role of spatial scale. It has been recognized for decades that the effects of vegetational diversity on biological control are likely to vary with spatial scale. Altieri and Whitcomb (1980) detected a positive influence of increased plant diversity on biological control of the fall armyworm, *Spodoptera frugiperda*, in corn at a site where experimental plots were separated by 50 m, but not at a different site where they were separated by only 8 m. This and other studies led Russell (1989) to conclude that experiments comparing small, adjacent plots that differed in plant diversity were likely to underestimate the effect of plant diversity since enemies could regard the collection of plots as a single diverse habitat, and other studies have supported this general view (Marino & Landis 1996; Thies & Tscharntke 1999; Thies et al. 2003; Lee &

Heimpel 2005; Schmidt & Tscharntke 2005; Gardiner et al. 2009b). However, a different set of studies has come to the exact opposite conclusion, namely, that plant diversity has a stronger influence on biological control at *smaller* spatial scales. Foremost among these studies is a meta-analysis of experiments on effects of habitat diversification done between 1980 and 1997 by Bommarco and Banks (2003). The analysis supported previous conclusions that herbivore abundance was lower in diversified than in monoculture plots overall and also that predator abundance was higher in diversified plots, but these results were strongly tempered by spatial scale. Both the negative effects on herbivores and the positive effects on predators disappeared at large spatial scales (i.e., when plots exceeded 256 m^2) and were strongest at small spatial scales (Figure 9.5). The scale-dependence of the results suggests that choice behavior by insects may drive experimental results and that insects are better able to choose between habitats when they are small and adjacent than when they are large or distant (Bergelson & Kareiva 1987).

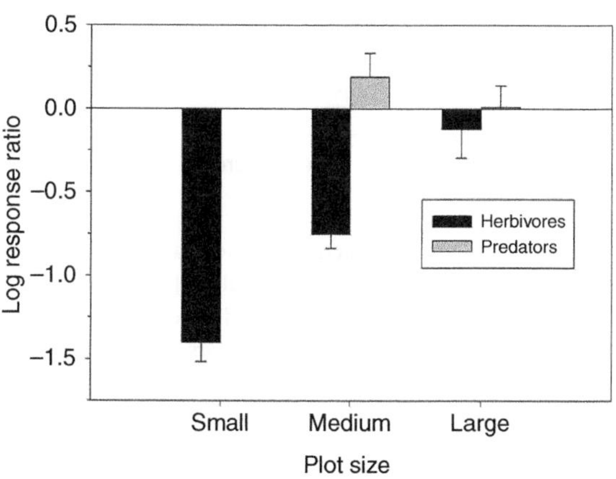

Figure 9.5. Results of a meta-analysis of experiments in which herbivore and predator abundances were compared in agricultural plots of various sizes that were subjected to plant diversification or not. The log response ratio is the average $\ln(X_{diversified}/X_{control})$, where X is the herbivore or predator abundance. Thus, negative values indicate that abundances are lower in diversified plots than control plots and positive values indicate that abundances are higher in diversified plots. No data were available for predators in small plots. Modified from Bommarco and Banks (2003).

The resolution to these divergent hypotheses and observations is clearly that different species of natural enemies (and herbivores) respond to plant diversity at different spatial scales. This point was made emphatically by Corbett and Plant (1993), who incorporated natural enemy dispersal rates into models of plant diversity and biological control, and is confirmed by a number of studies showing empirically that natural enemy species differ in the spatial scale at which plant diversification affects their activity (e.g., Russell 1989; Roland & Taylor 1997; Schmidt et al. 2008). More recently, Chapin-Kramer and colleagues (2011) have shown that specialist enemies (and pests) tend to respond more strongly to smaller spatial scales in landscape complexity studies than generalists do.

9.3.2 Relationship between Enemies and Resource Concentration Hypotheses

It is important to realize that the enemies and resource concentration hypotheses are not mutually exclusive and also that they may be related. Russell (1989) argued that the two hypotheses can be complementary, and he pointed to cases in which densities of different herbivore species attacking the same crop are reduced in polycultures differentially by mechanisms central to either the enemies or resource concentration hypotheses. Even for a single herbivore species, however, it is possible that both resource concentration and enemies contribute to lower herbivore abundance in diversified systems. For instance, Khan and coworkers (1997) found that interplanting molasses grass with maize on African farms led both to higher parasitism of lepidopteran stem borers by the native parasitoid *Cotesia sesamiae*, and to lower stem-borer densities independently of parasitism, resulting in higher maize yields. Laboratory assays showed that molasses grass had the dual effect of repelling stem borer females and attracting parasitoid females (Khan et al. 1997). In another example, Andow (1990) showed that the colonization rate of adult Mexican bean beetles, *Epilachna varivestris*, and the survivorship of Mexican bean beetle eggs were both lower in weedy than monoculture bean plots, consistent with both resource concentration and enemies leading to lower beetle density in the diversified plots. However, larval and pupal survivorship were *higher* in the diversified plots, effectively cancelling out the decreased egg survivorship. This study pointed out the benefits of investigating various life stages of the herbivore and also the usefulness of using formal demographic analysis to partition the effects of resource concentration and natural enemies in affecting herbivore densities.

An antagonistic relationship between resource concentration and enemies is possible if natural enemies are disfavored by vegetational diversity. This could be the case, for example, if searching for prey/hosts is more difficult in more diverse or complex habitats, providing herbivores with a refuge from enemy attack (e.g., Andow & Prokrym 1990; Gols et al. 2005; Bukovinszky et al. 2007; Randlkofer et al. 2007, 2010). Sheehan (1986) argued persuasively that just such a scenario should be *expected* for specialist natural enemies that rely on chemical or other cues to locate their hosts or prey. These foragers are aided by a simplified habitat in which cues are not masked by non-host and host-plant species. In essence, the enemies hypothesis is undermined by enemies that respond to polycultures in the same way that herbivores do under the resource concentration hypothesis. Sheehan's critique suggests that the enemies hypothesis should be restricted to generalist natural enemies. We return to this topic in Section 9.6.1.

Any changes in herbivore density that occur as a direct consequence of plant diversity (i.e., through resource concentration) have the potential to lead to indirect effects on natural enemies (Coll & Bottrell 1995; Costamagna et al. 2004; Heimpel & Jervis 2005). This is another form of interaction between the resource concentration and enemies hypotheses, and in this case, density-dependent attack rates on herbivores come into play. Specifically, in cases where herbivore density is lower in diversified systems, natural enemies can either enhance this effect with an inversely density-dependent response or they can attenuate it with a positively density-dependent response. A positively density-dependent response of natural enemies layered on top of a direct negative effect of plant diversity provides the herbivores with a low-density refuge from attack, and an inversely density-dependent attack rate leads to a synergistic effect of resource concentration and enemies on herbivore suppression. These effects are shown in Figure 9.6, in which overall natural enemy attack is

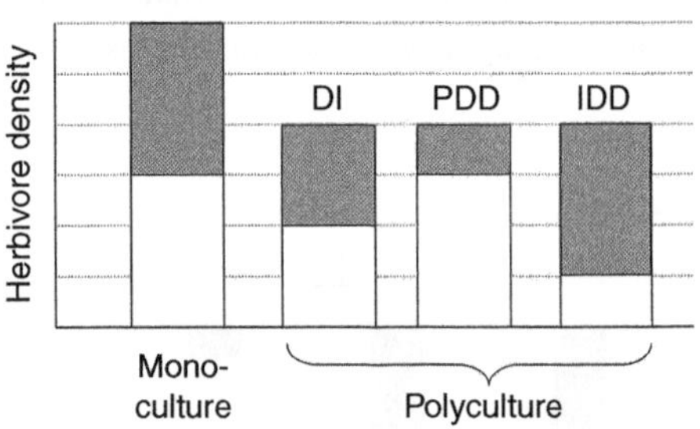

Figure 9.6. Potential interactions between a negative direct effect of plant diversity (polyculture) on herbivore density and the effect of this density difference on natural enemy attack rates. The height of the bars represents herbivore density and the hatched gray areas represent the herbivores attacked by biological control agents. In this scenario, the polyculture reduces herbivore density by one third independently of natural enemies, and the three polyculture bars represent the three possible responses to density differences (DI is density independence, PDD is positive density dependence and IDD is inverse density dependence). In the DI case, plant and enemy effects are additive, in the PDD a weaker response of natural enemies at lower densities eliminates the suppressive effect of the polyculture on herbivore densities, and in the IDD case, a stronger response of natural enemies at lower herbivore densities leads to synergism between plant and enemy effects on herbivore suppression. Adapted from Heimpel and Jervis (2005).

expressed as density-independent, positively density-dependent or inversely density-dependent. An example of synergism between resource concentration and natural enemy density dependence was provided by Coll and Bottrell in the eastern United States, who showed that resource dilution in maize-soybean dicultures (as opposed to soybeans alone) led to lower abundance of various herbivores (Coll & Bottrell 1994). Predators showed an inverse density-dependent response in the same study (Coll & Bottrell 1995), and the net result was a disproportionally high effect of predators in the diculture.

9.4 Mechanisms and Manipulations

In this section, we turn our attention more exclusively to the enemies hypothesis. While there is some question as to the primacy of the enemies hypothesis over resource concentration as we have just discussed, conservation biological control by habitat manipulation is primarily devoted to the elucidation of habitat manipulation strategies that enhance the efficacy of natural enemies. In other words, it is the goal of conservation biological control to create conditions that favor mechanisms that underlie the enemies hypothesis. We note also that mechanisms contributing to the resource concentration hypothesis can also be considered biological control (see Chapter 1 and Parolin et al. 2014). For this discussion though, we retain the traditional interpretation that only mechanisms underlying the enemies hypothesis constitute conservation biological control by habitat manipulation. Before discussing particular manipulations used for biological control, we consider here the requirements needed for habitat manipulations to lead to increased biological control, and the mechanisms by which successful conservation biological control can work.

For habitat manipulation to lead to increased biological control, three general requirements must be met. First, the manipulation itself must lead to increased diversity, abundance or activity of resident natural enemies. This implies that (i) natural enemies are present within dispersal range of the manipulation and (ii) that the resource being offered by the manipulation is in limiting supply. Based on these considerations, Tscharntke and colleagues (2005a) predicted that conservation biological control would be most effective within landscapes supporting intermediate levels of biodiversity, since particularly depauperate landscapes might not harbor appropriate natural enemies to conserve, and particularly diverse landscapes might have sufficient resources in the absence of the manipulation. This hypothesis was supported by studies showing that a positive effect of floral strips on biological control of aphids and caterpillars by parasitoids in kale was found in 'moderately simple' but not in 'highly complex' landscapes (Jonsson et al. 2012, 2015).

A second requirement is that higher diversity, abundance or activity of natural enemies must lead to increased pressure on target pests. This cannot be taken for granted because the natural enemies may feed on resources other than the target pests. Indeed, herbivore diversity and abundance in general are likely to increase with plant diversity, often providing opportunities for attacking alternative (non-pest) prey or hosts, as we discuss in Section 9.4.2. In this context, it is important to determine whether the habitat manipulation itself is a natural enemy source (as desired) or a natural enemy sink (Corbett & Plant 1993; Corbett 1998; Tscharntke et al. 2005b). For example, the activity and/or density of members of the ground beetle (carabid) genus *Amara* is enhanced by the addition of hedgerows, but these beetles rarely venture into agricultural fields (Thomas et al. 2001a). Other carabid species, on the other hand, are enhanced by the diversification of edge habitat and readily disperse into adjacent agricultural fields (Thomas et al. 1991, 2001a; Dennis & Fry 1992; Lee et al. 2001; Griffiths et al. 2008). Source-sink relationships are often mediated by timing, with natural enemies moving from areas experiencing declines in resource availability to areas experiencing increases in resource availability (Tscharntke et al. 2005b).

Effective manipulations take advantage of these natural movement patterns. In one example implemented in both the United States and Australia, alfalfa (lucerne) is harvested in strips every two weeks during the growing season, rather than all at once in two or three discrete harvesting events over the field season. In a variant of this system, narrower strips are left unharvested for longer periods than usual. In both cases, arthropod natural enemies move from the defaunated, harvested strips to the non-harvested strips that still harbor prey (Van den Bosch & Stern 1969; Summers 1976; Hossain et al. 2002). The alternative is to harvest the entire field at once, resulting in the emigration of all or most of the surviving natural enemies. Growing cotton adjacent to alfalfa can also take advantage of early- and mid-season alfalfa harvests since natural enemies move to the cotton fields, where they can help to suppress pests (Corbett et al. 1991; Mensah & Sequira 2004).

The last requirement is that pest density must be decreased in the presence of the manipulation. Note that this cannot be assumed even if the first two conditions are met, because the manipulation itself may increase the diversity or abundance of herbivores, potentially even exacerbating pest problems (see Section 9.6).

We now turn to two specific classes of mechanisms that are central to achieving conservation biological control by habitat manipulation – microhabitat improvement and resource supplementation, of which there are a number of variants. These are not the only mechanisms by which conservation biological control by habitat manipulation can be achieved, but they are the best-studied.

9.4.1 Improving Microhabitats for Natural Enemies

Increasing plant diversity in agricultural ecosystems can improve the microhabitat for natural enemies of arthropod pests and soil-inhabiting plant pathogens. These benefits can be through increased overwintering success, decreased mortality or increased attraction and/or retention of biological control agents in the general vicinity of pest organisms. Microhabitat effects are understood here to be restricted to abiotic effects and are often conceived of as being mediated by microclimate effects such as local temperature and humidity conditions, but they also include soil characteristics such as mineral content and pH.

For arthropod natural enemies of the pests of annual crops in temperate climates, overwintering typically occurs at field edges, and a number of studies have shown that diverse field edges increase the overwintering success of biological control agents (e.g., Sotherton 1985; Dennis et al. 1994; Wratten et al. 1998; Thies & Tscharntke 1999; Collins et al. 2003a, b; Griffiths et al. 2008; Geiger et al. 2009). Indeed, increased overwintering success is one of the main goals in the creation of 'beetle banks', as we discuss in Section 9.5.2. Similarly, incorporation of cover crops can provide moderate microclimates during the summer that improve the survival of arthropod natural enemies (Orr et al. 1997; Riechert 1998). Increased vegetational diversity can also provide important shelter from the wind or serve as a windbreak that concentrates natural enemies near pest populations (Corbett & Rosenheim 1996; Beane & Bugg 1998; Brandle et al. 2004).

A particularly stark microclimate effect of plant diversity is provided by differences between coffee bushes grown underneath a canopy of trees ('shade coffee') and coffee grown in areas cleared of such cover ('sun coffee'). In particular, Lin (2007) has demonstrated that sun coffee settings are associated which greater fluctuations of temperature and humidity than are shade coffee settings. Given these and other differences, neotropical coffee plantations have attracted a fair number of empirical studies on differences in pest control in coffee grown in shade versus sun conditions. A number of these studies have shown that diverse shade coffee plantations harbor a higher diversity and abundance of various predatory taxa, in particular ground-dwelling ants and insectivorous birds (Perfecto & Snelling 1995; Perfecto et al. 2004). Some studies have also presented data consistent with improved biological control under shade (Perfecto et al. 2004; Armbrecht & Gallego 2007; Johnson et al. 2009). Other studies, however, show that predation is not stronger under shade (Greenberg et al. 2000; De la Mora et al. 2008; Philpott et al. 2008; Van Bael et al. 2008; Gordon et al. 2009; Amaral et al. 2010; Johnson et al. 2010; Larsen & Philpott 2010). This lack of consistency was investigated for insectivorous birds in tropical ecosystems more generally using a meta-analysis approach by Van Bael and colleagues (2008) and Philpott and colleagues (2009), and these analyses failed to uncover a significant relationship between vegetational diversity and the strength of arthropod predation by birds. Thus, while studies like those reported by Perfecto and colleagues (2004) suggest that bird predation can be stronger under shade, the broader literature indicates that this can not necessarily be expected to occur.

Various cultural practices can be used to support biological control of soil-borne plant pathogens as well. In particular, the addition of compost to agricultural soils leads to numerous changes in soil characteristics, many of which can favor the activity of disease-suppressive microorganisms (Hoitink & Boehm 1999). A particularly well-documented example of such an effect involves the addition of a soil amendment called 'SF-21' to the soil of pine nurseries. This amendment includes milled pine bark and effectively suppresses damping-off disease in pine seedlings. Investigations of the mechanisms for this suppression revealed that a lowering of the soil pH caused by SF-21 led to an increase in the mycoparasitic fungi *Trichoderma harzianum* and *Penicillium oxalicum*, which suppressed the damping-off pathogen *Rhizoctonia solani* (Huang & Kuhlman 1991a, b). *Trichoderma* fungi were present in the soil naturally but ineffective at higher pH values, and SF-21 contained both an acidifying agent (aluminum sulfate) and a substrate for *Trichoderma* growth (the milled pine bark). The result of incorporating SF 21 into the soil was enhancement of the density and effectiveness of fungi capable of suppressing a devastating disease of pine seedlings. As we describe later, composts can increase disease suppression by other mechanisms as well (Sections 9.4.2, 9.5.3 and see Hoitink & Fahy 1986; Windels 1997; Hoitink & Boehm 1999; Boutler et al. 2000) and can be used in the context of both conservation and augmentative biological control (Section 9.7.2).

9.4.2 Resource Supplementation

Habitat manipulation can deliver critical resources to biological control agents. These resources can range from simple nutrients to populations of living organisms. Nutrient supplementation is exemplified by increases in usable nitrogen provided by composts and taken up by resident microbes that are involved in disease suppression and by the increased availability of nectar sugar to parasitoids in manipulations that add flowers to agroecosystems. Habitat manipulation can also lead to changes in species and community structure, and under the right circumstances this can translate into an increased abundance of alternative hosts or prey for arthropod biological control agents. In this section, we explore resource supplementation for conservation biological control by discussing three scenarios that can make habitat manipulation strategies function as successful conservation biological control tactics. These scenarios are: (i) enhanced nutrients for microbial agents of plant disease; (ii) sugar supplementation for increased biological control by parasitoids and hoverflies; and (iii) increasing the abundance of alternative prey or host species for predators or parasitoids of arthropod pests.

9.4.2.1 Nutrients for Microbial Agents for Plant Diseases

The addition of soil amendments such as compost can make limiting nutrients available to disease-suppressive microbes. The organic matter that develops through the composting process provides nutrients such as carbohydrates, chitin and cellulose that can boost the populations of microbial biological control agents (Hoitink & Boehm 1999). In essence, amendments function to fertilize the populations of biological control agents that are already present. As we have noted in Chapter 1 and will stress again later (Section 9.5.3), it can be difficult to separate these effects from direct fertilization of the plants and from increases in effectiveness of biological control agents that are introduced with the amendment itself. Nevertheless, providing limiting nutrients to resident beneficial microbes has been identified as an important mechanism leading to disease suppression. For example, the lignocellulosic substances present in composted tree bark serve as a growth substrate for resident populations of *Trichoderma* fungi that suppress *Rhizoctonia solani* – induced damping-off disease of seedlings in potting media (Nelson & Hoitink 1983; Hoitink & Boehm 1999).

9.4.2.2 Sugar Supplementation

One of the most attractive hypotheses in conservation biological control posits that natural biological control by arthropod pests can be improved by providing sugar to natural enemies. In particular, biological control agents that feed on pests in the larval stage but sugar in the mobile adult stage have been seen as promising targets for sugar supplementation. These agents include many if not most species of parasitoids, hoverflies and lacewings, among others (Jervis & Heimpel 2005). Adults of these insects often feed on sugar when it is available in the field (in floral or extrafloral nectar or in honeydew), and laboratory studies typically show a substantial longevity and fecundity boost associated with sugar feeding. These results, along with observations that many agroecosystems are sugar-poor, have led to an expectation that sugar supplementation could alleviate a resource that is both limiting in the field and critical to achieving successful biological control. In addition, feeding on sugar does not come at the direct expense of attacking pests, as can occur with feeding on alternative prey or host species.

Heimpel and Jervis (2005) formulated a list of conditions that need to be met for an experimental demonstration that sugar supplementation leads to increased biological control. While their discussion focused on using floral nectar to enhance parasitoids, it is broadly applicable to other sugar sources and to arthropod biological control agents beyond parasitoids. Their conditions can be summarized as follows. First, the biological control agents must be sugar-limited in the field in the absence of supplemental sugar. Second, they must actually feed on sugar that is supplemented. The most common habitat manipulation to supplement sugar involves incorporating flowering plants into agricultural settings. Extrafloral nectar can also be supplemented, in some cases by simply using different varieties of a crop or orchard plant (Stapel et al. 1997; Mathews et al. 2007). Sugar can also be supplemented directly by the application of sugar-rich sprays. Although it is debatable whether this truly constitutes habitat manipulation, this method has been an important experimental proof-of-principle that sugar limits biological control (Wade et al. 2008b; Evans et al. 2010). Definitive confirmation of sugar feeding can be achieved using either observations or one of several biochemical assays that can be applied to field-collected insects (Jervis et al. 1992; Heimpel et al. 2004b). The remaining requirements are that supplemental nectar leads to higher pest mortality and lower pest densities. Neither of these latter outcomes can be assumed even if sugar is limiting and fed on (Heimpel & Jervis 2005).

While a number of studies have documented higher levels of parasitism associated with supplemental floral plantings (e.g., English-Loeb et al. 2003; Tylianakis et al. 2004; Lee & Heimpel 2005; Balmer et al. 2014), unambiguous evidence that these increases are both due to sugar feeding and that they lead to lower pest densities is thin

(Harwood et al. 1994; Heimpel & Jervis 2005). This is despite evidence of sugar feeding by parasitoids in some field studies (e.g., Lavandero et al. 2005; Lee et al. 2006; Winkler et al. 2009b), including a demonstration of higher fecundity in sugar-fed individuals (Lee & Heimpel 2008). Demonstrating a link between sugar supplementation and biological control can be difficult because of methodological challenges and also because of some real limitations to the effectiveness of this strategy. As already noted (Section 9.3.1), an important methodological hurdle involves spatial scale – in particular if the dispersal range of the natural enemy greatly exceeds the distance between experimental plots (Corbett & Plant 1993; Lee & Heimpel 2005). In this case, an entire field experiment – composed of control plots and plots with flower strips, for instance – may be experienced as sugar-enhanced from the standpoint of foraging biological control agents. This phenomenon is exemplified by the literature on floral supplementation for syrphid flies (in which case use of pollen by adult flies in addition to nectar is likely). Here, the best results have been obtained either in studies done at relatively large spatial scales in the field (Hickman & Wratten 1996; Haenke et al. 2009) or in greenhouse trials where dispersal of adult syrphid flies between treatment plots is restricted (Pineda & Marcos-Garcia 2008) (Figure 9.7). Beyond these issues, however, there are real difficulties inherent in improving biological control via nectar supplementation. These difficulties include individual-based problems such as parasitoids not being able to access floral sugar due to either flower structure or competition (Heimpel & Jervis 2005; Russell 2015). There are also broad landscape-level considerations that may predetermine sugar-supplementation strategies toward success or failure. Indeed, conservation biological control through sugar supplementation appears to be a case-in-point for Tsharntke and colleagues' (2005a) hypothesis that habitat manipulation is most likely to be effective in landscapes of intermediate complexity, as noted earlier (Jonsson et al. 2012, 2015). Landscapes that harbor too little diversity may not supply enough natural enemies to benefit from sugar supplementation, whereas particularly diverse landscapes may be so sugar rich that natural enemies are not sugar limited in the first place. Last,

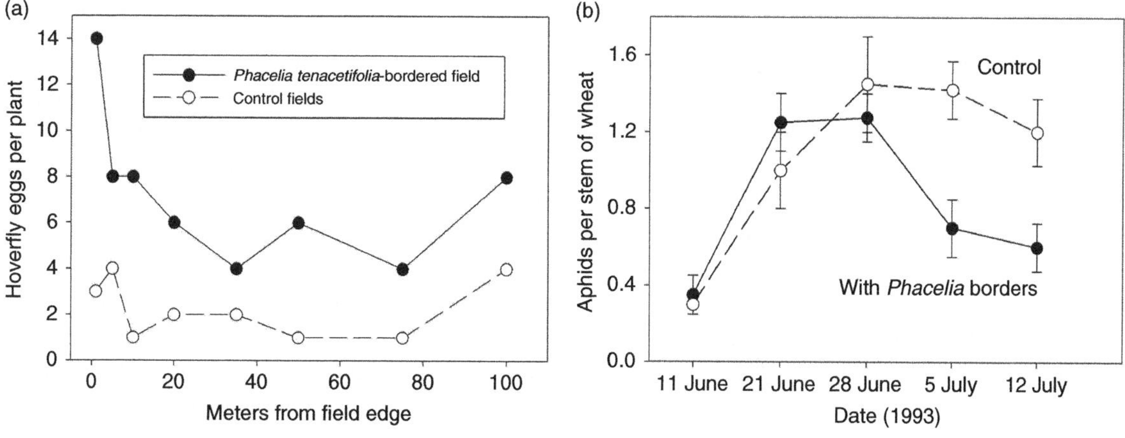

Figure 9.7. Effects of *Phacelia tanacetifolia* strips on the edge of wheat fields in southern England on the number of hoverfly eggs per plant within the field at various distances from the edge (a), and the numbers of grain aphids per stem of wheat (b). From Hickman and Wratten (1996).

an important limitation involves the risk that pests will use the supplemental sugar in addition to (or instead of) natural enemies, leading to increases in pest densities (see also Section 9.6.2). As we have discussed previously, the outcome of an increase in both pest and natural enemy abundance depends on both the magnitude of the responses and the relationship between pest density and attack rate by natural enemies (Section 9.4, Figure 9.6).

9.4.2.3 Increasing Abundance of Alternative Hosts or Prey for Arthropod Agents – Applied Apparent Competition

As discussed in Section 9.1, increasing vegetational diversity often increases the diversity and abundance of herbivores. Assuming that they do not become pests themselves, these additional herbivores can improve the efficacy of biological control agents in their role as alternative prey (or host) species. Whether this beneficial relationship will develop depends on a number of factors. The most important question involves source-sink considerations, as already discussed. If the alternative prey are highly suitable or preferred species, the biological control agent may concentrate its efforts on these species to the exclusion of the pest, leading to reduced pest control. In this scenario, the alternative prey is a sink for natural enemies, rather than a source (e.g., Bugg et al. 1987; Perrin 1975; Kemp & Barrett 1989; Bigger & Chaney 1998). Corbett and Plant (1993) modeled a number of scenarios under which agroecosystem diversification can function as a sink or a source of biological control agents. These analyses focused to a large extent on the dispersal patterns of biological control agents and suggested that conditions favoring diversified habitats functioning as a sink for agents include high mobility of agents, high attractiveness of diversified areas (vegetation 'strips') and close proximity of strips to one another. A major finding was also that supplemental vegetation that is available before the crop becomes available for pests can be much more beneficial (i.e., act as a source) than vegetation that is temporally synchronized with the crop. This prediction is supported by additional theoretical work (Wissinger 1997; Bianchi & Van der Werf 2004), as well as a number of studies showing benefits of perennial diversification of annual crops (e.g., Mensah 1999; Thies & Tscharntke 1999; Langer & Hance 2004; MacLeod et al. 2004; Pickett et al. 2004; but see Frere et al. 2007). In one of these studies, land that had been allowed to remain fallow for one versus six years was compared for its effect on biological control in adjacent oilseed rape (canola) fields. Parasitism of the rape pollen beetle, *Meligethes aeneus*, was more than twice as high in the center of fields adjacent to the longer-term fallow field, presumably because alternative prey were available for overwintering of pollen beetle parasitoids in these fields, allowing early colonization of the parasitoids (Thies & Tscharntke 1999).

At least two general scenarios are possible by which habitat manipulations in agriculture can improve biological control through the provision of alternative prey. In the first scenario, natural enemies feed on the alternative prey species in the manipulated habitat and then disperse to the crop to feed on the pest. An elegant example of this scenario involves overwintering hosts for the leafhopper parasitoid *Anagrus erythroneurae* (= *epos*) in the United States. This parasitoid must overwinter within host eggs, but the main pest that it attacks, the grape leafhopper, overwinters in the adult stage. Other leafhopper species do overwinter as eggs, and they do so on plants that often grow adjacent to vineyards, providing *Anagrus* parasitoids the opportunity to overwinter near vineyards and colonize grape leafhopper in spring (Doutt & Nakata 1973; Corbett & Rosenheim 1996; Williams & Martinson 2000; Prischmann et al. 2007). One of these plant species is prune trees, which has allowed for the development of a conservation biological control strategy in which prune trees are planted adjacent to vineyards as a means of improving biological control of grape leafhopper (Murphy et al. 1998).

In the second scenario, the alternative prey species themselves move from the manipulated habitat to the crop habitat, where they are consumed by natural enemies. Because of the higher overall abundance of food, this can lead to aggregation of natural enemies and, ultimately, greater pest suppression. This scenario has been demonstrated for weed-free versus weedy plots of sugarcane in the southeastern United States (Ali & Reagan 1985), weed strips in apple orchards in Switzerland (Wyss et al. 1995) and wheat interplanted with strips of diverse vegetation in France (Langer & Hance 2004). Results consistent with this scenario were also documented on golf courses in the eastern United States by Frank and Shrewsbury (2004). They planted 'conservation strips' of flowers and grasses adjacent to fairways and found that alternative prey species such as collembolans were augmented by these strips compared to grass controls. The predation rate on lepidopteran larvae experimentally placed onto the golf course adjacent to the strips was higher as well, suggesting an increased potential for biological control adjacent to the strips.

All cases of alternative prey or host species improving biological control (whether or not habitat manipulation is involved) are examples of apparent competition. In apparent competition the presence of one prey species negatively affects another (and vice versa) by increasing predator reproduction or aggregation. The term was coined in an expansive paper by Robert Holt (1977), who provided examples of biological control interactions in his initial development of the topic, citing cases in which predatory mites were maintained in orchards on a number of foods other than the pest mites which they controlled (see Collyer 1964). Holt noted that cases like these appear to be 'example[s] of pest control maintained by apparent competition among prey species', and Jeffries and Lawton (1984) have noted that the recognition of the importance of alternative hosts to biological control predates the term 'apparent competition'. A number of authors have reviewed evidence for apparent competition in biological control of arthropod pests (e.g., Van Veen et al. 2006; Evans 2008; Meyling & Hajek 2010; Chailleux et al. 2014; Kaser & Ode 2016), and Alhmedi and colleagues (2011) have shown how quantitative food web analysis can be used to demonstrate the beneficial effects of alternative prey species in biological control.

9.5 Case Studies in Conservation Biological Control

In this section, we provide brief accounts of three case studies of conservation biological control. All involve habitat manipulation, and there is evidence for some level of success for all of them.

9.5.1 Conserving Predatory Ants throughout History

We have already mentioned in Chapter 1 that conservation of *Oecophylla* weaver ants in orchard and other settings dates back at least 3,000 years. These practices continue to this day in tropical Asia, Australia and Africa, and they consist of habitat manipulations favoring the abundance and local establishment of weaver ants, that prey upon pest arthropods. Weaver ants are also favored as a delicacy for humans (Offenberg 2011), and their presence improves the quality and attractiveness of citrus fruits, possibly through direct effects of ant excretions on tree physiological processes (Barzman et al. 1996). Strategies to enhance *Oecophylla* spp. include moving nests to areas where they are needed (a practice more akin to augmentative biological control, as we noted in Chapter 8), feeding ants with fish or chicken intestines, constructing bridges between trees in citrus orchards (Figure 1.2), various forms of intercropping, and simply tolerance of colonies despite aggression to humans (Van Mele & Cuc 2000; Van Mele 2008; Sinzogan et al. 2008). These and other practices have been termed 'ant husbandry' by Barzman and colleagues (1996). The success of *Oecophylla* husbandry has been demonstrated in various settings, and it is clear that making use of these predators can in some cases greatly decrease the need for insecticide applications (e.g., Van Mele & Cuc 2000; Peng & Christian 2006; Van Mele et al. 2007 Sinzogan et al. 2008; Van Mele 2008; Crozier et al. 2010; Peng et al. 2011).

Weaver ants are particularly effective predators of arthropod pests in orchards because they construct large nests within the canopy and engage in extensive hunting raids that target herbivores on the trees. Members of other ant genera that co-occur with weaver ants often nest on or near the ground and tend to be less effective in suppressing pest arthropods. These ants can interfere with and severely disrupt the effectiveness of weaver ants, and a number of strategies have been developed to limit this competition as a means of increasing biological control (Barzman et al. 1996; Van Mele 2008). Habitat diversification is one class of these strategies, based on early observations of better biological control by weaver ants in more diversified orchard or plantation settings. These observations prompted a number of authors to speculate that a diverse understory would benefit *Oecophylla* spp. at the expense of competitor ant species that are less effective biological control agents (Rapp & Salum 1995). An experiment done in a citrus orchard in Tanzania was designed to test this hypothesis (Seguni et al. 2011). One half of a single grove was weeded to bare ground while vegetation was kept at approximately 10 cm in the other half. This procedure was maintained for 16 months, after which the treatments were switched for another two years of sampling. The ants of interest in this study were *Oecophylla longinoda* (a weaver ant) and the big-headed ant, *Pheidole megacephala*, a fierce competitor of *Oecophylla*. *Pheidole* were dominant on the initially bare side of the orchard, but both *Pheidole* and *Oecophylla* were present on the 'weedy' side for the first phase of the study. After the switch, *Oecophylla* moved into the side of the orchard that had previously been bare but in which weeds were allowed to grow, and largely abandoned the side of the orchard that was made bare. Because of the low replication of this study, these results cannot be seen as robust support for the hypothesis that understory vegetation tips the competitive balance toward *Oecophylla*. However, the strong response to switching the treatments halfway through the experiment increases confidence. The results suggest that bare ground increases contact between the two ant species, leading to local elimination of *Oecophylla*, the weaker competitor (Vanderplank 1960). Increased interactions are likely in bare orchards because an absence of undergrowth drives the ground-nesting *Pheidole* into the tree canopy in search of honeydew (Seguni et al. 2011). While Seguni and colleagues did not measure pest levels or strength of biological control in this experiment, previous work has shown that *Oecophylla* are more effective biological control agents than *Pheidole* are in the old-world tropics and also that *Oecophylla* are capable of suppressing arthropod pests in citrus and other orchard settings in Africa (e.g., Way 1951; Greenslade 1971; Van Mele 2008).

A counterpoint to this example involves strategies to disrupt the black ant, *Dolichoderus thoracicus*, a major antagonist of weaver ants in Vietnamese citrus groves. In this case, the appropriate strategy involves a *decrease* in plant diversity – namely, the avoidance of intercropping citrus with sapodilla fruit trees, which harbor black ant nests. The ability of weaver ants to suppress stinkbugs and caterpillars in citrus is severely reduced by the presence of black ants, which are themselves ineffective biological control agents in citrus. Limiting the ability of black ants to nest therefore improves biological control by weaver ants in citrus (Van Mele & Chien 2004).

Ants other than *Oecophylla* can be conserved for biological control purposes as well. Some of these interactions with an emphasis on Latin

America were reviewed by Perfecto and Castiñeras (1998), and an interesting study from eastern North America is described by Mathews and colleagues (2007, 2009). In this case, extrafloral nectaries on peach trees attracted and retained a number of predatory ant species, leading to lower herbivore densities, enhanced tree growth and less fruit injury. Hemipteran-tending by ants can also lead to improved biological control by ants, despite the fact that these hemipterans can also be pest insects. Styrsky and Eubanks (2010) showed that predation of caterpillars by the imported fire ant, *Solenopsis invicta*, was greater in the presence of cotton aphids than in their absence in cotton. Honeydew production by the aphids attracted the ants and resulted in ant tending of the aphids, as well as greatly increased predation on larvae of the beet armyworm, *Spodoptera exigua*. Since beet armyworms are more serious pests than cotton aphids are, the net effect of aphids on cotton production was positive. Similar observations have been reported by other authors (e.g., Nickerson et al. 1977; Messina 1981; Way et al. 1999; Altfeld & Stiling 2009; but see Fritz 1983). These kinds of results can lead to a very counterintuitive recommendation for conservation biological control – namely, that the presence of aphids should be encouraged under certain circumstances.

9.5.2 Beetle Banks in the United Kingdom and Beyond

The higher the ratio of field edge to interior is in agricultural ecosystems, the more likely it is for natural enemies to move from shelter habitat on the edge of fields into the interior. A number of studies have shown that the activity of some natural enemies extends only tens of meters (or even less) into annually cropped fields (e.g., Coombes & Sotherton 1986; Tylianakis et al. 2004). This dispersal limitation inspired the development of so-called beetle banks, in which critical elements of edge habitat were placed *within* fields in linear strips at a frequency great enough to allow natural enemy dispersal to the mid-point between strips. The first beetle banks were created by Thomas and colleagues (1991, 1992; see also MacLeod et al. 2004) within two wheat fields in the United Kingdom – they consisted of raised sections of soil 290 m in length and 1.5 m in width planted to various matt- and tussock-forming grasses known not to be invasive or weedy (Figure 9.8). Sampling done on and near these structures showed that they enhanced overwintering populations of predatory arthropods such as ground beetles, rove beetles and spiders, and also that some of these predators moved into the field centers in the spring and summer. Similar results have been reported from other studies investigating within-field

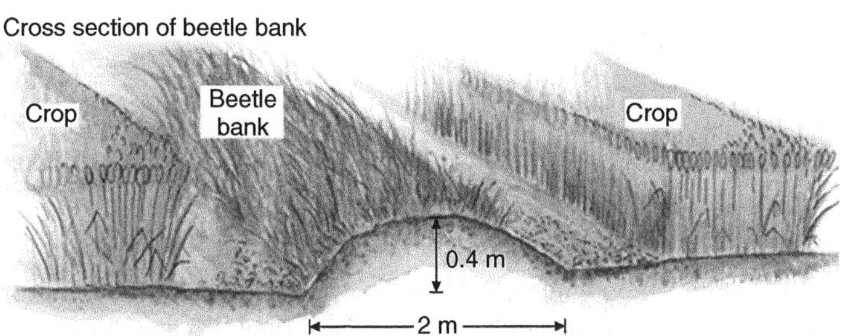

Figure 9.8. Diagram of a beetle bank from the Game & Wildlife Conservation Trust, Fordingbridge, Hampshire, the United Kingdom; http://www.gwct.org.uk/

strips of added vegetation within annual cropping systems in Europe (Lys & Nentwig 1992; Lys et al. 1994; Nentwig et al. 1998; Lemke & Poehling 2002), but movement of predators from strips to the adjacent crop habitat was not consistently seen in studies done in Michigan in the United States (Carmona & Landis 1999; Lee et al. 2001).

These early studies led to the promotion of beetle banks as components of 'agri-environment schemes' in the United Kingdom, and as a result they have been adopted throughout northern Europe and are now iconic features of the agricultural landscape there (Landis et al. 2000; Collins et al. 2003b; MacLeod et al. 2004). Prospective analyses suggested that the savings in pesticide use could outweigh costs attributable to taking land out of agricultural production (Sotherton 1995). Additionally, beetle banks can provide nesting sites for vertebrates of conservation interest (Thomas et al. 2001b; Bence et al. 2003; Ewald et al. 2010).

The first research group to address the effects of beetle banks on pest suppression was Collins and colleagues (2002), who used predator exclusion techniques at two distances from beetle banks to assess the usefulness of beetle banks in reducing cereal aphid populations. These studies showed that aphid suppression by predators was occurring in wheat fields with beetle banks and that this effect was stronger close to banks than farther from the banks (8 versus 83 m). Although these experiments did not incorporate control fields without beetle banks, they are consistent with the hypothesis that the beetle bank is contributing to increased biological control of cereal aphids. Collins and colleagues (2002, 2003a, b) suspected that the most important aphid predators enhanced by the beetle banks were certain species of ground beetles, rove beetles and lyniphiid spiders known to attack aphids, and also that the most important grass species from the beetle bank favoring these predators was the tussock-forming *Dactylis glomerata*.

Another study on the efficacy of beetle banks did compare beetle-banked fields with control fields and assessed predation in both types of fields. This study was done in the northwestern United States by Prasad and Snyder (2006), who found that while beetle banks (formed from *D. glomerata* grass) were effective in enhancing the abundance of predatory beetles within potato fields, predation of fly eggs (used as experimental surrogate pests) was unaffected. Experimental field cage studies revealed a likely explanation for this puzzling outcome: beetle banks led to higher levels of the large ground beetle *Pterostichus melanarius*, which turned out to be a more effective predator of smaller ground beetles than of the prey eggs. The smaller ground beetles, on the other hand, were effective egg predators, so enhanced *P. melanarius* populations likely led to intraguild predation that disrupted biological control. Prasad and Snyder's study also indicated that the presence of alternative prey (such as aphids) could disrupt biological control of other pest species by small ground beetles, but there was no indication that aphids were enhanced (or depressed) in fields with beetle banks. Indeed, we would generally expect lower aphid densities in these fields based on other studies (Collins et al. 2002).

As far as we are aware, the Prasad and Snyder (2006) study remains the only one that has compared pest suppression in fields with and without beetle banks. Thus, there is ample scope to build on the early beetle bank work, which showed positive effects on potential biological control agents. More studies are needed that employ proper controls and that assess effects on pest as well as predator populations.

9.5.3 Amendments to Create Suppressive Soils

It has long been known that organic amendments can improve biological control of soil-borne plant pathogens (Hoitink & Fahy 1986; Windels 1997; Hoitink & Boehm 1999; Mazzola 2004; and see Chapter 8). These amendments can include animal and green manures, or composts of various kinds, and they can either be used to improve naturally occurring biological control or they can involve the introduction of biological control agents themselves. As we have mentioned previously in Chapters 1 and 8, the biological control agents involved can suppress disease by a number of mechanisms including competition, antibiosis, parasitism and induced resistance. And the addition of organic amendments can aid these microbial biological control agents in various ways, including improvement of the abiotic conditions necessary for growth and activity and supplementation of critical resources.

Before describing a case study, it is worth relating the difficulties of disentangling various factors that may be involved in the facilitation of improved plant growth as a result of adding amendments to agricultural soils. These difficulties are due to the extreme complexity that reigns among the interaction of soil, plants, plant pathogens and other soil-borne microbes. First, amendments can lead to improved plant growth without suppressing plant pathogens. This can occur if composting increases the nutrient status of the soil, improving plant vigor (including tolerance to plant pathogens). Even if microbial diversity is increased by composting, improved plant growth may not be due to disease suppression at all, but instead to improvement of soil quality. These differences prompted the recommendation for a functional classification of rhizobacteria as either 'biocontrol plant growth-promoting bacteria' or simply 'plant growth-promoting bacteria' by Bashan and Holguin (1998), as discussed in Chapter 1. And this classification, while useful, may not even be sufficient since some species and strains of bacteria and fungi promote plant growth by a combination of both plant growth promotion and disease suppression (Hernandez-Rodriguez et al. 2008; Naik et al. 2008).

A second source of ambiguity in the context of conservation biological control involves the difference between *conserved* and *introduced* biological control agents. When the addition of a soil amendment leads to the suppression of plant pathogens via the activity of antagonistic microbes, this can be due to either increased activity of biological control agents already present in the soil or to agents introduced with the amendment (Lucas & Sarniguet 1998; Kowalchuk et al. 2003). Only the former of these mechanisms constitutes conservation biological control. As we discuss later in a section on combined conservation and augmentative biological control in these systems (Section 9.7.2), conserved and introduced biological control agents can combine to produce greater disease suppression than either can achieve alone.

Even if it can be clearly shown that a given soil amendment leads to biological control of a plant pathogen via resident microbial antagonists, the mechanism by which this occurs is often elusive. And while the mechanisms are often traits of particular organisms, they can also be context dependent. For instance, *Streptomyces* bacteria can suppress plant pathogens by the production of antibiotics or by resource competition (Neeno-Eckwall et al 2001). Which mechanism predominates in disease suppression is determined in large part by resource availability, with antibiosis dominating under high resource availability and competition predominating under low resource availability (Schlatter et al. 2009). Mazzola (2004) has reviewed many of these mechanisms as well as methods available to identify soil-borne antagonists of plant pathogens and the mechanisms by which they suppress plant disease.

Given the complexities inherent in demonstrating conservation biological control of plant pathogens due to the addition of soil amendments, we seek a case study that carefully shows how an amendment can lead to increased disease suppression through the enhanced action of antagonistic microbes that are not themselves introduced with the amendment. Improved control of soil-borne plant diseases by *Streptomyces* bacteria through the addition of seed meals and green manures fits these criteria. *Streptomyces* are ubiquitous soil organisms, and some species play important roles in suppression of a broad range of plant disease-causing organisms within the soil (e.g., Samac & Kinkel 2001; Xiao et al. 2002; Samac et al. 2003; Errakhi et al. 2007). In addition, a number of studies have produced data suggesting that soil amendments can improve the suppressive actions of *Streptomyces* against various important plant pathogens including scab, rot and wilt diseases in potato, alfalfa and apples (Cohen et al. 2005; Wiggins & Kinkel 2005a, b; Cohen & Mazzola 2006) (Figure 9.9). Despite strong circumstantial evidence that disease suppression by resident *Streptomyces* was improved by the addition of soil amendments in these studies, the mechanism(s) were unknown. *Streptomyces* bacteria are known to produce a variety of antimicrobial, antifungal and anti-nematicidal compounds, but can also stimulate plant resistance mechanisms and outcompete pathogenic bacteria (Liu et al. 1997; Schrey & Tarkka 2008; Chater et al. 2010).

The roles of various mechanisms contributing to amendment-mediated suppression by *Streptomyces* have been dissected by Mazzola and colleagues working with the seed meal of various *Brassica* species as amendments to soils containing apple seedlings plagued by *Rhizoctonia* root rot. *Brassica* seed meals (BSM) are a by-product of processing *Brassica* seeds for oil and they have a number of uses, including as soil amendments to increase suppressiveness of plant pathogens. The usefulness of BSM applications in suppressing *Rhizoctonia* root rot of apple seedlings was demonstrated in Washington State in the United States (Mazzola et al. 2001; Cohen et al. 2005; Mazzola & Mullinix 2005; see Fig. 9.9a), as were increases in density of *Streptomyces* in response to BSM (Cohen et al. 2005; Cohen & Mazzola 2006). The importance of *Streptomyces* in suppressing root rot was identified via a number of approaches. First, Cohen and Mazzola (2006) showed that pasteurizing the BSM amendment did not decrease its ability to suppress rot, and also that the amendment could not suppress rot within pasteurized soil samples. This showed that the amendment itself was not the source of suppressive agents but rather that the agents were present in the soil. Further studies showed that disease suppression could be recovered in pasteurized

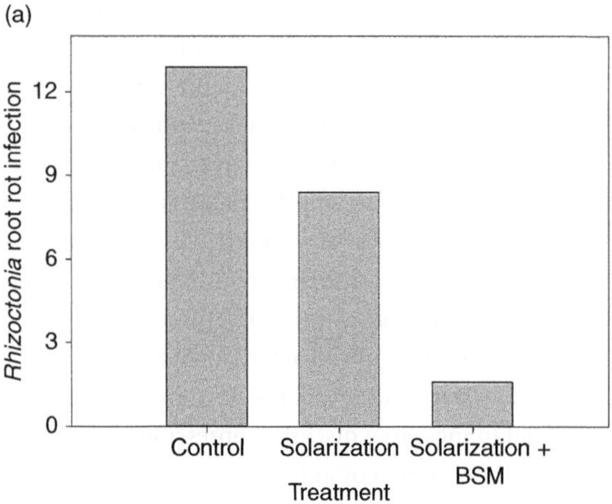

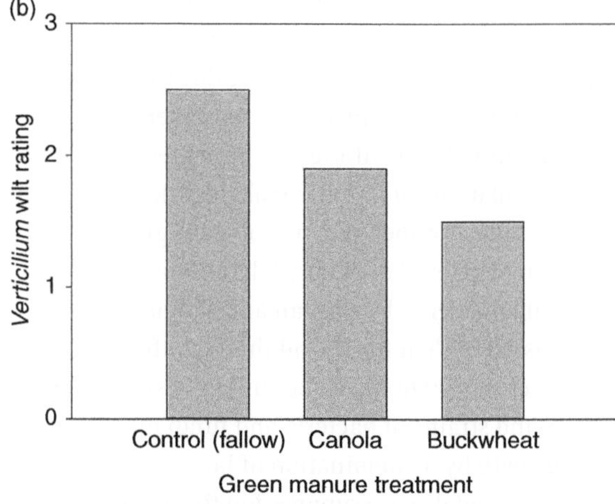

Figure 9.9. Effects of soil amendments on soil-borne plant pathogens in field experiments. (a) Incidence of *Rhizoctonia* root rot infection in apple seedlings in unamended soil (control), soil subjected to solarization with plastic tarps, and soils subjected to solarization and *Brassica napus* seed meal (BSM). From Mazzoli and Mullinix (2005). (b) *Verticillium* root rot rating on potatoes in plots subjected to no green manure (fallow control) and canola or buckwheat green manure treatment. From Wiggins and Kinkel (2005b).

soils by inoculation of *Streptomyces* that had been isolated from BSM-amended or non-amended soils (Cohen & Mazzola 2006). Together these studies showed that BSM amendments can protect apple seedlings from root rot, and that this suppression is correlated with increases in *Streptomyces* bacteria, which are capable of suppressing rot. However, the mechanism(s) by which BSM aids *Streptomycetes*, as well as the mechanism by which *Streptomyes* suppresses rot in this system have remained elusive (Mazzola et al. 2007; Mazzola & Manici 2012)

Additional insights on mechanisms has come from a similar system, however – one in which the use of green manures (i.e., plantings incorporated into the soil as the crop is sown) enhances suppression of potato scab, *Streptomyces scabies*, by beneficial *Streptomyces* bacteria. An early indication that conservation biological control may be possible against potato scab came from studies showing decreased incidence of disease following amendment of soils with grass clippings and *Actinomyces* (Millard & Taylor 1927). Millard and Taylor suspected their results depended on the grass clippings, but did not separate the effects of the grass itself and bacterial amendments co-inoculated with the clippings. Their interpretation got some support from later studies though, showing that the application of buckwheat, canola and sudangrass green manures could increase suppression of various soil-borne crop diseases (including potato scab) by *Streptomyces*, although the in-field levels of disease suppression were inconsistent, and the mechanism(s) remained unknown (Wiggins & Kinkel 2005a, b; Perez et al. 2008).

In studies performed on prairie soils from Minnesota in the United States, Schlatter and colleagues (2009) were able to shed light on the mechanism by which *Streptomyces* inhibited plant-pathogenic bacteria. These authors isolated soil from the Cedar Creek Ecosystem Science Reserve discussed in Section 9.1 and inoculated samples with carbon-based soil amendments, including cellulose and lignin at low and high levels. After incubating these mixtures for nine months, Schlatter and colleagues measured growth of *Streptomyces* populations that were originally present in the soil samples as well as the ability of these *Streptomyces* to inhibit plant-pathogenic bacteria. These studies showed that amendments of both cellulose and lignin increased *Streptomyces* densities. In addition to leading to increased densities of beneficial *Streptomyces*, the *Streptomyces* from the amended soils showed stronger inhibition of pathogens. Inhibition was tested by placing killed *Streptomyces* colonies from soils incubated with low and high levels of glucose and lignin onto petri dishes containing water agar along with live colonies of standard pathogenic bacteria (also in the genus *Streptomyces*). The 'inhibition zone', or physical area on the dishes containing killed pathogens, was then measured. Inhibition zones were consistently greater from soil samples receiving higher levels of glucose or lignin amendments. Schlatter and colleagues (2009) interpreted these results as suggesting that *Streptomyces* strains capable of producing higher amounts of antibiotics were favored by high levels of cellulose and lignin supplementation. The ecological basis for this hypothesis is that nutrient-rich environments cause increased interspecific interactions within the soil, which increase the value of antibiotic activity. At lower nutrient levels, antibiotic production is less critical and strains that are more efficient at utilizing nutrients are favored. If this hypothesis is correct, the usefulness of organic amendments in the conservation biological control of soil-borne plant pathogens is mediated at least in part by a fitness trade-off between antibiotic production and nutrient efficiency (Kinkel et al. 2011). Other hypotheses involving fitness trade-offs in biological control are discussed in Chapter 7.

The study by Schlatter and colleagues (2009) identified a mechanism by which resident beneficial *Streptomyces* can inhibit pathogenic bacteria in soil and how this suppression can be enhanced by carbon-based amendments. The data point to a double benefit – (i) higher densities of *Streptomyces* that show (ii) enhanced inhibitory qualities. The extent to which these mechanisms are operational (or dominant) in real agricultural settings is still not clear, however.

9.6 How Habitat Diversification Can *Decrease* Effectiveness of Biological Control

The notion that habitat diversification could exacerbate pest problems is by no means new, nor it is it far-fetched. H. van Emden warned about such effects more than 50 years ago (Van Emden 1965) while citing an older literature in which uncultivated vegetation was seen more as a source of pests than of biological control agents. Non-crop vegetation can indeed support pest populations and it can serve as a source of pest populations, leading in some cases to recommendations to *limit* rather than enhance plant diversity in order to decrease pest pressure (Herzog & Funderburk 1985; Coombs 2000; Jones et al. 2001; Power & Mitchell 2004; Schellhorn et al. 2010). Higher plant diversity can also lead to higher levels of humidity, which can in turn lead to increased pressure by plant pathogens (Johnson et al. 2009). In addition, although diverse field boundaries can act as sources of natural enemies, they can also serve as sinks of natural enemies instead, or as barriers to the movement of natural enemies into agricultural fields (Corbett & Plant 1993; Wratten et al. 2003). And although composting can be an effective way to enhance the suppression of plant pathogens as discussed previously, it can also enhance disease under certain conditions (Windels 1997). In this section, we focus on four specific mechanisms by which increased habitat diversity can interfere with biological control.

9.6.1 Resource Concentration for Specialist Natural Enemies

The resource concentration hypothesis holds that monocultures are easier to find and harder to lose than are polycultures for specialist herbivores (see Section 9.3). And as we have already noted, this same argument likely holds for specialist enemies of these herbivores (Section 9.3.2). Sheehan proposed this hypothesis in 1986 and cited a few cases of parasitoids in support of this general view. Since then a number of other studies have reported cases in which specialist natural enemies respond more positively to monoculture than polyculture settings (e.g., Coll & Bottrell 1996; Bezemer et al. 2010). In some cases, higher densities of biological control agents in monoculture settings are confounded with higher pest densities (e.g., Gold et al. 1989; Szendrei & Weber 2009). This could be due simply to aggregation of biological control agents to higher populations of pests rather than Sheehan's hypothesis that natural enemies are impeded by plant diversity. Controlled experiments can be useful in disentangling these and other confounding factors, as shown by Gols and colleagues (2005) and Bukovinsky and colleagues (2007), who used greenhouse and laboratory studies to show that the presence of companion plant species impeded host finding by *Diadegma semiclausum*, a specialist parasitoid of diamondback moth. This interference was not caused by a higher density of host or plants, but was enhanced when companion plants were taller than plants harboring hosts. In line with Sheehan's hypothesis, these authors concluded that chemical and structural diversity can provide a refuge to pests from their specialized parasitoids. For a similar set of studies, see Randlkofer and colleagues (2007, 2010).

9.6.2 Nectar Benefiting Pests

As we noted in Section 9.4.2, providing nectar to adult biological control agents that rely on sugar for survival is a common goal in conservation biological control. But what if the pests benefit from the nectar as well? This is a real danger for insect pests that use nectar as a sugar source, as was shown in an experiment on biological control of the potato tuber moth in Australia. Baggen and Gurr (1998) compared parasitism levels of tuber moth by

Copidosoma koehleri in plots of potatoes established at various distances from flowering coriander. While parasitism rates were higher the closer the plots were to the flowers, so was pest density and damage to the crop. Laboratory studies showed that both the tuber moth and parasitoid adults were able to use coriander nectar and that this led to increased longevity of both species. The conclusion was therefore that both insects availed themselves of nectar in this trial with a net effect of more serious crop damage. In a later study, Winkler and colleagues (2009b) used biochemical analyses of field-caught insects to show that both the diamondback moth (a pest) and its parasitoid, *D. semiclausum*, utilized supplemental nectar from various flowering plants in a conservation biological control field trial. Cautionary tales such as these have led to directed laboratory experiments comparing the ability of pests and parasitoids to utilize the nectar of plants contemplated for conservation biological control (e.g., Baggen et al. 1999; Begum et al. 2006; Lavandero et al. 2006; Winkler et al. 2009a; Nilsson et al. 2011; Geneau et al. 2012). These studies represent some of the most detailed attempts to define elements of the 'right forms of diversity' discussed in Section 9.2.

9.6.3 Enhancing Natural Enemies of Biological Control Agents

Habitat manipulation to enhance biological control agents can backfire if natural enemies of biological control agents are enhanced as we have already seen in studies on North American beetle banks (Prasad & Snyder 2006; see Section 9.5.2). The role of intraguild predators and secondary predators in interfering with biological control has received increased attention since the mid-1990s (e.g., Rosenheim et al. 1995; Rosenheim 1998; Brodeur & Boivin 2006; Straub et al. 2008; Chailleux et al. 2014). This recognition has highlighted the prospect of enhancing the 'wrong kind of diversity' at higher trophic levels through habitat diversification with a net effect of reducing pest suppression.

An example comes from orchards in New Zealand, where Stephens and colleagues (1998) found that the planting of flowering buckwheat enhanced the lacewing parasitoid *Anacharis* sp. Since lacewings can be beneficial insects in orchards as predators of aphids (but see Robinson et al. 2008), increased parasitism of these insects in the buckwheat-enhanced vineyard plots could in principle lead to higher aphid densities, but this was not measured by Stephens and colleagues. In a separate set of studies done in field cages, enhancement of *Anacharis* parasitism by buckwheat flowers did interfere with the ability of this same lacewing species to reduce aphid populations in alfalfa, but this occurred only when aphid densities were relatively high (Jonsson et al. 2009; Jacometti et al. 2010).

Enhancement of hyperparasitoids using habitat diversification constitutes a similar danger (Banks et al. 2008; Araj et al. 2009), particularly in cases where hyperparasitoids are less specialized than primary parasitoids are, as tends to be the case for parasitoids associated with aphids (Müller et al. 1999; Brodeur 2000). Three European studies evaluated the hypothesis that habitat diversity enhances hyperparasitism more than it enhances primary parasitism of aphids. One of these found support for such an effect (Rand et al. 2012) while two did not (Gagic et al. 2011; Plećaš et al. 2014). The pattern Rand and colleagues found is disconcerting for conservation biological control given the general pattern that generalist enemies are more likely to be positively affected by habitat diversification that we have already discussed (see Sections 9.3.2 and 9.6.1). Such a pattern may be illustrative of a broader trend if members of the fourth trophic level tend to have broader diet breadth than members of the third trophic level.

9.6.4 Disrupting Suppressive Soils

Finally, natural biological control of plant pathogens can be disrupted by diversity enhancing practices such as crop rotation when they interfere with the build-up of suppressive soils (e.g., Weller et al. 2002; Ryan et al. 2009). In these cases, the most effective way to achieve biological control can be to establish continuous monocultures of annual crops, a practice that would typically be considered anathema to conservation biological control. One example of this scenario is take-all decline in wheat. Take-all is a devastating pathogen of wheat caused by the fungus *Gaeumannomyces graminis* var. *tritici*. It had been noted for years that the incidence of take-all tended to drop precipitously after severe outbreaks of the disease. As we have already discussed in Chapter 8, these reductions were termed take-all decline (TAD) and tended to persist for many years under conditions of wheat monoculture, with disease returning to the original TAD sites after the initiation of crop rotation. This form of suppressiveness has been documented throughout the world and the agent responsible for TAD is an antibiotic-producing *Pseudomonas* bacterium (Ryan et al. 2009). A similar case has been documented for potato scab, a disease caused by the bacterium *Streptomyces scabies*. Two potato research plots in Minnesota became suppressive to potato scab after 28 and 8 years to the point where they were abandoned for scab research. Analysis of the soils in these fields showed that they had become suppressive by virtue of the predominance of various strains of *Streptomyces* that were not pathogenic to potatoes but instead produced antibiotics that suppressed *S. scabies* (Liu et al. 1997). In these and other cases, it appears that continuous monoculture facilitates the buildup of microbes or groups of microbes that confer suppressiveness.

9.7 Broader Applicability of Conservation Biological Control

We end this chapter with some comments on linkages between conservation biological control and other forms of biological control. A number of authors have suggested that conservation biological control strategies could be used to improve the effectiveness of both importation and augmentative biological control (e.g., Hopper & Roush 1993; Ehler 1998; Newman et al. 1998; Gurr & Wratten 1999; Heimpel & Asplen 2011; Fernanda Diaz et al. 2012). We agree with this sentiment and offer some examples of cases in which conservation and augmentative biological have been combined. While many conservation biological control programs are directed toward established importation biological control agents, we know of no cases in which manipulation of the habitat into which importation biological control agents are to be released has been explicitly evaluated as a means of increasing establishment. We therefore focus on the integration of conservation biological control (through habitat manipulation) and augmentation in this section. The topic of augmentation is dealt with more generally in Chapter 8, and in Chapter 10, we discuss ways in which augmentative biological control can be combined with conservation biological control through pesticide reduction (Section 10.3).

Two general strategies can be discerned that combine elements of conservation and augmentative biological control. In one, habitat manipulations are developed that improve the establishment or reproduction of augmentative biological control agents and in the other, biological control agents are added to manipulation tactics to add value to the manipulation. The first strategy is driven by augmentation and improved by conservation, and the second is driven by conservation and improved by augmentation. We briefly discuss examples of these two strategies.

9.7.1 Improving Augmentative Biological Control with Conservation Biological Control

Explicit attempts to improve the success of augmentative biological control through habitat manipulation are included in so-called banker plant systems. Banker plants consist of non-crop vegetation that is planted for the express purpose of establishing released natural enemies of herbivores adjacent to crop plantings. Herbivores are often released onto these plants as well in order to aid in the establishment of the biological control agents. Banker plants have been developed for both field and greenhouse use and are well-reviewed by Frank (2010) and Huang and colleagues (2011).

In a system similar to the banker plant approach, Corbett and colleagues (1991) experimented with using a 'nursery crop' of alfalfa (lucerne) adjacent to cotton as a means of conserving released predatory mites in California. Alfalfa is a perennial crop whereas cotton is planted as an annual and the idea was to establish the predatory mite *Galendromus* (= *Metaseiulus*) *occidentalis* earlier in the season so that phytophagous spider mites could not build up in the cotton. The predatory mites were expected to reproduce in the alfalfa on phytophagous mites that were themselves also introduced into the alfalfa and then to move into the cotton to suppress spider mites there. Because of this reproduction, more predatory mites could be 'released' into the cotton crop than by straight augmentation into the cotton. Corbett and colleagues' (1991) results suggest strongly that predatory mites moved from the alfalfa into the cotton and also that this led to suppression of spider mites in the cotton during most of the growing season.

A particularly clear example of combining augmentative and conservation biological control comes from attempts to improve within-field establishment of augmentative releases of the egg parasitoid *Trichogramma ostriniae* against the European corn borer in bell pepper fields in the southern United States. Russell and Bessin (2009) designed an experiment in which the release of *Trichogramma* wasps and the presence of flowering buckwheat, *Fagopyrum esculentum*, adjacent to small plots of bell peppers were both manipulated. Parasitism of European corn borer egg masses placed into the field as sentinels was the highest in plots that received both parasitoid releases and buckwheat addition, but the contribution of buckwheat to this trend was not statistically significant. However, pest damage to the peppers was also lowest in this treatment, and in this case the statistical analysis showed a significant interaction between release and buckwheat, suggesting synergism between the treatments (Figure 9.10). This difference suggests

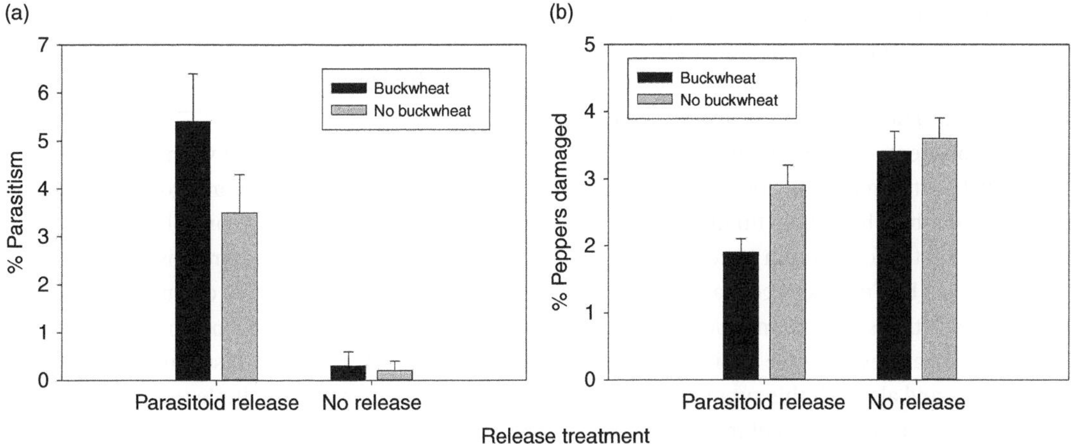

Figure 9.10. Parasitism of sentinel European corn borer egg masses (a) and percentage of pepper fruits damaged by European corn borer larvae (b) that were either subjected to releases of *Trichogramma ostriniae* or not and enhanced with plantings of buckwheat flowers or not. Modified from Russell and Bessin (2009).

a stronger effect of the treatments on parasitism of naturally produced corn borer egg masses than sentinels. The authors of this study suspected that *Trichogramma* wasps utilized buckwheat nectar and that this allowed higher parasitism rates.

9.7.2 Improving Conservation Biological Control by Using Augmentative Biological Control

The benefits of conservation biological control tactics can be improved upon by adding biological control agents to the habitat manipulation. Just such an approach is commonly applied in biological control of plant pathogens. Here, soil amendments used to conserve microbial biological control agents of soil-borne plant pathogens are often enriched with some of the biological control agents themselves, so that the incorporation of the amendment serves both to conserve the beneficial microbes already present and also to add new ones. Hoitink and Boehm (1999) have argued that adding various types of compost to agricultural systems serves to inoculate soils with antagonists of a number of soil-borne plant pathogens at the same time as providing nutrients for antagonists that are already present in the soil. This strategy is perhaps best studied with respect to the supplementation of various composts with *Trichoderma* fungi. Early studies showed that *Trichoderma* were among the biological control agents in some composts (e.g., Kwok et al. 1987; Cotxarrera et al. 2002), and this led to strategies by which these fungi were isolated and formulated for addition to composts. An example is given by Trillas and colleagues (2006), who showed that suppression of *Rhizoctonia* damping-off can be achieved by the addition of various composts, and also that this suppression can be greatly enhanced by a particular strain of *Trichoderma asperellum* (the strain T-34), depending on the compost it is associated with (Figure 9.11). The supplementation of composts with biological control agents has been developed in other contexts

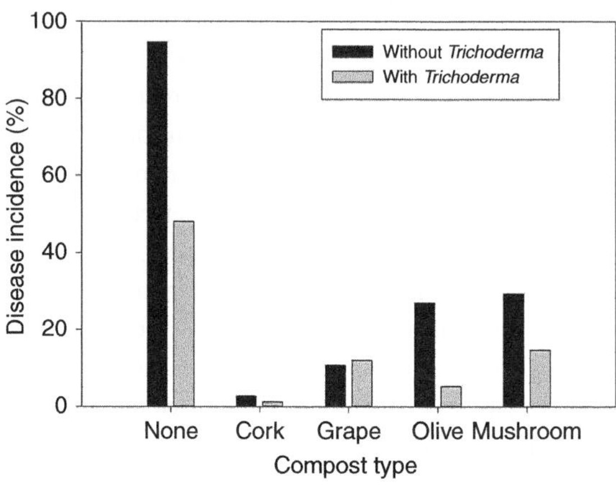

Figure 9.11. Percentage of cucumber seedlings showing symptoms of *Rhizoctonia solani* damping-off in the presence of four compost types and a peat control with and without addition of the fungus *Trichoderma asperellum* strain T-34. These results varied with age of the compost and concentration of *Trichoderma*. Modified from Trillas et al. (2006).

as well, and interest in this practice appears to be increasing (e.g., Postma et al. 2003; Siddiqui et al. 2008; Sant et al. 2010; Bernal-Vicente et al. 2012; St. Martin & Brathwaite 2012).

9.8 Conclusions

Conservation biological control via habitat manipulation is an explicit attempt to improve naturally available biological control services. Of all the classes of biological control discussed in this book, it is the most ecologically complex since it tends to involve the most interspecific interactions. The breadth of interactions is also particularly wide, including as it does multiple trophic levels and alterations of abiotic factors. Research into conservation biological control has historically had an optimistic orientation because it rests on the confrontation of ecological theories linking diversity and stability to observations of disturbance in managed systems. This confrontation leads to the clear hypothesis that if we can only ameliorate the disturbance associated with management practices, an increase in stability (i.e., fewer pest outbreaks) will follow. While this optimism has been rewarded in some cases, it has also been tempered more recently due to caveats that inevitably accompany the spiraling complexity found in these systems. Despite this trend, we remind readers that early researchers warned of just this complexity (Section 9.2) and pointed toward a search for the 'right kind of diversity'. This search continues as conservation biological control strategies become increasingly refined.

10 Conservation Biological Control II: Facilitating Natural Control through Pesticide Reduction or Selectivity

In Chapter 9, we discussed conservation biological control through habitat manipulation. Reducing pesticide use with the express goal of protecting biological control agents is traditionally also considered a conservation biological control strategy, and we turn our attention to these strategies here. We begin by discussing interactions between pesticides and biological control, some of which are paradoxical, and follow this with an outline of the major conservation biological control strategies that have been developed to protect biological control agents from pesticides. Last, and as we did in Chapter 9, we discuss ways in which this form of conservation biological control can be merged with importation and augmentative biological control.

10.1 Pest Resurgences and Secondary Pest Outbreaks

The use of broad-spectrum pesticides entails a sort of Faustian bargain: while broad-spectrum pesticides can be useful from the perspective that multiple species of pests are affected, they can also undermine pest control by killing natural enemies of pests. The application of broad-spectrum pesticides can lead paradoxically to outbreaks of the target pest itself (known as pest resurgence) or to outbreaks of other pests (known as secondary pest outbreaks) through the process of release from natural enemies (Dutcher 2007). Pest resurgences and secondary pest outbreaks form an important basis for understanding how pesticide reduction can lead to better pest control. We therefore begin this chapter by providing a brief discussion of these unintended consequences before outlining the conservation biological control tactics that are developed to overcome them.

10.1.1 Pest Resurgence

Pest resurgence occurs when a target pest population increases following a pesticide application – an unexpected outcome since the pesticide is ostensibly toxic to the pest! How can we explain this paradoxical effect? Hardin and colleagues (1995) reviewed five hypotheses to explain the resurgence of arthropod pests, and these are listed in Table 10.1. Two hypotheses involve negative effects of pesticides on the natural enemies of pests, and thus the release of pests controlled by these enemies, but these hypotheses differ with respect to the effect of the pesticide on the pest itself. In the first hypothesis, pesticides are more toxic to the natural enemies than they are to the pests. This disparity has been observed in many systems, and the most obvious mechanism is differential pesticide resistance. It is indeed the case that pests tend to evolve resistance to pesticides more rapidly than their natural enemies do (Tabashnik & Johnson 1999).

In the second hypothesis, however, pest resurgence via release from natural enemies occurs despite pesticides actually killing pests, and it is this case that requires further explanation. Sherratt and Jepson (1993) used a metapopulation model to look at the incidence of pest resurgence as a function of the relative toxicity of pesticides to predator and prey. Their analysis showed that while pest resurgence was most prevalent when toxicity was greater to predators than to pests, resurgence could occur at equal toxicities and even in cases where pesticide toxicity was higher for pests than predators (Figure 10.1). Such an outcome is possible in the context of metapopulation dynamics if pests re-invade the sprayed area more rapidly than predators do (Croft 1990; see also Trumper & Holt

Table 10.1. **Hypotheses and proposed mechanisms to explain resurgence of arthropod pests to insecticide (or acaricide) use. From Hardin and colleagues (1995).**

Hypothesis	Mechanism
Stronger effect of pesticide on natural enemy than on pest species	Release of pest from control by natural enemies
Strong effect of pesticides on both pests and their natural enemies	Release of pest from control by natural enemies
Alteration of plant quality by pesticides	Improved host plant quality for herbivorous pest species
Direct positive effects of pesticides on pest species	Enhanced fecundity, longevity or behavior
Negative effects of pesticides on non-target herbivores	Reduced competition

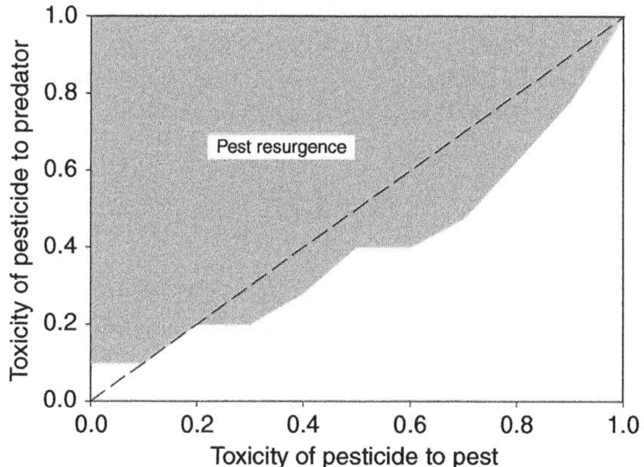

Figure 10.1. Combinations of toxicity of pesticides against prey and predators that lead to pest resurgence (gray area) or not from Sherratt and Jepson's (1993) metapopulation model. Parameter space above the diagonal line represents the case where toxicity is greater to predators than to pests.

1998). Alternatively, pests can be reduced to densities so low that natural enemies cannot find them and starve or emigrate as a result (Hardin et al. 1995).

But metapopulation dynamics are not necessary for pest resurgence in the face of effective pesticides. Hassell (1984a) used a host-parasitoid modeling approach to contrast four scenarios: (i) insecticides act prior to parasitism and only kill hosts; (ii) insecticides act after parasitism and only kill hosts; (iii) insecticides act after parasitism and kill unparasitized and parasitized hosts at the same rate; and (iv) insecticides act prior to parasitism and kill adult parasitoids as well as hosts. The first three scenarios all resulted in pest suppression, but in the fourth scenario, pest resurgence dynamics were possible (Figure 10.2). This analysis exposes the fact that pesticides and natural enemies can suppress pest densities additively when natural enemies are not killed, so that release from natural enemies can lead to resurgence even in the face of reasonably effective pesticides depending on the relative strength of suppression imposed by pesticides and natural enemies (see also Barclay 1982; Waage et al. 1985; Matsuoka & Seno 2008).

Godfray and Chan (1990) evaluated a similar resurgence scenario previously hypothesized for a number of tropical pests. In this scenario, insecticide applications synchronize pest populations at a particular life stage and thus interfere with the ability of parasitoids or other natural enemies that

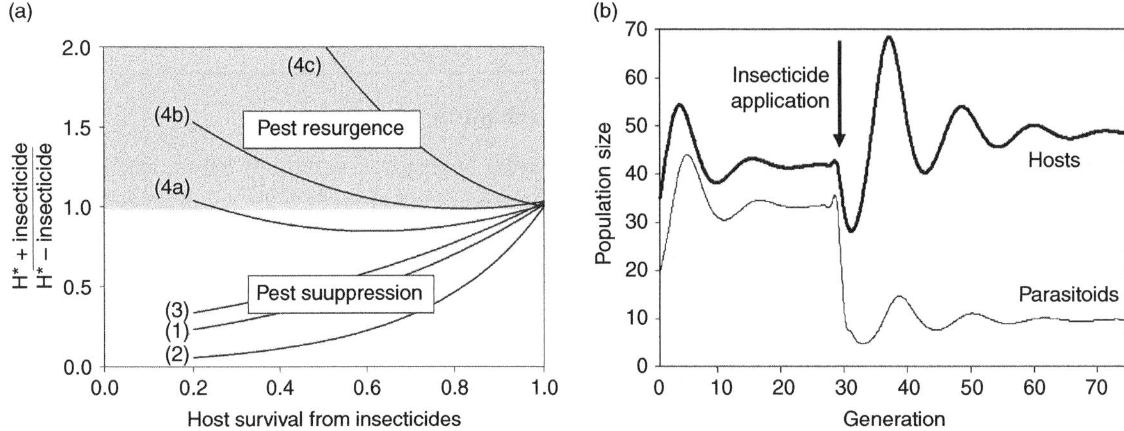

Figure 10.2. (a) Host suppression in four host-parasitoid models that differ in the way that insecticides kill hosts and their parasitoids. Host suppression is shown as a function of host survival from insecticides and expressed as the ratio of host equilibrium densities with and without insecticides. Values of this ratio above 1 indicate pest resurgence (gray), and values below 1 correspond to pest suppression (white). In scenarios (1) and (2) only hosts are killed, in scenario (3) both unparasitized and parasitized hosts are killed, while in scenario (4) unparasitized and parasitized hosts as well as adult parasitoids are killed. Three variants of scenario 4 are modeled, which differ in the extent to which adult parasitoids survive the insecticide application compared to their hosts. These values are (Host survival from insecticides)^m, with $m = 0.8$, 1.0 and 1.5 for (4a), (4b) and (4c), respectively. Modified from Hassell (1984a). (b) A simulation of scenario (4b) from Hassell (1984a) showing resurgence of hosts (thicker line) after application of insecticide at generation 30 (the thinner line indicates parasitoid populations). See also Waage and colleagues (1985).

require a different life stage of the pest to suppress it. Godfray and Chan referred to this hypothesis as 'catastrophic synchronization', and their analyses suggested that it is actually a relatively *unlikely* outcome of insecticide use, mainly because the stage structure of the host population should begin to shift from single to multiple stages soon after the insecticide application. Instead, they advocated a more nuanced scenario, 'dynamic synchronization', in which synchronization is imposed by natural enemies themselves – in particular parasitoids if predators are eliminated from the system. Previous models had shown that oscillating equilibria in host-parasitoid models have the tendency to synchronize host life stages in tropical environments (Godfray & Hassell 1987, 1989). Godfray and Chan posited that while a broad-based community of natural enemies including predators, parasitoids and pathogens may be capable of maintaining the pest population at a low stable density, the disruption of such control by insecticide application may leave a single specialist, which could very likely lead to oscillatory dynamics with peaks reminiscent of synchronized outbreaks. This scenario is in fact supported by a number of well-known tropical pest species, and Godfray and Chan's models confirm such dynamics are possible, at least theoretically. And while resurgence in these cases does depend on a form of natural enemy release, the critical insight for our purpose here is that it occurs despite the insecticides effectively killing pests.

Pest resurgence can also occur in the context of weed biological control, where it is in some ways even more paradoxical than in arthropod biological control because weed biological control agents are not expected to be susceptible to herbicides. While this is true for the most part, some herbicides do exhibit arthropod toxicity, including against some weed biological control agents (Messersmith & Adkins 1995; Ainsworth 2003). The mechanisms mediating weed resurgence in the face of herbicide use involve mainly removal of the primary resource of the biological control agent, which can lead to starvation or emigration of herbivores, or the loss of refuge from predators when the host plant is killed (Newman et al. 1998; Ainsworth 2003). Larson and colleagues (2007) studied the integration of herbicides and biological control of leafy spurge, *Euphorbia esula*, a Eurasian weed of pasturelands and natural areas in North America. Their study took place in a large swath of relatively pristine mixed-grass prairie within a national park in the United States that had been invaded by leafy spurge. Control efforts included releases of two species of flea beetles in the genus *Aphthona* and the aerial application of three kinds of herbicides, with some areas receiving only *Aphthona* beetles, some only herbicides and some both beetles and herbicides. Sampling results over a three-year period indicated that herbicide use did not improve weed control as compared to biological control. Instead, it increased seedling density in some years and interfered with biological control by (i) nullifying the year-to-year numerical response of flea beetles to spurge densities, and (ii) increasing interspecific competition between the two *Aphthona* flea beetle species. Larson and colleagues (2007) cautioned that herbicide use in this context was likely detrimental to the long-term control of leafy spurge within this national park.

To summarize, these analyses and examples show that pest resurgence via natural enemy release can occur in a number of complex ways. And as Hardin and colleagues (1995) and Morse (1998) have stressed (and as DeBach 1955 previously recognized), pest resurgence can occur even in the face of effective toxicity toward the pest. Before our discussion of conservation biological control strategies aimed at countering pest resurgence, we turn our attention to the topic of secondary pest outbreaks.

10.1.2 Secondary Pest Outbreaks

Indiscriminate mortality caused by broad-spectrum pesticides often leads to the emergence of pests that are quite unrelated to the original target pest. For example, fungicides applied to potato fields to eliminate a new strain of late blight, *Phytophthora infestans*, in the United States in the 1990s led to outbreaks of the green-peach aphid, *Myzus persicae*, because the fungicides interfered with entomopathogens that had previously kept the aphids far below pest levels (Lagnaoui & Radcliffe 1997, 1998) (Figure 10.3). These kinds of effects can also extend across agricultural habitats. Insecticides applied to Iranian apple orchards in the winter reportedly led to outbreaks of the wheat bug, *Eurygaster integriceps*, in neighboring grain fields by suppressing its egg parasitoids that overwinter in apple orchards as adults (Messenger et al. 1976a). Previously innocuous organisms that emerge as pests due to the effects of pesticide use against other species are typically called secondary pests. Most of

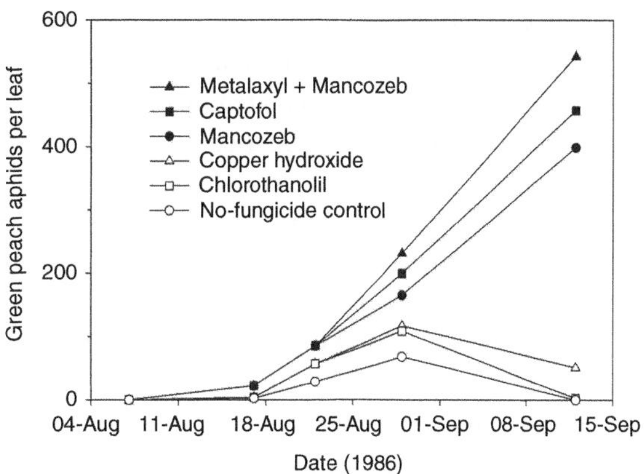

Figure 10.3. Green peach aphid density on potatoes under five different fungicidal regimes and a non-fungicidal control. The fungicides were applied to control late blight, *Phytophthora infestans*, and three of these treatments severely interfered with naturally occurring entomopathogens that are capable of controlling the aphids. Two other fungicides are compatible with natural biological control. From Lagnaoui and Radcliffe (1998).

the cases of secondary pests come from insecticide use impacting arthropod natural enemies of arthropod pests, although as noted earlier, fungicides can be implicated as well.

On the face of it, weeds would seem to be immune from secondary pest outbreaks associated with herbicide use since herbicides are not expected to have negative effects on herbivorous insects or plant pathogens, as discussed previously for pest resurgence. In some cases, however, herbicides can adversely affect arthropod biological control agents, and we are aware of one example that led to a secondary outbreak of an arthropod pest. Applications of the herbicide paraquat for weed control in apple orchards in the eastern United States had direct negative impacts on the predatory mite *Neoseiulus fallacis*, leading to release of the European red mite, *Panonychus ulmi* (Pfeiffer 1986). More generally, situations can be found in which weed management practices involve some kind of disturbance, such as cultivation, crop rotation, grazing and burning, that can interfere with biological control (Newman et al. 1998). Fire is sometimes used as an 'herbicide', and this can have a negative impact on herbivores that play a role in weed suppression. Briese (1996) reported a situation in which prescribed burns were used in Australia to reduce the risk of forest fires, and inadvertently suppressed populations of *Chrysolina quadrigemina*, an effective importation biological control agent of St. John's wort, *Hypericum perforatum*, leading to temporary outbreaks of this invasive weed within eucalyptus forests.

Of course, insecticides can impact arthropod weed control agents, and this could lead to weeds being secondary pests associated with insect control. Such an outcome can be a result of insecticide drift (Hoffmann & Moran 1995) or of insecticide application in areas of spatial overlap between an arthropod pest problem and weed biological control. Such a scenario may have occurred in the rangelands of northern New Mexico in the United States, where insecticides were applied annually between 1966 and 1994 to control rangeland grasshoppers and caterpillars. Pomerinke and colleagues (1995) suggested that these applications may have interfered with natural biological control of purple locoweed, *Astragalus mollissimus*, by a native weevil.

The emergence of secondary pests requires more than just indiscriminate killing of natural enemies. It is also necessary for the secondary pests themselves to be relatively unaffected by the sprays. The most common mechanism by which this is achieved is resistance to the pesticides on the part of the pests (e.g., Luck et al. 1977). Of course, secondary pest outbreaks also imply that the natural enemies are *not* resistant to the sprays (Tabashnik & Johnson 1999). The survival of secondary pests in the face of pesticide application can also be an outcome of the fact that the pesticides are applied with the primary pest in mind and so may end up missing secondary pests either temporally or spatially. In this case, secondary pest outbreaks occur as the unfortunate result of pesticides not reaching a potential pest but reaching its natural enemies instead.

10.1.3 Broader Implications

A point not lost on observers of biological control is that cases of pest resurgence and secondary pest outbreaks can constitute strong circumstantial evidence for the importance of natural biological control in field settings (Van den Bosch 1971; Messenger et al. 1976a; Hoffmann & Moran 1995). Insecticide-triggered pest outbreaks cannot be taken as absolute proof of natural biological control because of alternative hypotheses for insecticide-induced pest outbreaks such as pesticide-induced enhancement of pest fecundity, either directly or through positive effects on plant health (Luck et al. 1988; Hardin et al. 1995; Morse 1998; see Table 10.1). However, observations of pest resurgence and secondary pest outbreaks may well have been the inspiration for the development of the so-called insecticidal check method of evaluating the effectiveness of biological control agents, a method that provided experimental demonstrations of biological control by essentially duplicating the conditions that lead to secondary pest outbreaks (DeBach 1946, 1955).

10.2 Strategies to Counter Pesticide-Induced Disruption of Biological Control

It is difficult to understate the importance of pesticide-induced disruption of biological control. Luck and colleagues (1977) did an analysis of the top 25 arthropod pests of agricultural crops in California as of 1970 and found that all but one of these were either aggravated by insecticide use via pest resurgence or triggered secondary pest outbreaks. Clearly, there is great scope for strategies that can reestablish natural biological control in these and other systems. The main solution to pest resurgence and secondary pest outbreaks is the alleviation of natural enemy release by the reduction of the use of broad-spectrum pesticides, which can be achieved either by using more selective insecticides or by reducing the extent of pesticide use in general. The reduction of pesticide use is desirable to growers and land managers for economic, environmental and health reasons separately from any benefits that accrue to biological control, so it does not always constitute a true conservation biological control strategy. For instance, Coll (2004) has noted that 'precision agriculture', a set of practices aimed at applying insecticides and herbicides only where needed within fields, is used primarily to reduce input costs, rather than to conserve natural enemies per se. However, many pesticide-reduction strategies are designed with the explicit goal of conserving natural enemies, and these are clearly cases of conservation biological control.

We discuss conservation biological control in the context of pesticide selectively, covering first 'physiological selectivity', which is defined as the restriction in the taxonomic breadth of direct toxicity of pesticides, and then 'ecological selectivity', which encompasses a range of strategies that aim to restrict contact between the pesticide and non-target organisms. Both of these terms were coined by Ripper and colleagues (1951) and have been elaborated on by a number of authors (e.g., Newsom et al. 1976; Hull & Beers 1985; Croft 1990; Johnson & Tabashnik 1999).

10.2.1 Physiological Pesticide Selectivity

Increased *physiological selectivity* can be achieved via a number of insecticidal and fungicidal modes of action that we do not review here (interested readers can consult Croft 1990; Pfeiffer 2000; Klingen & Haukeland 2006; King et al. 2008). The Western Palearctic Regional Section of the International Organization of Biological Control (IOBC/WPRS) has provided very useful guidance in the use of compounds that show physiological selectivity with respect to natural enemies (Hassan 1989; Sterk et al. 1999). The long-standing Working Group on Pesticides and Beneficial Organisms of the IOBC/WPRS has periodically updated techniques used in testing pesticides for non-target effects and has catalogued more than 60 compounds in terms of their safety to biological control agents. The working group has gone a long way toward promoting and standardizing studies of effects of pesticides on biological control agents, but Thomson and Hoffman (2007) have called for increased refinements in terms of tailoring the list of agents tested to locally appropriate species, considering sublethal effects of pesticides and performing field evaluations.

A number of pest control programs have successfully used selective pesticides to solve problems of pest resurgence or secondary pest outbreaks. Stern and Van den Bosch (1959) were pioneers in this area and noted that an advantage of selective insecticides is that toxicity to the target pest need not be as high as needed from a broad-spectrum insecticide precisely because the conserved natural enemies provide additional mortality. They demonstrated this principle with aphids in alfalfa in California, and more recently Naranjo and Ellsworth (2009) showed that, while approximately five applications of conventional broad-spectrum insecticides were needed per season to control the sweet potato whitefly, *Bemisia tabaci*, in cotton fields in the Southwestern United States, only a single application of a more selective insect growth regulator is required. The reason is not that the insect growth regulator is more toxic or has a longer residual time. Rather, fewer applications are needed because natural enemies (mainly predatory insects) are conserved, and these keep the whitefly under control for the rest of the season. Naranjo (2001) called this continued presence and effect of natural enemies a 'bioresidual' produced by using the more selective insecticides (Figure 10.4).

Other cases in which selective insecticides have improved biological control have been reviewed by a number of authors (Poehling 1989; Croft 1990; Wright & Verkerk 1995; Johnson & Tabashnik 1999) and include improved biological control of arthropod pests in apples, corn, cotton and potatoes in North America and Australia (Agnello et al. 2003; Hewa-Kapuge et al. 2003; Koss et al. 2005; Naranjo & Ellsworth 2009), coffee in Hawaii (Vargas et al. 2001) and cabbage in Europe (Charleston et al. 2006). In one case, a selective mineral oil was applied against the black parlatoria scale, *Parlatoria ziziphi*, on grapefruit trees in Egypt to replace broad-spectrum insecticides that were killing *Encarsia citrina* and other parasitoids of the scale insects (Coll & Abd-Rabou 1998). The oils were effective in conserving *E. citrina*, and by pure serendipity retained toxicity to a hyperparasitoid of *E. citrina*, leading to even more effective biological control!

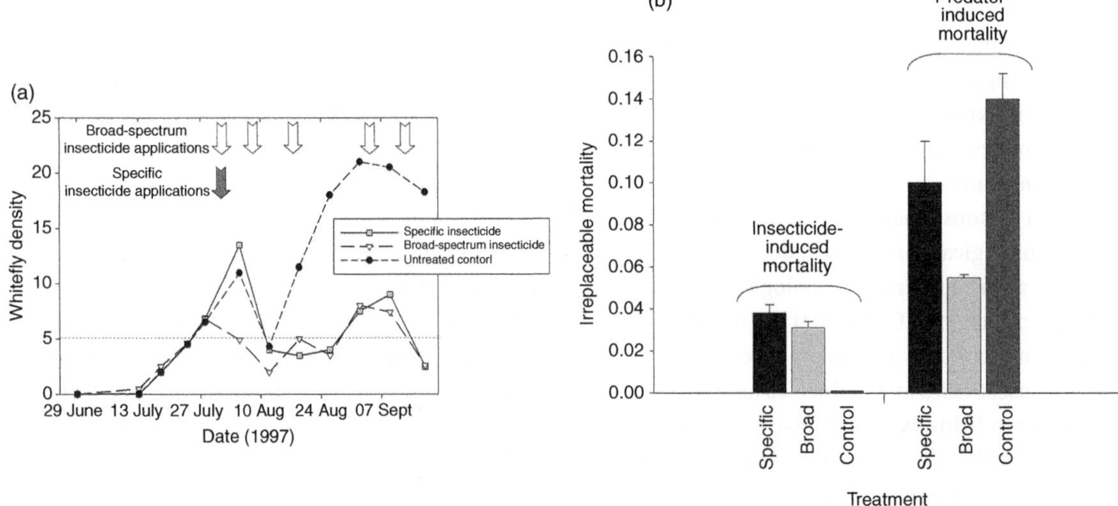

Figure 10.4. (a) Per-leaf density of the sweet potato whitefly, *Bemisia tabaci*, on cotton in Arizona in the United States under three management treatments: a single early season application of the specific insect growth regulators, buprofezin or pyriproxyfen, five applications of a mix of conventional broad-spectrum insecticides and an untreated control (from Naranjo 2001). The figure illustrates the point that only a single application of a selective insecticide is needed to provide the level of control achieved using five applications of the broad-spectrum insecticide. Panel (b) shows that use of the selective insecticide pyriproxyfen led to equivalent mortality of whiteflies as the broad-spectrum insecticide but increased predator-induced mortality (Naranjo & Ellsworth 2009), leading to a 'bioresidual' that prevents the need for additional insecticide application.

While these and other developments have been exciting, it is important to stress that some purportedly selective insecticides that are marketed as 'reduced risk' or 'IPM friendly' are in some cases no less toxic to natural enemies than broad-spectrum insecticides are (Gentz et al. 2010). Similarly, some organically approved compounds can be just as toxic as synthetic products (Biondi et al. 2012a, b). Attention has also been focused on sublethal effects of pesticides in addition to the more traditional concentration on acute toxicity (Desneux et al. 2007; Biondi et al. 2013), which necessitates more complex population models as components of ecological risk assessment for pesticide use (Stark et al. 2007; Banks et al. 2011; Forbes et al. 2011; Mills et al. 2016).

Conservation of predatory and parasitic insects has long been one of the acknowledged benefits of microbial pesticides, many of which have limited, if any, direct impacts on other biological control agents (Flexner et al. 1986). However, various sublethal and negative indirect effects of entomopathogens on arthropod natural enemies have been documented, and immature parasitoids within toxified hosts typically die along with their host (e.g., Flexner et al. 1986; Roy et al. 1998; Maniania et al. 2008). As with chemical pesticides, the key to the use of microbial agents for purposes of conservation biological control is their specificity. *Bacillus thuringiensis* (*Bt*) has been used as a selective biopesticide to replace broad-spectrum insecticides and conserve natural enemies of pest insects such as the diamondback moth and Colorado potato beetle (Hilbeck et al. 1998; Chilcutt & Tabashnik 1999).

10.2.2 Ecological Pesticide Selectivity

Ecological selectivity refers to the constellation of strategies aimed at reducing exposure of biological control agents to potentially harmful pesticides. Approaches range from an overall reduction in pesticide application to particular tactics in which the timing and/or placement of pesticides is manipulated to reduce exposure to biological control agents.

10.2.2.1 Reduced Application Frequency or Doses

We have already noted that lower toxicity can be tolerated in pesticides that conserve natural enemies because the decreased pest mortality can be compensated for by conserved natural enemies (see Figure 10.4). In similar fashion, reduced application frequency or doses of broad-spectrum insecticides may not lead to reduced pest control if natural enemies are conserved. A number of researchers have indeed noted that reducing the doses of pesticides from the recommended rates can maintain effective pest control while conserving natural enemies (Johnson & Tabashnik 1999; Acheampong & Stark 2004). This realization has led to improved control strategies of multiple-pest systems in apples in the United States and New Zealand with dosage reductions between 50% and 95% (Hull & Beers 1985). A number of studies of rice in Southeast Asia have similarly shown that reducing the frequency of insecticide applications can lead to improved control of insect pests such as the brown planthopper, *Nilaparvata lugens* (Matteson 2000). Insecticide-induced enemy release in these systems seems to be linked to disruption of detritivores early in the cropping cycle, which form the food base for generalist predators that can suppress pests later in the cropping cycle (Settle et al. 1996). Current recommendations in Asian rice production stress restraint in insecticide use, and, according to Matteson (2000), 'Farmers who do not drop insecticide use altogether are hard pressed to identify the infrequent occasions when it will be profitable to spray.'

The observation from these and other studies is that (i) a similar level of pest control can be achieved with low and high application rates (or doses) and (ii) natural enemies are conserved. However, it does not necessarily follow from this that conserved natural enemies are providing compensatory mortality at lower application rates and are therefore responsible for the maintenance of pest control. An alternative explanation is simply that the lower application rates or doses are sufficient to achieve adequate suppression. Establishing a causal relationship between reduced pesticide use, conservation biological control and resulting pest pressure takes a detailed approach that has not been attempted by many researchers. Ideally, a life-table approach is used in which mortality ascribed to pesticides and natural enemies is compared among different pesticide application treatments. Such an approach was used by Naranjo and Ellsworth (2009), who showed that fewer applications of more (physiologically) selective insecticides were required to achieve control of the sweet potato whitefly precisely because of increased predator effects (see Figure 10.4).

10.2.2.2 Spatial Placement of Pesticide Applications

Providing natural enemies a spatial refuge from insecticide contact can be an effective conservation biological control strategy. Tactics include spot-spraying parts of a field or infestation that have a lower abundance of natural enemies, and leaving a constant proportion of a given field or infestation unsprayed to conserve natural enemies. Systemic insecticides can also be seen as exhibiting spatial selectivity because they are present only on the inside of plants, meaning that only herbivores (or omnivores) receive direct exposure.

Precision treatment. The growing field of precision agriculture may lead to application strategies in which pesticide use is focused only in areas where it is needed, leading to conservation of natural enemies (e.g., Midgarden et al. 1997; Karimzadeh et al. 2011). Fleischer and colleagues (1999) coined the term *precision IPM* to refer specifically to the process of sampling and map-building that can lead to application of pesticides at aggregations of pests. The approach represents an updating of previous strategies in which 'spot treatments' were advocated for patchily distributed pests (Newsom et al. 1976). Another possibility is applying insecticides in areas that have a high ratio of pests to natural enemies and avoiding or reducing the use of insecticides in areas or fields with a more favorable ratio (e.g., Dowden 1952; Gonzalez & Wilson 1982; Hoffmann et al. 1991; Giles et al. 2003; Conway et al. 2006).

Strip treatment. While precision treatments tend to be sampling-intensive and spatially variable, strategies in which a fixed proportion of the target area is exempted from pesticide application in order to produce natural enemy refuges have been developed as well. For example, applying insecticides to alternating rows in citrus or apple orchards on a rotating schedule can conserve natural enemies while still achieving acceptable levels of pest mortality (DeBach & Landi 1961; Hull et al. 1983; Figure 10.5). Despite the practical successes of these programs, however, the critical size for a pesticide refuge is sometimes too large to be economically feasible. Lester and colleagues (1998) conducted a large-scale field trial in an apple orchard in Ontario, Canada, with the purpose of determining the size of a refuge from pyrethroid sprays necessary to avoid secondary outbreaks of phytophagous mites (especially the European red mite, *Panonychus ulmi*). Refuges were achieved by covering branches with plastic bags to protect 10%, 30% or 60% of the leaves during insecticide sprays. Significant reduction in secondary outbreaks of *P. ulmi* was indeed achieved, but only at the 60% refuge level, which was deemed economically unrealistic from the standpoint of lepidopteran control.

Spatial refugia from pesticide use can also be used in weed control, where herbicide use can lead to weed resurgence by depriving highly specialized biological control agents of their only food source, as noted earlier. In this case, leaving some fraction of the weed population untreated can maintain a

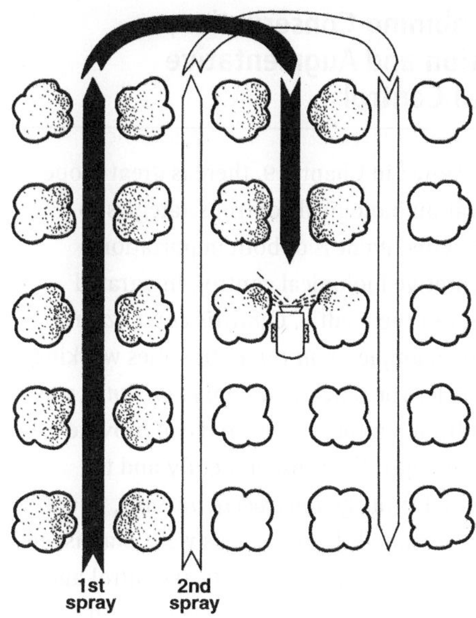

Figure 10.5. A diagram of the alternate-row spray technique used in apple orchards in Pennsylvania in the United States. Application of insecticide is primarily to the inside half of trees on either side of a given alley (denoted by stippling in the figure), leaving the outside half of the trees as a refuge for natural enemies. From Hull and Beers (1985).

population of biological control agents in the area that then have the potential to colonize weeds that have either survived the herbicide application or that recolonize the sprayed area. This tactic has been used effectively to maintain water hyacinth weevils in the genus *Neochetina* after herbicide applications against water hyacinth in the Southeastern United States (Haag & Habeck 1991). Newman and colleagues (1998) discussed other similar examples.

Systemic insecticides and insecticidal baits. Another way to protect natural enemies from pesticide applications is to develop application technologies that limit natural enemy exposure to the compounds. An example is systemic insecticides, which are present within, rather than on the surface of plants, eliminating the risk of topical application to natural enemies. Systemic insecticides as a group have been touted as safe for natural enemies for this reason (Ripper et al. 1951; Hull & Beers 1985; Johnson & Tabashnik 1999). Croft (1990), however, cautioned that while systemic insecticides have played an important role in conserving arthropod biological control agents, they are not always specific and they can find their way to natural enemies. One problem is that many arthropod natural enemies do feed on, or oviposit within, plant parts and can therefore be directly exposed to systemic insecticides (Lundgren 2009). Related to this is that biological control agents can be exposed by feeding on the nectar or pollen of plants treated with systemic insecticides (Smith & Krischik 1999; Krischik et al. 2007; Rogers et al. 2007). And second, the herbivores that have fed on the systemic insecticides can be toxic to natural enemies (Croft 1990).

So-called bait-sprays are similar to systemic insecticides in that topical application to natural enemies is minimized. Here, the insecticide is mixed with a bait substance that is attractive to pest herbivores and presumably not to their natural enemies. This has led to selective effects of insecticides on pest fruit flies, which feed on the baits to a greater extent than their parasitoids do (Vargas et al. 2002; Stark et al. 2004), opening the door to integrated control tactics (Vargas et al. 2001; McQuate et al. 2005). Again though, negative effects on natural enemies have been documented (Croft 1990).

Timing of pesticide applications. The timing of pesticide applications can be designed to protect natural enemies while still impacting the target pest or weed. Strategies of this kind have been developed both to protect natural enemies of the target pests themselves and natural enemies of potential secondary pests. In a number of arthropod pest/enemy systems, pests commence activity earlier in the season than their enemies do so that early sprays with short residual times can allow natural enemies to colonize (Hull & Beers 1985). However, this strategy can backfire if the absence of pests early in the season discourages immigration by natural enemies, leading to outbreaks later in the season (Johnson & Tabashnik 1999). In other systems, windows of opportunity have been identified within the growing season when broad-spectrum pesticides can be used with reduced impact on natural enemies (Bartlett 1964). For example, insecticide applications can be restricted to periods during which natural enemies are present in an invulnerable stage (e.g., Granett et al. 1976; Mensah & Madden 1993).

The proper timing of herbicide applications can be very effective in conserving weed biological control agents. As mentioned in Section 10.1.1, herbicide application can interfere with weed biological control by leading to starvation of weed biological control agents. In this case, the herbicide may control the weed in the short term, but the control agent is not available once the weed regrows or recolonizes the site. Applying herbicides during or before the feeding stage of herbivores causes the most disruption of this sort, and spray regimes aimed at plant stages or times of the year that follow feeding can improve biological control (Trumble & Kok 1982; Story & Stougaard 2006).

10.3 Combining Conservation, Importation and Augmentative Biological Control

As we have noted in Chapter 9, there is great scope for conservation biological control strategies to improve the effectiveness of both importation and augmentative biological control. Integrated management of the medfly, *Ceratitis capitata*, in Hawaii is an example of *all three* strategies working simultaneously. The parasitoid *Fopius arisanus* was introduced from Southeast Asia and can provide impressive biological control of medfly and the related oriental fruit fly, *Bactrocera dorsalis*, in coffee plantations. Although *F. arisanus* is well established in Hawaii as an importation biological control agent, site-specific effectiveness is improved through augmentation, and these releases are combined with bait sprays utilizing spinosad and phloxine B, insecticides that are less detrimental to populations of the parasitoid than malathion – the previous insecticide of choice (Vargas et al. 2001; McQuate et al. 2005). In this system, elements of conservation, augmentation and importation biological control are integrated in a way that improves overall suppression of a very serious coffee pest in Hawaii. A similar approach is under development for use on a broad geographic scale in Mexico (Ruiz et al. 2008).

REFERENCES

Abell KJ & Van Driesche RG (2012) Impact of latitude on synchrony of a scale (*Fiorinia externa*) (Hemiptera: Diaspididae) and its parasitoid (*Encarsia citrina*) (Hymenoptera: Aphelinidae) in the eastern United States. *Biological Control* **63**: 339–347.

Abrams PA (2015) Why ratio dependence is (still) a bad model of predation. *Biological Reviews* **90**: 794–814.

Abrams PA & Ginzburg LR (2000) The nature of predation: prey dependent, ratio dependent, or neither? *Trends in Ecology & Evolution* **15**: 337–341.

Abrantes J, Carmo CR, Matthee CA, Yamada F, Loo W & Esteves P (2011) A shared unusual genetic change at the chemokine receptor type 5 between *Oryctolagus, Bunolagus* and *Pentalagus*. *Conservation Genetics* **12**: 325–330.

Abu-Dieyeh MH & Watson AK (2007) Grass overseeding and a fungus combine to control *Taraxacum officinale*. *Journal of Applied Ecology* **44**: 115–124.

Acheampong S & Stark JD (2004) Can reduced rates of pymetrozine and natural enemies control the cabbage aphid, *Brevicoryne brassicae* (Homoptera: Aphididae) on broccoli? *International Journal of Pest Management* **50**: 275–279.

Acorn J (2007) *Ladybugs of Alberta*. Edmonton, University of Alberta Press.

Adachi-Hagimori T, Miura K & Stouthamer R (2008) A new cytogenetic mechanism for bacterial endosymbiont-induced parthenogenesis in Hymenoptera. *Proceedings of the Royal Society B-Biological Sciences* **275**: 2667–2673.

Adams JM, Fang W, Callaway RM, Cipollini D & Newell E (2009) A cross-continental test of the enemy release hypothesis: leaf herbivory on *Acer platanoides* (L.) is three times lower in North America than in its native Europe. *Biological Invasions* **11**: 1005–1016.

Adams PB (1990) The potential of mycoparasites for biological control of plant diseases. *Annual Review of Phytopathology* **28**: 59–72.

Agnello AM, Reissig WH, Kovach J & Nyrop JP (2003) Integrated apple pest management in New York State using predatory mites and selective pesticides. *Agriculture Ecosystems & Environment* **94**: 183–195.

Agrawal AA & Kotanen PM (2003) Herbivores and the success of exotic plants: a phylogenetically controlled experiment. *Ecology Letters* **6**: 712–715.

Agrawal AA, Kotanen PM, Mitchell CE, Power AG, Godsoe W & Klironomos J (2005) Enemy release? An experiment with congeneric plant pairs and diverse above- and below ground enemies. *Ecology* **86**: 2979–2989.

Agricola U & Fisher HU (1991) Hyperparasitism in two newly introduced parasitoids, *Epidinocarsis lopezi* and *Gyranosoidea tebygi* (Hymenoptera: Encyrtidae) after their establishment in Togo. *Bulletin of Entomological Research* **81**: 127–132.

Ainsworth N (2003) Integration of herbicides with arthropod biocontrol agents for weed control. *Biocontrol Science and Technology* **13**: 547–570.

Alabouvette C, Heilig U & Cordier C (2012) Microbial control of plant diseases. In *Beneficial Microorganisms in Agriculture, Food and the Environment* (Sundh I, Wilcks A & Goettel M, eds.) Wallingford, UK, CABI Publishing, pp. 96–111.

Alhmedi A, Haubruge E, D'Hoedt S & Francis F (2011) Quantitative food webs of herbivore and related beneficial community in non-crop and crop habitats. *Biological Control* **58**: 103–112.

Ali AD & Reagan TE (1985) Vegetation manipulation impact on predator and prey populations in Louisiana sugarcane ecosystems. *Journal of Economic Entomology* **78**: 1409–1414.

Allen SK, Thiery RG & Hagstrom NT (1986) Cytological evaluation of the likelihood that triploid grass carp will reproduce. *Transactions of the American Fisheries Society* **115**: 841–848.

Alphey L (2014) Genetic control of mosquitoes. *Annual Review of Entomology* **59**: 205–224.

Alstad DN & Edmonds JGF (1983) Selection, outbreeding depression, and the sex ratio of scale insects. *Science* **220**: 93–95.

Altfeld L & Stiling P (2009) Effects of aphid-tending Argentine ants, nitrogen enrichment and early-season herbivory on insects hosted by a coastal shrub. *Biological Invasions* **11**: 183–191.

Althoff DM (2003) Does parasitoid attack strategy influence host specificity? A test with new world braconids. *Ecological Entomology* **28**: 500–502.

Altieri MA & Letourneau DK (1982) Vegetation management and biological control in agroecosystems. *Crop Protection* **1**: 405–430.

Altieri MA & Whitcomb WH (1979) The potential use of weeds in the manipulation of beneficial insects. *HortScience* **14**: 12–18.

Altieri MA & Whitcomb WH (1980) Weed manipulation for insect pest management in corn. *Environmental Management* **4**: 483–489.

Alyokhin A & Sewell G (2004) Changes in a lady beetle community following the establishment of three alien species. *Biological Invasions* **6**: 463–471.

Amaral DS, Venzon M, Pallini A, Lima PC & DeSouza O (2010) Does vegetational diversification reduce coffee leaf miner *Leucoptera coffeella* (Guerin-Meneville) (Lepidoptera: Lyonetiidae) attack? *Neotropical Entomology* **39**: 543–548.

Amsellem Z, Cohen BA & Gressel J (2002) Engineering hypervirulence in a mycoherbicidal fungus for efficient weed control. *Nature Biotechnology* **20**: 1035–1039.

Anagnostakis SL (1982) Biological control of chestnut blight. *Science* **215**: 466–471.

Andersen MC, Ewald M & Northcott J (2005) Risk analysis and management decisions for weed biological control agents: ecological theory and modeling results. *Biological Control* **35**: 330–337.

Anderson G, Delfosse ES, Spencer N, Prosser C & Richard R (2000) Biological control of leafy spurge: an emerging success story. In *Proceedings of the X International Symposium on Biological Control of Weeds* (Spencer NR, ed.), Bozeman, Montana State University, pp. 15–25.

Anderson RM & May RM (1978) Regulation and stability of host-parasite population interactions. 1. Regulatory processes. *Journal of Animal Ecology* **47**: 219–247.

Anderson RM & May RM (1981) The population dynamics of microparasites and their invertebrate hosts. *Philosophical Transactions of the Royal Society of London, Series B* **291**: 451–524.

Anderson RM & May RM (1982) Coevolution of hosts and parasites. *Parasitology* **85**: 411–426.

Andow DA (1990) Population dynamics of an insect herbivore in simple and diverse habitats. *Ecology* **71**: 1006–1017.

Andow DA (1991) Vegetational diversity and arthropod population response. *Annual Review of Entomology* **36**: 561–586.

Andow DA & Hilbeck A (2004) Science-based risk assessment for non-target effects of transgenic crops. *BioScience* **54**: 637–649.

Andow DA & Imura O (1994) Specialization of phytophagous arthropod communities on introduced plants. *Ecology* **75**: 296–300.

Andow DA, Lane CP & Olson DM (1995) Use of *Trichogramma* in maize – estimating environmental risks. In *Biological Control: Benefits and Risks* (Hokkanen HMT & Lynch JM, eds.), Cambridge, UK, Cambridge University Press, pp. 101–118.

Andow DA & Prokrym DR (1990) Plant structural complexity and host-finding by a parasitoid. *Oecologia* **82**: 162–165.

Andres LA & Rees NE (1995) Musk thistle. In *Biological Control in the Western United States* (Nechols JR, ed.), Oakland, University of California Press, pp. 248–251.

Andrewartha HG & Birch LC (1954) *The Distribution and Abundance of Animals.* Chicago, University of Chicago Press.

Angalet GW & Fuester R (1977) The *Aphidius* parasites of the pea aphid *Acyrthosiphon pisum* in the eastern half of the United States. *Annals of the Entomological Society of America* **70**: 87–96.

Antolin MF (1999) A genetic perspective on mating systems and sex ratios of parasitoid wasps. *Researches on Population Ecology* **41**: 29–37.

Antolin MF, Bjorksten TA & Vaughn TT (2006) Host-related fitness trade-offs in a presumed generalist parasitoid, *Diaeretiella rapae* (Hymenoptera: Aphidiidae). *Ecological Entomology*, **31**: 242–254.

Antolin MF & Strand MR (1992) Mating system of *Bracon hebetor* (Hymenoptera: Braconidae). *Ecological Entomology* **17**: 1–7.

Antonovics J (1992) Toward community genetics. In *Plant Resistance to Herbivores and Pathogens: Ecology, Evolution, and Genetics* (Fritz RS & Simms EL, eds.), Chicago, University of Chicago Press, pp. 426–449.

Araj S-E, Wratten S, Lister A & Buckley H (2009) Adding floral nectar resources to improve biological control: potential pitfalls of the fourth trophic level. *Basic and Applied Ecology* **10**: 554–562.

Arditi R (1983) A unified model of the functional response of predators and parasitoids. *Journal of Animal Ecology* **52**: 293–303.

Arditi R & Akcakaya HR (1990) Underestimation of mutual interference in predators. *Oecologia* **83**: 358–361.

Arditi R & Ginzburg LR (1989) Coupling in predator–prey dynamics: ratio-dependence. *Journal of Theoretical Biology* **139**: 311–326.

Arditi R & Ginzburg LR (2012) *How Species Interact: Altering the Standard View on Trophic Ecology.* New York, Oxford University Press.

Argaud O, Croizier L, Lopez-Ferber M & Croizier G (1998) Two key mutations in the host-range specificity domain of the p143 gene of *Autographa californica* nucleopolyhedrovirus are required to kill *Bombyx mori* larvae. *Journal of General Virology* **79**: 931–935.

Armbrecht I & Gallego MC (2007) Testing ant predation on the coffee berry borer in shaded and sun coffee plantations in Colombia. *Entomologia Experimentalis et Applicata* **124**: 261–267.

Armbruster P, Bradshaw WE & Holzapfel CM (1997) Evolution of the genetic architecture underlying fitness in the pitcher-plant mosquito, *Wyeomyia smithii. Evolution* **51**: 451–458.

Ashburner M, Hoy MA & Peloquin JJ (1998) Prospects for the genetic transformation of arthropods. *Insect Molecular Biology* **7**: 201–213.

Asano S & Miyamoto K (2007) A laboratory method to evaluate the effectiveness of ultraviolet (UV) protectant for *Bacillus thuringiensis* product. *Japanese Journal of Applied Entomology and Zoology* **51**: 121–127.

Askew RR (1994) Parasitoids of leaf-mining Lepidoptera: what determines their host ranges? *Parasitoid Community Ecology* (Hawkins BA & Sheehan W, eds.), Oxford, UK, Oxford University Press, pp. 177–202.

Askew RR & Shaw MR (1986) Parasitoid communities: their size, structure and development. *Insect Parasitoids* (Waage JK & Greathead D, eds.), London, UK, Academic Press, pp. 225–264.

Aspi J (2000) Inbreeding and outbreeding depression in male courtship song characters in *Drosophila montana. Heredity* **84**: 273–282.

Asplen MK, Chacon JM & Heimpel GE (2016) Divergent sex-specific dispersal by a parasitoid wasp in the field. *Entomologia Experimentalis et Applicata* **159**: 252–259.

Asplen MK, Whitfield JB, de Boer JG & Heimpel GE (2009) Is single-locus complementary sex determination the ancestral mechanism for hymenopteran haplodiploidy? *Journal of Evolutionary Biology* **22**: 1762–1769.

Auerbach M & Simberloff D (1988) Rapid leaf-miner colonization of introduced trees and shifts in sources of herbivore mortality. *Oikos* **52**: 41–50.

Auslander DM, Oster GF & Huffaker CB (1974) Dynamics of interacting populations. *Journal of the Franklin Institute* **297**: 345–376.

Babendreier D, Kuske S & Bigler F (2003) Non-target host acceptance and parasitism by *Trichogramma brassicae* Bezdenko (Hymenoptera: Trichogrammatidae) in the laboratory. *Biological Control* **26**: 128–138.

Babendreier D, Bigler F & Kuhlmann U (2006) Current status and constraints in the assessment of non-target effects. *Environmental Impact of Invertebrates for Biological Control of Arthropods: Methods and Risk Assessment* (Bigler F, Babendreier D & Kuhlmann U, eds.), Wallingford, UK, CABI Publishing, pp. 1–14.

Baek J-M, Howell CR & Kenerley CM (1999) The role of an extracellular chitinase from *Trichoderma virens* Gv29-8 in the biocontrol of *Rhizoctonia solani. Current Genetics* **35**: 41–50.

Baer CF, Tripp DW, Bjorksten TA & Antolin MF (2004) Phylogeography of a parasitoid wasp (*Diaeretiella rapae*): no evidence of host-associated lineages. *Molecular Ecology* **13**: 1859–1869.

Baggen LR & Gurr GM (1998) The influence of food on *Copidosoma koehleri* (Hymenoptera: Encyrtidae), on the use of flowering plants as a habitat management tool to enhance biological control of potato moth, *Phythorimaea operculella*

(Lepidoptera: Gelechiidae). *Biological Control* **11**: 9–17.

Baggen LR, Gurr GM & Meats A (1999) Flowers in tri-trophic systems: mechanisms allowing selective exploitation by insect natural enemies for conservation biological control. *Entomologia Experimentalis et Applicata* **91**: 155–161.

Bai C, Shapiro-Ilan DI, Gaugler R & Hopper KR (2005) Stabilization of beneficial traits in *Heterorhabditis bacteriophora* through creation of inbred lines. *Biological Control* 32: 220–227.

Bailey EP (1992) Red foxes, *Vulpes vulpes*, as biological control agents for introduced arctic foxes, *Alopex lagopus*, on Alaskan islands. *Canadian Field-Naturalist* **106**: 200–205.

Baker DA, Loxdale HD & Edwards OR (2003). Genetic variation and founder effects in the parasitoid wasp, *Diaeretiella rapae* (M'intosh) (Hymenoptera: Braconidae: Aphidiinae), affecting its potential as a biological control agent. *Molecular Ecology* **12**: 3303–3311.

Baker KF (1987) Evolving concepts of the biological control of plant pathogens. *Annual Review of Plant Pathology* **25**:67–85.

Baker KF & Cook RJ 1974. *Biological Control of Plant Pathogens*. San Francisco, CA, Freeman.

Bakker PAHM, Pieterse CMJ & Van Loon LC (2007) Induced systemic resistance by fluorescent *Pseudomonas* spp. *Phytopathology* **97**: 239–243.

Bakker PAHM, Ran LX, Pieterse CMJ & Van Loon LC (2003) Understanding the involvement of rhizobacteria-mediated induction of systemic resistance in biocontrol of plant diseases. *Canadian Journal of Plant Pathology* **25**: 5–9.

Balciunas JK (2000) Code of best practices for classical biological control of weeds. In *Proceedings of the X International Symposium on Biological Control of Weeds* (Spencer NR, ed.), Bozeman, Montana State University, p. 435.

Balciunas JK (2004a) Are mono-specific agents necessarily safe? The need for pre-release assessment of probable impact of candidate biocontrol agents, with some examples. In *Proceedings of the XI Symposium on Biological Control of Weeds* (Cullen JM, Briese DT, Kriticos DJ, Lonsdale WM, Morin L & Scott JK, eds.), Canberra, Australia, CSIRO, pp. 252–257.

Balciunas JK (2004b) Four years of 'Code of Best Practices': has it had an impact? In *Proceedings of the XI Symposium on Biological Control of Weeds* (Cullen JM, Briese DT, Kriticos DJ, Lonsdale WM, Morin L & Scott JK, eds.), Canberra, Australia, CSIRO pp. 258–260.

Bale JS, Van Lenteren JC & Bigler F (2008) Biological control and sustainable food production. *Philosophical Transactions of the Royal Society B* **363**: 761–776.

Balmer O, Geneau CE, Belz E, Weishaupt B, Förderer G, Moos S, Ditner N, Juric I & Luka H (2014) Wildflower companion plants increase pest parasitation and yield in cabbage fields: experimental demonstration and call for caution. *Biological Control* **76**: 19–27.

Banks JE, Bommarco R & Ekbom B (2008) Population response to resource separation in conservation biological control. *Biological Control* **47**: 141–146.

Banks JE, Stark JD, Vargas RI & Ackleh AS (2011) Parasitoids and ecological risk assessment: can toxicity data developed for one species be used to protect an entire guild? *Biological Control* **59**: 336–339.

Barbercheck ME & Millar LC (2000) Environmental impacts of entomopathogenic nematodes used for biological control in soil. In *Nontarget Effects of Biological Control* (Follett PA & Duan JJ, eds.), Dordrecht, The Netherlands, Kluwer, pp. 287–308.

Barbosa P, ed. (1998) *Conservation Biological Control*. San Diego, CA, Academic Press.

Barclay HJ (1982) Models for pest control using predator release, habitat management and pesticide release in combination. *Journal of Applied Ecology* **19**: 337–348.

Barclay HJ, Otvos IS & Thomson AJ (1985) Models of periodic inundative of parasitoids for pest control. *Canadian Entomologist* **117**: 705–716.

Barlow ND (1994) Predicting the effect of a novel vertebrate biocontrol agent: a model for viral vectored immunocontraception of New Zealand opossums. *Journal of Applied Ecology* **31**: 454–462.

Barlow ND (1998) Biological control in New Zealand: new models from real systems. In *Proceedings of the VII International Congress of Ecology* (Farina A, Kennedy J & Bossu V, eds.), Cambridge, UK, Cambridge University Press, pp. 230–259.

Barlow ND (1999) Models in biological control: a field guide. In *Theoretical Approaches to Biological Control* (Hawkins BA & Cornell HV, eds.), Cambridge, UK, Cambridge University Press, pp. 43–68.

Barlow ND (2000) The ecological challenge of immunocontraception: editor's introduction. *Journal of Applied Ecology* **37**: 897–902.

Barlow ND, Barratt BIP, Ferguson CM & Barron MC (2004) Using models to estimate parasitoid impact on nontarget host abundance. *Environmental Entomology* **33**: 941–948.

Barlow ND, Barron MC & Parkes J (2002) Rabbit haemorrhagic disease in New Zealand: field test of a disease-host model. *Wildlife Research* **29**: 649–653.

Barlow ND & Dixon AFG (1980) *Simulation of Lime Aphid Population Dynamics.* Wageningen, The Netherlands, Centre for Agricultural Publishing.

Barlow ND & Goldson SL (1993) A modelling analysis of the successful biological control of *Sitona discoideus* (Coleoptera: Curculionidae) by *Microctonus aethiopoides* (Hymenoptera: Braconidae) in New Zealand. *Journal of Applied Ecology* **30**: 165–178.

Barlow ND, Goldson SL & McNeill R (1994) A prospective model for the phenology of *Microctonus hyperodae* (Hymenoptera: Braconidae), a potential biological control agent of Argentine stem weevil in New Zealand. *Biocontrol Science and Technology* **4**: 375–386.

Barlow ND & Kean JM (1998) Simple models for the impact of rabbit calicivirus disease (RCD) on Australasian rabbits. *Ecological Modelling* **109**: 225–241.

Barlow ND & Kean JM (2004) Resource abundance and invasiveness: a simple model. *Biological Invasions* **6**: 261–268.

Barlow ND, Moller AP & Beggs JR (1996) A model for the effect of *Sphecophaga vesparum vesparum* as a biological control agent of the common wasp in New Zealand. *Journal of Applied Ecology* **33**: 31–44.

Barratt BIP (2004) *Microctonus* parasitoids and New Zealand weevils: comparing laboratory estimates of host ranges to realized host ranges. In *Assessing Host Ranges for Parasitoids and Predators Used for Biological Control: A Guide to Best Practice* (Van Driesche RG & Reardon R, eds.), Morgantown, WV, USDA Forest Service, pp. 103–120.

Barratt BIP, Blossey B & Hokkanen H (2006) Post-release evaluation of non-target effects of biological control agents. In *Environmental Impact of Invertebrates for Biological Control of Arthropods* (Bigler F, Babendreier D & Kuhlmann U, eds.), Wallingford, UK, CABI Publishing, pp. 166–186.

Barratt BIP, Evans AA, Ferguson CM, Barker GM, McNeill MR & Phillips CB (1997) Laboratory nontarget host range of the introduced parasitoids *Microctonus aethiopoides* and *M. hyperodae* (Hymenoptera: Braconidae) compared with field parasitism in New Zealand. *Environmental Entomology* **26**: 694–702.

Barratt BIP, Ferguson CM, Bixley AS, Crook KE, Barton DM & Johnstone PD (2007) Field parasitism of nontarget weevil species (Coleoptera: Curculionidae) by the introduced biological control agent *Microctonus aethiopoides* Loan (Hymenoptera: Braconidae) over an altitude gradient. *Environmental Entomology* **36**: 826–839.

Barratt BIP, Howarth FG, Withers TM, Kean JM & Ridley GS (2010) Progress in risk assessment for classical biological control. *Biological Control* **52**: 245–254.

Barron MC (2007) Retrospective modelling indicates minimal impact of non-target parasitism by *Pteromalus puparum* on red admiral butterfly (*Bassaris gonerilla*) abundance. *Biological Control* **41**: 53–63.

Barron MC, Barlow ND & Wratten SD (2003) Non-target parasitism of the endemic New Zealand red admiral butterfly (*Bassaris gonerilla*) by the introduced biological control agent *Pteromalus puparum. Biological Control* **27**: 329–335.

Bartlett BR (1964) Integration of chemical and biological control. In *Biological Control of Insect Pests and Weeds* (DeBach P, ed.), New York, Reinhold, pp. 489–511.

Barton J (2004) How good are we at predicting the field host-range of fungal pathogens used for classical biological control of weeds? *Biological Control*, **31**: 99–122.

Barton Browne L & Withers TM (2002) Time-dependent changes in the host-acceptance threshold of insects: implications for host specificity testing of candidate biological control agents. *Biocontrol Science and Technology* 12: 677–693.

Barve N, Barve V, Jimenez-Valverde A, Lira-Noriega A, Mahera SP, Townsend Peterson A, Soberóna J & Villalobos F (2011) The crucial role of the accessible area in ecological niche modeling and species distribution modeling. *Ecological Modelling* 222: 1810–1819.

Barzman MS, Mills NJ & Cuc NTT (1996) Traditional knowledge and rationale for weaver ant husbandry in the Mekong Delta of Vietnam. *Agriculture and Human Values* 13: 2–9.

Bashan Y & Holguin G (1998) Proposal for the division of plant growth-promoting rhizobacteria into two classifications: Biocontrol-PGPB (Plant Growth-Promoting Bacteria) and PGPB. *Soil Biology & Biochemistry* 30: 1225–1228.

Bateman R, Carey M, Batt D Prior C, Abraham Y, Moore D, Jenkins N & Fenlon J (1996) Screening for virulent isolates of entomopathogenic fungi against the desert locust, *Schistocerca gregaria* (Forskal). *Biocontrol Science and Technology* 6: 549–560.

Bathon H (1996) Impact of entomopathogenic nematodes on non-target hosts. *Biocontrol Science and Technology* 6: 421–434.

Batta YA (2004) Postharvest biological control of apple gray mold by *Trichoderma harzianum* Rifai formulated in an invert emulsion. *Crop Protection* 23: 19–26.

Baudoin A, Abad RG, Kok LT & Bruckart WL (1993) Field evaluation of *Puccinia carduorum* for biological control of musk thistle. *Biological Control* 3: 53–60.

Baumhover AH (2002) A personal account of developing the sterile insect technique to eradicate the screwworm from Curacao, Florida and the Southeastern United States. *Florida Entomologist* 85: 666–673.

Bean DW, Dalin P & Dudley TL (2012) Evolution of critical day length for diapause induction enables range expansion of *Diorhabda carinulata*, a biological control agent against tamarisk (*Tamarix* spp.). *Evolutionary Applications* 5: 511–523.

Bean D, Dudley T & Hultine K (2013) Bring on the beetles! In *Tamarix: A Case Study of Ecological Change in the American West* (Sher A & Quigley MF, eds.), New York, Oxford University Press, pp. 377–403.

Bean DW, Dudley TL & Keller JC (2007) Seasonal timing of diapause induction limits the effective range of *Diorhabda elongata deserticola* (Coleoptera: Chrysomelidae) as a biological control agent for tamarisk (*Tamarix* spp.). *Environmental Entomology* 36: 15–25.

Beane K & Bugg RL (1998) Natural and artificial shelter to enhance arthropod biological control agents. In *Enhancing Biological Control: Habitat Management to Promote Natural Enemies of Agricultural Pests* (Pickett CH & Bugg RL, eds.), Berkeley, University of California Press, pp. 239–254.

Beckendorf SK & Hoy MA (1985) Genetic improvement of arthropod natural enemies through selection, hybridization or genetic engineering techniques. In *Biological Control in Agricultural IPM Systems* (Hoy MA & Herzog DC, eds.), Orlando, FL, Academic Press, pp. 167–187.

Becker J, Eisenhauer N, Scheu S & Jousset A (2012) Increasing antagonistic interactions cause bacterial communities to collapse at high diversity. *Ecology Letters* 15: 468–474.

Beddington JR (1974) Age distribution and the stability of simple discrete time population models. *Journal of Theoretical Biology* 47: 65–74.

Beddington JR (1975) Mutual interference between parasites or predators and its effects on searching efficiency. *Journal of Animal Ecology* 44: 331–340.

Beddington JR, Free CA & Lawton JH (1975) Dynamic complexity in predator–prey models framed in difference equations. *Nature* 255: 58–60.

Beddington JR, Free CA & Lawton JH (1978) Characteristics of successful natural enemies in models of biological control of insect pests. *Nature* 273: 513–519.

Beddington JR & Hammond PS (1977) On the dynamics of host-parasite-hyperparasite interactions. *Journal of Animal Ecology* 46: 811–821.

Beddington JR, Hassell MP & Lawton JH (1976) The components of arthropod predation II. The predator rate of increase. *Journal of Animal Ecology* 45: 165–185.

Begum M, Gurr GM, Wratten SD, Hedberg PR & Nicol HI (2006) Using selective food plants to maximize biological control of vineyard pests. *Journal of Applied Ecology* **43**: 547–554.

Behle R & Birthisel T (2014) Formulation of entomopathogens as bioinsecticides. In *Mass Production of Beneficial Organisms* (Morales-Ramos JA, Guadalupe Rojas M & Shapro-Ilan DL, eds.), Amsterdam, The Netherlands, Elsevier, pp. 483–517.

Beirne BP (1975) Biological control attempts by introductions against pest insects in the field in Canada. *Canadian Entomologist* **107**: 225–236.

Bellamy DE & Byrne DN (2001) Effects of gender and mating status on self-directed dispersal by the whitefly parasitoid *Eretmocerus eremicus*. *Ecological Entomology* **26**: 571–577.

Bellows TS, Paine TD, Gould JR, Bezark LG & Ball JC (1992) Biological control of ash whitefly: a success in progress. *California Agriculture* **46**: 24–28.

Bellows TS & Van Driesche RG (1999) Life table construction and analysis for evaluating biological control agents. In *Handbook of Biological Control* (Bellows TS & Fisher TW, eds.), San Diego, CA, Academic Press, pp. 199–223.

Belshaw R (1994) Life history characteristics of Tachinidae (Diptera) and their effect of polyphagy. In *Parasitoid Community Ecology* (Hawkins BA & Sheehan W, eds.), Oxford, UK, Oxford University Press, pp. 145–162.

Bence SL, Stander K & Griffiths M (2003) Habitat characteristics of harvest mouse nests on arable farmland. *Agriculture, Ecosystems and Environment* **99**: 179–186.

Benson J, Pasquale A, Van Driesche R & Elkinton J (2003b) Assessment of risk posed by introduced braconid wasps to *Pieris virginiensis*, a native woodland butterfly in New England. *Biological Control* **26**: 83–93.

Benson J, Van Driesche RG, Pasquale A & Elkinton J (2003a) Introduced braconid parasitoids and range reduction of a native butterfly in New England. *Biological Control* **28**: 197–213.

Benvenuto C, Tabone E, Vercken E, Sorbier N, Colombel E, Warot S, Fauverge X & Ris N (2012) Intraspecific variability in the parasitoid wasp *Trichogramma chilonis:* can we predict the outcome of hybridization? *Evolutionary Applications* **5**: 498–510.

Berberet RC, Zarrabi AA, Payton ME & Bisges AD (2003) Reduction in effective parasitism of *Hypera postica* (Coleoptera: Curculionidae) by *Bathyplectes curculionis* (Hymenoptera: Ichneumonidae) due to encapsulation. *Environmental Entomology* **32**: 1123–1130.

Bergelson J & Kareiva P (1987) Barriers to movement and the response of herbivores to alternative cropping patterns. *Oecologia* **71**: 457–460.

Bergstedt RA, McDonald RB, Twohey MB & Heinrich JW (2003) Reduction in sea lamprey hatching success due to release of sterilized males. *Journal of Great Lakes Research* **29**: 435–444.

Berkelhamer RC (1983) Intraspecific genetic variation and haplodiploidy, eusociality, and polygyny in the Hymenoptera. *Evolution* **37**: 540–545.

Bernal-Vicente A, Ros M & Antonio Pascual J (2012) Inoculation of *Trichoderma harzianum* during maturation of vineyard waste compost to control muskmelon *Fusarium* wilt. *Bioresources* **7**: 1948–1960.

Bernays EA & Graham M (1988) On the evolution of host specificity in phytophagous arthropods. *Ecology* **69**: 886–915.

Berryman AA & Hawkins BA (2006) The refuge as an integrating concept in ecology and evolution. *Oikos* **115**: 192–196.

Bess HA, Van den Bosch R & Haramoto FH (1961) Fruit fly parasites and their activities in Hawaii. *Proceedings of the Hawaiian Entomological Society* **17**: 367–378.

Best SM, Collins SV & Kerr PJ (2000) Coevolution of host and virus: cellular localization of virus in myxoma virus infection of resistant and susceptible European rabbits. *Virology* **277**: 76–91.

Best SM & Kerr PJ (2000) Coevolution of host and virus: the pathogenesis of virulent and attenuated strains of myxoma virus in resistant and susceptible European rabbits. *Virology* **267**: 36–48.

Beukeboom LW (2001) Single-locus complementary sex determination in the ichneumonid *Venturia canescens* (Gravenhorst) (Hymenoptera). *Netherlands Journal of Zoology* **51**: 1–15.

Bezemer TM, Harvey JA, Kamp AFD, Wagenaar R, Gols R, Kostenko O, Fortuna T, Engelkes T, Vet LEM, Van der Putten WH & Soler R. (2010) Behaviour of male and female parasitoids in the field: influence of patch size, host density, and habitat complexity. *Ecological Entomology* 35: 341–351.

Bianchi FJJA, Booij CJH & Tscharnkte T (2006) Sustainable pest regulation in agricultural landscapes: a review on landscape composition, biodiversity and natural pest control. *Proceedings of the Royal Society B-Biological Sciences* 273: 1715–1727.

Bianchi F, Goedhart PW & Baveco JM (2008) Enhanced pest control in cabbage crops near forest in The Netherlands. *Landscape Ecology* 23: 595–602.

Bianchi F & Van der Werf W (2004) Model evaluation of the function of prey in non-crop habitats for biological control by ladybeetles in agricultural landscapes. *Ecological Modelling* 171: 177–193.

Bigger DS & Chaney WE (1998) Effects of *Iberis umbellata* (Brassicaceae) on insect pests of cabbage and on potential biological control agents. *Environmental Entomology* 27: 161–167.

Bigler F, Babendreier D & Kuhlmann U (2006) *Environmental Impact of Invertebrates for Biological Control of Arthropods: Methods and Risk Assessment.* Wallingford, UK, CABI Publishing.

Bigler F & Kölliker-Ott UM (2006) Balancing environmental risks and benefits: a basic approach. In *Environmental Impact of Invertebrates for Biological Control of Arthropods* (Bigler F, Babendreier D & Kuhlmann U, eds.), Wallingford, UK, CABI Publishing, pp. 273–286.

Bilgrami AL, Gaugler R, Shapiro-Ilan DI & Adams BJ (2006) Source of trait deterioration in entomopathogenic nematodes *Heterorhabditis bacteriophora* and *Steinernema carpocapsae* during in vivo culture. *Nematology* 8: 397–409.

Biondi A, Desneux N, Siscaro G & Zappala L (2012a) Using organic-certified rather than synthetic pesticides may not be safer for biological control agents: selectivity and side effects of 14 pesticides on the predator *Orius laevigatus. Chemosphere* 87: 803–812.

Biondi A, Mommaerts V, Smagghe G, Viñuela E, Zappalà L & Desneux N (2012b) The non-target impact of spinosyns on beneficial arthropods. *Pest Management Science* 68: 1523–1536.

Biondi A, Zappala L, Stark JD & Desneux N (2013) Do biopesticides affect the demographic traits of a parasitoid wasp and its biocontrol services through sublethal effects? *PLoS One* 8: e76548.

Birnbaum MJ, Clem RJ & Miller LK (1994) An apoptosis-inhibiting gene from a nuclear polyhedrosis virus encoding a polypeptide with Cys/His sequence motifs. *Journal of Virology* 68: 2521–2528.

Bischoff JF, Rehner SA & Humber RA (2009) A multilocus phylogeny of the *Metarhizium anisopliae* lineage. *Mycologia* 101: 512–530.

Blackburn TM, Essl F, Evans T, Hulme PE, Jeschke JM, Kühn I, Kumschick S, Marková Z, Mrugała A, Nentwig W, Pergl J, Pyšek P, Rabitsch W, Ricciardi A, Richardson DM, Sendek A, Vilà M, Wilson JRU, Winter M, Genovesi P & Bacher S (2014) Towards a unified classification of alien species based on the magnitude of their environmental impacts. *PLoS Biology* 12: e1001850.

Blackburn TM, Prowse TAA, Lockwood JL & Cassey P (2013) Propagule pressure as a driver of establishment success in deliberately introduced exotic species: fact or artefact? *Biological Invasions* 15: 1459–1469.

Blackburn TM, Pysek P, Bacher S, Carlton JT, Duncan RP, Jarošík V, Wilson JR & Richardson DM (2011) A proposed unified framework for biological invasions. *Trends in Ecology & Evolution* 26: 333–339.

Blair AC, Hanson BD, Brunk GR, Marrs RA, Westra P, Nissen SJ, Hufbauer RA (2005) New techniques and findings in the study of a candidate allelochemical implicated in invasion success. *Ecology Letters* 8: 1039–1047.

Blair AC, Nissen SJ, Brunk GR & Hufbauer RA (2006) A lack of evidence for an ecological role of the putative allelochemical (+/-)-catechin in spotted knapweed invasion success. *Journal of Chemical Ecology* 32: 2327–2331.

Blair AC, Weston LA, Nissen SJ, Brunk GR & Hufbauer RA (2009) The importance of analytical techniques in allelopathy studies with the reported allelochemical catechin as an example. *Biological Invasions* 11: 325–332.

Blair AC & Wolfe LM (2004) The evolution of an invasive plant: an experimental study with *Silene latifolia. Ecology* 85: 3035–3042.

Blanford S, Thomas MB & Langewald J (1998) Behavioural fever in the Senegalese grasshopper, *Oedaleus senegalensis*, and its implications for biological control using pathogens. *Ecological Entomology* 23: 9–14.

Blossey B (2004) Monitoring in weed biological control programs. *Biological Control of Invasive Plants in the United States* (Coombs E, Clark JK, Piper GL & Cofrancesco Jr. AF, eds.), Corvallis, Oregon State University Press, pp. 95–105.

Blossey B, Casagrande R, Tewksbury L & Ellis DR (2001) Nontarget feeding of leaf-beetles introduced to control purple loosestrife (*Lythrum salicaria* L.). *Natural Areas Journal* **21**: 368–377.

Blossey B & Kamil J (1996) What determines the increased competitive ability of invasive non-indigenous plants? In *Proceedings of the IX Symposium on Biological Control* (Moran VC & Hoffmann JH, eds.), Stellenbosch, S. Africa, University of Cape Town, pp. 3–9.

Blossey B & Nötzold R (1995) Evolution of increased competitive ability in invasive nonindigenous plants: a hypothesis. *Journal of Ecology* **83**: 887–889.

Blows MW (1993) The genetics of central and marginal populations of *Drosophila serrata*. II Hybrid breakdown in fitness components as a correlated response to selection for dessication resistance. *Evolution* **47**: 1271–1285.

Boag B & Yeates GW (2001) The potential impact of the New Zealand flatworm, a predator of earthworms, in western Europe. *Ecological Applications* **11**: 1276–1286.

Boavida C, Neuenschwander P & Herren HR (1995) Experimental assessment of the impact of the introduced parastoid *Gyranusoidea tebygi* Noyes on the mango mealybug *Rastrococcus invadens* Williams, by physical exclusion. *Biological Control* **5**: 99–103.

Bock DG, Caseys C, Cousens RD, Hahn MA, Heredia SM, Hübner S, Turner KG, Whitney K & Rieseberg LH (2015) What we still don't know about invasion genetics. *Molecular Ecology* **24**: 2277–2297.

Boettner GH, Elkinton JS & Boettner CJ (2000) Effects of a biological control introduction on three nontarget native species of saturniid moths. *Conservation Biology* **14**: 1798–1806.

Boivin G, Kölliker-Ott UM, Bale J & Bigler F (2006) Assessing the establishment potential of inundative biological control agents. In *Environmental Impact of Invertebrates for Biological Control of Arthropods* (Bigler F, Babendreier D & Kuhlmann U, eds.), Wallingford, UK, CABI Publishing, pp. 98–113.

Bokonon-Ganta AH, De Groote H & Neuenschwander P (2002) Socio-economic impact of biological control of mango mealybug in Benin. *Agriculture, Ecosystems and Environment* **93**: 367–378.

Bokonon-Ganta AH & Neuenschwander P (1995) Impact of the biological control agent *Gyranusoidea tebygi* Noyes (Hymenoptera, Encyrtidae) on the mango mealybug, *Rastrococcus invadens* Williams (Homoptera, Pseudococcidae) in Benin. *Biocontrol Science and Technology* **5**: 95–107.

Bolwerk A, Lagopodi AL, Wijfjes AHM Lamers GE, Chin-A-Woeng TF, Lugtenberg BJ & Bloemberg GV (2003) Interactions in the tomato rhizosphere of two *Pseudomonas* biocontrol strains with the phytopathogenic fungus *Fusarium oxysporum* f. sp *radicis-lycopersici*. *Molecular Plant-Microbe Interactions* **16**: 983–993.

Bommarco R & Banks JE (2003) Scale as modifier in vegetation diversity experiments: effects on herbivores and predators. *Oikos* **102**: 440–448.

Bonning BC, Boughton AJ & Harrison RL (2002) Genetic enhancement of baculovirus insecticides. In *Advances in Microbial Control of Insect Pests* (Upadhyay RK, ed.), New York, Plenum, pp. 109–125.

Bonsall MB, French DR & Hassell MP (2002) Metapopulation structures affect persistence of predator–prey interactions. *Journal of Animal Ecology* **71**: 1075–1084.

Bonsall MB & Hassell MP (1999) Parasitoid-mediated effects: apparent competition and the persistence of host-parasitoid assemblages. *Researches on Population Ecology* **41**: 59–68.

Bonsall MB, O'Reilly DR, Cory JS & Hails RS (2005) Persistence and coexistence of engineered baculoviruses. *Theoretical Population Biology* **67**: 217–230.

Borer ET, Hosseini PR, Seabloom EW & Dobson AP (2007) Pathogen-induced reversal of native dominance in a grassland community. *Proceedings of the National Academy of Sciences of the United States of America* **104**: 5473–5478.

Bornemissza GF (1966) An attempt to control ragwort in Australia with the cinnabar moth, *Callimorpha jacobaeae* (L.) (Arctiidae: Lepidoptera). *Australian Journal of Zoology* **14**: 201–243.

Bossard CC (1991) The role of habitat disturbance, seed predation and ant dispersal on establishment of the exotic shrub *Cytisus scoparius* in California. *American Midland Naturalist* **126**: 1–13.

Bossdorf O, Prati D, Auge H & Schmid B (2004a) Reduced competitive ability in an invasive plant. *Ecology Letters* **7**: 346–353.

Bossdorf O, Schroder S, Prati D & Auge H (2004b) Palatability and tolerance to simulated herbivory in native and introduced populations of *Alliaria petiolata* (Brassicaceae). *American Journal of Botany* **91**: 856–862.

Boudouresque CF & Verlaque M (2002) Biological pollution in the Mediterranean Sea: invasive versus introduced macrophytes. *Biological Invasions* **44**: 32–38.

Boughton AJ & Pemberton RW (2008) Efforts to establish a foliage-feeding moth, *Austromusotima camptozonale*, against *Lygodium microphyllum* in Florida, considered in the light of a retrospective review of establishment success of weed biocontrol agents belonging to different arthropod taxa. *Biological Control* **47**: 28–36.

Boulter JI, Boland GJ & Trevors JT (2000) Compost: a study of the development process and end-product potential for suppression of turfgrass disease. *World Journal of Microbiology & Biotechnology* **16**: 115–134.

Bourchier RS and Smith SM (1996) Influence of environmental conditions and parasitoid quality on field performance of *Trichogramma minutum*. *Entomologia Experimentalis et Applicata* **80**: 461–468.

Bourdot GW, Baird D, Hurrell GA & de Jong MD (2006) Safety zones for a Sclerotinia sclerotiorum-based mycoherbicide: accounting for regional and yearly variation in climate. *Biocontrol Science and Technology* **16**: 345–358.

Bourdot GW, Hurrell GA, Saville DJ & de Jong MD (2001) Risk analysis of *Sclerotinia sclerotiorum* for biological control of *Cirsium arvense* in pasture: ascospore dispersal. *Biocontrol Science and Technology* **11**: 119–139.

Bourdot GW, Saville DJ, Hurrell GA, Harvey IC & De Jong MD (2000) Risk analysis of *Sclerotinia sclerotiorum* for biological control of *Cirsium arvense* in pasture: *sclerotiorum* survival. *Biocontrol Science and Technology* **10**: 411–425.

Braga GUL, Flint SD, Miller CD, Anderson AJ & Roberts DW (2001) Variability in response to UV-B among species and strains of *Metarhizium* isolated from sites at latitudes from 61degreeN to 54degreeS. *Journal of Invertebrate Pathology* **78**: 98–108.

Brandle JR, Hodges L & Zhou XH (2004) Windbreaks in North American agricultural systems. *Agroforestry Systems* **61–2**: 65–78.

Briese DT (1986a) Factors affecting the establishment and survival of *Anaitis efformata* (Lepidoptera: Geometridae) introduced into Australia for the biological control of St. John's Wort, *Hypericum perforatum*. II. Field trials. *Journal of Applied Ecology* **23**: 821–839.

Briese DT (1986b) Host resistance to microbial control agents. *Fortschritte der Zoologie* **32**: 233–256.

Briese DT (1986c) Insect resistance to baculoviruses. In *The Biology of Baculoviruses Volume II: Practical Application for Insect Control* (Granados RR & Federici BA, eds.), Boca Raton, FL, CRC Press, pp. 237–264.

Briese DT (1996) Biological control of weeds and fire management in protected natural areas: are they compatible strategies? *Biological Conservation* **77**: 135–141.

Briese DT (1997) Biological control of St. John's wort: past, present and future. *Plant Protection Quarterly* **12**: 73–80.

Briese DT (2005) Translating host-specificity test results into the real world: the need to harmonize the yin and yang of current testing procedures. *Biological Control* **35**: 208–214.

Briese DT & Walker A (2002) A new perspective on the selection of test plants for evaluating the host-specificity of weed biological control agents: the case of *Deuterocampta quadrijuga*, a potential insect control agent of *Heliotropium amplexicaule*. *Biological Control* **25**: 273–287.

Briese DT, Pettit WJ & Walker A (2004) Evaluation of the biological control agent, *Lixus cardui*, on *Onopordum* thistles: experimental studies on agent demography and impact. *Biological Control* **31**: 165–171.

Briese DT, Walker A, Pettit WJ & Sagliocco JL (2002a) Host-specificity of candidate agents for the biological control of *Onopordum* spp. thistles in Australia: an assessment of testing procedures. *Biocontrol Science and Technology* **12**: 149–163.

Briese DT, Zapater M, Andorno A & Perez-Camargo G (2002b) A two-phase open-field test to evaluate the host-specificity of candidate biological control agents for *Heliotropium amplexicaule*. *Biological Control* **25**: 259–272.

Briese DT, Zapater M & Walker A (2005) Implementation of the blue heliotrope biological control strategy: host-specificity testing of *Longitarsus* sp. In *A Report for the Rural Industries Research and Development Corporation*, Barton, Australia, Rural Industries Research and Development Corporation.

Briggs CJ (1993) Competition among parasitoid species on a stage-structured host and its effect on host suppression. *American Naturalist* **141**: 372–379.

Briggs CJ (2009) Host-parasitoid interactions. In *The Princeton Guide to Ecology* (Levin SA, ed.), Princeton NJ, Princeton University Press, pp. 213–219.

Briggs CJ & Borer ET (2005) Why short-term experiments may not allow long-term predictions about intraguild predation. *Ecological Applications* **15**: 1111–1117.

Briggs CJ & Godfray HCJ (1995a) Models of intermediate complexity in insect-pathogen interactions: population dynamics of the microsporidian pathogen, *Nosema pyrausta*, of the European corn borer, *Ostrinia nubilalis*. *Parasitology* **111**: S71–S89.

Briggs CJ & Godfray HCJ (1995b) The dynamics of insect-pathogen interactions in stage-structured populations. *American Naturalist* **145**: 855–887.

Briggs CJ & Godfray HCJ (1996) The dynamics of insect-pathogen interactions in seasonal environments. *Theoretical Population Biology* **50**: 149–177.

Briggs CJ, Hails ND, Barlow ND & Godfray HCJ (1995) The dynamics of insect-pathogen interactions. In *Ecology of Infectious Diseases in Natural Populations* (Grenfell B & Dobson A, eds.), Cambridge, UK, Cambridge University Press, pp. 295–306.

Briggs CJ & Hoopes MF (2004) Stabilizing effects in spatial parasitoid–host and predator–prey models: a review. *Theoretical Population Biology* **65**: 299–315.

Briggs CJ, Murdoch WW & Nisbet RM (1999) Recent developments in theory for biological control of insect pests by parasitoids. In *Theoretical Approaches to Biological Control* (Hawkins BA & Cornell HV, eds.), Cambridge, UK, Cambridge University Press, pp. 22–42.

Brodeur J (2000) Host specificity and trophic relationships of hyperparasitoids. In *Parasitoid Population Biology* (Hochberg ME & Ives AR, eds.), Princeton, NJ, Princeton University Press, pp. 163–183.

Brodeur J (2012) Host specificity in biological control: insights from opportunistic pathogens. *Evolutionary Applications* **5**: 470–480.

Brodeur J & Boivin G (2006) *Trophic and Guild Interactions in Biological Control*. Dordrecht, The Netherlands, Springer.

Brodeur J, Geervliet JBF & Vet LEM (1996) The role of host species, age and defensive behavior on ovipositional decisions in a solitary specialist and gregarious generalist parasitoid (*Cotesia* species). *Entomologia Experimentalis et Applicata* **81**: 125–132.

Brodeur J & Rosenheim JA (2000) Intraguild interactions in aphid parasitoids. *Entomologia Experimentalis et Applicata* **97**: 93–108.

Brodmann PA, Wilcox CV & Harrison S (1997) Mobile parasitoids may restrict the spatial spread of an insect outbreak. *Journal of Animal Ecology* **66**: 65–72.

Bronstein JL (1998) The contribution of ant plant protection studies to our understanding of mutualism. *Biotropica* **30**: 150–161.

Brown JH & Heske EJ (1990) Control of a desert-grassland transition by a keystone rodent guild. *Science* **250**: 1705–1707.

Brown PMJ, Adriaens T, Bathon H, Cuppen J, Goldarazena A, Hägg T, Klausnitzer BEM, Kovář I, Loomans AJM, Majerus MEN, Nedvěd O, Pedersen J, Rabitsch W, Roy HE, Ternois V, Zakharov IA, Roy DB (2008) *Harmonia axyridis* in Europe: spread and distribution of a non-native coccinellid. *BioControl* **53**: 5–21.

Brunner K, Zeilinger S, Ciliento R, Woo SL, Lorito M, Kubicek CP & Mach RL (2005) Improvement of the fungal biocontrol agent *Trichoderma atroviride* to enhance both antagonism and induction of plant systemic disease resistance. *Applied and Environmental Microbiology* **71**: 3959–3965.

Bucher GE & Harris P (1961) Food-plant spectrum and elimination of disease of cinnabar moth, *Hypocrita jacobaeae* (L.) (Lepidoptera: Arctiidae). *Canadian Entomologist* **93**: 931–936.

Buckingham GR (2001) Quarantine host range studies with *Lophyrotoma zonalis*, an Australian sawfly of interest for biological control of melaleuca, *Melaleuca quinquenervia*, in Florida. *BioControl* **46**: 363–386.

Buckley YM, Hinz HL, Matthies D & Rees M (2001) Interactions between density-dependent processes, population dynamics and control of an invasive plant species, *Tripleurospermum perforatum* (scentless chamomile). *Ecology Letters* **4**: 551–558.

Buckley YM, Rees M, Paynter Q & Lonsdale M (2004) Modelling integrated weed management of an invasive shrub in tropical Australia. *Journal of Applied Ecology* **41**: 547–560.

Buckley YM, Rees M, Sheppard AW & Smyth MJ (2005) Stable coexistence of an invasive plant and biocontrol agent: a parameterized coupled plant-herbivore model. *Journal of Applied Ecology* **42**: 70–79.

Buerger P, Hauxwell C & Murray D (2007) Nucleopolyhedrovirus introduction in Australia. *Virologica Sinica* **22**: 173–179.

Bugg RL, Ehler LE & Wilson LT (1987) Effect of common Knotweed (*Polygonum aviculare*) on abundance and efficiency of insect predators of crop pests. *Hilgardia* **55**: 1–53.

Bugg RL & Waddington C (1994) Using cover crops to manage arthropod pests of orchards: a review. *Agriculture, Ecosystems and Environment* **50**: 11–28.

Bukovinszky T, Gols R, Hemerik L, Van Lenteren JC & Vet LEM (2007) Time allocation of a parasitoid foraging in heterogeneous vegetation: implications for host-parasitoid interactions. *Journal of Animal Ecology* **76**: 845–853.

Bukovinszky T, Trefas H, Van Lenteren JC, Vet LEM & Fremont J (2004) Plant competition in pest-suppressive intercropping systems complicates evaluation of herbivore responses. *Agriculture, Ecosystems & Environment* **102**: 185–196.

Bull JJ (1994) Virulence. *Evolution* **48**: 1423–1437.

Burdon JJ, Groves RH & Cullen JM (1981) The impact of biological control on the distribution and abundance of *Chondrilla juncea* in south-eastern Australia. *Journal of Applied Ecology* **18**: 957–966.

Burdon JJ, Groves RH, Kaye PE & Speer SS (1984) Competition in mixtures of susceptible and resistant genotypes of *Chondrilla juncea* differentially infected with rust. *Oecologia* **64**: 199–203.

Burdon JJ & Marshall DR (1981) Biological control and the reproductive mode of weeds. *Journal of Applied Ecology* **18**: 649–658.

Burdon JJ & Thrall PH (2004) Genetic structure of natural plant and pathogen populations. In *Genetics, Evolution and Biological Control* (Ehler LE, Sforza R & Mateille T, eds.), Wallingford, UK, CABI Publishing, pp. 1–17.

Burges DH (1998) *Formulation of Microbial Pesticides*. Dordrecht, The Netherlands, Kluwer.

Burges DH & Jones KA (1998) Formulation of bacteria, viruses and Protozoa to control insects. In *Formulation of Microbial Pesticides* (Burges DH, ed.), Dordrecht, The Netherlands, Kluwer, pp. 33–127.

Bürgi LP & Mills NJ (2014) Lack of enemy release for an invasive leafroller in California: temporal patterns and influence of host plant origin. *Biological Invasions* **16**: 1021–1034.

Bürgi LP, Roltsch WJ & Mills NJ (2015) Allee effects and population regulation: a test for biotic resistance against an invasive leafroller by resident parasitoids. *Population Ecology* **57**: 215–225.

Burnett T (1958) A model of host-parasite interactions. *Proceedings of the 10th International Congress of Entomology* **2**: 679–686.

Burrows BE & Balciunas JK (1997) Biology, distribution and host-range of the sawfly, *Lophyrotoma zonalis* (Hym., Pergidae), a potential biological control agent for the paperbark tree, *Melaleuca quinquenervia*. *Entomophaga* **42**: 299–313.

Bush GL (1969) Sympatric host race formation and speciation in frugivorous flies of the genus *Rhagoletis* (Diptera, Tephritidae). *Evolution* **23**: 237–251.

Butcher RDJ, Whitfield WGF & Hubbard SF (2000). Complementary sex determination in the genus *Diadegma* (Hymenoptera: Ichneumonidae). *Journal of Evolutionary Biology* **13**: 593–606.

Callaway RM, DeLuca TH & Belliveau WM (1999) Biological-control herbivores may increase competitive ability of the noxious weed *Centaurea maculosa*. *Ecology* **80**: 1196–1201.

Callaway RM & Ridenour WM (2004) Novel weapons: invasive success and the evolution of increased competitive ability. *Frontiers in Ecology and the Environment* **2**: 436–443.

Caltagirone LE (1981) Landmark examples in classical biological control. *Annual Review of Entomology* **26**: 213–232.

Caltagirone LE & Doutt RL (1989) The history of the vedalia beetle importation to California and its impact on the development of biological control. *Annual Review of Entomology* **34**:1–16.

Cameron E (1935) A study of the natural control of ragwort (*Senecio jacobaea*) L. *Journal of Ecology* **23**: 265–322.

Cameron PJ, Hill RL, Bain J & Thomas WP (1993) Analysis of importations for biological control of insect pests and weeds in New Zealand. *Biocontrol Science and Technology* **3**: 387–404.

Campbell MM (1976) Colonisation of *Aphytis melinus* DeBach (Hymenoptera, Aphelinidae) in *Aonidiella aurantii* (Mask.) (Hemiptera, Coccidae) on citrus in South Australia. *Bulletin of Entomological Research* **65**: 659–668.

Cao YQ, Peng GX, He ZB, Wang ZK, Yin YP, Xia YX (2007) Transformation of *Metarhizium anisopliae* with benomyl resistance and green fluorescent protein genes provides a tag for genetically engineered strains. *Biotechnology Letters* **29**: 907–911.

Carabajal Paladino L, Muntaabski I, Lanzavecchia S, Le Bagousse-Pinguet Y, Viscarret M, Juri M, Fueyo-Sánchez L, Papeschi A, Cladera J, Bressa MJ (2015) Complementary sex determination in the parasitic wasp *Diachasmimorpha longicaudata*. *PLoS One* **10**: e0119619.

Carey JR (1989) The multiple decrement life table: A unifying framework for cause-of-death analysis in ecology. *Oecologia* **78**: 131–137.

Carmona DM & Landis DA (1999) Influence of refuge habitats and cover crops on seasonal activity-density of ground beetles (Coleoptera: Carabidae) in field crops. *Environmental Entomology* **28**: 1145–1153.

Carisse O & Rolland D (2004) Effect of timing of application of the biological control agent *Microsphaeropsis ochracea* on the production and ejection pattern of ascospores by *Venturia inaequalis*. *Phytopathology* **94**: 1305–1314.

Carruthers RI & D'Antonio CM (2005) Science and decision making in biological control of weeds: benefits and risks of biological control. *Biological Control* **35**: 181–182.

Carson WP, Hovick SM, Baumert AJ, Bunker DE & Pendergast TH (2008) Evaluating the post-release efficacy of invasive plant biocontrol by insects: a comprehensive approach. *Arthropod-Plant Interactions* **2**: 77–86.

Carter N, Dixon AFG & Rabbinge R (1982) *Cereal Aphid Populations: Biology, Simulation and Prediction.* Wageningen, The Netherlands, Centre for Agricultural Publishing and Documentation.

Carton Y & Kitano H (1981) Evolutionary relationships to parasitism by seven species of the *Drosophila melanogaster* subgroup. *Biological Journal of the Linnean Society* **16**: 227–241.

Carvalheiro LG, Buckley YM, Ventim R, Fowler SV & Memmott J (2008) Apparent competition can compromise the safety of highly specific biological control agents. *Ecology Letters* **11**: 690–700.

Case CM & Crawley MJ (2000) Effect of intraspecific competition and herbivory on the recruitment of an invasive alien plant: *Conyza sumatrensis*. *Biological Invasions* **2**: 103–110.

Cassani JR & Caton WE (1985) Induced triploidy in grass garp, *Ctenopharyngodon idella* Val. *Aquaculture* **46**: 37–44.

Castells E, Morante M, Blanco-Moreno JM, Sans FX, Vilatersana R & Blasco A (2013) Reduced seed predation after invasion supports enemy release in a broad biogeographical survey. *Oecologia* **173**: 1397–1409.

Caughley G & Lawton JH (1981) Plant-herbivore systems. In *Theoretical Ecology: Principles and Applications*, 2nd edition (May RM, ed.), Oxford, UK, Blackwell, pp. 132–166.

Causton CE (2009) Success in biological control: the scale and the ladybird. In *Galapagos: Preserving Darwin's Legacy* (De Roy T, ed.), Auckland, New Zealand, David Bateman, pp. 184–190.

Cavallini P & Serafini P (1995) Winter diet of the small Indian mongoose, *Herpestes auropunctatus*, on an Adriatic island. *Journal of Mammalogy* **76**: 569–574.

Ceballo FA, Walter GH & Rochester W (2010) The impact of climate on the biological control of citrus mealybug [*Planococcus citri* (Risso)] by the parasitoid *Coccidoxenoides perminutus* Girault as predicted by the climate-matching program CLIMEX. *Philippine Agricultural Scientist* **93**: 317–328.

Center TD & Dray FA (2010) Bottom-up control of water hyacinth weevil populations: do the plants regulate the insects? *Journal of Applied Ecology* **47**: 329–337.

Center TD, Dray FA, Jubinsky GP & Grodowitz MJ (1999) Biological control of water hyacinth under conditions of maintenance management: can herbicides and insects be integrated? *Environmental Management* **23**: 241–256.

Center TD, Purcell MF, Pratt PD, Rayamajhi M, Tipping PW, Wright SA & Dray A (2012) Biological control of *Melaleuca quinquenervia*: an Everglades invader. *BioControl* **57**: 151–165.

Chaboudez P & Burdon JJ (1995) Frequency-dependent selection in a wild plant-pathogen system. *Oecologia* **102**: 490–493.

Chaboudez P & Sheppard AW (1995) Are particular weeds more amenable to biological control? A reanalysis of reproduction of life history. In *Proceedings of the Eighth International Symposium on Biological Control of Weeds* (Delfosse ES & Scott RR, eds.), Melbourne, Australia, DSIR/CSIRO, pp. 95–102.

Chace AB (1979) *The Rhind Mathematical Papyrus*. Oberlin, OH, Mathematical Association of America.

Chacon J & Heimpel GE (2010) Density-dependent intraguild predation of an aphid parasitoid. *Oecologia* **164**: 213–220.

Chaddick PR & Leek FF (1972) Further specimens of stored product insects found in an ancient Egyptian tomb. *Journal of Stored Product Research* **8**: 83–86.

Chailleux A, Mohl EK, Alves MT, Messelink GJ & Desneux N (2014) Natural enemy-mediated indirect interactions among prey species: potential for enhancing biocontrol services in agroecosystems. *Pest Management Science* **70**: 1769–1779.

Channer AGdR & Gowen SR (1992) Selection for increased host resistance and increased pathogen specificity in the *Meloidogyne-Pasteuria penetrans* interaction. *Fundamental and Applied Nematology* **15**: 331–339.

Chaplin-Kramer R, O'Rourke ME, Blitzer EJ & Kremen C (2011) A meta-analysis of crop pest and natural enemy response to landscape complexity. *Ecology Letters* **14**: 922–932.

Charleston DS, Kfir R, Dicke M & Vet LEM (2006) Impact of botanical extracts derived from *Melia azedarach* and *Azadirachta indica* on populations of Plutella xylostella and its natural enemies: a field test of laboratory findings. *Biological Control* **39**: 105–114.

Charudattan R (2001) Biological control of weeds by means of plant pathogens: significance for integrated weed management in modern agro-ecology. *BioControl* **46**: 229–260.

Charudattan R (2005) Ecological, practical, and political inputs into selection of weed targets: what makes a good biological control target? *Biological Control* **35**: 183–196.

Charudattan R, Chandramohan S & Wyss GS (2002) Biological control. In *Pesticides in Agriculture and the Environment* (Wheeler WB, ed.), New York, Marcel Dekker, pp. 25–58.

Chaston JM, Dillman AR, Shapiro-Ilan DI, Bilgrami AL, Gaugler R, Hopper KR, Adams BJ (2011) Outcrossing and crossbreeding recovers deteriorated traits in laboratory cultured *Steinernema carpocapsae* nematodes. *International Journal for Parasitology* **41**: 801–809.

Chater KF, Biro S, Lee KJ, Palmer T & Schrempf H (2010) The complex extracellular biology of *Streptomyces*. *Fems Microbiology Reviews* **34**: 171–198.

Chavalle S, Buhl PN, Censier F & De Proft M (2015) Comparative emergence phenology of the orange wheat blossom midge, *Sitodiplosis mosellana* (Gehin) (Diptera: Cecidomyiidae) and its parasitoids (Hymenoptera: Pteromalidae and Platygastridae) under controlled conditions. *Crop Protection* **76**: 114–120.

Chen C-J & Thiem SM (1997) Differential infectivity of two *Autographa californica* nucleopolyhedrovirus mutants on three permissive cell lines is the result of lef-7 deletion. *Virology* **227**: 88–95.

Cherry AJ, Jenkins NE, Heviefo G, Bateman R & Lomer CJ (1999) Operational and economic analysis of a West African pilot-scale production plant for aerial conidia of *Metarhizium* spp. for use as a mycoinsecticide against locusts and grasshoppers. *Biocontrol Science and Technology* **9**: 35–51.

Chesson PL & Murdoch WW (1986) Aggregation of risk relationships among host-parasitoid models. *American Naturalist* **127**: 696–715.

Chilcutt CF & Tabashnik BE (1999) Simulation of integration of *Bacillus thuringiensis* and the parasitoid *Cotesia plutellae* (Hymenoptera: Braconidae) for control of susceptible and resistant diamondback moth (Lepidoptera: Plutellidae). *Environmental Entomology* **28**: 505–512.

Childs MR (2006) Comparison of gila topminnow and western mosquitofish as biological control agents of mosquitoes. *Western North American Naturalist* **66**: 181–190.

Chobot V, Huber C, Trettenhahn G & Hadacek F (2009) (+/−)-Catechin: chemical weapon, antioxidant, or stress regulator? *Journal of Chemical Ecology* **35**: 980–996.

Chun YJ, Van Kleunen M & Dawson W (2010) The role of enemy release, tolerance and resistance in plant invasions: linking damage to performance. *Ecology Letters* **13**: 937–946.

Cilliers CJ & Neser S (1991) Biological control of *Lantana camara* (Verbenaceae) in South Africa. *Agriculture, Ecosystems and Environment* **37**: 57–75.

Civeyrel L & Simberloff D (1996) A tale of two snails: is the cure worse than the disease? *Biodiversity and Conservation* **5**: 1231–1252.

Clarke B, Murray J & Johnson MS (1984) The extinction of endemic species by a program of biological control. *Pacific Science* **38**: 97–104.

Clausen CP (1978) *Introduced Parasites and Predators of Arthropod Pests and Weeds: A World Review*. Washington, DC, USDA Agriculture Research Service.

Clay K (2014) Defensive symbionts: a microbial perspective. *Functional Ecology* **28**: 293–298.

Clay K & Kover PX (1996) The red queen hypothesis and plant/pathogen interactions. *Annual Review of Phytopathology* **34**: 29–50.

Clem RJ, Fechheimer M & Miller LK (1991) Prevention of apoptosis by a baculovirus gene during infection of insect cells. *Science* **254**: 1388–1390.

Clem RJ & Miller LK (1993) Apoptosis reduces both the in vitro replication and the in vivo infectivity of a baculovirus. *Journal of Virology* **67**: 3730–3738.

Clewley GD, Eschen R, Shaw RH & Wright DJ (2012) The effectiveness of classical biological control of invasive plants. *Journal of Applied Ecology* **49**: 1287–1295.

Coatzee JA, Hill MP, Byrne MJ & Bownes AA (2011) A review of the biological control programmes on *Eichhornia crassipes* (C.Mart.) *Solms* (Pontederiaceae), *Salvinia molesta* DS Mitch. (Salviniaceae), *Pistia stratiotes* L. (Araceae), *Myriophyllum aquaticum* (Vell.) Verdc. (Haloragaceae) and *Azolla filiculoides* Lam. (Azollaceae) in South Africa. *African Entomology* **19**: 451–468.

Cock MJW (1985) *A Review of Biological Control of Pests in the Commonwealth Caribbean and Bermuda up to 1982*. Slough, UK, Commonwealth Agricultural Bureau.

Cock MJW (1986) Requirements for biological control: an ecological perspective. *Biocontrol News and Information* **7**: 7–17.

Cock MJW, van Lenteren JC, Brodeur J, Barratt BIP, Bigler F, Bolckmans K, Consoli FL, Haas F, Mason PG & Parra JRP (2010) Do new access and benefit sharing procedures under the Convention on Biological Diversity threaten the future of biological control? *BioControl* **55**:199–218.

Cohen AC (2004) *Insect Diets: Science and Technology*. Boca Raton, FL, CRC Press.

Cohen AC & Smith LK (1998) A new concept in artificial diets for *Chrysoperla rufilabris*: the efficacy of solid diets. *Biological Control* **13**: 49–54.

Cohen MF & Mazzola M (2006) Resident bacteria, nitric oxide emission and particle size modulate the effect of *Brassica napus* seed meal on disease incited by *Rhizoctonia solani* and *Pythium* spp. *Plant and Soil* **286**: 75–86.

Cohen MF, Yamasaki H & Mazzola M (2005) *Brassica napus* seed meal soil amendment modifies microbial community structure, nitric oxide production and incidence of *Rhizoctonia* root rot. *Soil Biology & Biochemistry* **37**: 1215–1227.

Colautti RI, Ricciardi A, Grigorovich IA & MacIsaac HJ (2004) Is invasion success explained by the enemy release hypothesis? *Ecology Letters* **7**: 721–733.

Coll M (2004) Precision agriculture approaches in support of ecological engineering for pest management. In *Ecological Engineering for Pest Management* (Gurr GM, ed.), Ithaca, NY, Cornell University Press, pp. 113–142.

Coll M & Abd-Rabou S (1998) Effect of oil emulsion sprays on parasitoids of the black parlatoria, *Parlatoria ziziphi*, in grapefruit. *BioControl* **43**: 29–37.

Coll M & Bottrell DG (1994) Effects of nonhost plants on an insect herbivore in diverse habitats. *Ecology* **73**: 723–731.

Coll M & Bottrell DG (1995) Predator–prey association in monocultures and dicultures – effect of maize and bean vegetation. *Agriculture Ecosystems & Environment* **54**: 115–125.

Coll M & Bottrell DG (1996) Movement of an insect parasitoid in simple and divers plant assemblages. *Ecological Entomology* **21**: 141–149.

Coll M, De Mendoza LG & Roderick GK (1994) Population structure of a predatory beetle: the importance of gene flow for intertrophic level interactions. *Heredity* **72**: 228–236.

Coll M & Guershon M (2002) Omnivory in terrestrial arthropods: mixing plant and prey diets. *Annual Review of Entomology* **47**: 267–298.

Collier TR, Murdoch WW & Nisbet RM (1994) Egg load and the decision to host-feed in the parasitoid, *Aphytis melinus*. *Journal of Animal Ecology* **63**: 299–306.

Collier T & Van Steenwyk R (2004) A critical evaluation of augmentative biological control. *Biological Control* **31**: 245–256.

Collier T & Van Steenwyk R (2006) How to make a convincing case for augmentative biological control. *Biological Control* **39**: 119–120.

Collins DP & Jacobsen BJ (2003) Optimizing a *Bacillus subtilis* isolate for biological control of sugar beet cercospora leaf spot. *Biological Control* **26**: 153–161.

Collins KL, Boatman ND, Wilcox A & Holland JM (2003a) A 5-year comparison of overwintering polyphagous predator densities within a beetle bank and two conventional hedgebanks. *Annals of Applied Biology* **143**: 63–71.

Collins KL, Boatman ND, Wilcox A & Holland JM (2003b) Effects of different grass treatments used to create overwintering habitat for predatory arthropods on arable farmland. *Agriculture Ecosystems & Environment* **96**: 59–67.

Collins KL, Boatman ND, Wilcox A, Holland JM & Chaney K (2002) Influence of beetle banks on cereal aphid predation in winter wheat. *Agriculture Ecosystems & Environment* **93**: 337–350.

Collyer E (1964) The effect of an alternative food supply between two *Typhlodromus* species and *Panonychus ulmi* (Kock) (Acarina). *Entomologia Experimentalis et Applicata* **7**: 120–124.

Connor EF, Faeth SH, Simberloff D & Opler PA (1980) Taxonomic isolation and accumulation of herbivorous insects: a comparison of introduced and native trees. *Ecological Entomology* **5**: 205–211.

Conway HE, Steinkraus DC, Ruberson JR & Kring TJ (2006) Experimental treatment threshold for the cotton aphid (Homoptera: Aphididae) using natural enemies in Arkansas cotton. *Journal of Entomological Science* **41**: 361–373.

Cook JM (1993). Inbred lines as reservoirs of sex alleles in parasitoid rearing programs. *Environmental Entomology* **22**: 1213–1216.

Cook RJ (2007) Take-all decline: model system in the science of biological control and clue to the success of intensive cropping. In *Biological Control: A Global Perspective* (Vincent C, Goettel M & Lazarovits G, eds.), Wallingford, UK, CABI Publishing, pp. 399–413.

Coombes DS & Sotherton NW (1986) The dispersal and distribution of polyphagous predatory Coleoptera in cereals. *Annals of Applied Biology* **108**: 461–474.

Coombs MT (2000) Seasonal phenology, parasitism, and evaluation of mowing as a control measure for *Nezara viridula* (Hemiptera: Pentatomidae) in Australian pecans. *Environmental Entomology* **29**: 1027–1033.

Coombs M (2004) Estimating the host range of the tachinid *Trichopoda giacomellii*, introduced into Australia for biological control of the green vegetable bug. In *Assessing Host Ranges of Parasitoids and Predators Used for Classical Biological Control: A Guide to Best Practice* (Van Driesche RG & Reardon R, eds.), Morgantown, WV, USDA Forest Service, pp. 143–151.

Coquillard P, Thibaut T, Hill DRC, Gueugnot J, Mazel C & Coquillard Y (2000) Simulation of the mollusc Ascoglossa *Elysia subornata* population dynamics: application to the potential biocontrol of *Caulerpa taxifolia* growth in the Mediterranean Sea. *Ecological Modelling* **135**: 1–16.

Corbett A (1998) The importance of movement in the response of natural enemies to habitat manipulation. In *Enhancing Biological Control* (Pickett CH & Bugg RL, eds.), Berkeley, University of California Press, pp. 25–48.

Corbett A, Leigh TF & Wilson LT (1991) Interplanting alfalfa as a source of *Mataseiulus occidentalis* (Acari: Phytoseiidae) for managing spider mites in cotton. *Biological Control* **1**: 188–196.

Corbett A & Plant RE (1993) Role of movement in the response of natural enemies to agroecosystem diversification: a theoretical evaluation. *Environmental Entomology* **22**: 519–531.

Corbett A & Rosenheim JA (1996) Impact of a natural enemy overwintering refuge and its interaction with the surrounding landscape. *Ecological Entomology* **21**: 155–164.

Cornell HV & Hawkins BA (1993) Accumulation of native parasitoid species on introduced herbivores: a comparison of hosts as natives and hosts as invaders. *American Naturalist* **141**: 847–865.

Cornwallis LJ, Stewart A, Bourdot GW, Gaunt RE, Harvey IC & Saville DJ (1999) Pathogenicity of *Sclerotinia sclerotiorum* on *Ranunculus acris* is in dairy pasture. *Biocontrol Science and Technology* **9**: 365–377.

Cory JS (2003) Ecological impacts of virus insecticides: host range and non-target organisms. In *Environmental Impact of Microbial Insecticides* (Hokkanen HMT & Hajek AE, eds.), Dordrecht, The Netherlands, Kluwer, pp. 73–92.

Cory JS (2007) Field tests in the UK of a genetically modified virus. In *Biological Control: A Global Perspective* (Vincent C, Goettel MS & Lazarovotis G, eds.), Wallingford, UK, CABI Publishing, pp. 362–373.

Cory JS & Franklin MT (2012) Evolution and microbial control of insects. *Evolutionary Applications* **5**: 455–469.

Cory JS, Hirst ML, Williams T, Hails RS, Goulson D, Green BM, Carty TM, Possee RD, Cayley PJ & Bishop DHL (1994) Field trial of a genetically improved baculovirus insecticide. *Nature* **370**: 138–140.

Costa A & Stary P (1988) *Lysiphlebus testaceipes*, an introduced aphid parasitoid in Portugal (Hym.: Aphidiidae). *Entomophaga* **33**: 403–412.

Costamagna AC, Menalled FD & Landis DA (2004) Host density influences parasitism of the armyworm *Pseudaletia unipuncta* in agricultural landscapes. *Basic and Applied Ecology* **5**: 347–355.

Costanza R, d'Arge R, de Groot R, Farber S, Grasso M, Hannon B, Limburg K, O'Neill RV, Paruelo J, Raskin RG, Naeem S, Sutton PC & Van Den Belt M (1997) The value of the world's ecosystem services and natural capital. *Nature* **387**: 253–260.

Costello MJ & Daane KM (2003) Spider and leafhopper (*Erythroneura* spp.) response to vineyard ground cover. *Environmental Entomology* **32**: 1085–1098.

Cottrell TE (2004) Suitability of exotic and native lady beetle eggs (Coleoptera: Coccinellidae) for development of lady beetle larvae. *Biological Control* **31**: 362–371.

Cottrell TE (2005) Predation and cannibalism of lady beetle eggs by adult lady beetles. *Biological Control* **34**: 159–164.

Cottrell TE & Yeargan KV (1998) Intraguild predation between an introduced lady beetle, *Harmonia axyridis*(Coleoptera: Coccinellidae), and a native lady beetle, *Coleomegilla maculata* (Coleoptera: Coccinellidae). *Journal of the Kansas Entomological Society* **71**: 159–163.

Cotxarrera L, Trillas-Gay MI, Steinberg C & Alabouvette C (2002) Use of sewage sludge compost and *Trichoderma asperellum* isolates to suppress Fusarium wilt of tomato. *Soil Biology & Biochemistry* **34**: 467–476.

Coupland R & Baker G (2007) Search for biological control agents of invasive Mediterranean snails. In *Biological Control: A Global Perspective* (Vincent C, Goettel MS & Lazarovits G, eds.), Wallingford, UK, CABI Publishing, pp. 7–12.

Courchamp F, Berec L & Gasciogne J (2008) *Allee Effects in Ecology and Conservation*. Oxford, UK, Oxford University Press.

Courchamp F, Chapuis J-L & Pascal M (2003) Mammal invaders on islands: impact, control and control impact. *Biological Reviews* **78**: 347–383.

Courchamp F & Cornell SJ (2000) Virus-vectored immunocontraception to control feral cats on islands: a mathematical model. *Journal of Applied Ecology* **37**: 903–913.

Courchamp F & Sugihara G (1999) Modeling the biological control of an alien predator to protect island species from extinction. *Ecological Applications* **9**: 112–123.

Courtenay WR & Meffe GK (1989) Small fishes in strange places: a review of introduced poeciliids. In *Ecology and Evolution of Livebearing Fishes* (Meffe GK & Snelson FF, eds.), Englewood Cliffs, NJ, Prentice Hall, pp. 319–331.

Cousens R & Croft AM (2000) Weed populations and pathogens. *Weed Research* **40**: 63–82.

Cowan DP & Stahlhut JK (2004) Functionally reproductive diploid and haploid males in an inbreeding hymenopteran with complementary sex determination. *Proceedings of the National Academy of Sciences, USA*, **101**: 10374–10379.

Cowie RH (2001) Can snails ever be effective and safe biocontrol agents? *International Journal of Pest Management* **47**: 23–40.

Crawley MJ (1986) The population biology of invaders. *Philosophical Transactions of the Royal Society of London, Series B* **314**: 711–731.

Crawley MJ (1987) What makes a community invasible? In *Colonization, Succession and Stability* (Gray AJ, Crawley MJ & Edwards PJ, eds.), Oxford, UK, Blackwell, pp. 429–453.

Crawley MJ (1989a) Insect herbivores and plant population dynamics. *Annual Review of Entomology* **34**: 531–564.

Crawley MJ (1989b) The successes and failures of weed biocontrol using insects. *Biocontrol News and Information* **10**: 213–223.

Crawley MJ (1997) Plant-herbivore dynamics. In *Plant Ecology* (Crawley MJ, ed.), Oxford, UK, Blackwell, pp. 401–474.

Croft BA (1990) *Arthropod Biological Control Agents and Pesticides*. New York, John Wiley & Sons.

Croizier G, Croizier L, Argaud O & Poudevigne D (1994) Extension of *Autographa californica* nuclear polyhedrosis virus host range by interspecific replacement of a short DNA sequence in the p143 helicase gene. *Proceedings of the National Academy of Sciences of the United States of America* **91**: 48–52.

Crone EE, Menges ES, Ellis MM, Bell T, Bierzychudek P, Ehrlén J, Kaye TN, Knight TM, Lesica P, Morris WF, Oostermeijer G, Quintana-Ascensio PF, Stanley A, Ticktin T, Valverde T & Williams JL (2011) How do plant ecologists use matrix population models? *Ecology Letters* **14**: 1–8.

Cronin JT & Reeve JD (2005) Host-parasitoid spatial ecology: a plea for a landscape-level synthesis. *Proceedings of the Royal Society B-Biological Sciences* **272**: 2225–2235.

Cronin JT & Reeve JD (2014) An integrative approach to understanding host-parasitoid population dynamics in real landscapes. *Basic and Applied Ecology* **15**: 101–113.

Cronin JT, Reeve JD, Wilkens R & Turchin P (2000) The pattern and range of movement of a checkered beetle predator relative to its bark beetle prey. *Oikos* **90**: 127–138.

Crook NE, Clem RJ & Miller LK (1993) An apoptosis-inhibiting baculovirus gene with a zinc finger-like motif. *Journal of Virology* **67**: 2168–2174.

Cross JV, Burgess CM & Hanks GR (1996) Integrating insecticide use with biological control of two spotted spider mite (*Tetranychus urticae*) by *Phytoseiulus persimilis* on strawberry in the UK. In *Brighton Crop Protection Conference: Pests and Diseases, Volume 3*. Alton, UK, British Crop Protection Council, pp. 899–906.

Crowder DW (2007) Impact of release rates on the effectiveness of augmentative biological control agents. *Journal of Insect Science* **7**: 15.

Crowley PH & Martin EK (1989) Functional responses and interference within and between year classes of a dragonfly population. *Journal of the North American Benthological Society* **8**: 211–221.

Crozier RH, Newey PS, Schluns EA & Robson SKA (2010) A masterpiece of evolution – *Oecophylla* weaver ants (Hymenoptera: Formicidae). *Myrmecological News* **13**: 57–71.

Crump NS, Cother EJ & Ash GJ (1999) Clarifying the nomenclature in microbial weed control. *Biocontrol Science and Technology* **9**: 89–97.

Cullen JM & Delfosse ES (1985) *Echium plantanineum*: catalyst for conflict and change in Australia. In *Proceedings of the VI International Symposium on Biological Control* (Delfosse ES, ed.), Ottawa, Canada, Agriculture Canada, pp. 249–292.

Cullen JM, Kable PF & Catt M (1973) Epidemic spread of a rust imported for biological control. *Nature* **244**: 462–464.

Cullen JM, McFadyen REC & Julien MH (2011) One hundred years of biological control of weeds in Australia. *Proceedings of the XIII International Symposium of Weed Biological Control* (Wu Y, Johnson TD, Singh S, Raghu S, Wheeler G, Pratt P, Warner K, Center T, Goolsby J & Reardon R, eds.), Washington, DC, U.S. Forest Service, pp. 360–367.

Cullen J & Sheppard AW (2012) *Carduus nutans* L. – nodding thistle. In *Biological Control of Weeds in Australia* (Julien M, McFadyen R & Cullen J, eds.), Canberra, Australia, CSIRO, pp. 118–130.

Culliney TW (2005) Benefits of classical biological control for managing invasive plants. *Critical Reviews in Plant Sciences* **24**: 131–150.

Culver JJ (1919) A study of *Compsilura concinnata*, an imported tachinid parasite of the gipsy moth and the brown-tail moth. *U.S. Department of Agriculture Bulletin* **766**: 1–27.

Cunniffe NJ & Gilligan CA (2011) A theoretical framework for biological control of soil-borne plant pathogens: identifying effective strategies. *Journal of Theoretical Biology* **278**: 32–43.

Currie CR, Scott JA, Summerbell RC & Malloch D (1999) Fungus-growing ants use antibiotic-producing bacteria to control garden parasites. *Nature* **398**: 701–704.

Currie GA & Fyfe RV (1938) The fate of certain European insects introduced into Australia for the control of weeds. *Journal of the Council for Scientific and Industrial Research* **11**: 289–301.

Curtis CF (2000) The case for deemphasizing genomics in malaria control. *Science* **230**: 1508.

Cutting KJ & Hough-Goldstein J (2013) Integration of biological control and native seeding to restore invaded plant communities. *Restoration Ecology* **21**: 648–655.

Da Ros N, Ostermayer R, Roques A & Raimbault JP (1993) Insect damage to cones of exotic conifer species introduced in arboreta I. Interspecific variations within the genus *Picea*. *Journal of Applied Entomology* **115**: 113–133.

Daehler CC & Strong DR (1997) Reduced herbivore resistance in an introduced smooth cordgrass (*Spartina alterniflora*) after a century of herbivore-free growth. *Oecologia* **110**: 99–108.

Dambroski HR, Lin JC, Berlocher SH, Forbes AA, Roelofs W & Feder JL (2005) The genetic basis for fruit odor discrimination in *Rhagoletis* flies and its significance for sympatric host shifts. *Evolution* **59**: 1953–1964.

D'Antonio CM (1993) Mechanisms controlling invasion of coastal plant communities by the alien succulent *Carpobrotus edulis*. *Ecology* **74**: 83–95.

D'Antonio CM & Dudley TL (1995) Biological invasions as agents of change on islands versus mainlands. In *Islands: Biological Diversity and Ecosystem Function* (Vitousek PM, Loope LL & Adsersen H, eds.), Berlin, Germany, Springer, pp. 103–122.

D'Antonio CM, Dudley TL & Mack M (1999) Disturbance and biological invasions: direct effects and feedbacks. In *Ecosystems of the World: Ecosystems of Disturbed Ground* (Walker LR, ed.), Amsterdam, The Netherlands, Elsevier, pp. 413–452.

Danyk TP & Mackauer M (1996) An extraserosal envelope in eggs of *Praon pequodorum* (Hymenoptera: Aphidiidae), a parasitoid of pea aphid. *Biological Control* **7**: 67–70.

Darwin C (1859) *On the Origin of Species*. Middlesex, UK, Penguin Books.

Dauer JT, McEvoy PB & Van Sickle J (2012) Controlling a plant invader by targeted disruption of its life cycle. *Journal of Applied Ecology* **49**: 322–330.

Daugherty MP, Harmon JP & Briggs CJ (2007) Trophic supplements to intraguild predation. *Oikos* **11**: 662–677.

David AS, Kaser JM, Morey AC, Roth AM & Andow DA (2013) Release of genetically engineered insects: a framework to identify potential ecological effects. *Ecology and Evolution* 3: 4000–4015.

Davies AP, Takashino K, Watanabe M & Miura K (2009) Parental genetic traits in offspring from inter-specific crosses between introduced and indigenous *Diadegma* Foerster (Hymenoptera: Ichneumonidae): possible effects on conservation genetics. *Applied Entomology and Zoology* **44**: 535–541.

Davis MA (2009) *Invasion Biology*. Oxford, UK, Oxford University Press.

Davis AS, Landis DA, Nuzzo V, Blossey B, Hinz H & Gerber E (2006) Demographic models inform selection of biocontrol agents for garlic mustard (*Alliaria petiolata*). *Ecological Applications* **16**: 2399–2410.

Dawkins R & Krebs JR (1979) Arms races between and within species. *Proceedings of the Royal Society of London, Series B* **205**: 489–511.

Day WH (1970) The survival value of its jumping cocoons to *Bathyplectes anurus*, a parasite of the alfalfa weevil. *Journal of Economic Entomology* **63**: 586–589.

Day WH (1981) Biological control of the alfalfa weevil in the northeastern United States. In *Biological Control in Crop Production* (Papavizas GC, ed.), London, UK, Allanheld, Osmun, & Co, pp. 361–374.

Day WH (2005) Changes in abundance of native and introduced parasites (Hymenoptera: Braconidae), and of the target and non-target plant bug species (Hemiptera: Miridae), during two classical biological control programs in alfalfa. *Biological Control* 33: 368–374.

Day WH, Eaton AT, Romig RF, Tilmon KJ, Mayer M & Dorsey T (2003) *Peristenus digoneutis* (Hymenoptera: Braconidae), a parasite of *Lygus lineolaris* (Hemiptera: Miridae) in northeastern United States alfalfa, and the need for research on other crops. *Entomological News* **114**: 105–111.

Day WH, Prokrym DR, Ellis DR & Chianese RJ (1994) The known distribution of the predator *Propylea quattuordecimpunctata* (Coleoptera, Coccinellidae) in the United States, and thoughts on the origin of this species and 5 other exotic lady beetles in eastern North America. *Entomological News* **105**: 244–256.

DeAngelis DL, Goldstein RA & O'Neill RV (1975) A model for trophic interaction. *Ecology* **56**: 881–892.

de Boer JG & Dicke M (2005) Information use by the predatory mite *Phytoseiulus persimilis* (Acari: Phytoseiidae), a specialised natural enemy of herbivorous spider mites. *Applied Entomology and Zoology* **40**: 1–12.

de Boer JG, Groenen MAM, Pannebakker BA, Beukeboom LW & Kraus RHS (2015) Population-level consequences of complementary sex determination in a solitary parasitoid. *BMC Evolutionary Biology* **15**: 98.

de Boer JG, Kuijper B, Heimpel GE & Beukeboom LW (2012) Sex determination meltdown upon biological control introduction of the parasitoid *Cotesia rubecula? Evolutionary Applications* 5: 444–454.

de Boer JG, Ode PJ, Rendahl AK, Vet LEM, Whitfield JB & Heimpel GE (2008) Experimental support for *multiple-locus* complementary sex determination in the parasitoid *Cotesia vestalis. Genetics* **180**: 1525–1535.

de Boer, Ode PJ, Vet LEM, Whitfield J & Heimpel GE (2007) Diploid males sire triploid daughters and sons in the parasitoid wasp *Cotesia vestalis. Heredity* **99**: 288–294.

de Briano AEE, Acciaresi HA & Briano JA (2013) Establishment, dispersal, and prevalence of *Rhinocyllus conicus* (Coleoptera: Curculionidae), a biological control agent of thistles, *Carduus* species (Asteraceae), in Argentina, with experimental information on its damage. *Biological Control* **67**: 186–193.

De Clercq P, Mason PG & Babendreier D (2011) Benefits and risks of exotic biological control agents. *BioControl* **56**: 681–698.

de Jong MD, Aylor DE & Bourdot GW (1999) A methodology for risk analysis of plurivorous fungi in biological weed control: *Scleroinia sclerotiorum* as a model. *BioControl* **43**: 397–419.

de Jong MD, Bourdot GW, Powell J & Gourdiaan J (2002) A model of the escape of *Sclerotinia sclerotiorum* ascospores from pasture. *Ecological Modelling* **150**: 83–105.

De La Fuente L, Mavrodi DV, Landa BB, Thomashow LS & Weller DM (2006) *phlD*-based genetic diversity and detection of genotypes of 2,4-diacetylphloroglucinol-producing *Pseudomonas fluorescens*. *FEMS Microbiology Ecology* **56**: 64–78.

De la Mora A, Livingston G & Philpott SM (2008) Arboreal ant abundance and leaf miner damage in coffee agroecosystems in Mexico. *Biotropica* **40**: 742–746.

De Nardo EAB, Grewal PS, McCartney D & Stinner BR (2006) Non-target effects of entomopathogenic nematodes on soil microbial community and nutrient cycling processes: a microcosm study. *Applied Soil Ecology* **34**: 250–257.

De Nardo EAB & Hopper KR (2004) Using the literature to evaluate parasitoid host ranges: a case study of *Macrocentrus grandii* (Hymenoptera: Braconidae) introduced into North America to control *Ostrinia nubilalis* (Lepidoptera: Crambidae). *Biological Control* **31**: 280–295.

de Oliveira ACS, Martins SGF & Zacarias MS (2013) An individual-based model for the interaction of the mite *Tetranychus urticae* (Koch, 1836) with its predator *Neoseiulus californicus* (McGregor, 1954) (Acari: Tetranychidae, Phytoseiidae). *Ecological Modelling* **255**: 11–20.

de Roos AM, McCauley E & Wilson WG (1991) Mobility versus density-limited predator–prey dynamics on different spatial scales. *Proceedings of the Royal Society of London B* **246**: 117–122.

de Roos AM, McCauley E & Wilson WG (1998) Pattern formation and the spatial scale of interaction between predators and their prey. *Theoretical Population Biology* **53**: 108–130.

DeBach P (1946) An insecticidal check method for measuring the efficacy of entomophagous insects. *Journal of Economic Entomology* **39**: 695–697.

DeBach P (1955) Validity of the insecticidal check method as a measure of the effectiveness of natural enemies of diaspine scale insects. *Journal of Economic Entomology* **48**: 584–588.

DeBach P (1964) *Biological Control of Insect Pests and Weeds*. New York, Reinhold.

DeBach P (1965) Some biological and ecological phenomena associated with colonizing enomophagous insects. In *The Genetics of Colonizing Species* (Baker HG & Stebbins GL, eds.), New York, Academic Press, pp. 287–303.

DeBach P (1971) The use of imported natural enemies in insect pest management ecology. *Proceedings of the Tall Timbers Conference on Ecological Animal Control by Habitat Management* **3**: 211–233.

DeBach P (1972) The use of imported natural enemies in insect pest management ecology. *Proceedings of the Tall Timbers Conference on Ecological Animal Control by Habitat Management* **3**: 211–233.

DeBach P (1974) *Biological Control by Natural Enemies*. Cambridge, UK, Cambridge University Press.

DeBach P & Hagen KS (1964) Manipulation of entomophagous species. In *Biological Control of Insect Pests and Weeds* (DeBach P, ed.), New York, Reinhold, pp. 429–458.

DeBach P & Landi J (1961) The introduced purple scale parasite, *Aphytis lepidosaphes* Compere, and a method of integrating chemical with biological control. *Hilgardia* **31**: 459–496.

DeBach P & Rosen D (1991) *Biological Control by Natural Enemies*. Second edition. Cambridge, UK, Cambridge University Press.

DeBach P & Sundby RA (1963) Competitive displacement between ecological homologues. *Hilgardia* **34**: 105–166.

Decou CG (1994) Biological control of the two-spotted spider mite (Acarina: Tetranychidae) on commercial strawberries in Florida with *Phytoseiulus persimilis* (Acarina: Phytoseiidae). *Florida Entomologist* **77**: 33–41.

Delfosse ES (1985) *Echium plantagineum* in Australia: effects of a major conflict of interest. In *Proceedings of the VI International Symposium on Biological Control* (Delfosse ES, ed.), Ottawa, Canada, Agriculture Canada, pp. 293–299.

Delfosse ES (2005) Risk and ethics in biological control. *Biological Control* **35**: 319–329.

Dempster JP (1983) The natural control of populations of butterflies and moths. *Biological Reviews* **58**: 461–481.

Dennis P & Fry GLA (1992) Field margins – can they enhance natural enemy population densities and general arthropod diversity on farmland? *Agriculture Ecosystems & Environment* **40**: 95–115.

Dennis P, Thomas MB & Sotherton NW (1994) Structural features of field boundaries which influence the overwintering densities of beneficial arthropod predators. *Journal of Applied Ecology* **31**: 361–370.

Denno RF & Finke DL (2006) Multiple predator interactions and foodweb connectance implications for biological control. In *Trophic and Guild Interactions in Biological Control* (Brodeur J & Boivin G, eds.), Dordrecht, The Netherlands, Springer, pp. 21–44.

Denoth M, Frid L & Myers JH (2002) Multiple agents in biological control: improving the odds? *Biological Control* **24**: 20–30.

Denoth M & Myers JH (2005) Variable success of biological control of *Lythrum salicaria* in British Columbia. *Biological Control* **32**: 269–279.

Denslow JS (2003) Weeds in paradise: thoughts on the invasibility of tropical islands. *Annals of the Missouri Botanical Garden* **90**: 119–127.

Desneux N, Blahnik R, Delebecque CJ & Heimpel GE (2012) Host phylogeny and specialisation in parasitoids. *Ecology Letters* **15**: 453–460.

Desneux N, Decourtye A & Delpuech JM (2007) The sublethal effects of pesticides on beneficial arthropods. *Annual Review of Entomology* **52**: 81–106.

DeWalt SJ, Denslow JS & Ickes K (2004) Natural-enemy release facilitates habitat expansion of the invasive tropical shrub, *Clidemia hirta*. *Ecology* **85**: 471–483.

Dhileepan K (2003) Evaluating the effectiveness of weed biocontrol at the local scale. In *Improving the Selection, Testing and Evaluation of Weed Biological Control Agents* (Jacob HS & Briese DT, eds.), Osmond, Australia, CRC, pp. 51–60.

Dhileepan K, Taylor DBJ, McCarthy J, King A & Shabbir A (2013) Development of cat's claw creeper leaf-tying moth *Hypocosmia pyrochroma* (Lepidoptera: Pyralidae) at different temperatures: Implications for establishment as a biological control agent in Australia and South Africa. *Biological Control* **67**: 194–202.

Dhileepan K, Trevino M, Donnelly GP & Raghu S (2005) Risk to non-target plants from *Charidotis auroguttata* (Chrysomelidae: Coleoptera), a potential biocontrol agent for cat's claw creeper *Macfadyena unguis-cati* (Bignoniaceae) in Australia. *Biological Control* **32**: 450–460.

Di Giallonardo F & Holmes EC (2015a) Exploring host-pathogen interactions through biological control. *PLoS Pathogens* **11**: e1004865.

Di Giallonardo F & Holmes EC (2015b) Viral biocontrol: grand experiments in disease emergence and evolution. *Trends in Microbiology* **23**: 83–90.

Di Giusto B, Anstett MC, Dounias E & McKey DB (2001) Variation in the effectiveness of biotic defence: the case of an opportunistic ant-plant protection mutualism. *Oecologia* **129**: 367–375.

Diamond JM (1997) *Guns, Germs and Steel: The Fates of Human Societies*. New York, Norton & Co.

Didham RK, Tylianakis JM, Hutchison MA, Ewers RM & Gemmell NJ (2005) Are invasive species the drivers of ecological change? *Trends in Ecology and Evolution* **20**: 470–474.

Dieckhoff C (2011) Host acceptance behavior in the soybean aphid parasitoid *Binodoxys communis* (Hymenoptera: Braconidae) – the role of physiological state in biological control. Thesis, University of Minnesota, Minneapolis, U.S.A.

Dieckhoff C, Theobald JC, Wäckers FL & Heimpel GE (2014) Egg load dynamics and the risk of egg and time limitation experienced by an aphid parasitoid in the field. *Ecology and Evolution* **4**: 1739–1750.

Diehl JK, Holliday NJ, Lindgren CJ & Roughley RE (1997) Insects associated with purple loosestrife, *Lythrum salicaria* L., in southern Manitoba. *Canadian Entomologist* **129**: 937–948.

Dijkerman HJ (1990) Suitability of 8 *Yponomeuta* species as hosts of *Diadegma armillata*. *Entomologia Experimentalis et Applicata* **54**: 173–180.

Dixon AFG (2000) *Insect Predator–Prey Dynamics: Ladybird Beetles & Biological Control*. Cambridge, UK, Cambridge University Press.

Djonovic S, Vittone G, Mendoza-Herrera A & Kenerley CM (2007) Enhanced biocontrol activity of *Trichoderma virens* transformants constitutively coexpressing beta-1,3- and beta-1,6-glucanase genes. *Molecular Plant Pathology* **8**: 469–480.

Dobson AP (1988) Restoring island ecosystems: the potential of parasites to control introduced mammals. *Conservation Biology* **2**: 31–39.

Dobson SL (2003) Reversing *Wolbachia*-based population replacement. *Trends in Parasitology* **19**: 128–133.

Dobson SL, Fox CW & Jiggins FM (2002) The effect of *Wolbachia*-induced cytoplasmic incompatibility on host population size in natural and manipulated systems. *Proceedings of the Royal Society of London, B* **269**: 437–445.

Dodd AP (1940) *The Biological Campaign Against the Prickly-Pear.* Brisbane, Australia, Commonwealth Prickly Pear Board.

Domenech J, Reddy MS, Kloepper JW, Ramos B & Gutierrez-Manero J (2006) Combined application of the biological product LS213 with *Bacillus, Pseudomonas* or *Chryseobacterium* for growth promotion and biological control of soil-borne diseases in pepper and tomato. *BioControl* **51**: 245–258.

Dormann CF, Schymanski SJ, Cabral J, Chuine I, Graham C, Hartig F, Kearney M, Morin X, Römermann C & Singer A (2012) Correlation and process in species distribution models: bridging a dichotomy. *Journal of Biogeography* **39**: 2119–2131.

Doutt RL (1964) The historical development of biological control. In *Biological Control of Insect Pests and Weeds* (DeBach P, ed.), New York, Halsted, pp. 21–44.

Doutt RL & Nakata J (1973) The rubus leafhopper and its egg parasitoid: an endemic biotic system useful in grape-pest management. *Environmental Entomology* **2**: 381–386.

Dowden PB (1952) The importance of coordinating applied control and natural control of forest insects. *Journal of Economic Entomology* **45**: 481–483.

Driver F, Milner RJ & Trueman JWH (2000) A taxonomic revision of *Metarhizium* based on a phylogenetic analysis of rDNA sequence data. *Mycological Research* **104**: 134–150.

Drukker B, Janssen A, Ravensberg, W & Sabelis MW (1997) Improved control capacity of the mite predator *Phytoseiulus persimilis* (Acari: Phytoseiidae) on tomato. *Experimental and Applied Acarology* **21**: 507–518.

Drukker B, Yaninek JS & Herren HR (1993) A packaging and delivery system for aerial release of Phytoseiidae for biological control. *Experimental and Applied Acarology* **17**: 129–143.

Duan JJ, Bauer LS, Abell KJ, Ulyshen MD & Van Driesche RG (2015) Population dynamics of an invasive forest insect and associated natural enemies in the aftermath of invasion: implications for biological control. *Journal of Applied Ecology* **52**: 1246–1254.

Duan JJ, Lundgren JG, Naranjo S & Marvier M (2010) Extrapolating non-target risk of Bt crops from laboratory to field. *Biology Letters* **6**: 74–77.

Duan JJ & Messing RH (2000) Evaluating nontarget effects of classical biological control: fruit fly parasitoids in Hawaii as a case study. In *Nontarget Effects of Biological Control* (Follett PA & Duan JJ, eds.), Dordrecht, The Netherlands, Kluwer, pp. 95–110.

Dubuffet A, Dupas S, Frey F, Drezen J-M, Poirié M & Carton Y (2007) Genetic interactions between the parasitoid wasp *Leptopilina boulardi* and its *Drosophila* host. *Heredity* **98**: 21–27.

Dudley TL & DeLoach CJ (2004) Saltcedar (*Tamarix* spp.), endangered species, and biological weed control – Can they mix? *Weed Technology* **18**: 1542–1551.

Dudley TL & Kazmer DJ (2005) Field assessment of the risk posed by *Diorhabda elongata*, a biocontrol agent for control of saltcedar (*Tamarix* spp.), to a nontarget plant, *Frankenia salina. Biological Control* **35**: 265–275.

Duffy BK, Simon A & Weller DM (1996) Combination of *Trichoderma koningii* with fluorescent pseudomonads for control of take-all on wheat. *Phytopathology* **86**: 188–194.

Dugaw CJ, Hastings A, Preisser EL & Strong DR (2004) Parasitoid-mediated effects on apparent competition and the persistence of host–parasitoid assemblages. *Bulletin of Mathematical Biology* **66**: 583–594.

Dunn RR (2005) Modern insect extinctions, the neglected majority. *Conservation Biology* **19**: 1030–1036.

Dupas S & Carton Y (1999) Two non-linked genes for specific virulence of *Leptopilina boulardi* against *Drosophila melanogaster* and *D. yakuba*. *Evolutionary Ecology* **13**: 211–220.

Durrant WE & Dong X (2004) Systemic acquired resistance. *Annual Review of Phytopathology* **42**: 185–209.

Dushoff J & Dwyer G (2001) Evaluating the risks of engineered viruses: modeling pathogen competition. *Ecological Applications* **11**: 1602–1609.

Dutcher JD (2007) A review of resurgence and replacement causing pest outbreaks in IPM. In *General Concepts in Integrated Pest and Disease Management* (Ciancio A & Mukerji KG, eds.), Dordrecht, The Netherlands, Springer, pp. 27–43.

Dwyer G, Levin SA & Buttel L (1990) A simulation model of the population dynamics and evolution of myxomatosis. *Ecological Monographs* **60**: 423–447.

Eckberg JO, Tenhumberg B & Louda SM (2012) Insect herbivory and propagule pressure influence *Cirsium vulgare* invasiveness across the landscape. *Ecology* **93**: 1787–1794.

Eckberg JO, Tenhumberg B & Louda SM (2014) Native insect herbivory limits population growth rate of a non-native thistle. *Oecologia* **175**: 129–138.

Ehler LE (1990) Environmental impact of introduced biological-control agents: implications for agricultural biotechnology. In *Risk Assessment in Agricultural Biotechnology* (Marois JJ & Bruening G, eds.), Wallingford, UK, CABI Publishing, pp. 85–96.

Ehler LE (1998) Invasion biology and biological control. *Biological Control* **13**: 127–133.

Ehler LE (2000) Critical issues related to nontarget effects in classical biological control of insects. In *Nontarget Effects of Biological Control* (Follett PA & Duan JJ, eds.), Dordrecht, The Netherlands, Kluwer, pp. 3–13.

Ehler LE (2007) Impact of native predators and parasites on *Spodoptera exigua*, an introduced pest of alfalfa hay in northern California. *Biocontrol* **52**: 323–338.

Ehler LE, Sforza R & Mateille T (2004) *Genetics, Evolution and Biological Control*. Wallingford, UK, CABI Publishing.

Ehlers R-U (2003) Biocontrol nematodes. In *Environmental Impacts of Microbial Insecticides: Need and Methods for Risk Assessment* (Hokkannen HMT & Hajek AE, eds.), Dordrecht, The Netherlands, Kluwer, pp. 177–220.

Eilenberg J, Hajek A & Lomer C (2001) Suggestions for unifying the terminology in biological control. *BioControl* **46**: 387–400.

Elias J, Mazzi D & Dorn S (2009) No need to discriminate? Reproductive diploid males in a parasitoid with complementary sex determination. *PLoS One* **4**: e6024.

Elkinton JS & Boettner GH (2012) Benefits and harm caused by the introduced generalist tachinid, *Compsilura concinnata*, in North America. *Biocontrol* **57**: 277–288.

Elkinton JS, Buonaccorsi JP, Bellows TS & Van Driesche RG (1992) Marginal attack rate, k-values and density dependence in the analysis of contemporaneous mortality factors. *Researches on Population Ecology* **34**: 29–44.

Elkinton JS, Liebhold AM & Muzika RM (2004) Effects of alternative prey on predation by small mammals on gypsy moth pupae. *Population Ecology* **46**: 171–178.

Elkinton JS, Parry D & Boettner GH (2006) Implicating an introduced generalist parasitoid in the invasive browntail moth's enigmatic demise. *Ecology* **87**: 2664–2672.

Elliot NC, Kieckhefer R & Kauffman W (1996) Effects of an invading coccinellid on native coccinellids in an agricultural landscape. *Oecologia* **105**: 537–544.

Ellison CA, Evans HC & Ineson J (2004) The significance of intraspecific pathogenicity in the selection of a rust pathotype for the classical biological control of *Mikania micrantha* (mile-a-minute weed) in Southeast Asia. In *Proceedings of the XI International Symposium on Biological Control of Weeds* (Cullen JM, Briese TD, Kriticos DJ Lonsdale WM, Morin L & Scott JK, eds.), Canberra, Australia, CSIRO Entomology, pp. 102–107.

Ellner SP, McCauley E, Kendall BE, Briggs CJ, Hosseini PR, Wood SN, Janssen A, Sabelis MW, Turchin P, Nisbet RM & Murdoch WW (2001) Habitat structure and population persistence in an experimental community. *Nature* **412**: 538–543.

El-Tarabily KA, Hardy G, Sivasithamparam K, Hussein AM & Kurtboke DI (1997) The potential for the biological control of cavity-spot disease of carrots, caused by *Pythium coloratum*, by streptomycete and non-streptomycete actinomycetes. *New Phytologist* **137**: 495–507.

Elton CS (1958) *The Ecology of Invasions by Animals and Plants.* Chicago, IL, Chicago University Press.

Emge RG, Melching JS & Kingsolver CH (1981) Epidemiology of *Puccinia chondrillina*, a rust pathogen for the biological control of rush skeletonweed in the United States. *Phytopathology* **71**: 839–843.

English-Loeb GM, Rhainds M, Martinson T & Ugine T (2003) Influence of flowering cover crops on *Anagrus* parasitoids (Hymenoptera: Mymaridae) and *Erythroneura* leafhoppers (Homoptera: Cicadellidae) in New York vineyards. *Agricultural and Forest Entomology* **5**: 173–181.

Epstein AH & Hill JH (1999) Status of rose rosette disease as a biological control for multiflora rose. *Plant Disease* **83**: 92–101.

Errakhi R, Bouteau F, Lebrihi A & Barakate M (2007) Evidences of biological control capacities of *Streptomyces* spp. against *Sclerotium rolfsii* responsible for damping-off disease in sugar beet (*Beta vulgaris* L.). *World Journal of Microbiology & Biotechnology* **23**: 1503–1509.

Espiau C, Riviere D, Burdon JJ Gartner S, Daclinat B, Hasan S & Chaboudez P (1998) Host-pathogen diversity in a wild system: *Chondrilla juncea-Puccinea chondrillina. Oecologia* **113**: 133–139.

Evans EW (1991) Intra versus interspecific interactions of ladybeetles (Coleoptera: Coccinellidae) attacking aphids. *Oecologia* **87**: 401–408.

Evans EW (2000) Morphology of invasion: body size patterns associated with establishment of *Coccinella septempunctata* (Coleoptera: Coccinellidae) in western North America. *European Journal of Entomology* **97**: 469–474.

Evans EW (2004) Habitat displacement of North American ladybirds by an introduced species. *Ecology* **85**: 637–647.

Evans EW (2008) Multitrophic interactions among plants, aphids, alternate prey and shared natural enemies – a review. *European Journal of Entomology* **105**: 369–380.

Evans EW, Anderson MR & Bowling PD (2010) Targeted sugar provision promotes parasitism of the cereal leaf beetle *Oulema melanopus. Agricultural and Forest Entomology* **12**: 41–47.

Evans EW, Carlile NR, Innes MB & Pitigala N (2013) Warm springs reduce parasitism of the cereal leaf beetle through phenological mismatch. *Journal of Applied Entomology* **137**: 383–391.

Evans EW, Karren JB & Israelsen CE (2006) Interactions over time between cereal leaf beetle (Coleoptera: Chrysomelidae) and larval parasitoid *Tetrastichus julis* (Hymenoptera: Eulophidae) in Utah. *Journal of Economic Entomology* **99**: 1967–1973.

Evans JA, Davis AS, Raghu S, Ragavendran A, Landis DA & Schemske DW (2012) The importance of space, time, and stochasticity to the demography and management of *Alliaria petiolata. Ecological Applications* **22**: 1497–1511.

Evans KJ & Gomez DR (2004) Genetic markers in rust fungi and their application to weed biocontrol. In *Genetics, Evolution and Biological Control* (Ehler LE, Sforza R & Mateille T, eds.), Wallingford, UK, CABI Publishing, pp. 73–96.

Evenhuis NL (2002) *Hawaii's Extinct Species – Insects.* Website: http://hbs.bishopmuseum.org/endangered/ext-insects.html.

Ewald JA, Aebischer NJ, Richardson SM, Grice PV & Cooke AI (2010) The effect of agri-environment schemes on grey partridges at the farm level in England. *Agriculture, Ecosystems and Environment* **138**: 55–63.

Facon B, Crespin L, Loiseau A, Lombaert E, Magro A & Estoup A (2010) Can things get worse when an invasive species hybridizes? The harlequin ladybird *Harmonia axyridis* in France as a case study. *Evolutionary Applications* **4**: 71–88.

Facon B, Hufbauer RA, Tayeh A, Loiseau A, Lombaert E, Vitalis R, Guillemaud T, Lundgren JG & Estoup A (2011) Inbreeding depression is purged in the invasive insect *Harmonia axyridis. Current Biology* **21**: 424–427.

Fagan WF & Bishop JG (2000) Trophic interactions during primary succession: herbivores slow a plant reinvasion at Mount St. Helens. *American Naturalist* **155**: 238–251.

Fagan WF, Lewis MA, Neubert MG & Van den Driesche P (2002) Invasion theory and biological control. *Ecology Letters* **5**: 148–157.

Fagan WF, Lewis M, Neubert MG, Aumann C, Apple JL & Bishop JG (2005) When can herbivores slow or reverse the spread of an invading plant? A test case from Mount St. Helens. *American Naturalist* **166**: 669–685.

Fan Y, Fang W, Guo S, Pei X, Zhang Y, Xiao Y, Li D, Jin K, Bidochka MJ & Pei Y (2007) Increased insect virulence in *Beauveria bassiana* strains overexpressing an engineered chitinase. *Applied and Environmental Microbiology* **73**: 295–302.

Farrar RR, Shapiro M & Javaid I (2003) Photostabilized titanium dioxide and a fluorescent brightener as adjuvants for a nucleopolyhedrovirus. *BioControl* **48**: 543–560.

Fauvergue X & Hopper KR (2009) French wasps in the New World: experimental biological control introductions reveal a demographic Allee effect. *Population Ecology* **51**: 385–397.

Fauvergue X, Malausa J-C, Giuge L & Courchamp F (2007) Invading parasitoids suffer no Allee effect: a manipulative field experiment. *Ecology* **88**: 2392–2403.

Fauvergue X, Vercken E, Malausa T & Hufbauer RA (2012) The biology of small, introduced populations, with special reference to biological control. *Evolutionary Applications* **5**: 424–443.

Federici BA (2005) Insecticidal bacteria: an overwhelming success for invertebrate pathology. *Journal of Invertebrate Pathology* **89**:30–38.

Federici BA & Maddox JV (1996) Host specificity in microbe-insect interactions. *BioScience* **46**: 410–421.

Felker-Quinn E, Schweitzer JA & Bailey JK (2013) Meta-analysis reveals evolution in invasive plant species but little support for evolution of increased competitive ability (EICA). *Ecology and Evolution* **3**: 739–751.

Fellowes MDE & Godfray HCJ (2000) The evolutionary ecology of resistance to parasitoids in *Drosophila*. *Heredity* **84**: 1–8.

Fenner F (1953) Changes in the mortality rate due to myxomatosis in the Australian wild rabbit. *Nature* **172**: 228–230.

Fenner F & Ratcliffe FN (1965) *Myxomatosis*. Cambridge, UK, Cambridge University Press.

Fenner F & Fantini B (1999) *Biological Control of Vertebrate Pests: The History of Myxomatosis – An Experiment in Evolution*. Wallingford, UK, CABI Publishing.

Fernanda Diaz M, Ramirez A & Poveda K (2012) Efficiency of different egg parasitoids and increased floral diversity for the biological control of noctuid pests. *Biological Control* **60**: 182–191.

Fernandes EKK, Rangel DEN, Braga GUL & Roberts DW (2015) Tolerance of entomopathogenic fungi to ultraviolet radiation: a review on screening of strains and their formulation. *Current Genetics* **61**: 427–440.

Ferrari J, Darby AC, Daniel TJ, Godfray HCJ & Douglas AE (2004) Linking the bacterial community in pea aphids with host-plant use and natural enemy resistance. *Ecological Entomology* **29**: 60–65.

Ferrari J, Müller CB, Kraaijeveld AR & Godfray HCJ (2001) Clonal variation and covariation in aphid resistance to parasitoids and a pathogen. *Evolution* **55**: 1805–1814.

Ferrero-Serrano A, Collier TR, Hild AL, Mealor BA & Smith T (2008) Combined impacts of native grass competition and introduced weevil herbivory on Canada thistle (*Cirsium arvense*). *Rangeland Ecology & Management* **61**: 529–534.

Fine PVA (2002) The invasibility of tropical forests by exotic plants. *Journal of Tropical Ecology* **18**: 687–705.

Finlayson CJ, Landry KN & Alyokhin AV (2008) Abundance of native and non-native lady beetles (Coleoptera: Coccinellidae) in different habitats in Maine. *Annals of the Entomological Society of America* **101**: 1078–1087.

Fisher AJ, Smith L & Woods DM (2011) Climatic analysis to determine where to collect and release *Puccinia jaceae* var. *solstitialis* for biological control of yellow starthistle. *Biocontrol Science and Technology* **21**: 333–351.

Flanagan GJ, Hills A & Wilson CG (2000) The successful biological control of spineyhead sida, *Sida acuta* (Malvaceae), by *Calligrapha pantherina* (Col.: Chrysomelidae) in Australia's Northern Territory. In *Proceedings of the 10th International Symposium on Biological Control of Weeds* (Spencer NR, ed.), Bozeman, Montana State University, pp. 35–41.

Flanders SE (1930) Mass production of egg parasites of the genus *Trichogramma*. *Hilgardia* **4**: 465–501.

Flanders SE (1951) Mass culture of California red scale and its golden chalcid parasites. *Hilgardia* **21**: 1–42.

Fleischer SJ, Blom PE & Weisz R (1999) Sampling in precision IPM: When the objective is a map. *Phytopathology* **89**: 1112–1118.

Fletcher JP, Hughes JP & Harvey I (1994) Life expectancy and egg load affects oviposition decisions of a solitary parasitoid. *Proceedings of the Royal Society of London, B* **258**: 63–167.

Flexner JL, Lighthart B & Croft BA (1986) The effects of microbial pesticides on non-target beneficial arthropods. *Agriculture, Ecosystems and Environment* **16**: 203–254.

Flinn PW & Hagstrum D (1995) Simulation model of *Cephalonomyia waterstoni* (Hymenoptera: Bethylidae) parasitizing the rusty grain beetle (Coleoptera: Cucujidae). *Environmental Entomology* **24**: 1608–1615.

Flint ML & Dreistadt SH (1998) *Natural Enemies Handbook*. Berkeley, University of California Press.

Flower CE, Long LC, Knight KS, Rebbeck J & Brown JS (2014) Native bark-foraging birds preferentially forage in infected ash (*Fraxinus* spp.) and prove effective predators of the invasive emerald ash borer (*Agrilus planipennis* Fairmaire). *Forest Ecology and Management* **313**: 300–306.

Follett PA & Duan JJ (2000) *Nontarget Effects of Biological Control*. Dordrecht, The Netherlands, Kluwer.

Follett PA, Johnson MT & Jones VP (2000) Parasitoid drift in Hawaiian pentatomids. In *Nontarget Effects of Biological Control* (Follett PA & Duan JJ, eds.), Dordrecht, The Netherlands, Kluwer, pp. 77–94.

Forbes VE, Calow P, Grimm V, Hayashi TI, Jager T, Katholm A, Palmqvist A, Pastorok R, Salvito D, Sibly R, Spromberg J, Stark J, Stillman RA (2011) Adding value to ecological risk assessment with population modeling. *Human and Ecological Risk Assessment* **17**: 287–299.

Force DC (1967) Genetics in the colonization of natural enemies for biological control. *Annals of the Entomological Society of America* **60**: 722–728.

Fowler SV, Syrett P & Hill RL (2000) Success and safety in the biological control of environmental weeds in New Zealand. *Australian Ecology* **25**: 553–562.

Frank SD (2010) Biological control of arthropod pests using banker plant systems: past progress and future directions. *Biological Control* **52**: 8–16.

Frank SD & Shrewsbury PM (2004) Effect of conservation strips on the abundance and distribution of natural enemies and predation of *Agrotis ipsilon* (Lepidoptera: Noctuidae) on golf course fairways. *Environmental Entomology* **33**: 1662–1672.

Franks SJ, Pratt PD & Tsutsui ND (2011) The genetic consequences of a demographic bottleneck in an introduced biological control insect. *Conservation Genetics* **12**: 201–211.

Fraser SM & Lawton JH (1994) Host range expansion by British moths onto introduced conifers. *Ecological Entomology* **19**: 127–137.

Frazer BD & Van den Bosch R (1973) Biological control of the walnut aphid in California: the interrelationship of the aphid and its parasite. *Environmental Entomology* **2**: 561–568.

Freestone AL, Ruiz GM & Torchin ME (2013) Stronger biotic resistance in tropics relative to temperate zone: effects of predation on marine invasion dynamics. *Ecology* **94**: 1370–1377.

Frere I, Fabry J & Hance T (2007) Apparent competition or apparent mutualism? An analysis of the influence of rose bush strip management on aphid population in wheat field. *Journal of Applied Entomology* **131**: 275–283.

Fridley JD, Stachowicz JJ, Naeem S, Sax DF, Seabloom EW, Smith MD, Stohlgren TJ, Tilman D & Von Holle B (2007) The invasion paradox: reconciling pattern and process in species invasions. *Ecology* **88**: 3–17.

Fritz RS (1983) Ant protection of a host plant's defoliator – consequence of an ant-membracid mutualism. *Ecology* **64**: 789–797.

Fuester R, Kenis M, Swain KS, Kingsley PC, Lopez-Vaamonde C & Herard F (2001) Host range of *Aphantorhaphopsis samarensis* (Diptera: Tachinidae), a larval parasite of the gypsy moth (Lepidoptera: Lymantriidae). *Environmental Entomology* **30**: 605–611.

Fuester R, Swan KS, Kenis M & Herard F (2004) Determining the host range of *Aphantorhaphopsis samarensis*, a specialized tachinid introduced against gypsy moth. In *Assessing Host Ranges of Parasitoids and Predators Used in Biological Control: A Guide to Best Practice* (Van Driesche RG & Reardon R, eds.), Morgantown, WV, USDA Forest Service, pp. 177–194.

Fuller EB, Elderd D & Dwyer G (2012) Persistence in the environment and insect-baculovirus interactions: disease-density thresholds, epidemic burnout, and insect outbreaks. *American Naturalist* **179**: E70–E96.

Funasaki GY, Lai P-Y, Nakahara LM, Beardsley JW & Ota AK (1988) A review of biological control introductions in Hawaii: 1890–1985. *Proceedings of the Hawaiian Entomological Society* **28**: 105–160.

Futuyma DJ & Moreno G (1988) The evolution of ecological specialization. *Annual Review of Ecology and Systematics* **19**: 207–233.

Fuxa JR (1998) Environmental manipulation for microbial control of insects. In *Conservation Biological Control* (Barbosa P, ed.), San Diego, CA, Academic Press, pp. 255–268.

Gagic V, Tscharnkte T, Dormann CF, Gruber B, Wilstermann A & Thies C (2011) Food web structure and biocontrol in a four-trophic level system across a landscape complexity gradient. *Proceedings of the Royal Society B* **278**: 2946–2953.

Gagne WC & Howarth FG (1985) Conservation status of endemic Hawaiian Lepidoptera. In *Proceedings of 3rd Congress of European Lepidoptera*, Cambridge, UK, Societas Europaea Lepidopterologica, pp. 74–84.

Gagnon AE, Heimpel GE & Brodeur J (2011) The ubiquity of intraguild predation among predatory arthropods. *PLoS One* **6**: e28061.

Galat DL & Robertson B (1992) Response of endangered *Poeciliopsis occidentalis* Sonoriensis in the Rio-Yaqui Drainage, Arizona, to introduced *Gambusia affinis*. *Environmental Biology of Fishes* **33**: 249–264.

Gamradt SC & Kats LB (1996) Effect of introduced crayfish and mosquitofish on California newts. *Conservation Biology* **10**: 1155–1162.

Gan-Mor S & Matthews GA (2003) Recent developments in sprayers for application of biopesticides – an overview. *Biosystems Engineering* **84**: 119–125.

Gao L, Sun MH, Liu XZ & Che YS (2007) Effects of carbon concentration and carbon to nitrogen ratio on the growth and sporulation of several biocontrol fungi. *Mycological Research* **111**: 87–92.

Garcia R, Caltagirone LE & Gutierrez AP (1988) Comments on a redefinition of biological control. *BioScience* **38**: 692–694.

Garcia R & Legner EF (1999) Biological control of medical and veterinary pests. In *Handbook of Biological Control: Principles and Applications of Biological Control* (Bellows TS & Fisher TW, eds.), San Diego, CA, Academic Press, pp. 935–954.

Garcia-Rossi D, Rank N & Strong DR (2003). Potential for self-defeating biological control? Variation in herbivore vulnerability among invasive *Spartina* genotypes. *Ecological Applications*, **13**: 1640–1649.

Gardiner MM, Landis DA, Gratton C, Schmidt N, O'Neal M, Mueller E, Chacon J, Heimpel GE & DiFonzo CD (2009a) Landscape composition influences patterns of native and exotic lady beetle abundance. *Diversity and Distributions* **15**: 554–564.

Gardiner MM, Landis DA, Gratton C, DiFonzo CD, O'Neal M, Chacon JM, Wayo MT, Schmidt NP, Mueller EE & Heimpel GE (2009b) Landscape diversity enhances biological control of an introduced crop pest in the north-central USA. *Ecological Applications* **19**: 143–154.

Gardner DE, Smith CW & Markin GP (1995) Biological control of alien plants in natural areas of Hawaii. In *Proceedings of the 8th International Symposium on Biological Control of Weeds* (Delfosse ES & Scott RR, eds.), Canberra, Australia, CSIRO, pp. 2–7.

Gardner J & Giles K (1997) Mechanical distribution of *Chrysoperla rufilabris* and *Trichogramma pretiosum*: survival and uniformity of discharge after spray dispersal in an aqueous suspension. *Biological Control* **8**: 138–142.

Garrett SD (1965) Toward biological control of soil-borne pathogens. In *Ecology of Soil-Borne Pathogens* (Baker KF & Snyder WC, eds.), Berkeley, University of California Press, pp. 4–16.

Gaskin JF, Bon MC, Cock JW, Cristofaro M, De Biase A, De Clerck-Floate R, Ellison CA, Hinz HL, Hufbauer RA, Julien MH & Sforza R (2011) Applying molecular-based approaches to classical biological control of weeds. *Biological Control* **58**: 1–21.

Gassmann A & Louda SM (2001) *Rhynocillus conicus*: Initial evaluation and subsequent ecological impacts in North America. In *Evaluating Indirect Ecological Effects of Biological Control* (Wajnberg E, Scott JK & Quimby PC, eds.), Wallingford, UK, CABI Publishing, pp. 147–184.

Gaugler R (1993) Ecological genetics of entomopathogenic nematodes. In *Nematodes and the Biological Control of Insect Pests* (Bedding R, Akhurst R & Kaya H, eds.), East Melbourne, Australia, CSIRO, pp. 89–95.

Gaugler R, Lewis E & Stuart RJ (1997a) Ecology in the service of biological control: the case of entomopathogenic nematodes. *Oecologia* **109**: 483–489.

Gaugler R, Wilson M & Shearer P (1997b) Field release and environmental fate of a transgenic entomopathogenic nematode. *Biological Control* **9**: 75–80.

Geiger F, Wäckers FL & Bianchi FJJA (2009) Hibernation of predatory arthropods in semi-natural habitats. *BioControl* **54**: 529–535.

Gelernter WD & Lomer CJ (2000) Success in biological control of above-ground insects by pathogens. In *Biological Control: Measures of Success* (Gurr G & Wratten S, eds.), Dordrecht, The Netherlands, Kluwer, pp. 297–322.

Geneau CE, Wäckers FL, Luka H, Daniel C & Balmer O (2012) Selective flowers to enhance biological control of cabbage pests by parasitoids. *Basic and Applied Ecology* **13**: 85–93.

Gentz MC, Murdoch G & King GF (2010) Tandem use of selective insecticides and natural enemies for effective, reduced-risk pest management. *Biological Control* **52**: 208–215.

Gerlach J (2001) Predator, prey and pathogen interactions in introduced snail populations. *Animal Conservation* **4**: 203–209.

Getz WM (1984) Population dynamics: a resource per capita approach. *Journal of Theoretical Biology* **108**: 623–644.

Getz WM (1996) A hypothesis regarding the abruptness of density dependence and the growth rate of populations. *Ecology* **77**: 2014–2026.

Getz WM & Mills NJ (1996) Host-parasitoid coexistence and egg-limited encounter rates. *The American Naturalist* **148**: 333–347.

Ghorbani R, Leifert C & Seel W (2005) Biological control of weeds with antagonistic plant pathogens. *Advances in Agronomy* **86**: 191–225.

Giles DK & Wunderlich LR (1998) Electronically-controlled delivery system for beneficial insect eggs in liquid suspensions. *Transactions of the American Society of Agricultural Engineers* **41**: 839–847.

Giles DK, Gardner J & Studer HE (1995) Mechanical release of predacious mites for biological pest control in strawberries. *Transactions of the American Society of Agricultural Engineers* **38**: 1289–1296.

Giles KL, Jones DB, Royer TA, Elliot NC & Kindler SD (2003) Development of a sampling plant in winter wheat that estimates cereal aphid parasitism levels and predicts population suppression. *Journal of Economic Entomology* **96**: 975–982.

Gillespie DR & Raworth DA (2004) Biological control of two-spotted spider mites on greenhouse vegetable crops. In *Biocontrol in Protected Culture* (Heinz KM, Van Driesche RG & Parrella MP, eds.), Batavia, IL, Ball, pp 201–220

Gilligan CA (2002) An epidemiological framework for disease management. *Advances in Botanical Research* **38**: 1–64.

Giorgini M, Monti MM, Caprio E, Stouthamer R & Hunter MS (2009) Feminization and the collapse of haplodiploidy in an asexual parasitoid wasp harboring the bacterial symbiont *Cardinium*. *Heredity* **102**: 365–371.

Glare TJ, Caradus W, Gelernter T, Jackson T, Keyhani N, Köhl J, Marrone P, Morin L & Stewart A (2012) Have biopesticides come of age? *Trends in Biotechnology* **30**: 250–258.

Glaser RW & Farrell CC (1935) Field experiments with the Japanese beetle and its nematode parasite. *Journal of the New York Entomological Society* **43**: 345–371.

Goddard JH, Torchin ME, Kuris AM & Lafferty KD (2005) Host specificity of *Sacculina carcini*, a potential biological control agent of introduced European green crab *Carcinus maenas* in California. *Biological Invasions* **7**: 895–912.

Godfray HCJ (1994) *Parasitoids: Behavioral and Evolutionary Ecology*. Princeton, NJ, Princeton University Press.

Godfray HCJ (1995) Field experiments with genetically manipulated insect viruses – ecological issues. *Trends in Ecology & Evolution* **10**: 465–469.

Godfray HCJ, Briggs CJ, Barlow ND, O'Callaghan MO, Glare TR & Jackson TA (1999) A model of insect-pathogen dynamics in which a pathogenic bacterium can also reproduce saprophytically. *Proceedings of the Royal Society of London B* **266**: 233–240.

Godfray HCJ & Chan MS (1990) How insecticides trigger single-stage outbreaks in tropical pests. *Functional Ecology* **4**: 329–337.

Godfray HCJ & Hassell MP (1987) Natural enemies may be a cause of discrete generations in tropical insects. *Nature* **327**: 144–147.

Godfray HCJ & Hassell MP (1989) Discrete and continuous insect populations in tropical environments. *Journal of Animal Ecology* **58**: 153–174.

Godfray HCJ, Hassell MP & Holt RD (1994) The population dynamic consequences of phenological asynchrony between parasitoids and their hosts. *Journal of Animal Ecology* **63**: 1–10.

Godfray HCJ & Waage JK (1991) Predictive modelling in biological control: the mango mealy bug (*Rastrococcus invadens*) and its parasitoids. *Journal of Applied Ecology* **28**: 434–453.

Godsoe W, Murray R & Plank MJ (2015) Information on biotic interactions improves transferability of distribution models. *American Naturalist* **185**: 281–290.

Goeden RD (1978) Biological control of weeds. In *Introduced Parasites and Predators of Arthropod Pests and Weeds; USDA Handbook No. 480* (Clausen CP, ed.), Washington, DC, USDA Agricultural Research Service, pp. 357–413.

Goeden RD & Kok LT (1986) Comments on a proposed "new" approach for selecting agents for the biological control of weeds. *Canadian Entomologist* **118**: 51–58.

Goeden RD & Louda SM (1976) Biotic interference with insects imported for weed control. *Annual Review of Entomology* **21**: 325–341.

Goettel MS & Hajek A (2001) Evaluation of non-target effects of pathogens used for management of arthopods. In *Evaluating Indirect Ecological Effects of Biological Control* (Wajnberg E, Scott JK & Quimby PC, eds.), Wallingford, UK, CABI Publishing, pp. 81–97.

Goettel MS, Hajek AE, Siegel JP & Evans HC (2001) Safety of fungal biocontrol agents. In *Fungi as Biocontrol Agents* (Butt TM, Jackson CW & Magan N, eds.), Wallingford, UK, CABI Publishing, pp. 347–375.

Goettel MS, Poprawski TJ, Vandenberg JD, Li Z & Roberts DW (1990) Safety to nontarget invertebrates of fungal biocontrol agents. In *Safety of Microbial Insecticides* (Laird M, Lacey LA & Davidson EW, eds.), Boca Raton, FL, CRC Press, pp. 210–231.

Gold CS, Altieri MA & Bellotti AC (1989) The effects of intercropping and mized varieties of predators and parasitoids of cassava whiteflies (Hemiptera: Aleyrodidae) in Colombia. *Bulletin of Entomological Research* **79**: 115–121.

Goldson SL, McNeill MR & Proffitt JR (2003) Negative effects of strain hybridisation on the biocontrol agent *Microctonus aethiopoides. New Zealand Plant Protection*, **56**, 138–142.

Goldson SL, Proffitt JR & McNeil MR (1990) Seasonal biology and ecology in New Zealand of *Microtonus aethiopoides* (Hymenoptera: Braconidae), a parasitoid of *Sitona* spp. (Coleoptera: Curculionidae), with special emphasis on atypical behaviour. *Journal of Applied Ecology* **27**: 703–722.

Goldson SL, Wratten SD, Ferguson CM, Gerard PJ, Barratt BIP, Hardwick S, McNeill MR, Phillips CB, Popay AJ, Tylianakis JM & Tomasetto F (2014) If and when successful classical biological control fails. *Biological Control* **72**: 76–79.

Goldsworthy CA & Bettoli PW (2006) Growth, body condition, reproduction and survival of stocked Barrens topminnows, *Fundulus julisia* (Fundulidae). *American Midland Naturalist* **156**: 331–343.

Gols R, Bukovinszky T, Hemerik L, Harvey JA, van Lenteren JC & Vet LEM (2005) Reduced foraging efficiency of a parasitoid under habitat complexity: implications for population stability and species coexistence. *Journal of Animal Ecology* **74**: 1059–1068.

Gonthier DJ, Ennis KK, Philpott SM, Vandermeer J & Perfecto I (2013) Ants defend coffee from berry borer colonization. *BioControl* **58**: 815–820.

Gonzalez D & Wilson LT (1982) A food-web approach to economic thresholds: a sequence of prests/predaceous arthropods in California cotton. *Entomophaga* 31–43.

Gonzalez D, Hagen KS, Stary P, Bishop GW, Davis DW & Pike KS (1995) Pea aphid and blue alfalfa aphid. In *Biological Control in the Western United States* (Nechols JR, ed.), Oakland, University of California Press, pp. 129–135.

Goodsell JA & Kats LB (1999) Effect of introduced mosquitofish on pacific treefrogs and the role of alternative prey. *Conservation Biology* **13**: 921–924.

Goolsby JA, De Barro PJ, Kirk AA, Sutherst RW, Canas L & Ciomperlik MA (2005). Post-release evaluation of biological control of *Bemisia tabaci* biotype 'B' in the USA and the development of predictive tools to guide introductions for other countries. *Biological Control* **32**: 70–77.

Goolsby JA, De Barro PJ, Makinson JR, Pemberton RW, Hartley DM & Frohlich DR (2006a) Matching the origin of an invasive weed for selection of a herbivore haplotype for a biological control programme. *Molecular Ecology* **15**: 287–297.

Goolsby JA, Van Klinken RD & Palmer WA (2006b) Maximising the contribution of native-range studies towards the identification and prioritisation of weed biocontrol agents. *Australian Journal of Entomology* **45**: 276–286.

Gordon CE, McGill B, Ibarra-Nunez G, Greenberg R & Perfecto I (2009) Simplification of a coffee foliage-dwelling beetle community under low-shade management. *Basic and Applied Ecology* **10**: 246–254.

Gould F (1994) Potential and problems with high-dose strategies for pesticidal engineered crops. *Biocontrol Science and Technology* **4**: 451–461.

Gould F & Schliekelman P (2004) Population genetics of autocidal control and strain replacement. *Annual Review of Entomology* **49**: 193–217.

Gould JR, Bellows TS & Paine TD (1992) Population dynamics of *Siphoninus phillyreae* in California in the presence and absence of a parasitoid *Encarsia partenopea*. *Ecological Entomology* **17**: 127–134.

Gould JR, Elkinton JS & Wallner WE (1990) Density-dependent suppression of experimentally created gypsy moth, *Lymantria dispar* (Lepidoptera, Lymantriidae), populations by natural enemies. *Journal of Animal Ecology* **59**: 213–233.

Graham GL, Peng G, Bailey KL & Holm FA (2006) Effect of dew temperature, post-inoculation condition, and pathogen does on suppression of scentless chamomile by *Colletrichum truncatum*. *Biocontrol Science and Technology* **16**: 271–280.

Granett J, Dunbar DM & Weseloh RM (1976) Gypsy moth control with Dimilin sprays timed to minimize effects on the parasite *Apanteles melanoscelus*. *Journal of Economic Entomology* **69**: 403–404.

Grasman J, Van Herwaarden OA, Hemerik L & Van Lenteren JC (2001) A two-component model of host-parasitoid interactions: determination of the size of inundative releases of parasitoids in biological pest control. *Mathematical Biosciences* **169**: 207–216.

Graur D (1985) Gene diversity in Hymenoptera. *Evolution* **39**: 190–199.

Gray RH, Lorimer CG, Tobin PC & Raffa KF (2008) Preoutbreak dynamics of a recently established invasive herbivore: roles of natural enemies and habitat structure in stage-specific performance of gypsy moth (Lepidoptera: Lymantriidae) populations in northeastern Wisconsin. *Environmental Entomology* **37**: 1174–1184.

Greathead D (1971) *A Review of Biological Control in the Ethiopian Region. Commonwealth Institute of Biological Control, Technical Communication No. 5*. Slough, UK, Commonwealth Institute of Biological Control.

Greathead DJ (1986) Parasitoids in classical biological control. In *Insect Parasitoids* (Waage J & Greathead DJ, eds.), London, UK, Academic Press, pp. 290–318.

Greathead DJ & Greathead AH (1992) Biological control of insects pests by insect parasitoids and predators: the BIOCAT database. *Biocontrol News and Information* **13**: 61 N–67 N.

Greaves MP, Pring RJ & Lawrie J (2001) A proposed mode of action of oil-based formulations of a microbial herbicide. *Biocontrol Science and Technology* **11**: 273–281.

Greenberg R, Bichier P, Angon AC, MacVean C, Perez R & Cano E (2000) The impact of avian insectivory on arthropods and leaf damage in some Guatemalan coffee plantations. *Ecology* **81**: 1750–1755.

Greenslade PJM (1971) Interspecific competition and frequency changes among ants in Solomon Islands coconut plantations. *Journal of Applied Ecology* **8**: 323–352.

Gressel J (2001) Potential failsafe mechanisms against the spread and introgression of transgenic hypervirulent biocontrol fungi. *Trends in Biotechnology* **19**: 149–154.

Gressel J (2002) *Molecular Biology of Weed Control.* London, UK, Taylor & Francis.

Grevstad FS (1999a) Experimental invasions using biological control introductions: the influence of release size on the chance of population establishment. *Biological Invasions* **1**: 313–323.

Grevstad FS (1999b) Factors influencing the chance of population establishment: implications for release strategies in biocontrol. *Ecological Applications* **9**: 1439–1447.

Grewal PS, Bornstein-Forst S, Burnell AM, Glazer I and Jagdale GB (2006) Physiological, genetic, and molecular mechanisms of chemoreception, thermobiosis, and anhydrobiosis in entomopathogenic nematodes. *Biological Control* **38**: 54–65.

Grewal PS & Peters A (2005) Formulation and quality. In *Nematodes as Biocontrol Agents* (Grewal PS, Ehlers R-U & Shapiro-Ilan DI, eds.), Wallingford, UK, CABI Publishing, pp. 79–90.

Grewal PS, Selvan S & Gaugler R (1994) Thermal adaptation of entomopathogenic nematodes – niche breadth for infection, establishment and reproduction. *Journal of Thermal Biology* **19**: 245–253.

Griffiths GJK, Holland JM, Bailey A & Thomas MB (2008) Efficacy and economics of shelter habitats for conservation biological control. *Biological Control* **45**: 200–209.

Griffiths O, Cook A & Wells SM (1993) The diet of the introduced carnivorous snail *Euglandina rosea* in Mauritius and its implications for threatened island gastropod faunas. *Journal of Zoology, London* **229**: 79–89.

Groenteman R, Kelly D, Fowler SV & Bourdot GW (2011) Abundance, phenology and impact of biocontrol agents on nodding thistle (*Carduus nutans*) in Canterbury 35 years into a biocontrol programme. *New Zealand Journal of Agricultural Research* **54**: 1–13.

Gruner DS (2005) Biotic resistance to an invasive spider conferred by generalist insectivorous birds on Hawai'i Island. *Biological Invasions* **7**: 547–552.

Guetsky R, Shtienberg D, Elad Y, Fischer E & Dinoor A (2002) Improving biological control by combining biocontrol agents each with several mechanisms of disease suppression. *Phytopathology* **92**: 976–985.

Guo QF, Sax DF, Qian H & Early R (2012) Latitudinal shifts of introduced species: possible causes and implications. *Biological Invasions* **14**: 547–556.

Guretzky JA & Louda SM (1997) Evidence for natural biological control: insects decrease survival and growth of a native thistle. *Ecological Applications* **7**: 1330–1340.

Gurney WSC, Nisbet RM & Lawton JH (1983) The systematic formulation of tractable single species population models incorporating age structure. *Journal of Animal Ecology* **52**: 479–496.

Gurr GM & Wratten S (1999) Integrated biological control: a proposal for enhancing success in biological control. *International Journal of Pest Management* **45**: 81–84.

Gurr GM, Wratten SD & Barbosa P (2000) Success in conservation biological control of arthropods. In *Biological Control: Measures of Success* (Gurr GM & Wratten SD, eds.), Dordrecht, The Netherlands, Kluwer, pp. 105–132.

Gurr GM, Wratten SD & Altieri MA (2004) *Ecological Engineering for Pest Management.* Ithaca, NY, Comstock.

Gurr GM, Wratten SD & Luna JM (2003) Multi-function agricultural biodiversity: pest management and other benefits. *Basic and Applied Ecology* **4**: 107–116.

Gurr GM, Wratten SD, Tylianakis J, Kean J & Keller M (2005) Providing plant foods for natural enemies in farming systems: balancing practicalities and theory. In *Plant-Provided Food for Carnivorous Insects: A Protective Mutualism and Its Applications* (Wäckers F, Van Rijn PCJ & Bruin J, eds.), Cambridge, UK, Cambridge University Press, pp. 326–345.

Gutierrez AP (1996) *Applied Population Ecology: A Supply-Demand Approach.* New York, Wiley.

Gutierrez AP & Baumgärtner JU (1984) Multitrophic models of predator–prey energetics: 1. Age specific energetics models – pea aphid *Acyrthosiphon pisum* (Harris) (Homoptera: Aphididae) as an example. *Canadian Entomologist* **116**: 924–932.

Gutierrez AP, Caltagirone LE & Meikle W (1999) Evaluation of results, economics of biological control. In *Handbook of Biological Control* (Bellows TS & Fisher TW, eds.), San Diego, CA, Academic Press, pp. 243–252.

Gutierrez AP, Daane KM, Ponti L, Walton VM & Ellis CK (2008) Prospective evaluation of the biological control of vine mealybug: refuge effects and climate. *Journal of Applied Ecology* **45**: 524–536.

Gutierrez AP, Mills NJ, Schreiber SJ & Ellis CK (1994) A physiologically based tritrophic perspective on bottom-up-top-down regulation of populations. *Ecology* **75**: 2227–2242.

Gutierrez AP, Neuenschwander P, Schulthess F, Herren HR, Baumgaertner JU, Wermelinger B, Lohr B & Ellis CK (1988a) Analysis of biological control of cassava pests in Africa. II. Cassava mealybug *Phenacoccus manihoti. Journal of Applied Ecology* **25**: 921–940.

Gutierrez AP, Neuenschwander P & Van Alphen JJM (1993) Factors affecting biological control of cassava mealybug by exotic parasitoids: A ratio-dependent supply-demand driven model. *Journal of Applied Ecology* **30**: 706–721.

Gutierrez AP, Pitcairn MJ, Ellis CK, Carruthers N & Ghazelbash R (2005) Evaluating biological control of yellow starthistle (*Centaurea solstitialis*) in California: a GIS based supply-demand demographic model. *Biological Control* **34**: 115–131.

Gutierrez AP & Ponti L (2013) Deconstructing the control of the spotted alfalfa aphid *Therioaphis maculata. Agricultural and Forest Entomology* **15**: 272–284.

Gutierrez AP, Ponti L, Hoddle M, Almeida RPP & Irvin NA (2011) Geographic distribution and relative abundance of the invasive glassy-winged sharpshooter: effects of temperature and egg parasitoids. *Environmental Entomology* **40**: 755–769.

Gutierrez AP, Villacorta A, Cure JR & Ellis CK (1998a) Tritrophic analysis of the coffee (*Coffee arabica*): Coffee berry borer (*Hypothenemus hampei* (Ferrari)): parasitoid system. *Anais da Sociedade Entomologica do Brasil* **27**: 357–385.

Gutierrez AP, Wemelinger B, Schulthess F, Baumgaertner JU, Herren HR, Ellis CK & Yaninek JS (1988b) Analysis of biological control of cassava pests in Africa. I. Simulation of carbon, nitrogen and water dynamics in cassava. *Ecology* **25**: 901–920.

Gutierrez AP, Yaninek JS, Neuenschwander P & Ellis CK (1999) A physiologically-based tritrophic metapopulation model of the African cassava food web. *Ecological Modelling* **123**: 225–242.

Gutierrez AP, Yaninek JS, Wermelinger B, Herren HR & Ellis CK (1988c) Analysis of biological control of cassava pests in Africa. III. Cassava green mite *Mononychellus tanajoa. Journal of Applied Ecology* **25**: 941–950.

Gwynn DM, Callaghan A, Gorham J, Walters KFA & Fellowes MDE (2005) Resistance is costly: trade-offs between immunity, fecundity and survival in the pea aphid. *Proceedings of the Royal Society of London B* **272**: 1803–1808.

Haag KH & Habeck DH (1991) Enhanced biological control of water hyacinth following limited herbicide application. *Journal of Aquatic Plant Management* **29**: 24–28.

Haas D & Défago F (2005) Biological control of soil-borne pathogens by fluorescent pseudomonads. *Nature Reviews Microbiology* **3**: 307–319.

Haddad NM, Crutsinger GM, Gross K, Haarstad J, Knops JMH & Tilman D (2009) Plant species loss decreases arthropod diversity and shifts trophic structure. *Ecology Letters* **12**: 1029–1039.

Haddad NM, Crutsinger GM, Gross K, Haarstad J & Tilman D (2011) Plant diversity and the stability of food webs. *Ecology Letters* **14**: 42–46.

Haddad NM, Haarstad J & Tilman D (2000) The effects of long-term nitrogen loading on grassland insect communities. *Oecologia* **124**: 73–84.

Haddad NM, Tilman D, Haarstad J, Ritchie M & Knops JMH (2001) Contrasting effects of plant richness and composition on insect communities: a field experiment. *American Naturalist* **158**: 17–35.

Hadfield MG & Mountain BS (1980) A field study of a vanishing species, *Achatinella mustelina* (Gastropoda, Pulmonata), in the Waianae mountains of Oahu. *Pacific Science* **34**: 345–358.

Hadfield MG, Miller SE & Carwille AH (1993) The decimation of endemic Hawai'ian tree snails by alien predators. *American Zoologist* 33: 610–622.

Haenke S, Scheid B, Schaefer M, Tscharntke T & Thies C (2009) Increasing syrphid fly diversity and density in sown flower strips within simple vs. complex landscapes. *Journal of Applied Ecology* **46**: 1106–1114.

Hagen KS, Mills NJ, Gordh G & McMurtry JA (1999) Terrestrial arthropod predators of insect and mite pests. In *Handbook of Biological Control* (Bellows TS & Fisher TW, eds.), San Diego, CA, Academic Press, pp. 383–504.

Hails RS, Hernandez-Crespo P, Sait SM, Donnelly CA, Green BM & Cory JS (2002) Transmission patterns of natural and recombinant baculoviruses. *Ecology* **83**: 906–916.

Hajek AE (2004) *Natural Enemies: An Introduction to Biological Control*. Cambridge, UK, Cambridge University Press.

Hajek AE & Butler L (2000) Predicting the host range of entomopathogenic fungi. In *Nontarget Effects of Biological Control* (Follett PA & Duan JJ, eds.), Dordrecht, The Netherlands, Kluwer, pp. 263–276.

Hajek AE, Butler L & Wheeler MM (1995) Laboratory bioassays testing the host range of the gypsy-moth fungal pathogen *Entomophaga maimaiga*. *Biological Control* **5**: 530–544.

Hajek AE, Butler L, Walsh SRA, Silver JC, Hain FP, Hastings FL, ODell TM & Smitley DR (1996) Host range of the gypsy moth (Lepidoptera: Lymantriidae) pathogen *Entomophaga maimaiga* (Zygomycetes: Entomophthorales) in the field versus laboratory. *Environmental Entomology* **25**: 709–721.

Hajek AE, McManus DP & Delalibera Jr. I (2005) *Catalogue of Introductions of Pathogens and Nematodes for Classical Biological Control of Insects and Mites*. Washington, DC, USA, U.S. Forest Service.

Hajek AE, McManus DP & Delalibera Jr. I (2007) A review of introductions of pathogens and nematodes for classical biological control of insects and mites. *Biological Control* **41**: 1–13.

Hall RW & Ehler LE (1979) Rate of establishment of natural enemies in classical biological control. *Bulletin of the Entomological Society of America* **25**: 280–282.

Hall RW, Ehler LE & Bisabri-Ershadi B (1980) Rate of success in classical biological control of arthropods. *Bulletin of the Entomological Society of America* **26**: 111–114.

Hallett RH, Bahlai CA, Xue YG & Schaafsma AW (2014) Incorporating natural enemy units into a dynamic action threshold for the soybean aphid, *Aphis glycines* (Homoptera: Aphididae). *Pest Management Science* **70**: 879–888.

Hallett SG (2005) Where are the bioherbicides? *Weed Science* **53**: 404–415.

Hamer AJ, Lane SJ & Mahony MJ (2002) The role of introduced mosquitofish (*Gambusia holbrooki*) in excluding the native green and golden bell frog (*Litoria aurea*) from original habitats in south-eastern Australia. *Oecologia* **132**: 445–452.

Hamilton WD (1967). Extraordinary sex ratios. *Science* **156**: 477–488.

Hammock B (1992) Virus release evaluation. *Nature* **355**: 119.

Hance T (1988) The demographic parameters of *Phytoseiulus persimilis* Athias-Henriot (Acari: Phytoseiidae) in relation to the possibilities of using it to control populations of *Tetranychus urticae* (Acari: Tetranychidae). *Annales de la Societe Royale Zoologique de Belgique* **118**: 161–170.

Hance T, Van Baaren J, Vernon P & Boivin G (2007) Impact of extreme temperatures on parasitoids in a climate change perspective. *Annual Review of Entomology* **52**: 107–126.

Handley LJL, Estoup A, Evans D, Thomas CE, Lombaert E, Facon B, Aebi A & Roy HE (2011) Ecological genetics of invasive alien species. *BioControl* **56**: 409–428.

Hanlon SG, Hoyer MV, Cichra CE & Canfield Jr. DE (2000) Evaluation of macrophyte control in 38 Florida lakes using triploid grass carp. *Journal of Aquatic Plant Management* **39**: 48–54.

Hanna R, Onzo A, Lingeman R, Yaninek JS & Sabelis MW (2005) Seasonal cycles and persistence in an acarine predator–prey system on cassava in Africa. *Population Ecology* **47**: 107–117.

Hanski I (1999) *Metapopulation Ecology*. Oxford, UK, Oxford University Press.

Hanski I & Simberloff D (1997) The metapopulation approach, its history, conceptual domain, and application to conservation. In *Metapopulation Biology: Ecology, Genetics and Evolution* (Hanski I & Gilpin ME, eds.) San Diego, CA, Academic Press, pp. 5–26.

Hardin MR, Benrey B, Coll M, Lamp WO, Roderick GK & Barbosa P (1995) Arthropod pest resurgence: an overview of potential mechanisms. *Crop Protection* **14**: 3–18.

Hardman JM, Van der Werf W & Blatt SE (2013) Simulating effects of environmental factors on biological control of *Tetranychus urticae* by *Typhlodromus pyri* in apple orchards. *Experimental and Applied Acaraology* **60**: 181–203.

Hardy CM, Hinds LA, Kerr PJ, Lloyd M, Redwood A, Shellam G & Strive T (2006) Biological control of vertebrate pests using virally vectored immunocontraception. *Journal of Reproductive Immunology* **71**: 102–111.

Hare JD & Morgan DJW (1997) Mass-priming *Aphytis*: behavioral improvement of insectary-reared biological control agents. *Biological Control* **10**: 207–214.

Hare JD, Morgan DJW & Nguyun T (1997) Increased parasitization of California red scale in the field after exposing its parasitoid, *Aphytis melinus*, to a synthetic kairomone. *Entomologia Experimentalis et Applicata* **82**: 73–81.

Harley KLS & Forno IW (1992) *Biological Control of Weeds: A Handbook for Practitioners and Students*. Melbourne, Australia, Inkata Press.

Harman GE (2006) Overview of mechanisms and uses of *Trichoderma* spp. *Phytopathology* **96**: 190–194.

Harman GE, Howell CR, Viterbo A, Chet I & Loritto M (2004) *Trichoderma* species – opportunistic, avirulent plant symbionts. *Nature Reviews Microbiology* **2**: 43–56.

Harmon JP & Andow DA (2004) Indirect effects between shared prey: predictions for biological control. *BioControl* **49**: 605–626.

Harpur BA, Subhani M & Zayed A (2012) A review of the consequences of complementary sex determination and diploid male production on mating failures in Hymenoptera. *Entomologia Experimentalis et Applicata* **146**: 156–164.

Harris P (1981) Stress as a strategy in the biological control of weeds. In *Biological Control and Crop Protection* (Papavizas GC, ed.), Totowa, Allanheld Osmun, pp. 333–340.

Harris P (1988) Environmental impact of weed-control insects. *BioScience* **38**: 542–548.

Harris P (1991) Classical biocontrol of weeds: its definitions, selection of effective agents, and administrative-political problems. *Canadian Entomologist* **123**: 827–849.

Harris P, Wilkinson ATS, Neary ME & Thompson LS (1971) *Senecio jacobaea* L., tansy ragwort (Compositae). In *Biological Control Programmes Against Insects and Weeds in Canada: 1959–1968* (Kelleher JS & Hulme MA, eds.), Slough, UK, Commonwealth Agricultural Bureau, pp. 97–104.

Harris P, Wilkinson ATS, Neary ME, Thompson LS & Finnamore D (1975) Establishment in Canada of the cinnabar moth, *Tyria jacobaeae* (Lepidoptera: Arctiidae). *Canadian Entomologist* **107**: 913–917.

Harris P & Zwölfer H (1968) Screening of phytophagous insects for biological control of weeds. *Canadian Entomologist* **100**: 295–303.

Harrison S & Taylor AD (1997) Empirical evidence for metapopulation dynamics. In *Metapopulation Biology: Ecology, Genetics and Evolution* (Hanski I & Gilpin ME, eds.), San Diego, CA, Academic Press, pp. 27–44.

Harrison RL & Bonning BC (2000) Genetic engineering of biocontrol agents for insects. In *Biological and Biotechnological Control of Insect Pests* (Rechcigl JE & Rechcigl NA, eds.), Boca Raton, FL, CRC Press, pp. 243–280.

Hart AD (1978) The onslaught against Hawaii's tree snails. *Natural History* **87**: 46–57.

Hart AJ, Bale JS, Tullett AG, Worland MR & Walters KFA (2002a) Effects of temperature on the establishment potential of the predatory mite *Amblyseius californicus* McGregor (Acari: Phytoseiidae) in the UK. *Journal of Insect Physiology* **48**: 593–599.

Hart AJ, Tullett AG, Bale JS & Walters KFA (2002b) Effects of temperature on the establishment potential in the UK of the non-native glasshouse biocontrol agent *Macrolophus caliginosus*. *Physiological Entomology* **27**: 112–123.

Harvey IC & Bourdot GW (2001) Giant buttercup (*Ranunculus acris* L.) control in pasture using a mycoherbicide based on *Sclerotinia sclerotiorum*. *New Zealand Plant Protection* **54**: 120–124.

Harwood RJW, Hickman JM, MacLeod A, Sherratt T & Wratten SD (1994) Managing field margin for hover flies. In *Field Margins: Integrating Agriculture and Conservation* (Boatman N, ed.), Alton, UK, British Crop Protection Society, pp. 147–152.

Hasan S, Chaboudez P & Espiau C (1995) Isozyme patterns and susceptibility of North American forms of *Chondrilla juncea* to European strains of the rust fungus *Puccinia chondrillina*. In *Proceedings of the Eighth International Symposium on Biological Control of Weeds* (Delfosse ES & Scott RR, eds.), Canterbury, New Zealand, DSIR/CSIRO, pp. 367–373.

Hasan S & Delfosse ES (1995) Susceptibility of the Australian native, *Heliotropium crispatum*, to the rust fungus *Uromyces heliotropii* introduced to control common heliotrope, *Heliotropium europaeum*. *Biocontrol Science and Technology* **5**: 165–174.

Hashmi S, Hashmi G, Glazer I, Gaugler R (1998) Thermal response of *Heterorhabditis bacteriophora* transformed with the *Caenorhabditis elegans hsp70* encoding gene. *Journal of Experimental Zoology* **281**: 164–170.

Hassan SA (1989) Testing methodology and the concept of the IOBC/WPRS working group. In *Pesticides and Non-target Invertebrates* (Jepson PC, ed.), Andover, UK, Intercept, pp. 1–18.

Hassell MP (1978) *The Dynamics of Arthropod Predator–Prey Systems*. Princeton, NJ, Princeton University Press.

Hassell MP (1980) Foraging strategies, population models and biological control: a case study. *Journal of Animal Ecology* **49**: 603–628.

Hassell MP (1984a) Insecticides in host parasitoid interactions. *Theoretical Population Biology* **26**: 378–386.

Hassell MP (1984b) Parasitism in patchy environments: inverse density dependence can be stabilizing. *IMA Journal of Mathematics Applied in Medicine and Biology* **1**: 123–133.

Hassell MP (2000) *The Spatial and Temporal Dynamics of Host-Parasitoid Interactions*. Oxford, UK, Oxford University Press.

Hassell MP & Comins HN (1978) Sigmoid functional responses and population stability. *Theoretical Population Biology* **14**: 62–67.

Hassell MP, Lawton JH & Beddington JR (1976) The components of arthropod predation I. The prey death-rate. *Journal of Animal Ecology* **45**: 135–164.

Hassell MP & May RM (1973) Stability in insect host-parasite models. *Journal of Animal Ecology* **42**: 693–726.

Hassell MP & Varley GC (1969) New inductive population model for insect parasites and its bearing on biological control. *Nature* **223**: 1133–1136.

Hastings A & Higgins K (1994) Persistence of transients in spatially structured ecological models. *Science* **263**: 1133–1136.

Hatherly IS, Bale JS, Walters KFA & Worland MR (2004) Thermal biology of *Typhlodromips montdorensis*: implications for its introduction as a glasshouse biological control agent in the UK. *Entomologia Experimentalis et Applicata* **111**: 97–109.

Hatherly IS, Hart AJ, Tullett AG & Bale JS (2005) Use of thermal data as a screen for the establishment potential of non-native biological control agents in the UK. *BioControl* **50**: 687–698.

Hautier L, Gregoire JC, de Schauwers J, San Martin G, Callier P, Jansen J-P & de Biseau J-C (2008) Intraguild predation by *Harmonia axyridis* on coccinellids revealed by exogenous alkaloid sequestration. *Chemoecology* **18**: 191–196.

Havill NP, Davis G, Mausel DL, Klein J, McDonald R, Jones C, Fischer M, Salom S & Caccone A (2012) Hybridization between a native and introduced predator of Adelgidae: an unintended result of classical biological control. *Biological Control* **63**: 359–369.

Hawkes RB (1973) Natural mortality of cinnabar moth in California. *Annals of the Entomological Society of America* **66**: 137–146.

Hawkins BA & Cornell HV (1994) Maximum parasitism rates and successful biological control. *Science* **266**: 1886.

Hawkins BA, Cornell HV & Hochberg ME (1997) Predators, parasitoids, and pathogens as mortality agents in phytophagous insect populations. *Ecology* **78**: 2145–2152.

Hawkins BA & Marino PC (1997) The colonization of native phytophagous insects in North America by exotic parasitoids. *Oecologia* **112**: 566–571.

Hawkins BA, Mills NJ, Jervis MA & Price PW (1999) Is the biological control of insects a natural phenomenon? *Oikos* **86**: 493–506.

Hawkins BA, Thomas MB & Hochberg ME (1993) Refuge theory and biological control. *Science* **262**: 1429–1432.

Haye T, Goulet H, Mason PG & Kuhlmann U (2005) Does fundamental host range match ecological host range? A retrospective case study of a *Lygus* plant bug parasitoid. *Biological Control* **35**: 55–67.

Haye T, Kuhlmann U, Goulet H & Mason PG (2006) Controlling *Lygus* plant bugs (Heteroptera: Miridae) with European *Peristenus relictus* (Hymenoptera: Braconidae) in Canada – risky or not? *Bulletin of Entomological Research* **96**: 187–196.

Hayes KR & Barry SC (2008) Are there any consistent predictors of invasion success? *Biological Invasions* **10**: 483–506.

Hayes L, Fowler SV, Paynter Q, Groenteman R, Peterson P, Dodd S & Bellgard S (2013) Biocontrol of weeds: achievements to date and future outlook. In *Ecosystem Services in New Zealand: Conditions and Trends* (Dymond JR, ed.), Lincoln, New Zealand, Manaaki Whenua, pp. 375–385.

Hector A, Dobson K, Minns A, Bazely-White E & Lawton JH (2001) Community diversity and invasion resistance: an experimental test of a grassland ecosystem and a review of comparable studies. *Ecological Research* **16**: 819–831.

Heimpel GE (2000) Effects of clutch size on host-parasitoid population dynamics. In *Parasitoid Population Biology* (Hochberg ME & Ives AR, eds.), Princeton, NJ, Princeton University Press, pp. 27–40.

Heimpel GE, Antolin MF, Franqui RA & Strand MR (1997) Reproductive isolation and genetic variation between two 'strains' of *Bracon hebetor* (Hymenoptera: Braconidae). *Biological Control* **9**: 149–156.

Heimpel GE, Antolin MR & Strand MR (1999). Diversity of sex-determining alleles in *Bracon hebetor*. *Heredity* **82**: 282–291.

Heimpel GE & Asplen MK (2011) A 'goldilocks' hypothesis for dispersal of biological control agents. *BioControl* **56**: 441–450.

Heimpel GE & Casas J (2008) Parasitoid foraging and oviposition behaviour in the field. In *Behavioural Ecology of Insect Parasitoids* (Wajnberg E, Bernstein C & Van Alphen JJM, eds.), Oxford, UK, Blackwell, pp. 51–70.

Heimpel GE & Collier TR (1996) The evolution of host-feeding behavior in insect parasitoids. *Biological Reviews* **71**: 373–400.

Heimpel GE & de Boer JG (2008) Sex determination in the Hymenoptera. *Annual Review of Entomology* **53**: 209–230.

Heimpel GE, Frelich LE, Landis DA, Hopper KR, Hoelmer KA, Sezen Z, Asplen MK & Wu K (2010) European buckthorn and Asian soybean aphid as components of an extensive invasional meltdown in North America. *Biological Invasions* **12**: 2913–2931.

Heimpel GE & Jervis MA (2005) Does floral nectar improve biological control by parasitoids? In *Plant-Provided Food and Plant-Carnivore Mutualism* (Wäckers F, Van Rijn P & Bruin J, eds.), Cambridge, UK, Cambridge University Press, pp. 267–304.

Heimpel GE, Lee JC, Wu Z, Weiser L, Wackers F & Jervis MA (2004b) Gut sugar analysis in field-caught parasitoids: adapting methods originally developed for biting flies. *International Journal of Pest Management* **50**: 193–198.

Heimpel GE & Lundgren JG (2000) Sex ratios of commercially reared biological control agents. *Biological Control* **19**: 77–93.

Heimpel GE, Mangel M & Rosenheim JA (1998) Effects of egg and time limitation on lifetime reproductive success of a parasitoid in the field. *American Naturalist* **152**: 273–289.

Heimpel GE, Neuhauser C & Andow DA (2005) Natural enemies and the evolution of resistance to transgenic insecticidal crops by pest insects: the role of egg mortality. *Environmental Entomology* **34**: 512–526.

Heimpel GE, Neuhauser C & Hoogendoorn M (2003) Effects of parasitoid fecundity and host resistance on indirect interactions among hosts sharing a parasitoid. *Ecology Letters* **6**: 556–566.

Heimpel GE, Ragsdale DW, Venette R, Hopper KR, O'Neil RJ, Rutledge CE & Wu Z (2004a) Prospects for importation biological control of the soybean aphid: anticipating potential costs and benefits. *Annals of the Entomological Society of America* **97**: 249–258.

Heimpel GE & Rosenheim JA (1998) Egg limitation in parasitoids: a review of the evidence and a case study. *Biological Control* **11**: 160–168.

Heimpel GE, Rosenheim JA & Mangel M (1996) Egg limitation, host quality, and dynamic behavior by a parasitoid in the field. *Ecology* **77**: 2410–2420.

Heimpel GE, Yang Y, Hill JD & Ragsdale DW (2013) Environmental consequences of invasive species: greenhouse gas emissions of insecticide use and the role of biological control in reducing emissions. *PLoS ONE* **8**: e72293.

Hein S, Poethke H-J & Dorn S (2009) What stops the 'diploid male vortex'? A simulation study for species with single locus complementary sex determination. *Ecological Modelling* **220**: 1663–1669.

Heinz KM, Nunney L and Parrella MP (1993) Toward predictable biological control of *Liriomyza trifolii* (Diptera: Agromyzidae) infesting greenhouse cut chrysanthemums. *Environmental Entomology* **22**: 1217–1233.

Heinz KM, Van Driesche R & Parrella MP (2005) *Biocontrol in Protected Culture*. Batavia, IL, Ball.

Hennemann ML & Memmott J (2001) Infiltration of a Hawaiian community by introduced biological control agents. *Science* **293**: 1314–1316.

Henning J, Meers J, Davies PR & Morris RS (2005) Survival of rabbit haemorrhagic disease virus (RHDV) in the environment. *Epidemiology and Infection* **113**: 719–730.

Henry LM, May N, Acheampong S, Gillespie DR & Roitberg BD (2010) Host-adapted parasitoids in biological control: does source matter? *Ecological Applications* **20**: 242–250.

Henry LM, Roitberg BD & Gillespie DR (2008) Host-range evolution in *Aphidius* parasitoids: fidelity, virulence and fitness trade-offs on an ancestral host. *Evolution* **62**: 689–699.

Henter HJ (1995). The potential for coevolution in a host-parasitoid system. II. Genetic variation within a population of wasps in the ability to parasitize an aphid host. *Evolution* **49**: 439–445.

Henter HJ (2003). Inbreeding depression and haplodiploidy: experimental measures in a parasitoid and comparisons across diploid and haplodiploid insect taxa. *Evolution* **57**: 1793–1803.

Henter HJ & Via S (1995) The potential for coevolution in a host-parasitoid system. I. Genetic variation within an aphid population in susceptibility to a parasitic wasp. *Evolution* **49**: 427–438.

Hernandez-Rodriguez A, Heydrich-Perez M, Acebo-Guerrero Y, Velazquez-del Valle MG & Hernandez-Lauzardo AN (2008) Antagonistic activity of Cuban native rhizobacteria against *Fusarium verticillioides* (Sacc.) Nirenb. in maize (*Zea mays* L.). *Applied Soil Ecology* **39**: 180–186.

Herzog DC & Funderburk JE (1985) Ecological bases for habitat management and pest cultural control. In *Ecological Theory and Integrated Pest Management Practice* (Kogan M, ed.), New York, Wiley, pp. 217–250.

Hewa-Kapuge S, McDougall S & Hoffmann AA (2003) Effects of methoxyfenozide, indoxacarb, and other insecticides on the beneficial egg parasitoid *Trichogramma* nr. *brassicae* (Hymenoptera: Trichogrammatidae) under laboratory and field conditions. *Journal of Economic Entomology* **96**: 1083–1090.

Hickling GJ (2000) Success in biological control of vertebrate pests. In *Biological Control: Measures of Success* (Gurr GM & Wratten SD, eds.), Dordrecht, The Netherlands, Kluwer, pp. 341–368.

Hickman JM & Wratten SD (1996) Use of *Phacelia tanacetifolia* strips to enhance biological control of aphids by hoverfly larvae in cereal fields. *Journal of Economic Entomology* **89**: 832–840.

Hight SD, Carpenter JE, Bloem S & Bloem KA (2005) Developing a sterile insect release program for *Cactoblastis cactorum* (Berg) (Lepidoptera: Pyralidae): effective overflooding ratios and release-recapture field studies. *Environmental Entomology* **34**: 850–856.

Hilbeck A, Eckel C & Kennedy GC (1998) Impact of *Bacillus thuringiensis* – insecticides on population dynamics and egg predation of the Colorado potato beetle in North Carolina potato plantings. *BioControl* **43**: 65–75.

Hilborn R (1975) The effect of spatial heterogeneity on the persistence of predator–prey interactions. *Theoretical Population Biology* **8**: 346–355.

Hill G & Greathead D (2000) Economic evaluation in classical biological control. In *The Economics of Biological Invasions* (Perrings C, Williamson M & Dalmazzone S, eds.), Cheltenham, UK, Edward Elgar, pp. 208–223.

Hill MP & Hulley PE (1995) Host-range expansion by native parasitoids to weed biocontrol agents introduced into South Africa. *Biological Control* **5**: 297–302.

Hochberg ME & Hawkins BA (1992) Refuges as a predictor of parasitoid diversity. *Science* **255**: 973–976.

Hochberg ME & Hawkins BA (1993) Predicting parasitoid species richness. *American Naturalist* **142**: 671–693.

Hochberg ME & Hawkins BA (1994) The implications of population dynamics theory to parasitoid diversity and biological control. In *Parasitoid Community Ecology* (Hawkins BA & Sheehan W, eds.), Oxford, UK, Oxford University Press, pp. 451–471.

Hochberg ME & Holt RD (1995) Refuge evolution and the population dynamics of coupled host-parasitoid associations. *Evolutionary Ecology* **9**: 633–661.

Hochberg ME & Waage JK (1991) A model for the biological control of *Oryctes rhinoceros* (Coleoptera: Scarabaeidae) by means of pathogens. *Journal of Applied Ecology* **28**: 514–531.

Hoddle MS (1999) Biological control of vertebrates. In *Handbook of Biological Control* (Bellows TS & Fisher TW, eds.), San Diego, CA, Academic Press pp. 955–974.

Hoddle M (2004b) Analysis of fauna in the receiving area for the purpose of identifying native species that exotic natural enemies may potentially attack. In *Assessing Host Ranges of Parasitoids and Predators Used for Classical Biological Control: A Guide to Best Practice* (Van Driesche RG & Reardon R, eds.), Morgantown, WV, U.S. Forest Service, pp. 24–39.

Hoddle MS (2004a) Restoring balance: using exotic species to control invasive exotic species. *Conservation Biology* **18**: 38–49.

Hoddle MS (2006) Historical review of control programs for *Levuana iridescens* (Lepidoptera: Zygaenidae) in Fiji and examination of possible extinction of this moth by *Bessa remota* (Diptera: Tachinidae). *Pacific Science* **60**: 439–453.

Hoddle MS, Van Driesche RG & Sanderson JP (1998) Biology and use of the whitefly parasitoid *Encarsia formosa*. *Annual Review of Entomology* **43**: 645–659.

Hoelmer KA & Kirk AA (2005) Selecting arthropod biological control agents against arthropod pests: can the science be improved to decrease the risk of releasing ineffective agents? *Biological Control* **34**: 255–264.

Hoffmann AA (2014) Facilitating *Wolbachia* invasions. *Austral Entomology* **53**: 125–132.

Hoffmann AA, Montgomery BL, Popovici J, Iturbe-Ormaetxe I, Johnson PH, Muzzi F, Greenfield M, Durkan M, Leong YS, Dong Y, Cook H, Axford J, Callahan AG, Kenny N, Omodei C, McGraw EA, Ryan PA, Ritchie SA, Turelli M & O'Neill SL (2011) Successful establishment of *Wolbachia* in *Aedes* populations to suppress dengue transmission. *Nature* **476**: 454–457.

Hoffmann JH (1995) Biological control of weeds: the way forward, a South African perspective. Proceedings of the BCPC Symposium. *Weeds in a Changing World* **64**: 77–89.

Hoffmann JH, Impson FAC & Volchansky CR (2002) Biological control of cactus weeds: implications of hybridization between control agent biotypes. *Journal of Applied Ecology* **39**: 900–908.

Hoffmann JH & Moran VC (1995) Localized failure of a weed biological control agent attributed to insecticide drift. *Agriculture Ecosystems & Environment* **52**: 197–203.

Hoffmann JH & Moran VC (1998) The population dynamics of an introduced tree, *Sesbania punicea*, in South Africa, in response to long-term damage caused by different combinations of three species of biological control agents. *Oecologia* **114**: 343–348.

Hoffmann MP, Wilson LT, Zalom FG & Hilton RJ (1990) Parasitism of *Heliothis zea* (Lepidoptera:

Noctuidae) eggs: effect on pest managment decision rules for processing tomatoes in the Sacramento Valley of California. *Environmental Entomology* **19**: 753–763.

Hoffmann MP, Wilson LT, Zalom FG & Hilton RJ (1991) Dynamic sequential sampling plan for *Helicoverpa zea* (Lepidoptera: Noctuidae) eggs in processing tomatoes: parasitism and temporal patterns. *Environmental Entomology* **20**: 1005–1012.

Hoffmeister T (1992) Factors determining the structure and diversity of parasitoid complexes in Tephritid fruit flies. *Oecologia* **89**: 288–297.

Hoffmeister TS, Babendreier D & Wajnberg E (2006) Statistical tools to improve the quality of experiments and data analysis for assessing non-target effects. In *Environmental Impact of Invertebrates for Biological Control of Arthropods* (Bigler F, Babendreier D & Kuhlmann U, eds.), Wallingford, UK, CABI Publishing, pp. 222–240.

Hoitink HAJ & Boehm MJ (1999) Biocontrol within the context of soil microbial communities: a substrate-dependent phenomenon. *Annual Review of Phytopathology* **37**: 427–446.

Hoitink HAJ & Fahy PC (1986) Basis for the control of soilborne plant pathogens with composts. *Annual Review of Phytopathology* **24**: 93–114.

Hokkanen H (1985a) Exploiter-victim relationships of major plant diseases: implications for biological weed control. *Agriculture, Ecosystems & Environment* **14**: 63–76.

Hokkanen HMT (1985b) Success in classical biological control. *CRC Critical Reviews in Plant Sciences* **3**: 35–72.

Hokkanen HMT & Hajek A (2003) *Environmental Impacts of Microbial Insecticides*. Dordrecht, The Netherlands, Kluwer.

Hokkanen HMT & Lynch JM (1995) *Biological Control: Benefits and Risks*. Cambridge, UK, Cambridge University Press.

Hokkanen H & Pimentel D (1984) New approach for selecting biological control agents. *Canadian Entomologist* **116**: 1109–1121.

Hokkanen H & Pimentel D (1989) New associations in biological control: theory and practice. *Canadian Entomologist*. **121**: 829–840.

Holling CS (1959) The components of predation as revealed by a study of small-mammal predation of the European pine sawfly. *The Canadian Entomologist* **41**: 293–320.

Holloway AK, Heimpel GE, Strand MR & Antolin MF (1999) Survival of diploid males in *Bracon* sp. near *hebetor* (Hymenoptera: Braconidae). *Annals of the Entomological Society of America* **92**: 110–116.

Holmes PM (1990) Dispersal and predation in alien *Acacia*. *Oecologia* **83**: 288–290.

Holst N & Ruggle P (1997) A physiologically based model of pest-natural enemy interactions. *Experimental and Applied Acarology* **62**: 325–341.

Holt RD (1977) Predation, apparent competition and the structure of prey communities. *Theoretical Population Biology* **12**: 197–229.

Holt RD & Hassell MP (1993) Environmental heterogeneity and the stability of host-parasitoid interactions. *Journal of Animal Ecology* **62**: 89–100.

Holt RD & Hochberg ME (1997) When is biological control evolutionarily stable (or is it?). *Ecology* **78**: 1673–1683.

Holt RD & Hochberg ME (2001) Indirect interactions: community modules and biological control: a theoretical perspective. In *Evaluating Indirect Ecological Effects of Biological Control* (Wajnberg E, Scott JK & Quimby PC, eds.), Wallingford, UK, CABI Publishing, pp. 13–38.

Holt RD, Hochberg ME & Barfield M (1999) Population dynamics and the evolutionary stability of biological control. In *Theoretical Approaches to Biological Control* (Hawkins BA & Cornell HV, eds.), Cambridge, UK, Cambridge University Press, pp. 219–230.

Honegger RE (1981) List of amphibians and reptiles either known or thought to have become extinct since 1600. *Biological Conservation* **19**: 141–158.

Hood GM, Chesson P & Pech RP (2000) Biological control using sterilizing viruses: host suppression and competition between viruses in non-spatial models. *Journal of Applied Ecology* **37**: 914–925.

Hoogendoorn M & Heimpel GE (2002) Indirect interactions between an introduced and a native ladybird species mediated by a shared parasitoid. *Biological Control* **25**: 224–230.

Hoogendoorn M & Heimpel GE (2004) Competitive interactions between an exotic and a native ladybeetle: a field cage study. *Entomologia Experimentalis et Applicata* **111**: 19–28.

Hoopes MF, Holt RD & Holyoak M (2005) The effects of spatial processes on two species interactions. In *Spatial Dynamics and Ecological Communities* (Holyoak M, Leibhold MA & Holt RD, eds.), Chicago, IL, University of Chicago Press, pp. 35–67.

Hoover K, Schultz CM, Lane SS, Bonning BC, Duffey SS, McCutchen BF & Hammock BD (1995) Reduction in damage to cotton plants by recombinant baculovirus that knocks moribund larvae of *Heliothis virescens* off the plant. *Biological Control* **5**: 419–426.

Hopper KR, Britch SC & Wajnberg E (2006) Risks of interbreeding between species used in biological control and native species, and methods for evaluating their occurrence and impact. In *Environmental Impact of Invertebrates for Biological Control of Arthropods* (Bigler F, Babendreier D & Kuhlmann U, eds.), Wallingford, UK, CABI Publishing, pp. 78–97.

Hopper KR, Prager SM & Heimpel GE (2013) Is parasitoid acceptance of different host species dynamic? *Functional Ecology* **27**: 1201–1211.

Hopper KR & Roush RT (1993) Mate finding, dispersal, number released, and the success of biological control introductions. *Ecological Entomology* **18**: 321–331.

Hopper KR, Roush RT & Powell W (1993). Management of genetics of biological control introductions. *Annual Review of Entomology* **38**: 27–51.

Hornby D (1983) Suppressive soils. *Annual Review of Phytopathology* **21**: 65–85.

Hossain Z, Gurr GM, Wratten SD & Raman A (2002) Habitat manipulation in lucerne *Medicago sativa*: arthropod population dynamics in harvested 'refuge' crop strips. *Journal of Applied Ecology* **39**: 445–454.

Hougardy E & Mills NJ (2006) The influence of host deprivation and egg expenditure on the rate of dispersal of a parasitoid following field release. *Biological Control* **37**: 206–213.

House HL (1967) The decreasing occurrence of diapause in the fly *Pseudosarcophaga affinis* through laboratory-reared generations. *Canadian Journal of Zoology* **45**: 149–153.

Howarth FG (1983) Classical biocontrol: panacea or Pandora's box? *Proceedings of the Hawaiian Entomological Society* **24**: 239–244.

Howarth FG (1991) Environmental impacts of classical biological control. *Annual Review of Entomology* **36**: 485–509.

Hoy MA (1985) Recent advances in genetics and genetic improvement of the Phytoseiidae. *Annual Review of Entomology* **30**: 345–370.

Hoy MA (2000a) Deploying transgenic arthropods in pest management programs: risks and realities. In *Insect Transgenesis: Methods and Applications* (Handler AM & James AA, eds.), Boca Raton, FL, CRC Press, pp. 335–368.

Hoy MA (2000b) Transgenic arthropods for pest management programs: risks and realities. *Experimental and Applied Acarology* **24**: 463–495.

Hoy MA (2003) *Insect Molecular Genetics: An Introduction to Principles and Applications.* Second edition. San Diego, CA, Academic Press.

Hoy MA (2008) Augmentative biological control. In *Encyclopedia of Entomology, Second Edition* (Capinera JL, ed.), Dordrecht, The Netherlands, Springer, pp. 327–334.

Hsiao TH (1996) Studies of interaction between alfalfa weevil strains, *Wolbachia* endosymbionts and parasitoids. In *The Ecology of Agricultural Pests* (Symondson WOC & Liddell JE, eds.) London, UK, Chapman & Hall, pp. 51–71.

Hu G & St. Leger J (2002) Field studies using a recombinant mycoinsecticide (*Metarhizium anisopliae*) reveal that it is rhizosphere competent. *Applied and Environmental Microbiology* **68**: 6383–6387.

Huang JW & Kuhlman EG (1991a) Formulation of a soil amendment to control damping-off of slash pine-seedlings. *Phytopathology* **81**: 163–170.

Huang JW & Kuhlman EG (1991b) Mechanisms inhibiting damping-off pathogens of slash pine seedlings with a formulated soil amendment. *Phytopathology* **81**: 171–177.

Huang NX, Enkegaard A, Osborne LS, Ramakers PMJ, Messelink GJ, Pijnakker J & Murphy G (2011) The banker plant method in biological control. *Critical Reviews in Plant Sciences* **30**: 259–278.

Huang Z, Bonsall RF, Mavrodi DV, Weller DM & Thomashow LS (2004) Transformation of *Pseudomonas fluorescens* with genes for biosynthesis of phenazine-1-carboxylic acid improves biocontrol of rhizoctonia root rot and

in situ antibiotic production. *FEMS Microbiology Ecology* **49**: 243–251.
Hufbauer RA (2001) Pea aphid-parasitoid interactions: have parasitoids adapted to differential resistance? *Ecology* **82**: 717–725.
Hufbauer RA (2002a) Aphid population dynamics: does resistance to parasitism influence population size? *Ecological Entomology* **27**: 25–32.
Hufbauer RA (2002b) Evidence for nonadaptive evolution in parasitoid virulence following a biological control introduction. *Ecological Applications* **12**: 66–78.
Hufbauer RA, Bogdanowicz SM & Harrison RG (2004) The population genetics of biological control introduction: mitochondrial DNA and microsatellite variation in native and introduced populations of *Aphidius ervi*, a parasitoid wasp. *Molecular Ecology* **13**: 337–348.
Hufbauer RA & Roderick GK (2005) Microevolution in biological control: mechanisms, patterns, and processes. *Biological Control* **35**: 227–239.
Hufbauer RA & Via S (1999) Evolution of an aphid-parasitoid interaction: variation in resistance to parasitism among aphid populations specialized on different plants. *Evolution* **53**: 1435–1445.
Huffaker CB (1957) Fundamentals of biological control of weeds. *Hilgardia* **27**: 101–157.
Huffaker CB & Kennett CE (1956) Experimental studies on predation: predation and cyclamen-mite populations on strawberries in California. *Hilgardia* **26**: 191–222.
Huffaker CB & Kennett CE (1959) A ten-year study of vegetational changes associated with biological control of Klamath weed. *Journal of Range Management* **12**: 69–82.
Huffaker CB & Kennett CE (1966) Biological control of olive scale *Parlatoria oleae* (Colvee) through the compensatory action of two introduced parasites. *Hilgardia* **37**: 283–335.
Huffaker CB, Simmonds FJ & Laing JE (1976) The theoretical and empirical basis of biological control. In *Theory and Practice of Biological Control* (Huffaker CB & Messenger PS, eds.) New York, Academic Press, pp. 41–78.
Hughes RD & Bryce MA (1984) Biological characterization of two biotypes of pea aphid, one susceptible and the other resistant to fungal pathogens, coexisting on lucerne in Australia. *Entomologia Experimentalis et Applicata* **36**: 225–229.
Hull LA & Beers EH (1985) Ecological selectivity: modifying chemical control practices to preserve natural enemies. In *Biological Control in Agricultural IPM Systems* (Hoy MA & Herzog DC, eds.), Orlando, FL, Academic Press pp. 103–122.
Hull LA, Hickey KD & Kanour WW (1983) Pesticide usage patterns and associated pest damage in commercial apple orchards of Pennsylvania. *Journal of Economic Entomology* **76**: 577–583.
Hultine KR, Bean DW, Dudley TL & Gehring CA (2015) Species introductions and their cascading impacts on biotic interactions in desert riparian ecosystems. *Integrative and Comparative Biology* **55**: 587–601.
Hunter DM (2005) Mycopesticides as part of integrated pest management of locusts and grasshoppers. *Journal of Orthoptera Research* **14**: 197–201.
Hunter DM, Milner RJ, Scanlan JC & Spurgin PA (1999) Aerial treatment of the migratory locust, *Locusta migratoria* (L.) (Orthoptera: Acrididae) with *Metarhizium anisopliae* (Deuteromycotina: Hyphomycetes) in Australia. *Crop Protection* **18**: 699–704.
Hunter MD & Price PW (1992) Playing chutes and ladders – heterogeneity and the relative roles of bottom-up and top-down forces in natural communities. *Ecology* **73**: 724–732.
Hunter MS (1999) Genetic conflict in natural enemies: a review, and consequences for the biological control of arthropods. In *Theoretical Approaches to Biological Control* (Hawkins BA & Cornell HV, eds.), Cambridge, UK, Cambridge University Press, pp. 231–258.
Hunter MS & Woolley JB (2001) Evolution and behavioral ecology of heteronomous aphelinid parasitoids. *Annual Review of Entomology* **46**: 251–290.
Hunt-Joshi TR, Root RB & Blossey B (2005) Disruption of weed biological control by an opportunistic mirid predator. *Ecological Applications* **15**: 861–870.
Hussey NW & Scopes N (1985) *Biological Pest Control: The Glasshouse Experience*. Ithaca, NY, Cornell University Press.

Hutchinson GE (1959) Homage to Santa Rosalia or why are there so many kinds of animals? *American Naturalist* 93: 145–159.

Hwang AS, Northrup SL, Alexander JK, Vo T & Edmands S (2011) Long-term experimental hybrid swarms between moderately incompatible *Tigriopus californicus* populations: hybrid inferiority in early generations yields to hybrid superiority in later generations. *Conservation Genetics* 12: 895–909.

Hynes RK & Boyetchko SM (2006) Research initiatives in the art and science of biopesticide formulations. *Soil Biology & Biochemistry* 38: 845–849.

Ikeda M, Reimbold EA & Thiem SM (2005) Functional analysis of the baculovirus host range gene, hrf-1. *Virology* 332: 602–613.

Imms AD (1937) *Recent Advances in Entomology, Second edition.* Philadelphia, PA, P. Blackiston's Son & Co.

Inceoglu AB, Kamita SG & Hammock BD (2006) Genetically modified baculoviruses: a historical overview and future outlook. *Advances in Virus Research* 68: 323–360.

IPPC (2006) *Guidelines for the Export, Shipment, Import and Release of Biological Control Agents and Other Beneficial Organisms.* Rome, Italy, International Plant Protection Convention, Food and Agriculture Organization.

Ireson JE, Gourlay AH, Holloway RJ, Chatterton WS, Foster SD & Kwong RM (2008) Host specificity, establishment and dispersal of the gorse thrips, *Sericothrips staphylinus* Haliday (Thysanoptera: Thripidae), a biological control agent for gorse, *Ulex europaeus* L. (Fabaceae), in Australia. *Biological Control* 45: 460–471.

Irvin NA, Hoddle MS, O'Brochta DA, Carey B & Atkinson PW (2004) Assessing fitness costs for transgenic *Aedes aegypti* expressing the GFP marker and transposase genes. *Proceedings of the National Academy of Sciences* 101: 891–896.

Irvin NA, Suarez Espinosa J & Hoddle MS (2014) Maximum realised lifetime parasitism and occurrence of time limitation in *Gonatocerus ashmeadi* (Hymenoptera: Mymaridae) foraging in citrus orchards. *Biocontrol Science and Technology* 24: 662–679.

Isaacs R, Tuell J, Fiedler A, Gardiner M & Landis D (2009) Maximizing arthropod-mediated ecosystem services in agricultural landscapes: the role of native plants. *Frontiers in Ecology and the Environment* 7: 196–203.

Ives WGH & Muldrew JA (1981) *Pristiphora erichsonii* (Hartig), Larch Sawfly (Hymenoptera: Tenthredinidae). In *Biological Control Programmes against Insects and Weeds in Canada, 1969–1980* (Kelleher JS & Hulme MA, eds.), Slough, UK, Commonwealth Agricultural Bureau, pp. 369–380.

Ives AR & Carpenter SR (2007) Stability and diversity of ecosystems. *Science* 317: 58–62.

Ivlev VS (1961) *Experimental Ecology of the Feeding of Fish.* New Haven, CT, Yale University Press.

Jackson TA, Alves SB & Pereira RM (2000) Success in biological control of soil-dwelling insects by pathogens and nematodes. In *Biological Control: Measures of Success* (Gurr G & Wratten S, eds.), Dordrecht, The Netherlands, Kluwer, pp. 271–296.

Jackson TA, Crawford AM & Glare TR (2005) *Oryctes* virus – time for a new look at a useful biocontrol agent. *Journal of Invertebrate Pathology* 89: 91–94.

Jacobsen BJ, Zidack NK & Larson BJ (2004) The role of *Bacillus*-based biological control agents in integrated pest management systems: plant diseases. *Phytopathology* 94: 1272–1275.

Jacobson, RJ, Croft P & Fenlon J (2001) Suppressing establishment of *Frankliniella occidentalis* Pergande (Thysanoptera: Thripidae) in cucumber crops by prophylactic release of *Amblyseius cucumeris* Oudemans (Acarina: Phytoseiidae). *Biocontrol Science and Technology* 11: 27–34.

Jacometti M, Jorgensen N & Wratten S (2010) Enhancing biological control by an omnivorous lacewing: floral resources reduce aphid numbers at low aphid densities. *Biological Control* 55: 159–165.

Jaksić FM & Yáñez JL (1983) Rabbit and fox introductions in Tierra del Fuego: history and assessment of the attempts at biological control of the rabbit infestation. *Biological Conservation* 26: 367–374.

James RR, McEvoy PB & Cox CS (1992) Combining the cinnabar moth (*Tyria jacobaeae*) and the ragwort leaf beetle (*Longitarsus jacobaeae*) for control of ragwort (*Senecio jacobaea*): an experimental analysis. *Journal of Applied Ecology* 29: 589–596.

Janisiewicz WJ, Tworkoski TJ & Sharer C (2000) Characterizing the mechanism of biological control of postharvest diseases on fruits with a simple method to study competition for nutrients. *Phytopathology* **90**: 1196–1200.

Janssen A (1999) Plants with spider-mite prey attract more predatory mites than clean plants under greenhouse conditions. *Entomologia Experimentalis et Applicata* **90**: 191–198.

Janssen A, Montserrat M, HilleRisLambers R, De Roos AM, Pallini A & Sabelis MW (2006) Intraguild predation usually does not disrupt biological control. In *Trophic and Guild Interactions in Biological Control* (Brodeur J & Boivin G, eds.), Dordrecht, The Netherlands, Springer, pp. 21–44.

Janssen A & Sabelis MW (1992) Phytoseiid life histories, local predator–prey dynamics, and strategies for control of tetranychid mites. *Experimental and Applied Acarology* **14**: 233–250.

Jansson JK (2003) Marker and reporter genes: illuminating tools for environmental microbiologists. *Current Opinion in Microbiology* **6**: 310–316.

Jaronski ST (2012) Microbial control of invertebrate pests. In *Beneficial Microorganisms in Agriculture, Food and the Environment* (Sundh I, Wilcks A & Goettel M, eds.), Wallingford, UK, CABI Publishing, pp. 72–95.

Jaronski ST & Jackson MA (2012) Mass production of entomopathogenic *Hypocreales*. In *Manual of Techniques in Invertebrate Pathology* (Lacey LA, ed.), San Diego, CA, Academic Press, pp. 255–284.

Jarvis PJ, Fowler SV, Paynter Q & Syrett P (2006) Predicting the economic benefits and costs of introducing new biological control agents for Scotch broom *Cytisus scoparius* into New Zealand. *Biological Control* **39**: 135–146.

Jedlicka JA, Greenberg R & Letourneau DK (2011) Avian conservation practices strengthen ecosystem services in California vineyards. *PLoS One* **6**: e27347

Jeffries MJ & Lawton JH (1984) Enemy free space and the structure of ecological communities. *Biological Journal of Linnean Society* **23**: 269–286.

Jeger MJ, Jeffries P, Elad Y. & Xu X.-M. (2009) A generic theoretical model for biological control of foliar plant diseases. *Journal of Theoretical Biology* **256**: 201–214.

Jenner WH, Kuhlmann U, Miall JH, Cappuccino N & Mason PG (2014) Does parasitoid state affect host range expression? *Biological Control* **78**: 15–22.

Jenkins NE & Grzywacz D (2000) Quality control of fungal and viral biocontrol agents: assurance of product performance. *Biocontrol Science and Technology* **10**: 753–777.

Jenkins NE & Grzywacz D (2003) Towards the standardisation of quality control of fungal and viral biocontrol agents. In *Quality Control and Production of Biological Control Agents: Theory and Testing Procedures* (Van Lenteren JC, ed.), Wallingford, UK, CABI Publishing, pp. 247–263.

Jensen GL, Shelton WL, Yang SL & Wilken LO (1983) Sex reversal of gynogenetic grass carp by implantation of methyltestosterone. *Transactions of the American Fisheries Society* **112**: 79–85.

Jervis MA & Heimpel GE (2005) Phytophagy. In *Insects as Natural Enemies: A Practical Perspective* (Jervis MA, ed.), Dordrecht, The Netherlands, Springer, pp. 525–550.

Jervis MA & Kidd NAC (1986) Host-feeding strategies in hymenopteran parasitoids. *Biological Reviews* **61**: 396–434.

Jervis MA, Kidd NAC & Walton M (1992) A review of methods for determining dietary range in adult parasitoids. *Entomophaga* **37**: 565–574.

Jervis MA, Lee JC & Heimpel GE (2004) Use of behavioural and life-history studies to understand the effects of habitat manipulation. In *Ecological Engineering for Pest Management* (Gurr GM & Wratten SD, eds.), Collingwood, Australia, CSIRO, pp. 65–100.

Jervis MA, Moe A & Heimpel GE (2012) The evolution of parasitoid fecundity: a paradigm under scrutiny. *Ecology Letters* **15**: 357–364.

Jeschke JM, Kopp M & Tollrian R (2002) Predator functional responses: discriminating between handling and digesting prey. *Ecological Monographs* **72**: 95–112.

Jetter K (2005) Economic framework for decision making in biological control. *Biological Control* **35**: 348–357.

Jezorek H, Stiling P & Carpenter J (2011) Ant predation on an invasive herbivore: can an extrafloral nectar-producing plant provide associational resistance to *Opuntia* individuals? *Biological Invasions* **13**: 2261–2273.

Jobin A, Schaffner U & Nentwig W (1996) The structure of the phyophagous insect fanua on the introduced weed *Solidago altissima* in Switzerland. *Entomologia Experimentalis et Applicata* **79**: 33–42.

Johnson DM & Stiling P (1996) Host specificity of *Cactoblastis cactorum* (Lepidoptera: Pyralidae), an exotic *Opuntia*-feeding moth, in Florida. *Environmental Entomology* **25**: 743–748.

Johnson DM & Stiling PD (1998) Distribution and dispersal of *Cactoblastis cactorum* (Lepidoptera: Pyralidae), an exotic *Opuntia*-feeding moth, in Florida. *Florida Entomologist* **81**: 12–22.

Johnson KB (1994) Dose-response relationships and inundative biological control. *Phytopathology* **84**: 780–784.

Johnson KB (1999) Dose-response relationships in biocontrol of plant disease and their use to define pathogen refuge size. In *Theoretical Approaches to Biological Control* (Hawkins BA & Cornell HV, eds.), Cambridge, UK, Cambridge University Press, pp. 385–392.

Johnson KB (2010) Pathogen refuge: a key to understanding biological control. *Annual Review of Phytopathology* **48**: 141–160.

Johnson KB & DiLeone JA (1999) Effect of antibiosis on antagonist dose-plant disease response relationships for the biological control of crown gall of tomato and cherry. *Phytopathology* **89**: 974–980.

Johnson MD, Kellermann JL & Stercho AM (2010) Pest reduction services by birds in shade and sun coffee in Jamaica. *Animal Conservation* **13**: 140–147.

Johnson MD, Levy NJ, Kellermann JL & Robinson DE (2009) Effects of shade and bird exclusion on arthropods and leaf damage on coffee farms in Jamaica's Blue Mountains. *Agroforestry Systems* **76**: 139–148.

Johnson MT, Follett PA, Taylor AD & Jones VP (2005) Impacts of biological control and invasive species on a non-target native Hawaiian insect. *Oecologia* **142**: 529–540.

Johnson MW & Tabashnik BE (1999) Enhanced biological control through pesticide selectivity. In *Handbook of Biological Control* (Bellows TS & Fisher TW, eds.), San Diego, CA, Academic Press, pp. 297–317.

Johnston J (2001) *Pesticides and Wildlife*. Washington, DC, American Chemical Society.

Jones CD (2005) The genetics of adaptation in *Drosophila seychellia*. *Genetica* **123**: 137–145.

Jones DA, Ryder MH, Clare BG, Farrand SK & Kerr A (1988) Construction of a Tra-deletion mutant of pAgK84 to safeguard the biological control of crown gall. *Molecular and General Genetics* **212**: 207–214.

Jones KA & Burges D (1998) Technology of formulation and application. In *Formulation of Microbial Pesticides* (Burges DH, ed.), Dordrecht, The Netherlands, Kluwer, pp. 7–30.

Jones VP (1995) Reassessment of the role of predators and *Trissolcus basalis* in biological control of southern green stink bug (Hemiptera, Pentatomidae) in Hawaii. *Biological Control* **5**: 566–572.

Jones VP, Westcott DM, Finson NN & Nishimoto RK (2001) Relationship between community structure and southern green stink bug (Heteroptera: Pentatomidae) damage in macadamia nuts. *Environmental Entomology* **30**: 1028–1035.

Jonsen ID, Bourchier RS & Roland J (2007) Influence of dispersal, stochasticity, and an Allee effect on the persistence of weed biocontrol introductions. *Ecological Modelling* **203**: 521–526.

Jonsson M, Buckley HL, Case BS, Wratten SD, Hale RJ & Didham RK (2012) Agricultural intensification drives landscape-context effects on host-parasitoid interactions in agroecosystems. *Journal of Applied Ecology* **49**: 706–714.

Jonsson M, Straub CS, Didham RK, Buckley HL, Case BS, Hale RJ, Gratton C & Wratten SD (2015) Experimental evidence that the effectiveness of conservation biological control depends on landscape complexity. *Journal of Applied Ecology* **52**: 1274–1282.

Jonsson M, Wratten SD, Landis DA & Gurr GM (2008) Recent advances in conservation biological control of arthropods by arthropods. *Biological Control* **45**: 172–175.

Jonsson M, Wratten SD, Landis DA, Tompkins JML & Cullen R (2010) Habitat manipulation to mitigate the impacts of invasive arthropod pests. *Biological Invasions* **12**: 2933–2945.

Jonsson M, Wratten SD, Robinson KA & Sam SA (2009) The impact of floral resources and omnivory on a four trophic level food web. *Bulletin of Entomological Research* **99**: 275–285.

Joshi J & Vrieling K (2005) The enemy release and EICA hypothesis revisited: incorporating the fundamental difference between specialist and generalist herbivores. *Ecology Letters* **8**: 704–714.

Julien MH (1982) *Biological Control of Weeds: A World Catalogue of Agents and Their Target Weeds.* Slough, UK, Commonwealth Agricultural Bureaux.

Julien MH (1987) *Biological Control of Weeds: A World Catalogue of Agents and Their Target Weeds*, 2nd edition. Wallingford, UK, CABI Publishing.

Julien MH (1992) *Biological Control of Weeds: A World Catalogue of Agents and Their Target Weeds*, 3rd edition. Wallingford, UK, CABI Publishing.

Julien MH & Griffiths MW (1998) *Biological Control of Weeds: A World Catalogue of Agents and Their Target Weeds*, 4th edition. Wallingford, UK, CABI Publishing.

Kabaluk T & Gazdik Z (2007) *Directory of Microbial Pesticides for Agricultural Crops in OECD Countries.* Agriculture and AgriFood Canada, http://publications.gc.ca/site/eng/359060/publication.html

Kadir JB, Charudattan R, Stall WM & Brecke BJ (2000) Field efficacy of *Dactylaria higginsii* as a bioherbicide for the control of purple nutsedge (*Cyperus rotundus*). *Weed Technology* **14**: 1–6.

Kajita Y, Takano F, Yasuda H & Agarwala BK (2000) Effects of indigenous ladybird species (Coleoptera: Coccinellidae) on the survival of an exotic species in relation to prey abundance. *Applied Entomology and Zoology* **35**: 473–479.

Kajita Y, Takano F, Yasuda H & Evans EW (2006a) Interactions between introduced and native predatory ladybirds (Coleoptera, Coccinellidae): factors influencing the success of species introductions. *Ecological Entomology* **31**: 58–67.

Kajita Y, Yasuda H & Evans EW (2006b) Effects of native ladybirds on oviposition of the exotic species, *Adalia bipunctata* (Coleoptera: Coccinellidae), in Japan. *Applied Entomology and Zoology* **41**: 57–61.

Kaplan I & Thaler JS (2010) Plant resistance attenuates the consumptive and non-consumptive impacts of predators on prey. *Oikos* **119**: 1105–1113.

Kapranas A, Morse JG, Pacheco P, Forster LD & Luck RF (2007) Survey of brown soft scale *Coccus hesperidum* L. parasitoids in southern California citrus. *Biological Control* **42**: 288–299.

Kapuscinski AR & Patronski TJ (2005) *Genetic Methods for Biological Control of Non-native Fish in the Gila River Basin.* St. Paul, University of Minnesota, Institute for Social Economic and Ecological Sustainability.

Karban R & Baldwin IT (2000) *Induced Responses to Herbivory.* Chicago, IL, University of Chicago Press.

Karban R, English-Loeb G & Hougen-Eitzman D (1997) Mite vaccinations for sustainable management of spider mites in vineyards. *Ecological Applications* **7**: 183–193.

Kareiva P (1990) Establishing a foothold for theory in biocontrol practice: using models to guide experimental design and release protocols. In *New Directions in Biological Control: Alternatives for Suppressing Agricultural Pests and Diseases* (Baker RR & Dunn PE, eds.), New York, Alan R. Liss, pp. 65–81.

Kareiva P (1996) Contributions of ecology to biological control. *Ecology* **77**: 1963–1964.

Karimzadeh R, Hejazi MJ, Helali H, Iranipour S & Mohammadi SA (2011) Assessing the impact of site-specific spraying on control of *Eurygaster integriceps* (Hemiptera: Scutelleridae) damage and natural enemies. *Precision Agriculture* **12**: 576–593.

Kaser JM & Heimpel GE (2015) Linking risk and efficacy in biological control host-parasitoid models. *Biological Control* **90**: 49–60.

Kaser JM & Ode PJ (2016) Hidden risks and benefits of natural enemy mediated indirect effects *Current Opinion in Insect Science* **14**: 105–111.

Kassa A, Stephan D, Vidal S & Zimmermann G (2004) Laboratory and field evaluation of different formulations of *Metarhizium anisopliae* var. *acridum* submerged spores and aerial conidia for the control of locusts and grasshoppers. *BioControl* **49**: 63–81.

Kaya HK & Gaugler R (1993) Entomopathogenic nematodes. *Annual Review of Entomology* **38**: 181–206.

Kean JM & Barlow ND (2000a) Can host-parasitoid metapopulations explain successful biological control? *Ecology* **81**: 2188–2197.

Kean JM & Barlow ND (2000b) Effects of dispersal on local population increase. *Ecology Letters* **3**: 479–482.

Kean JM & Barlow ND (2000c) Long-term assessment of the biological control of *Sitona discoideus* by *Microctonus aethiopoides* and test of a model. *Biocontrol Science and Technology* **10**: 215–221.

Kean JM & Barlow ND (2001) A spatial model for the successful biological control of *Sitona discoideus* by *Microctonus aethiopoides*. *Journal of Applied Ecology* **38**: 162–169.

Kean J, Wratten S, Tylianakis J & Barlow N (2003) The population consequences of natural enemy enhancement, and implications for conservation biological control. *Ecology Letters* **6**: 604–612.

Keane RM & Crawley MJ (2002) Exotic plant invasions and the enemy release hypothesis. *Trends in Ecology & Evolution* **17**: 164–170.

Kearney MR & Porter WP (2009) Mechanistic niche modelling: combining physiological and spatial data to predict species' ranges. *Ecology Letters* **12**: 334–350.

Kelch DG & McClay A (2004) Putting the phylogeny into the centrifugal phylogenetic method. In *XI International Symposium on Biological Control of Weeds* (Cullen JM, Briese DT, Kriticos DJ, Lonsdale WM, Morin L & Scott JK, eds.), Canberra, Australia, CSIRO, pp. 287–291.

Keller S, Schweizer C & Shah P (1999) Differential susceptibility of two *Melolontha populations* to infections by the fungus *Beauveria brongniartii*. *Biocontrol Science and Technology* **9**: 441–446.

Kellogg SK, Fink LS & Brower LP (2003) Parasitism of native luna moths, *Actias luna* (L.) (Lepidoptera: Saturniidae) by the introduced *Compsilura concinnata* (Meigen) (Diptera: Tachinidae) in central Virginia, and their hyperparasitism by trigonalid wasps (Hymenoptera: Trigonalidae). *Environmental Entomology* **32**: 1019–1027.

Kemp JC & Barrett GW (1989) Spatial patterning – impact of uncultivated corridors on arthropod populations within soybean agroecosystems. *Ecology* **70**: 114–128.

Kennedy TA, Naeem S, Howe KM, Knops MH, Tilman D & Reich P (2002) Biodiversity as a barrier to ecological invasion. *Nature* **417**: 636–638.

Kermack WO & McKendrick AG (1927) A contribution to the mathematical theory of epidemics. *Proceedings of the Royal Society, A* **115**: 700–721.

Kerr PJ, Ghedin E, DePasse JV, Fitch A, Cattadori IM, Hudson PJ, Tscharke DC, Read AF & Holmes EC (2012) Evolutionary history and attenuation of myxoma virus on two continents. *PLoS Pathogens* **8**: e1002950.

Kerr P & McFadden G (2002) Immune responses to myxoma virus. *Viral Immunology* **15**: 229–246.

Kerr PJ, Rogers MB, Fitch A, Depasse JV, Cattadori IM, Twaddle AC, Hudson PJ, Tscharke DC, Read AF, Holmes EC & Ghedin E (2013) Genome scale evolution of myxoma virus reveals host-pathogen adaptation and rapid geographic spread. *Journal of Virology* **87**: 12900–12915.

Kessel GJT, De Haas BH, Van der Werf W & Kohl J (2002) Competitive substrate colonisation by *Botrytis cinerea* and *Ulocladium atrum* in relation to biological control of *B. cinerea* in cyclamen. *Mycological Research* **106**: 716–728.

Kettenring KM & Mock KE (2012) Genetic diversity, reproductive mode, and dispersal differ between the cryptic invader, *Phragmites australis*, and its native conspecific. *Biological Invasions* **14**: 2489–2504.

Kevan PG, Sutton J & Shipp L (2007) Pollinators as vectors of biocontrol agents – the B52 story. In *Biological Control: A Global Perspective* (Vincent C, Goettel M & Lazarovits G, eds.), Wallingford, UK, CABI Publishing, pp. 319–327.

Khan ZR, Ampong Nyarko K, Chiliswa P, Hassanali A, Kimani S, Lwande W, Overholt WA, Pickett JA, Smart LE, Wadhams LJ & Woodcock CM (1997) Intercropping increases parasitism of pests. *Nature* **388**: 631–632.

Kidd D & Amaresekare P (2012) The role of transient dynamics in biological pest control: insights from a host-parasitoid community. *Journal of Animal Ecology* **81**: 47–57.

Kimberling DN (2004) Lessons from history: predicting successes and risks of intentional introductions for arthropod biological control. *Biological Invasions* **6**: 301–318.

Kimbro DL, Cheng BS & Grosholz ED (2013) Biotic resistance in marine environments. *Ecology Letters* **16**: 821–833.

Kindlmann P & Dixon AFG (1999) Strategies of aphidophagous predators: lessons from modelling insect predator–prey dynamics. *Journal of Applied Entomology* **123**: 397–399.

Kindlmann P & Dixon AFG (2001) When and why top-down regulation fails in arthropod predator–prey systems. *Basic and Applied Ecology* **2**: 333–340.

King C & Rubinoff D (2008) First record of fossorial behavior in Hawaiian leafroller moth larvae, *Omiodes continuatalis* (Lepidoptera: Crambidae). *Pacific Science* **62**: 147–150.

King GF, Escoubas P & Nicholson GM (2008) Peptide toxins that selectively target insect Na-V and Ca-V channels. *Channels* **2**: 100–116.

King JR & Tschinkel WR (2008) Experimental evidence that human impacts drive fire ant invasions and ecological change. *Proceedings of the National Academy of Sciences of the United States of America* **105**: 20339–20343.

Kinkel LL, Bakker MG & Schlatter DC (2011) A coevolutionary framework for managing disease-suppressive soils. *Annual Review of Phytopathology* **49**: 47–67.

Kinzie RA (1992) Predation by the introduced carnivorous snail *Euglandina rosea* (Ferussac) on endemic aquatic lymnaeid snails in Hawaii. *Biological Conservation* **60**: 149–155.

Kirkpatrick JF & Frank KM (2005) Contraception in free-ranging wildlife. In *Wildlife Contraception: Issues, Methods, and Applications* (Asa CS & Porton IJ, eds.), Baltimore, MD, Johns Hopkins University Press, pp. 195–221.

Klingen I & Haukeland S (2006) The soil as a reservoir for natural enemies of pest insects and mites with emphasis on fungi and nematodes. In *An Ecological and Societal Approach to Biological Control* (Eilenberg J & Hokkanen HMT, eds.), Dordrecht, The Netherlands, Springer, pp. 145–211.

Kloepper J, Tuzun S & Kuć J (1992) Proposed definitions related to induced disease resistance. *Biocontrol Science and Technology* **2**: 347–349.

Knipling EF (1992) *Principles of Insect Parasitism Analyzed from New Perspectives: Practical Implications for Regulating Insect Populations by Biological Means.* Washington, DC, USDA-ARS Agriculture Handbook No. 693.

Knops JMH, Tilman D, Haddad NM, Naeem S, Mitchell CE, Haarstad J, Ritchie ME, Howe KM, Reich PB, Siemann E & Groth J (1999) Effects of plant species richness on invasion dynamics, disease outbreaks, insect abundances and diversity. *Ecology Letters* **2**: 286–293.

Knutson AE (2003) Release of *Trichogramma pretiosum* in cotton with a novel ground sprayer. *Southwestern Entomologist* **28**: 11–17.

Kolbe JJ, Glor RE, Rodríguez Schettino L, Lara AC, Larson A & Losos JB (2004) Genetic variation increases during biological invasion by a Cuban lizard. *Nature* **431**: 177–181.

Konishi M & Ito Y (1973) Early entomology in East Asia. In *History of Entomology* (Smith RF, Mittler TE & Smith CN, eds.), Palo Alto, CA, Annual Reviews Inc., pp. 1–20.

Koss AM, Jensen AS, Schreiber A, Pike KS & Snyder WE (2005) Comparison of predator and pest communities in Washington potato fields treated with broad-spectrum, selective, or organic insecticides. *Environmental Entomology* **34**: 87–95.

Kovaliski J (1998) Monitoring the spread of rabbit hemorrhagic disease virus as a new biological agent for control of wild European rabbits in Australia. *Journal of Wildlife Diseases* **34**: 421–428.

Kowalchuk GA, Os GJ, Aartrijk J & Veen JA (2003) Microbial community responses to disease management soil treatments used in flower bulb cultivation. *Biology and Fertility of Soils* **37**: 55–63.

Kraaijeveld AR (2004) Experimental evolution in host-parasitoid interactions. In *Genetics, Evolution and Biological Control* (Ehler LE, Sforza R & Mateille T, eds.), Wallingford, UK, CABI Publishing, pp. 163–181.

Kraaijeveld AR, Ferrari J & Godfray HCJ (2002) Costs of resistance in insect-parasite and insect-parasitoid interactions. *Parasitology* **125**: S71–S82.

Kraaijeveld AR & Van Alphen JJM (1995) Geographical variation in encapsulation ability of *Drosophila melanogaster* larvae and evidence for parasitoid-specific components. *Evolutionary Ecology* **9**: 10–17.

Krafsur ES (1998) Sterile insect technique for suppressing and eradicating insect populations: 55 years and counting. *Journal of Agricultural Entomology* **15**: 303–317.

Kramer AM, Dennis B, Liebhold AM & Drake JM (2009) The evidence for Allee effects. *Population Ecology* **51**: 341–354.

Kremer RJ & Li JM (2003) Developing weed-suppressive soils through improved soil quality management. *Soil & Tillage Research* **72**: 193–202.

Krenek L & Rudolph VHW (2014) Allometric scaling of indirect effects: body size ratios predict non-consumptive effects in multi-predator systems. *Journal of Animal Ecology* **83**: 1461–1468.

Krips OE, Willems PEL & Dicke M (1999) Compatibility of host plant resistance and biological control of the two-spotted spider mite *Tetranychus urticae* in the ornamental crop gerbera. *Biological Control* **16**: 155–163.

Krischik VA, Landmark AL & Heimpel GE (2007) Soil-applied imidacloprid is translocated to nectar and kills nectar-feeding *Anagyrus pseudococci* (Girault) (Hymenoptera: Encyrtidae). *Environmental Entomology* **36**: 1238–1245.

Kriticos DJ (2003) The roles of ecological models in evaluating weed biological control agents and projects. In *Improving the Selection, Testing and Evaluation of Weed Biological Control Agents* (Spafford Jacob H & Briese DT, eds.), Osmond, Australia, CRC, pp. 69–74.

Kriticos DJ, Stuart RM & Ash JE (2004) Exploring interactions between cultural and biological control techniques: modelling bitou bush (*Chrysantehmoides monilifera* spp. *rotundata*) and a seed fly (*Mesoclanis polana*). In *Proceedings of the XI International Symposium on Biological Control of Weeds* (Cullen JM, Briese DT, Kriticos DJ, Lonsdale WM, Morin L & Scott JK, eds.), Canberra, Australia., CSIRO, p. 559.

Kruess A & Tscharntke T (1994) Habitat fragmentation, species loss, and biological control. *Science* **264**: 1581–1584.

Kuhlmann U, Mason PG, Hinz HL, Blossey B, De Clerck-Floate RA, Dosdall LM, McCaffrey JP, Schwarzlaender M, Olfert O, Brodeur J, Gassmann A, McClay AS & Wiedenmann RN (2006a) Avoiding conflicts between insect and weed biological control: selection of non-target species to assess host specificity of cabbage seedpod weevil parasitoids. *Journal of Applied Entomology* **130**: 129–141.

Kuhlmann U, Schaffner U & Mason PG (2006b) Selection of non-target species for host specificity testing. In *Environmental Impact of Invertebrates for Biological Control of Arthropods: Methods and Risk Assessment* (Bigler F, Babendreier D & Kuhlmann U, eds.), Wallingford, UK, CABI Publishing, pp. 15–37.

Kumschick S, Gaertner M, Vilá M, Essl F, Jeschke JM, Pysek P, Ricciardi A, Bacher S, Blackburn TM, Dick JTA, Evans T, Hulme PE, Kuhn I, Mrugala A, Pergl J, Rabitsch W, Richardson DM, Sendek A & Winter M (2015) Ecological impacts of alien species: quantification, scope, caveats, and recommendations. *BioScience* **65**: 55–63.

Kuris AM (2003) Did biological control cause extinction of the coconut moth, *Levuana iridescens*, in Fiji? *Biological Invasions* **5**: 133–141.

Kuske S, Babendreier D, Edwards PJ, Turlings TCJ & Bigler F (2004) Parasitism of non-target lepidoptera by mass-released *Trichogramma brassicae* and its implication for the larval parasitoid *Lydella thomposoni*. *BioControl* **49**: 1–19.

Kuske S, Widmer F, Edwards JP, Turlings TCJ, Babendreier D & Bigler F (2003) Dispersal and persistence of mass released *Trichogramma brassicae* (Hymenoptera: Trichogrammatidae) in non-target habitats. *Biological Control* **27**: 181–193.

Kwak Y-S & Weller DM (2013) Take-all of wheat and natural disease suppression: a review. *Plant Pathology Journal* **29**: 125–135.

Kwok OCH, Fahy PC, Hoitink HAJ & Kuter GA (1987) Interactions between bacteria and *Trichoderma hamatum* in suppression of *Rhizoctonia* damping-off in bark compost media. *Phytopathology* **77**: 1206–1212.

Lacey LA & Arthurs SP (2005) New method for testing solar sensitivity of commercial formulations of the granulovirus of codling moth (*Cydia pomonella*,

Tortricidae: Lepidoptera). *Journal of Invertebrate Pathology* **90**: 85–90.

Lacey LA, Grzywacz D, Shapiro-Ilan DI, Frutos R, Brownbridge M, Goettel MS (2015) Insect pathogens as biological control agents: back to the future. *Journal of Invertebrate Pathology* **132**: 1–41.

Lacey LA, Kaya HK & Vail P (2001) Insect pathogens as biological control agents: do they have a future? *Biological Control* **21**: 230–248.

Lacey LA, Thomson D, Vincent C & Arthurs SP (2008) Codling moth granulovirus: a comprehensive review. *Biocontrol Science and Technology* **18**: 639–663.

Lafferty KD & Kuris AM (1996) Biological control of marine pests. *Ecology* **77**: 1989–2000.

Lagnaoui A & Radcliffe EB (1997) Interference of fungicides with entomopathogens: effects on entomophthoran pathogens of green peach aphids. In *Ecological Interactions and Biological Control* (Andow DA, Ragsdale DW & Nyvall RF, eds.), Boulder, CO, Westview, pp. 301–315.

Lagnaoui A & Radcliffe EB (1998) Potato fungicides interfere with entomopathogenic fungi impacting population dynamics of green peach aphid. *American Journal of Potato Research* **75**: 19–25.

Laha M & Mattingly HT (2007) Ex situ evaluation of impacts of invasive mosquitofish on the imperiled Barrens topminnow. *Environmental Biology of Fishes* **78**: 1–11.

Lambrinos JG (2004) How interactions between ecology and evolution influence contemporary invasion dynamics. *Ecology* **85**: 2061–2070.

Landis DA, Wratten SD & Gurr GM (2000) Habitat management to conserve natural enemies of arthropod pests in agriculture. *Annual Review of Entomology* **45**: 175–201.

Lane SD, Mills NJ & Getz WM (1999) The effects of parasitoid fecundity and host taxon on the biological control of insect pests: the relationship between theory and data. *Ecological Entomology* **24**: 181–190.

Lane SD, St. Mary CM & Getz WM (2006) Coexistence of attack-limited parasitoids sequentially exploiting the same resource and its implications for biological control. *Annales Zoologici Fennici* **43**: 17–34.

Langellotto GA & Denno RF (2004) Responses of invertebrate natural enemies to complex-structured habitats: a meta-analytical synthesis. *Oecologia* **139**: 1–10.

Langer A & Hance T (2004) Enhancing parasitism of wheat aphids through apparent competition: a tool for biological control. *Agriculture Ecosystems & Environment* **102**: 205–212.

Langewald J & Kooyman C (2007) Green Muscle, a fungal biopesticide for control of locusts and grasshoppers in Africa. In *Biological Control: A Global Perspective* (Vincent C, Goettel M & Lazarovits G, eds.), Wallingford, UK, CABI Publishing, pp 311–318.

Larkin RP & Fravel DR (1999) Mechanisms of action and dose-response relationships governing biological control of fusarium wilt of tomato by nonpathogenic *Fusarium* spp. *Phytopathology* **89**: 1152–1161.

Larsen A & Philpott SM (2010) Twig-nesting ants: the hidden predators of the coffee berry borer in Chiapas, Mexico. *Biotropica* **42**: 342–347.

Larson DL, Grace JB, Rabie PA & Andersen P (2007) Short-term disruption of a leafy spurge (*Euphorbia esula*) biocontrol program following herbicide application. *Biological Control* **40**: 1–8.

Lau JA, Puliafico KP, Kopshever JA, Steltzer H, Jarvis EP, Schwarzländer M, Strauss SY & Hufbauer RA (2008) Inference of allelopathy is complicated by effects of activated carbon on plant growth. *New Phytologist* **178**: 412–423.

Lavandero BI, Wratten SD, Didham RK & Gurr GM (2006) Increasing floral diversity for selective enhancement of biological control agents: a double-edged sward? *Basic and Applied Ecology* **7**: 236–243.

Lavandero B, Wratten S, Shishehbor P & Worner S (2005) Enhancing the effectiveness of the parasitoid *Diadegma semiclausum* (Helen): movement after use of nectar in the field. *Biological Control* **34**: 152–158.

Laven H (1967) Eradication of *Culex pipiens fatigans* through cytoplasmic incompatibility. *Nature* **216**: 383–384.

Lavergne S & Molofsky J (2007) Increased genetic variation and evolutionary potential drive the success of an invasive grass. *Proceedings of the National Academy of Sciences, USA* **104**: 3883–3888.

Lawton JH (1985) Ecological theory and choice of biological control agents. *Proceedings VI International Symposium on the Biological Control of Weeds* (Delfosse S, ed.), Ottawa, Canada, Agriculture Canada, pp. 13–26.

le Masurier AD & Waage JK (1993) A comparison of attack rates in a native and an introduced population of the parasitoid *Cotesia glomerata*. *Biocontrol Science and Technology* **3**: 467–474.

Lee CE (2002) Evolutionary genetics of invasive species. *Trends in Ecology & Evolution*, **17**: 386–391.

Lee JC, Andow DA & Heimpel GE (2006) Influence of floral resources on sugar feeding and nutrient dynamics of a parasitoid in the field. *Ecological Entomology* **31**: 470–480.

Lee JC & Heimpel GE (2005) Impact of flowering buckwheat on lepidopteran cabbage pests and their parasitoids at two spatial scales. *Biological Control* **34**: 290–301.

Lee JC & Heimpel GE (2008) Floral resources impact longevity and oviposition rate of a parasitoid in the field. *Journal of Animal Ecology* **77**: 565–572.

Lee JC, Menalled FD & Landis DA (2001) Refuge habitats modify impact of insecticide disturbance on carabid beetle communities. *Journal of Applied Ecology* **38**: 472–483.

Legner EF (1972) Observations on hybridization and heterosis in parasitoids of synanthropic flies. *Annals of the Entomological Society of America* **65**: 254–263.

Legner EF (2008) Biological pest control: a history. In *Encyclopedia of Pest Management* (Pimentel D, ed.), London, UK, Taylor & Francis.

Leland JE, Mullins DE, Vaughan LJ and Warren HL (2005) Effects of media composition on submerged culture spores of the entomopathogenic fungus, *Metarhizium anisopliae* var. *acridum*, Part 1: Comparison of cell wall characteristics and drying stability among three spore types. *Biocontrol Science and Technology* **15**: 379–392.

Lemke A & Poehling HM (2002) Sown weed strips in cereal fields: overwintering site and 'source' habitat for *Oedothorax apicatus* (Blackwall) and *Erigone atra* (Blackwall) (Araneae: Erigonidae). *Agriculture Ecosystems & Environment* **90**: 67–80.

Leonardo TE (2004) Removal of a specialization-associated symbiont does not affect aphid fitness. *Ecology Letters* **7**: 461–468.

Lester PJ, Thistlewood HMA & Harmsen R (1998) The effects of refuge size and number on acarine predator–prey dynamics in a pesticide-disturbed apple orchard. *Journal of Applied Ecology* **35**: 323–331.

Letourneau DK & Bothwell SG (2008) Comparison of organic and conventional farms: challenging ecologists to make biodiversity functional. *Frontiers in Ecology and the Environment* **6**: 430–438.

Lever C (1985) *Naturalized Mammals of the World*. London, UK, Longman.

Levin S & Pimentel D (1981) Selection of intermediate rates of increase in parasite-host systems. *American Naturalist* **117**: 308–315.

Levine JM, Adler PB & Yelenik SG (2004) A meta-analysis of biotic resistance to exotic plant invasions. *Ecology Letters* **7**: 975–989.

Levine JM & D'Antonio CM (1999) Elton revisited: a review of evidence linking diversity and invasibility. *Oikos* **87**: 15–26.

Levine JM, Vila M, D'Antonio CM, Dukes JS, Grigulis K & Lavorel S (2003) Mechanisms underlying the impacts of exotic plant invasions. *Proceedings of the Royal Society B-Biological Sciences* **270**: 775–781.

Levins R (1969) Some demographic and genetic consequences of environmental heterogeneity for biological control. *Bulletin of the Entomological Society of America* **15**: 237–240.

Levins R (1975) Evolution in communities near equilibrium. *Ecology and Evolution of Communities* (Cody ML & Diamond JM, eds.), Cambridge, MA, Harvard University Press, pp. 16–50.

Lewis EE, Campbell JF & Gaugler R (1998) A conservation approach to using entomopathogenic nematodes in turf and landscapes. In *Conservation Biological Control* (Barbosa P, ed.), San Diego, CA, Academic Press, pp. 235–254.

Lewis MA & Kareiva P (1993) Allee dynamics and the spread of invading organisms. *Theoretical Population Biology* **43**: 141–158.

Leyse KE, Lawler SP & Strange T (2004) Effects of an alien fish, *Gambusia affinis*, on an endemic California fairy shrimp, *Linderiella occidentalis*: implications for conservation of diversity in fishless waters. *Biological Conservation* **118**: 57–65.

Li S, Falabella P, Giannanonio S, Fanti P, Battaglia D, Digilio MC, Völkl W, Sloggett JJ, Weisser W & Pennacchio F (2002) Pea aphid clonal resistance to the endophagous parasitoid *Aphidius ervi. Journal of Insect Physiology* **48**: 971–980.

Liebhold AM & Elkinton JS (1989) Elevated parasitism in artificially augmented populations of *Lymantria dispar* (Lepidoptera: Lymantriidae). *Environmental Entomology* **18**: 986–995.

Liebhold AM & Tobin PC (2008) Population ecology of insect invasions and their management. *Annual Review of Entomology* **53**: 387–408.

Liljesthroom G & Rabinovich J (2004) Modeling biological control: the population regulation of *Nezara viridula* by *Trichopoda giacomellii. Ecological Applications* **14**: 254–267.

Lin BB (2007) Agroforestry management as an adaptive strategy against potential microclimate extremes in coffee agriculture. *Agricultural and Forest Meteorology* **144**: 85–94.

Lincango MP, Causton CE, Alvarez CC & Jimenez-Uzcategui G (2011) Evaluating the safety of *Rodolia cardinalis* to two species of Galapagos finch; *Camarhynchus parvulus* and *Geospiza fuliginosa. Biological Control* **56**: 145–149.

Lindow SE & Leveau JE (2002) Phyllosphere microbiology. *Current Opinions in Biotechnology* **13**: 238–243.

Liu D, Kinkel LL, Eckwall EC, Anderson NA & Schottel JL (1997) Biological control of plant disease using antagonistic *Streptomyces*. In *Ecological Interactions and Biological Control* (Andow DA, Ragsdale DW & Nyvall RF, eds.), Boulder, CO, Westview, pp. 224–239.

Liu H & Stiling P (2006) Testing the enemy release hypothesis: a review and meta-analysis. *Biological Invasions* **8**: 1535–1545.

Liu H, Stiling P & Pemberton RW (2007) Does enemy release matter for invasive plants? Evidence from a comparison of insect herbivore damage among invasive, non-invasive and native congeners. *Biological Invasions* **9**: 773–781.

Liu XX, Chen M, Collins HL, Onstad DW, Roush RT, Zhang QW, Earle ED & Shelton AM (2014) Natural enemies delay insect resistance to Bt crops. *PLoS One* **9**: e90366.

Lloyd CJ, Hufbauer RA, Jackson A, Nissen SJ & Norton AP (2005). Pre- and post-introduction patterns in neutral genetic diversity in the leafy spurge gall midge, *Spurgia capitigena* (Bremi) (Diptera: Cecidomyiidae). *Biological Control* **33**: 153–164.

Lockwood JA (1993) Environmental issues involved in biological control of rangeland grasshoppers (Orthoptera: Acrididae) with exotic agents. *Environmental Entomology* **22**: 503–518.

Lockwood JA, Cassey P & Blackburn T (2005) The role of propagule pressure in explaining species invasions. *Trends in Ecology & Evolution* **20**: 223–228.

Lockwood JL, Hoopes MF & Marchetti MP (2013) *Invasion Ecology, 2nd edition.* Malden, MA, Blackwell.

Lockwood JL, Simberloff D, McKinney ML & Von Holle B (2001) How many, and which, plants will invade natural areas? *Biological Invasions* **3**: 1–8.

Lombaert E, Guillemaud T, Lundgren J, Koch R, Facon B, Grez A, Loomans A, Malausa T, Nedved O, Rhule E, Staverlokk A, Steenberg T & Estoup A (2014) Complementarity of statistical treatments to reconstruct worldwide routes of invasion: the case of the Asian ladybird *Harmonia axyridis. Molecular Ecology* **23**: 5979–5997.

Lomer CJ, Bateman RJ, Dent D, de Groote H, Duoro-Kpindou O-K, Kooyman C, Langewald J, Ouambama Z, Peveling R & Thomas M (1999) Development of strategies for the incorporation of biological pesticides into the integrated pest management of locusts and grasshoppers. *Agricultural and Forest Entomology* **1**: 71–88.

Lomer CJ, Bateman RP, Johnson DL, Langewald J & Thomas M (2001) Biological control of locusts and grasshoppers. *Annual Review of Entomology* **46**: 667–702.

Lonsdale WM (1999) Global patterns of plant invasions and the concept of invasibility. *Ecology* **80**: 1522–1536.

Lonsdale WM, Briese DT & Cullen JM (2001) Risk analysis and weed biological control. In *Evaluating Indirect Ecological Effects of Biological Control* (Wajnberg E, Scott JK & Quimby PC, eds.), Wallingford, UK, CABI Publishing, pp. 185–210.

Lonsdale WM, Farrell G & Wilson CG (1995) Biological control of a tropical weed: a population model and experiment for *Sida acuta*. *Journal of Applied Ecology* **32**: 391–399.

Loper JE, Kobayashi DY & Paulsen IT (2007) The genomic sequence of *Pseudomonas fluorescens* Pf-5: insights into biological control. *Phytopathology* **97**: 233–238.

Loreau M, Mouquet N & Gonzalez A (2003) Biodiversity as spatial insurance in heterogeneous landscapes. *Proceedings of the National Academy of Sciences of the United States of America* **100**: 12765–12770.

Losey JE & Denno RF (1998) Positive predator-predator interactions: enhanced predation rates and synergistic suppression of aphid populations. *Ecology* **79**: 2143–2152.

Losey JE & Vaughan M (2006) The economic value of ecological services provided by insects. *BioScience* **56**: 311–323.

Lotka AJ (1923) Contribution to quantitative parasitology. *Journal of the Washington Academy of Sciences* **13**: 152–158.

Louda SM (1998) Population growth of *Rhinocyllus conicus* (Coleoptera: Curculionidae) on two species of native thistles in Prairie. *Environmental Entomology* **27**: 834–841.

Louda SM & Arnett AE (2000) Predicting non-target ecological effects of biological control agents: evidence from *Rhinocyllus conicus*. In *Proceedings of the X International Symposium on Biological Control of Weeds* (Spencer NR, ed.), Bozeman, Montana State University, pp. 551–567.

Louda SM, Kendall D, Connor J & Simberloff D (1997) Ecological effects of an insect introduced for the biological control of weeds. *Science* **277**: 1088–1090.

Louda SM & O'Brien CW (2002) Unexpected ecological effects of distributing the exotic weevil, *Larinus planus* (F.), for the biological control of Canada thistle. *Conservation Biology* **16**: 717–727.

Louda SM, Pemberton RW, Johnson MT & Follett PA (2003) Nontarget effects: the Achilles' heel of biological control? *Annual Review of Entomology* **48**: 365–396.

Louda SM & Potvin MA (1995) Effect of inflorescence-feeding insects on the demography and lifetime firness of a native plant. *Ecology* **76**: 229–245.

Louda SM, Potvin MA & Collinge SK (1990) Predispersal seed predation, postdispersal seed predation and competition in the recruitment of seedlings of a native thistle in sandhills prairie. *The American Midland Naturalist* **124**: 105–113.

Louda SM, Rand TA, Arnett AE, McClay AS, Shea K & McEachern AK (2005a) Evaluation of ecological risk to populations of a threatened plant from an invasive biocontrol insect. *Ecological Applications* **15**: 234–249.

Louda SM, Rand TA, Russell FL & Arnett AE (2005b) Assessment of ecological risks in weed biocontrol: input from retrospective ecological analyses. *Biological Control* **35**: 253–264.

Lovei GL, Andow DA & Arpaia S (2009) Transgenic insecticidal crops and natural enemies: a detailed review of laboratory studies. *Environmental Entomology* **38**: 293–306.

Lozier JD & Mills NJ (2011) Predicting the potential invasive range of light brown apple moth (*Epiphyas postvittana*) using biologically informed and correlative species distribution models. *Biological Invasions* **13**: 2409–2421.

Lu A & Miller LK (1995) Differential requirements for baculovirus late expression factor genes in two cell lines. *Journal of Virology* **69**: 6265–6272.

Lu YH, Wu KM, Jiang YY, Guo YY & Desneux N (2012) Widespread adoption of Bt cotton and insecticide decrease promotes biocontrol services. *Nature* **487**: 362–365.

Lucas AM (1969) The effect of population structure on the success of insect introductions. *Heredity* **24**: 151–154.

Lucas P & Sarniguet A (1998) Biological control of soil-borne pathogens with resident versus introduced antagonists: should diverging approaches become strategic convergence? In *Conservation Biological Control* (Barbosa P, ed.), San Diego, CA, Academic Press, pp. 351–370.

Luck RF & Forster LD (2003) Quality of augmentative biological control agents: a historical perspective and lessons learned from evaluating *Trichogramma*. In *Quality Control and Production of Biological Control Agents: Theory and Testing Procedures* (Van Lenteren JC, ed.), Wallingford, UK, CABI Publishing, pp. 231–246.

Luck RF, Messenger PS & Barbieri JF (1981) The influence of hyperparasitism on the performance of biological control agents. In *The Role of Hyperparasitism in Biological Control* (Rosen D, ed.), Berkeley, University of California Press, pp. 34–42.

Luck RF & Podoler H (1985) Competitive exclusion of *Aphytis lingnanensis* by *A. melinus*: potential role of host size. *Ecology* **66**: 904–913.

Luck RF, Shepard BM & Kenmore PE (1988) Experimental methods for evaluating arthropod natural enemies. *Annual Review of Entomology* **33**: 367–391.

Luck RF, Shepard BM & Kenmore PE (1999) Evaluation of biological control with experimental methods. In *Handbook of Biological Control* (Bellows TS & Fisher TW, eds.), San Diego, CA, Academic Press, pp. 225–242.

Luck RF, Van den Bosch R & Garcia R (1977) Chemical insect control – a troubled pest management strategy. *BioScience* **27**: 606–611.

Lundgren JG (2009) *Relationships of Natural Enemies and Non-prey Foods*. Dordrecht, The Netherlands, Springer.

Lundgren JG & Fergen JK (2010) The effects of a winter cover crop on *Diabrotica virgifera* (Coleoptera: Chrysomelidae) populations and beneficial arthropod communities in no-till maize. *Environmental Entomology* **39**: 1816–1828.

Lundgren JG, Gassmann AJ, Bernal J, Duan JJ & Ruberson J (2009) Ecological compatibility of GM crops and biological control. *Crop Protection* **28**: 1017–1030.

Lynch M (1991) The genetic interpretation of inbreeding depression and outbreeding depression. *Evolution* **45**: 622–629.

Lys J-A & Nentwig W (1992) Augmentation of beneficial arthropods by strip-management. 4. Surface activity, movements and activity density of abundant carabid beetles in a cereal field. *Oecologia* **92**: 373–382.

Lys J-A, Zimmermann M & Nentwig W (1994) Increase in activity density and species number of carabid beetles in cereals as a result of strip-management. *Entomologia Experimentalis et Applicata* **73**: 1–9.

Ma WJ, Kuijper B, de Boer JG, van de Zande L, Beukeboom LW, Wertheim B & Pannebakker BA (2013) Absence of complementary sex determination in the parasitoid wasp genus *Asobara* (Hymenoptera: Braconidae). *PLoS One* **8**: e60459.

MacArthur RH & Wilson EO (1967) *The Theory of Island Biogeography*. Princeton, NJ, Princeton University Press.

MacDonald IAW & Cooper J (1995) Insular lessons for global biodiversity conservation with particular reference to alien invasions. In *Islands: Biological Diversity and Ecosystem Function* (Vitousek PM, Loope LL & Adsersen H, eds.), Berlin, Germany, Springer, pp. 189–203.

Macen JL, Graham KA, Lee SF, Schreiber M, Boshkov LK & McFadden G (1996) Expression of the myxoma virus tumor necrosis factor receptor homologue and M11 L genes is required to prevent virus-induced apoptosis in infected rabbit T lymphocytes. *Virology* **218**: 232–237.

Mack RN (1996) Biotic barriers to plant naturalization. In *Proceedings of the IX International Symposium on Biological Control of Weeds* (Moran VC & Hoffmann JH, eds.), Stellenbosch, South Africa, University of Cape Town, pp. 39–46.

Mack RN (2002) Natural barriers to plant naturalizations and invasions in the Sonoran Desert. In *Invasive Exotic Species in the Sonoral Region* (Tellman B, ed.), Tucson, University of Arizona Press, pp. 63–76.

Mack RN, Simberloff D, Lonsdale WM, Evans H, Clout M & Bazzaz FA (2000) Biotic invasions: causes, epidemiology, global consequences, and control. *Ecological Applications* **10**: 698–710.

Mackauer M (1976) Genetic problems in the production of biological control agents. *Annual Review of Entomology* **21**: 369–385.

Mackauer M. & Volkl W (1993) Regulation of aphid populations by aphidiid wasps: does parasitoid foraging behaviour or hyperparasitism limit impact? *Oecologia* **94**: 339–350.

MacLeod A, Wratten SD, Sotherton NW & Thomas MB (2004) 'Beetle banks' as refuges for beneficial arthropods in farmland: long-term changes in predator communities and habitat. *Agricultural and Forest Entomology* **6**: 147–154.

Maddox DM (1982) Biological control of diffuse knapweed (*Centaurea diffusa*) and spotted knapweed (*C. maculosa*). *Weed Science* **30**: 76–82.

Maeda S, Kamita SG & Kondo A (1993) Host-range expansion of *Autographa californica* nuclear polyhedrosis virus (Npv) following recombination of a 0.6-kilobase-pair DNA fragment originating from *Bombyx mori* Npv. *Journal of Virology* **67**: 6234–6238.

Mafokoane LD, Zimmermann HG & Hill MP (2007) Development of *Cactoblastis cactorum* (Berg) (Lepidoptera: pyralidae) on six north American *Opuntia* species. *African Entomology* **15**: 295–299.

Majerus MEN (1994) *Ladybirds*. London, UK, Harper Collins.

Majerus MEN & Hurst GDD (1997) Ladybirds as a model system for the study of male-killing endosymbionts. *Entomophaga* **42**: 13–20.

Malecki RA, Blossey B, Hight SD, Schroeder D, Kok LT & Coulson JR (1993) Biological control of purple loosestrife. *BioScience* **43**: 680–686.

Malmstrom CM, McCullough AJ, Johnson HA, Newton LA & Borer ET (2005) Invasive annual grasses indirectly increase virus incidence in California native perennial bunchgrasses. *Oecologia* **145**: 153–164.

Maniania NK, Bugeme DM, Wekesa VW, Delalibera I & Knapp M (2008) Role of entomopathogenic fungi in the control of *Tetranychus evansi* and *Tetranychus urticae* (Acari: Tetranychidae), pests of horticultural crops. *Experimental and Applied Acarology* **46**: 259–274.

Manrique V, Diaz R, Erazo L, Reddi N, Wheeler GS, Williams D & Overholt WA (2014) Comparison of two populations of *Pseudophilothrips ichini* (Thysanoptera: Phlaeothripidae) as candidates for biological control of the invasive weed *Schinus terebinthifolia* (Sapindales: Anacardiaceae). *Biocontrol Science and Technology* **24**: 518–535.

Mansfield S & Mills NJ (2004) A comparison of methodologies for the assessment of host preference of the gregarious egg parasitoid *Trichogramma platneri*. *Biological Control* **29**: 332–340.

Marchetto KM, Shea K, Kelly D, Groenteman R, Sezen Z & Jongejans E (2014) Unrecognized impact of a biocontrol agent on the spread rate of an invasive thistle. *Ecological Applications* **24**: 1178–1187.

Marino PC & Landis DA (1996) Effect of landscape structure on parasitoid diversity and parasitism in agroecosystems. *Ecological Applications* **6**: 276–284.

Maron JL & Harrison S (1997) Spatial pattern formation in an insect host-parasitoid system. *Science* **278**: 1619–1621.

Maron JL & Marler M (2008) Effects of native species diversity and resource additions on invader impact. *American Naturalist* **172**: S18–S33.

Maron JL & Vilá M (2001) When do herbivores affect plant invasion? Evidence for the natural enemies and biotic resistance hypotheses. *Oikos* **95**: 361–373.

Maron JL, Vilá M & Arnason J (2004) Loss of enemy resistance among introduced populations of St. John's Wort (*Hypericum perforatum*). *Ecology* **85**: 3243–3253.

Marrone PG (2002) An effective biofungicide with novel modes of action. *Pesticide Outlook* **13**: 193–194.

Marsden JS, Martin GE, Parham DJ, Risdill-Smith TJ & Johnson BG (1980) *Returns on Australian Agricultural Research*. Canberra, Australian CSIRO Division of Entomology.

Marsh FL (1937) Ecological observations upon the enemies of *Cecropia*, with particular reference to its hymenopterous parasites. *Ecology* **18**: 106–112.

Marsh FL (1941) A few life-history details of *Samia cecropia* within the southwestern limits of Chicago. *Ecology* **22**: 331–337.

Marsh PM (1977) Notes on the taxonomy and nomenclature of *Aphidius* species (Hym.: Aphidiidae) parasitic on the pea aphid in North America. *Entomophaga* **22**: 365–372.

Marshall ID & Fenner F (1958) Studies in the epidemiology of infectious myxomatosis of rabbits. V. Changes in the innate resistance of Australian wild rabbits exposed to myxomatosis. *Journal of Hygiene* **56**: 288–302.

Marsico TD, Burt JW, Espeland EK, Gilchrist GW, Jamies MA, Linsstrom L, Roderick GK, Swope S, Szucs M & Tsutsui ND (2010) Underutilized resources for studying the evolution of invasive species during their introduction, establishment, and lag phases. *Evolutionary Applications* **3**: 203–219.

Mason PG & Hopper KR (1997) Temperature dependence in locomotion of the parasitoid *Aphelinus asychis* (Hymenoptera: Aphelindae) from geographical regions with different climates. *Environmental Entomology* **26**: 1416–1423.

Mason RR & Torgerson TR (1987) Dynamics of a nonoutbreak population of the douglas-fir tussock moth (Lepidoptera: Lymantriidae) in southern Oregon. *Environmental Entomology* **16**: 1217–1227.

Massad E (1987) Transmission rates and the evolution of pathogenicity. *Evolution* **41**: 1127–1130.

Massei G & Cowan D (2014) Fertility control to mitigate human-wildlife conflicts: a review. *Wildlife Research* **41**: 1–21.

Mathenge CW, Holford P, Hoffmann JH, Zimmermann HG, Spooner-Hart R & Beattie GAC (2010) Hybridization between *Dactylopius tomentosus* (Hemiptera: Dactylopiidae) biotypes and its effects on host specificity. *Bulletin of Entomological Research* **100**: 331–338.

Mathews CR, Bottrell DG & Brown MW (2009) Extrafloral nectaries alter arthropod community structure and mediate peach (*Prunus persica*) plant defense. *Ecological Applications* **19**: 722–730.

Mathews CR, Brown MW & Bottrell DG (2007) Leaf extrafloral nectaries enhance biological control of a key economic pest, *Grapholita molesta* (Lepidoptera: Tortricidae), in peach (Rosales: Rosaceae). *Environmental Entomology* **36**: 383–389.

Matsumoto T, Itioka T, Nishida T & Inoue T (2003) Introduction of parasitoids has maintained a stable population of arrowhead scales at extremely low levels. *Entomologia Experimentalis et Applicata* **106**: 115–125.

Matsuo T, Sugaya S, Yasukawa J, Aigaki T & Fuyama Y (2007) Odorant-binding proteins OBP57d and OBP57e affect taste perception and host-plant preference in *Drosophila seychellia*. *PLoS Biology* **5**: e118.

Matsuoka T & Seno H (2008) Ecological balance in the native population dynamics may cause the paradox of pest control with harvesting. *Journal of Theoretical Biology* **252**: 87–97.

Matteson PC (2000) Insect pest management in tropical Asian irrigated rice. *Annual Review of Entomology* **45**: 549–574.

Mausel DL, Salom SM, Kok LT & Fidgen JG (2008) Propagation, synchrony, and impact of introduced and native *Laricobius* spp. (Coleoptera: Derodontidae) on hemlock woolly adelgid in Virginia. *Environmental Entomology* **37**: 1498–1507.

Mausel DL, Van Driesche RG & Elkinton JS (2011) Comparative cold tolerance and climate matching of coastal and inland *Laricobius nigrinus* (Coleoptera: Derodontidae), a biological control agent of hemlock woolly adelgid. *Biological Control* **58**: 96–102.

May RM (1973) *Stability and Complexity in Model Ecosystems*. Princeton, NJ, Princeton University Press.

May RM (1978) Host-parasitoid systems in patchy environments: a phenomenological model. *Journal of Animal Ecology* **47**: 833–844.

May RM & Hassell MP (1981) The dynamics of multiparasitoid-host interactions. *American Naturalist* **117**: 234–261.

May RM, Hassell MP, Anderson RM & Tonkyn DW (1981) Density dependence in host-parasitoid models. *Journal of Animal Ecology* **50**: 855–865.

Mayer VE, Frederickson ME, McKey D & Blatrix R (2014) Current issues in the evolutionary ecology of ant-plant symbioses. *New Phytologist* **202**: 749–764.

Maynard Smith J, Smith NH, O'Rourke M & Spratt BG (1993) How clonal are bacteria? *Proceedings of the National Academy of Sciences, USA* **90**: 4384–4388.

Mazzola M (2004) Assessment and management of soil microbial community structure for disease suppression. *Annual Review of Phytopathology* **42**: 35–59.

Mazzola M, Brown J, Izzo AD & Cohen MF (2007) Mechanism of action and efficacy of seed meal-induced pathogen suppression differ in a Brassicaceae species and time-dependent manner. *Phytopathology* **97**: 454–460.

Mazzola M, Fujimoto DK, Thomashow LS & Cook RJ (1995) Variation in sensitivity of *Gaeumannomyces graminis* to antibiotics produced by fluorescent *Pseudomonas* spp. and effect on biological control of take-all of wheat. *Applied and Environmental Microbiology* **61**: 2554–2559.

Mazzola M, Granatstein DM, Elfving DC & Mullinix K (2001) Suppression of specific apple root pathogens by *Brassica napus* seed meal amendment regardless of glucosinolate content. *Phytopathology* **91**: 673–679.

Mazzola M & Manici LM (2012) Apple replant disease: role of microbial ecology in cause and control. *Annual Review of Phytopathology* **50**: 45–65.

Mazzola M & Mullinix K (2005) Comparative field efficacy of management strategies containing *Brassica napus* seed meal or green manure for the control of apple replant disease. *Plant Disease* **89**: 1207–1213.

McBride CS (2007) Rapid evolution of smell and taste receptor genes during host specialization in *Drosophila seychellia. Proceedings of the National Academy of Sciences, USA* **104**: 4996–5001.

McCallum H (1996) Immunocontraception for wildlife population control. *Trends in Ecology & Evolution* **11**: 491–493.

McClay AS & Balciunas JK (2005) The role of pre-release efficacy assessment in selecting classical biological control agents for weeds – applying the Anna Karenina principle. *Biological Control* **35**: 197–207.

McClure M (1986) Population dynamics of Japanese hemlock scales: a comparison of endemic and exotic communities. *Ecology* **67**: 1411–1421.

McConnachie AJ, Hill KM & Byrne MJ (2004) Field assessment of a frond-feeding weevil, a successful biological control agent of red waterfern, *Azolla filiculoides*, in southern Africa. *Biological Control* **28**: 25–32.

McCullough DG, Mercader RJ & Siegert NW (2015) Developing and integrating tactics to slow ash (Oleaceae) mortality caused by emerald ash borer (Coleoptera: Buprestidae). *Canadian Entomologist* **147**: 349–358.

McDonald RC & Kok LT (1991) Hyperparasites attacking *Cotesia glomerata* (L.) and *Cotesia rubecula* (Marshall) (Hymenoptera: Braconidae) in southwestern Virginia. *Biological Control* **1**: 170–175.

McEvoy PB & Coombs EM (1999) Biological control of plant invaders: regional patterns, field experiments, and structured population models. *Ecological Applications* **9**: 387–401.

McEvoy PB, Higgs KM, Coombs EM, Karacetin E & Starcevich LA (2012) Evolving while invading: rapid adaptive evolution in juvenile development time for a biological control organism colonizing a high-elevation environment. *Evolutionary Applications* **5**: 524–536.

McEvoy PB, Rudd NT, Cox CS & Huso M (1993) Disturbance, competition, and herbivory effects on ragwort *Senecio jacobaea* populations. *Ecological Monographs* **63**: 55–75.

McFadden G (2005) Poxvirus tropism. *Nature Reviews Microbiology* **3**: 201–213.

McFadyen REC (1998) Biological control of weeds. *Annual Review of Entomology* **43**: 369–393.

McFadyen REC (2000) Successes in biological control of weeds. In *Proceedings of the X International Symposium on Biological Control of Weeds* (Spencer NR, ed.), Bozeman, Montana State University, pp. 3–14.

McFadyen RE (2003) Does ecology help in the selection of biocontrol agents? In *Improving the Selection, Testing and Evaluation of Weed Biological Control Agents* (Spafford Jacob H & Briese DT, eds.), Glen Osmond, Australia, CRC, pp. 5–9.

McFadyen R (2008) Return on investment: determining the economic impact of biological control programs. In *Proceedings of the XII International Symposium on Biological Control of Weeds* (Julien MH, Sforza R, Bon MC, Evans HC, Hatcher PE, Hinz HL & Rector BG, eds.), Wallingford, UK, CABI Publishing, pp. 67–74.

McFadyen R & Spafford Jacob H (2004) Insects for the biocontrol of weeds: predicting parasitism levels in the new country. In *Proceedings of the XI International Symposium on Biological Control of Weeds* (Cullen JM, Briese DT, Kriticos DJ, Lonsdale WM, Morin L & Scott JK, eds.), Canberra, Australia, CSIRO, pp. 135–140.

McMurtry JA (1991) Augmentative releases to control mites in agriculture. In *Modern Acarology, Volume 1* (Dusbabeck F & Bukva V, eds.), The Hague, The Netherlands, SPB Academic, pp. 151–157.

McMurtry JA & Croft BA (1997) Life-styles of phytoseiid mites and their roles in biological control. *Annual Review of Entomology* **42**: 291–321.

McMurtry JA, Moraes GJ & Famah Sourassou N (2013) Revision of lifestyles of phytoseiid mites (Acari: Phytoseiidae) and implications for biological control strategies. *Systematic and Applied Acarology* **18**: 297–320.

McQuate GT, Sylva CD & Jang EB (2005) Mediterranean fruit fly (Dipt., Tephritidae) suppression in persimmon through bait sprays in adjacent coffee plantings. *Journal of Applied Entomology* **129**: 110–117.

McSpadden Gardener BB (2004) Ecology of *Bacillus* and *Paenibacillus* in agricultural systems. *Phytopathology* **94**: 1252–1258.

McSpadden Gardener BB (2007) Diversity and ecology of biocontrol *Pseudomonas* spp. in agricultural systems. *Phytopathology* **97**: 221–226.

Mead AR (1979) *Pulmonates Volume 2B, Economic Malacology with Particular Reference to Achatina fulica.* London, UK, Academic Press.

Meffe GK (1983) Attempted chemical renovation of an Arizona springbrook for management of the endangered Sonoran topminnow. *North American Journal of Fisheries Management* **3**: 315–321.

Meffe GK (1984) Effects of abiotic disturbance on coexistence of predator–prey fish species. *Ecology* **65**: 1525–1534.

Meffe GK (1985) Predation and species replacement in American southwestern fishes: a case study. *The Southwestern Naturalist* **30**: 173–187.

Meffe GK, Hendrickson DA, Minckley WL & Rinne JN (1983) Factors resulting in decline of the endangered Sonoran topminnow *Poeciliopsis occidentalis* (Atheriniformes, Poeciliidae) in the United-States. *Biological Conservation* **25**: 135–159.

Memmott J, Craze PG, Harman HM, Syrett P & Fowler SV (2005) The effect of propagule size on the invasion of an alien insect. *Journal of Animal Ecology* **74**: 50–62.

Memmott J, Fowler SV & Hill RL (1998) The effect of the release size on the probability of establishment of biological control agents: gorse thrips (*Sericothrips staphylinus*) released against gorse (*Ulex europaeus*) in New Zealand. *Biocontrol Science and Technology* **8**: 103–115.

Mensah RK (1999) Habitat diversity: implications for the conservation and use of predatory insects of *Helicoverpa* spp. in cotton systems in Australia. *International Journal of Pest Management* **45**: 91–100.

Mensah RK & Madden JL (1993) Development and application of an integrated pest management program for the psyllid, *Ctenarytaina thysanura* on *Boronia megastigma* in Tasmania. *Entomologia Experimentalis et Applicata* **66**: 59–74.

Mensah RK & Sequeira RV (2004) Habitat manipulation for insect pest management in cotton cropping systems. In *Ecological Engineering for Pest Management* (Gurr GM, Wratten SD & Altieri MA, eds.), Ithaca, NY, Comstock, pp. 187–198.

Mercado-Blanco, J. 2015. *Pseudomonas* strains that exert biocontrol of plant pathogens. In *Pseudomonas, Volume 7, New Aspects of Pseudomonas Biology* (Ramos J-L, Goldberg JB & Filloux A, eds.), Dordrecht, The Netherlands, Springer, pp. 121–172.

Messelink GJ, Bloemhard CMJ, Cortes JA, Sabelis MW & Janssen A (2011) Hyperpredation by generalist predatory mites disrupts biological control of aphids by the aphidophagous gall midge *Aphidoletes aphidimyza. Biological Control* **57**: 246–252.

Messelink GJ & Janssen A (2014) Increased control of thrips and aphids in greenhouses with two species of generalist predatory bugs involved in intraguild predation. *Biological Control* **79**: 1–7.

Messenger PS (1970) Bioclimatic inputs to biological control and pest management programs. In *Concepts of Pest Management* (Rabb RL & Guthrie FE, eds.), Raleigh, University of North Carolina Press.

Messenger PS, Biliotti E & Van den Bosch R (1976a) The importance of natural enemies in integrated control. In *Theory and Practice of Biological Control* (Huffaker CB & Messenger PS, eds.), New York, Academic Press, pp. 543–654.

Messenger PS, Wilson F & Whitten MJ (1976b) Variation, fitness, and adaptability of natural enemies. In *Theory and Practice in Biological Control* (Huffaker CB & Messenger PS, eds.), New York, Academic Press, pp. 209–232.

Messersmith CG & Adkins SW (1995) Integrating weed-feeding insects and herbicides for weed control. *Weed Technology* **9**: 199–208.

Messina FJ (1981) Plant protection as a consequence of an ant-membracid mutualism – Interactions on Goldenrod (*Solidago* sp). *Ecology* **62**: 1433–1440.

Messing RH (2000) The impact of nontarget concerns on the practice of biological control. In *Nontarget Effects of Biological Control* (Follett PA & Duan JJ, eds.) Dordrecht, The Netherlands, Kluwer, pp. 45–58.

Messing RH (2001) Centrifugal phylogeny as a basis for non-target host testing in biological control: is it relevant for parasitoids? *Phytoparasitica* **29**: 187–190.

Messing RH, Roitberg BD & Brodeur J (2006) Measuring and predicting indirect impacts of biological control: competition, displacement and secondary interactions. In *Environmental Impact of Invertebrates for Biological Control of Arthropods* (Bigler F, Babendreier D & Kuhlmann U, eds.), Wallingford, UK, CABI Publishing, pp. 64–77.

Messing RH & Wright MG (2006) Biological control of invasive species: solution or pollution? *Frontiers in Ecology and the Environment* **4**: 132–140.

Meyer G, Clare R & Weber E (2005) An experimental test of the evolution of increased competitive ability hypothesis in goldenrod, *Solidago gigantea*. *Oecologia* **144**: 299–307.

Meyling NV & Eilenberg J (2007) Ecology of the entomopathogenic fungi *Beauveria bassiana* and *Metarhizium anisopliae* in temperate agroecosystems: potential for conservation biological control. *Biological Control* **43**: 145–155.

Meyling NV & Hajek AE (2010) Principles from community and metapopulation ecology: application to fungal entomopathogens. *Biocontrol* **55**: 39–54.

Meyling NV, Lubeck M, Buckley EP, Eilenberg J & Rehner SA (2009) Community composition, host range and genetic structure of the fungal entomopathogen *Beauveria* in adjoining agricultural and seminatural habitats. *Molecular Ecology* **18**: 1282–1293.

Michaud JP (1999) Sources of mortality in colonies of brown citrus aphid, *Toxoptera citricida*. *Biocontrol* **44**: 347–367.

Michaud JP (2002) Invasion of the Florida citrus ecosystem by *Harmonia axyridis* (Coleoptera: Coccinellidae) and asymmetric competition with a native species, *Cycloneda sanguinea*. *Environmental Entomology* **31**: 827–835.

Midgarden D, Fleischer SJ, Weisz R & Smilowitz Z (1997) Site-specific integrated pest management impact on development of esfenvalerate resistance in Colorado potato beetle (Coleoptera: Chrysomelidae)and on densities of natural enemies. *Journal of Economic Entomology* **90**: 855–867.

Milgroom MG & Cortesi P (2004) Biological control of chestnut blight with hypovirulence: a critical analysis. *Annual Review of Phytopathology* **42**: 311-338.

Millar LC & Barbercheck ME (2001) Interaction between endemic and introduced entomopathogenic nematodes in conventional-till and no-till corn. *Biological Control* **22**: 235–245.

Millard WA & Taylor CB (1927) Antagonism of micro-organisms as the controlling factor in the inhibition of scab by green-manuring. *Annals of Applied Biology* **14**: 202–216.

Miller D (1936) Biological control of noxious weeds. *New Zealand Journal of Science and Technology* **18**: 581–584.

Miller LK, Lingg AJ & Bulla LA (1983) Bacterial, viral and fungal insecticides. *Science* **219**: 715–721

Miller TEX, Shaw AK, Inouye BD & Neubert MG (2011) Sex-biased dispersal and the speed of two-sex invasions. *American Naturalist* **177**: 549–561.

Miller-Rushing AJ, Hoye TT, Inouye DW & Post E (2010) The effects of phenological mismatches on demography. *Philosophical Transactions of the Royal Society B: Biological Sciences* **365**: 3177–3186.

Mills LS, Doak DF & Wisdom MJ (1999) Reliability of conservation actions based on elasticity analysis of matrix models. *Conservation Biology* **13**: 815–829.

Mills MD, Rader RB & Belk MC (2004) Complex interactions between native and invasive fish: the simultaneous effects of multiple negative interactions. *Oecologia* **141**: 713–721.

Mills NJ (1982) Satiation and the functional response: a test of a new model. *Ecological Entomology* **7**: 305–315.

Mills NJ (1992) Parasitoid guilds, life-styles, and host ranges in the parasitoid complexes of torticoid hosts (Lepidoptera: Torticoidea). *Environmental Entomology* **21**: 230–239.

Mills NJ (1994) Biological control: some emerging trends. In *Individuals, Populations and Patterns in Ecology* (Leather SR, Watt AD, Mills NJ & Walters KFA, eds.), Andover, UK, Intercept, pp. 213–222.

Mills NJ (1998) *Trichogramma*: the field efficacy of inundative biological control of the codling moth in Californian orchards. In *Proceedings of the 1st California Conference on Biological Control* (Hoddle MS, ed.), Berkeley, University of California, pp. 10–11.

Mills NJ (2000) Biological control: the need for realistic models and experimental approaches to parasitoid introductions. In *Parasitoid Population Biology* (Hochberg ME & Ives AR, eds.), Princeton, NJ, Princeton University Press, pp. 217–234.

Mills NJ (2003) Augmentation in orchards: improving the efficacy of *Trichogramma* inundation. In *Proceedings of 1st International Symposium on Biological Control of Arthropods* (Van Driesche RG, ed.), Honolulu, USDA Forest Service, pp. 130–135.

Mills NJ (2005) Selecting effective parasitoids for biological control introductions: codling moth as a case study. *Biological Control* **34**: 274–282.

Mills NJ (2006a) Accounting for differential success in the biological control of homopteran and lepidopteran pests. *New Zealand Journal of Ecology* **30**: 61–72.

Mills NJ (2006b) Interspecific competition among natural enemies and single versus multiple introductions in biological control. In *Trophic and Guild Interactions in Biological Control* (Brodeur J & Boivin G, eds.), Dordrecht, The Netherlands, Springer, pp. 191–220.

Mills NJ (2008) Can matrix models guide the selection of parasitoids for biological control introductions? LBAM in California as a case study. In *Proceedings of the 3rd International Symposium on Biological Control of Arthropods* (Mason PG, Gillespie DR & Vincent C, eds.), Christchurch, New Zealand, USDA Forest Service, pp. 89–97.

Mills NJ (2010) Egg parasitoids in biological control and integrated pest management. In *Egg Parasitoids in Agroecosystems with Emphasis on Trichogramma* (Consoli FL, Parra JRP & Zucchi RA, eds.), Dordrecht, The Netherlands, Springer, pp. 389–411.

Mills NJ, Babendreier D & Loomans AJM (2006) Methods for monitoring the dispersal of natural enemies from point source releases associated with augmentative biological control. In *Environmental Impact of Invertebrates for Biological Control of Arthropods* (Bigler F, Babendreier D & Kuhlmann U, eds.), Wallingford, UK, CABI Publishing, pp. 114–131.

Mills NJ, Beers EH, Shearer PW, Unruh TR & Amarasekare KG (2016) Comparative analysis of pesticide effects on natural enemies in western orchards: a synthesis of laboratory bioassay data. *Biological Control* **102**: 17–25.

Mills NJ & Getz WM (1996) Modelling the biological control of insect pests: a review of host-parasitoid models. *Ecological Modelling* **92**: 121–143.

Mills NJ & Gutierrez AP (1996) Prospective modelling in biological control: an analysis of the dynamics of heteronomous hyperparasitism in a cotton-whitefly-parasitoid system. *Journal of Applied Ecology* **33**: 1379–1394.

Mills NJ & Gutierrez AP (1999) Biological control of insect pests: a tritrophic perspective. In *Theoretical Approaches to Biological Control* (Hawkins BA & Cornell HV, eds.), Cambridge, UK, Cambridge University Press, pp. 89–102.

Mills NJ & Lacan I (2004) Ratio dependence in the functional response of insect parasitoids: evidence from *Trichogramma minutum* foraging for eggs in small host patches. *Ecological Entomology* **29**: 208–216.

Mills, NJ, Pickel C, Mansfield S, McDougall S, Buchner R, Caprile J, Edstrom J, Elkins R, Hasey J, Kelley K, Krueger W, Olson W & Stocker R (2000) Mass releases of Trichogramma wasps can reduce damage from codling moth. *California Agriculture* **54**: 22–25.

Mills NJ & Wajnberg E (2008) Optimal foraging behavior and efficient biological control methods. In *Behavioral Ecology of Insect Parasitoids* (Wajnberg E, Bernstein C & Van Alphen JJM, eds.), Malden, MA, Blackwell Publishing, pp. 3–30.

Milner RJ & Hunter DM (2001) Recent developments in the use of fungi as biopesticides against locusts and grasshoppers in Australia. *Journal of Orthoptera Research* **10**: 271–276.

Minckley WL (1999) Ecological review and management recommendations for recovery of the endangered Gila topminnow. *Great Basin Naturalist* **59**: 230–244.

Minckley WL & Deacon JE (1968) Southwestern fishes and the enigma or 'endangered species'. *Science* **159**: 1424–1432.

Minkenberg OPJM, Tatar M & Rosenheim JA (1992) Egg load as a major source of variability in insect foraging and oviposition behavior. *Oikos* **65**: 134–142.

Mitchell AJ & Kelly AM (2006) The public sector role in establishment of grass carp in the United States. *Fisheries* **31**: 113–122.

Mitchell CE & Power AG (2003) Release of invasive plants from fungal and viral pathogens. *Nature* **421**: 625–627.

Mkoji GMB, Hofkin BV, Kuris AM, Stewart-Oaten A, Mungai BN, Kihara JH, Mungai F, Yundu J, Mbui J, Rashid JR, Kariuki CH, Ouma JH, Koech DK & Loker ES (1999) Impact of the crayfish *Procambarus clarkii* on *Schistosoma haematobium* transmission in Kenya. *American Journal of Tropical Medicine and Hygiene* **61**: 751–759.

Mochiah MB, Ngi-Song AJ, Overholt WA & Stouthamer R (2002) *Wolbachia* infection in *Cotesia sesamiae* (Hymenoptera: Braconidae) causes cytoplasmic incompatability: implications for biological control. *Biological Control* **25**: 74–80.

Moeed A, Hickson R & Barratt BIP (2006) Principles of environmental risk assessment with emphasis on the New Zealand perspective. In *Environmental Impact of Invertebrates for Biological Control of Arthropods* (Bigler F, Babendreier D & Kuhlmann U, eds.), Wallingford, UK, CABI Publishing, pp. 241–253.

Mols CMM & Visser ME (2007) Great tits (*Parus major*) reduce caterpillar damage in commercial apple orchards. *PLoS One* **2**: e202.

Momanyi C, Lohr B & Gitonga L (2006) Biological impact of the exotic parasitoid, *Diadegma semiclausum* (Hellen), of diamondback moth, *Plutella xylostella* L., in Kenya. *Biological Control* **38**: 254–263.

Moon RD (1980) Biological control through interspecific competition. *Environmental Entomology* **9**: 723–728

Moore D, Douro-Kpindou O-K, Jenkins NE & Lomer CJ (1996) Effects of moisture content and temperature on storage of *Metarhizium flavoviride* conidia. *Biocontrol Science and Technology* **6**: 51–61.

Moore NW (1987) *The Bird of Time*. Cambridge, UK, Cambridge University Press.

Morales Ramos JA, Summy KR & King EG (1996) ARCASIM, a model to evaluate augmentation strategies of the parasitoid *Catolaccus grandis* against boll weevil populations. *Ecological Modelling* **93**: 221–235.

Moran VC, Hoffmann JH & Hill MP (2011) A context for the 2011 compilation of reviews on the biological control of invasive alien plants in South Africa. *African Entomology* **19**: 177–185.

Moreau G, Eveleigh ES, Lucarotti CJ & Quiring DT (2006) Stage-specific responses to ecosystem alteration in an eruptive herbivorous insect. *Journal of Applied Ecology* **43**: 28–34.

Morehead SA & Feener Jr. DH (2000) An experimental test of potential host range in the ant parasitoid *Apocephalus paraponerae*. *Ecological Entomology* **25**: 332–340.

Morin L, Evans KJ & Sheppard AW (2006) Selection of pathogen agents in weed biological control: critical issues and peculiarities in relation to arthropod agents. *Australian Journal of Entomology* **45**: 349–365.

Morin L, Reid AM, Sims-Chilton NM, Buckley YM, Dhileepan K, Hastwell GT, Nordblom TL & Raghu S (2009) Review of approaches to evaluate the effectiveness of weed biological control agents. *Biological Control* **51**: 1–15.

Morse JG (1998) Agricultural implications of pesticide-induced hormesis of insects and mites. *Human and Experimental Toxicology* **17**: 266–269.

Mossman K, Lee SF, Barry M, Boshkov L & McFadden G (1996) Disruption of M-T5, a novel myxoma virus gene member of the poxvirus host range superfamily, results in dramatic attenuation of myxomatosis in infected European rabbits. *Journal of Virology* **70**: 4394–4410.

Mouquet N, Thomas JA, Elmes GW, Clarke RT & Hochberg ME (2005) Population dynamics and conservation of a specialized predator: a case study of *Maculinea arion*. *Ecological Monographs* **75**: 525–542.

Muesebeck CFW & Dohanian SM (1927) *A study in hyperparasitism, with particular reference to the parasites of Apanteles melanoscelus (Ratzeburg)*. USDA Bulletin 1487.

Mukherjee A, Christman MC, Overholt WA & Cuda JP (2011) Prioritizing areas in the native range of *Hygrophila* for surveys to collect biological control agents. *Biological Control* **56**: 254–262.

Muldrew JA (1953) The natural immunity of the larch sawfly (*Pristiphora erichsonii* (Htg.)) to the introduced parasite *Mesoleius tenthredinis* Morley, in Manitoba and Saskatchewan. *Canadian Journal of Zoology* **31**: 313–332.

Müller CB, Adriaanse ICT, Belshaw R & Godfray HCJ (1999) The structure of an aphid-parasitoid community. *Journal of Animal Ecology* **68**: 346–370.

Müller-Schärer H (1991) The impact of root herbivory as a function of plant density and competition: survival, growth and fecundity of *Centaurea maculosa* in field plots. *Journal of Applied Ecology* **28**: 759–776.

Müller-Schärer H & Schaffner U (2008) Classical biological control: exploiting enemy escape to manage plant invasions. *Biological Invasions* **10**: 859–874.

Mumm R & Dicke M (2010) Variation in natural plant products and the attraction of bodyguards involved in indirect plant defense. *Canadian Journal of Zoology* **88**: 628–667.

Munro VMW & Henderson IM (2002) Nontarget effects of entomophagous biocontrol: shared parasitism between native lepidopteran parasitism and the biocontrol agent *Trigonospila brevifacies* (Diptera: Tachinidae) in forest habitats. *Environmental Entomology* **31**: 388–396.

Münster-Swendsen M & Nachman G (1978) Asynchrony in insect host–parasite interaction and its effect on stability, studied by a simulation model. *Journal of Animal Ecology* **47**: 159–171.

Murdoch WW (1990) The relevance of pest-enemy models to biological control. In *Critical Issues in Biological Control* (Mackauer M, Ehler LE & Roland J, eds.), Andover, UK, Intercept, pp. 1–24.

Murdoch WW (2009) Biological control: theory and practice. In *The Princeton Guide to Ecology* (Levin SA, ed.), Princeton, NJ, Princeton University Press, pp. 683–688.

Murdoch WW & Bence J (1987) General predators and unstable prey populations. In *Predation: Direct and Indirect Impacts on Aquatic Communities* (Kerfoot WC & Sih A, eds.), Hanover, NH, University Press of New England, pp. 17–30.

Murdoch WW & Briggs CJ (1996) Theory for biological control: recent developments. *Ecology* **77**: 2001–2013.

Murdoch WW, Briggs CJ & Nisbet RM (1996a) Competitive displacement and biological control in parasitoids: a model. *American Naturalist* **148**: 807–826.

Murdoch WW, Briggs CJ & Nisbet (2003) *Consumer-Resource Dynamics*. Princeton, NJ, Princeton University Press.

Murdoch WW, Briggs CJ & Swarbrick S (2005) Host suppression and stability in a parasitoid-host system: experimental demonstration. *Science* **309**: 610–613.

Murdoch WW, Chesson J & Chesson PL (1985) Biological control in theory and practice. *American Naturalist* **125**: 344–366.

Murdoch WW, Evans FC & Peterson CH (1972) Diversity and pattern in plants and insects. *Ecology* **53**: 819–829.

Murdoch WW, Luck RF, Swarbrick SL, Walde S, Yu DS & Reeve JD (1995) Regulation of an insect population under biological control. *Ecology* **76**:206–217.

Murdoch WW, Luck RF, Walde SJ, Reeve JD & Yu DS (1989) A refuge for red scale under control by *Aphytis*: structural aspects. *Ecology* **70**: 1707–1714.

Murdoch WW, Nisbet RM, Blythe SP, Gurney WSC & Reeve JD (1987) An invulnerable age class and stability in delay-differential parasitoid-host models. *American Naturalist* **129**: 263–282.

Murdoch WW, Nisbet RM, Luck RF, Godfray HCJ & Gurney WSC (1992) Size-selective sex-allocation and host feeding in a parasitoid-host model. *Journal of Animal Ecology* **61**: 533–541.

Murdoch WW & Oaten A (1975) Predation and population stability. *Advances in Ecological Research* **9**: 1–125

Murdoch WW & Oaten A (1989) Aggregation by parasitoids and predators: effects on equilibrium and stability. *American Naturalist* **123**: 371–392.

Murdoch WW, Swarbrick SL & Briggs CJ (2006b) Biological control: lessons from a study of California red scale. *Population Ecology* **48**: 297–305.

Murdoch WW, Swarbrick SL, Luck, RF, Walde S & Yu DS (1996b) Refuge dynamics and metapopulation dynamics: an experimental test. *American Naturalist* **147**: 424–444.

Murphy SM (2004) Enemy-free space maintains swallowtail butterfly host shift. *Proceedings of the National Academy of Sciences, USA* **101**: 18048–18052.

Murphy BC, Rosenheim JA, Dowell RV & Granett J (1998) Habitat diversification for improving biological control: parasitism of the western grape leafhopper. *Entomologia Experimentalis et Applicata* **87**: 225–235.

Murray E (1993) The sinister snail. *Endeavour* **17**: 78–83.

Murray J, Murray E, Johnson MS & Clarke B (1988) The extinction of *Partula* on Moorea. *Pacific Science* **42**: 150–153.

Murray TJ, Withers TM & Mansfield S (2010) Choice versus no-choice test interpretation and the role of biology and behavior in parasitoid host specificity tests. *Biological Control* **52**: 153–159.

Myers JH (1985) How many insect species are necessary for successful biocontrol of weeds? *Proceedings of the 6th International Symposium on the Biological Control of Weeds* (Delfosse ES, ed.), Ottawa, Canada, Agriculture Canada, pp. 77–82.

Myers JH & Bazely DR (2003) *Ecology and Control of Introduced Plants*. Cambridge, UK, Cambridge University Press.

Myers JH & Harris P (1980) Distribution of *Urophora* galls in flower heads of diffuse and spotted knapweed in British Columbia. *Journal of Applied Ecology* **17**: 359–367.

Myers JH, Higgins C & Kovacs E (1989) How many insect species are necessary for the biological control of insects? *Environmental Entomology* **18**: 541–547.

Myers JH, Risley C & Eng R (1989) The ability of plants to compensate for insect attack: why biological control of weeds with insects is so difficult. In *Proceedings of the 7th International Symposium on the Biological Control of Weeds* (Delfosse ES, ed.), Rome, Italy, Istituto Sperimentale per la Patalogia Vegetale, pp. 67–73.

Myers JH & Sabath MD (1981) Genetic and phenotypic variability, genetic variance, and the success of establishment of insect introductions for the biological control of weeds. In *Proceedings of the V International Symposium on Biological Control of Weeds* (Delfosse ES, ed.), Brisbane, Australia, Commonwealth Scientific and Industrial Research Organization, pp. 91–102.

Myers K, Marshall ID & Fenner F (1954) Studies in the epidemiology of infectious myxomatosis in rabbits. *Journal of Hygiene* **52**: 337–360.

Naeem S & Li SB (1997) Biodiversity enhances ecosystem reliability. *Nature* **390**: 507–509.

Nachman G (1987) Systems analysis of acarine predator–prey interactions. I. A stochastic simulation model of spatial processes. *Journal of Animal Ecology* **56**: 247–265.

Nachman G & Zemek R (2003) Interactions in a tritrophic acarine predator–prey metapopulation system V: within-plant dynamics of *Phytoseiulus persimilis* and *Tetranychus urticae* (Acari: Phytoseiidae, Tetranychidae). *Experimental and Applied Acarology* **29**:35–68.

Nagayama K, Watanabe S, Kumakura K, Ichikawa T & Makino T (2007) Development and commercialization of *Trichoderma asperellum* SKT-1 (Ecohope (R)), a microbial pesticide. *Journal of Pesticide Science* **32**: 141–142.

Naik PR, Raman G, Narayanan KB & Sakthivel N (2008) Assessment of genetic and functional diversity of phosphate solubilizing fluorescent pseudomonads isolated from rhizospheric soil. *BMC Microbiology* **8**: 230

Naka H, Mitsunaga T & Mochizuki A (2005) Laboratory hybridization between the introduced and the indigenous green lacewings (Neuroptera: Chrysopidae: Chrysoperla) in Japan. *Environmental Entomology* **34**: 727–731.

Nakasuji F, Yamanaka H & Kiritani K (1973) The disturbing effect of mycriphantid spiders on the larval aggregation of the tobacco cutworm, *Spodoptera litura* (Lepidoptera: Noctuidae). *Kontyu* **41**: 220–227.

Nakayama S, Seko T, Takatsuki J-I, Miura K & Miyatake T (2010) Walking activity of flightless *Harmonia axyridis* (Coleoptera: Coccinellidae) as a biological control agent. *Journal of Economic Entomology* **103**: 1564–1568.

Naranjo SE (2001) Conservation and evaluation of natural enemies in IPM systems for *Bemisia tabaci*. *Crop Protection* **20**: 835–852.

Naranjo SE & Ellsworth PC (2009) The contribution of conservation biological control to integrated control of *Bemisia tabaci* in cotton. *Biological Control* **51**: 458–470.

Naranjo SE, Ellsworth PC & Frisvold GB (2015) Economic value of biological control in integrated pest management of managed plant systems. *Annual Review of Entomology* **60**: 621–645.

Navarro-Llopis V, Ayala I, Sanchis J, Primo J & Moya P (2015) Field efficacy of a *Metarhizium anisopliae*-based attractant-contaminant device to control *Ceratitis capitata* (Diptera: Tephritidae). *Journal of Economic Entomology* **108**: 1570–1578.

Nechols JR, Obrycki JJ, Tauber CA & Tauber MJ (1996) Potential impact of native natural enemies on *Galerucella* spp (Coleoptera: Chrysomelidae) imported for biological control of purple loosestrife: a field evaluation. *Biological Control* **7**: 60–66.

Neeno-Eckwall EC, Kinkel LL & Schottel JL (2001) Competition and antibiosis in the biological control of potato scab. *Canadian Journal of Microbiology* **47**: 332–340.

Nei M, Taruyama T & Chakraborty R (1975) The bottleneck effect and genetic variability in populations. *Evolution* **29**: 1–10.

Nelson EB & Hoitink HAJ (1983) The role of microorganisms in the suppression of *Rhizoctonia solani* in container media amended with composted hardwood bark. *Phytopathology* **73**: 274–278.

Nelson EH, Matthews CE & Rosenheim JA (2004) Predators reduce prey population growth by inducing changes in prey behavior. *Ecology* **85**: 1853–1858.

Nentwig W, Frank T & Lethmayer C (1998) Sown weed strips: artificial ecological compensation areas as an important tool in conservation biological control. In *Conservation Biological Control* (Barbosa P, ed.), San Diego, CA, Academic Press, pp. 133–154.

Neuenschwander P (2001) Biological control of the cassava mealybug in Africa: a review. *Biological Control* **21**: 214–229.

Neuenschwander P, Schulthess F & Madojemu E (1986) Experimental evaluation of the efficiency of *Epidinocarsis lopezi*, a parasitoid introduced into Africa against the cassava mealybug *Phenacoccus manihoti*. *Entomologia Experimentalis et Applicata* **42**: 133–138.

Neuhauser C, Andow DA, Heimpel GE, May G, Shaw RG & Wagenius S (2003) Community genetics: expanding the synthesis of ecology and genetics. *Ecology* **84**: 545–558.

Newman RM (2004) Biological control of Eurasian watermilfoil by aquatic insects: basic insights from an applied problem. *Archiv Fur Hydrobiologie* **159**: 145–184.

Newman RM, Borman ME & Castro SW (1997) Developmental performance of the weevil *Euhrychiopsis lecontei* on native and exotic watermilfoil host plants. *Journal of the American Benthological Society* **16**: 627–634.

Newman RM, Thompson DC & Richman DB (1998) Conservation strategies for the biological control of weeds. In *Conservation Biological Control* (Barbosa P, ed.), New York, Academic Press, pp. 371–396.

Newsom LD, Smith RF & Whitcomb WH (1976) Selective pesticides and selective use of pesticides. In *Theory and Practices of Biological Control* (Huffaker CB & Messenger PS, eds.), New York, Academic Press, pp. 565–692.

Nicholson AJ (1933) The balance of animal populations. *Journal of Animal Ecology* **2**: 132–178.

Nicholson AJ & Bailey VA (1935) The balance of animal populations. Part 1. *Proceedings of the Zoological Society of London* **3**: 551–598.

Nickerson JC, Rolphkay CA, Buschman LL & Whitcomb WH (1977) Presence of *Spissistilus festinus* (Homoptera: Membracidae) as a factor affecting egg predation by ants (Hymenoptera: Formicidae) in soybeans. *Florida Entomologist* **60**: 193–199.

Nielsen C, Keena M & Hajek AE (2005) Virulence and fitness of the fungal pathogen *Entomophaga maimaiga* in its host *Lymantria dispar*, for pathogen and host strains originating from Asia, Europe, and North America. *Journal of Invertebrate Pathology* **89**: 232–242.

Nilsson U, Rannback LM, Anderson P, Eriksson A & Ramert B (2011) Comparison of nectar use and preference in the parasitoid *Trybliographa rapae* (Hymenoptera: Figitidae) and its host, the cabbage root fly, *Delia radicum* (Diptera: Anthomyiidae). *Biocontrol Science and Technology* **21**: 1117–1132.

Nimkingrat P, Khanam S, Strauch O & Ehlers R-U (2013) Hybridisation and selective breeding for improvement of low temperature activity of the entomopathogenic nematode *Steinernema feltiae*. *BioControl* **58**: 417–426.

Nisbet RM, Diehl S, Wilson WG, Cooper SD, Donaldson DD & Kratz K (1997) Primary-productivity gradients and short-term population dynamics in open systems. *Ecological Monographs* **67**: 535–553.

Niyibigira EI, Overholt WA & Stouthamer R (2004a) *Cotesia flavipes* Cameron and *Cotesia sesamiae* (Cameron) (Hymenoptera: Braconidae) do not exhibit complementary sex determination: evidence from field populations. *Applied Entomology and Zoology*, **39**: 705–715.

Niyibigira EI, Overholt WA & Stouthamer R (2004b) *Cotesia flavipes* Cameron (Hymenoptera: Braconidae) does not exhibit complementary sex determination (ii) evidence from laboratory experiments. *Applied Entomology and Zoology* **39**: 717–725.

Noonburg EG & Byers JE (2005) More harm than good: when invader vulnerability to predators enhances impact on native species. *Ecology* **86**: 2555–2560.

Nordblom TL, Smyth MJ, Swirepik A, Sheppard AW & Briese DT (2002) Spatial economics of biological control: investing in new releases of insects for earlier limitation of Patterson's curse in Australia. *Agricultural Economics* **27**: 403–424.

Norris RF (1997) Impact of leaf mining on the growth of *Portulaca oleracea* (common purslane) and its competitive interaction with *Beta vulgaris* (sugarbeet). *Journal of Applied Ecology* **34**: 349–362.

Novotny V, Miller SE, Cizek L, Leps J, Janda M, Basset Y, Weiblen GD & Darrow K (2003) Colonising aliens: caterpillars (Lepidoptera) feeding on *Piper aduncum* and *P-umbellatum* in rainforests of Papua New Guinea. *Ecological Entomology* **28**: 704–716.

NRC (2002) *Predicting Invasions of Nonindigenous Plants and Plant Pests*. Washington, DC, National Academy of Sciences.

Nuñez MA, Simberloff D & Relva MA (2008) Seed predation as a barrier to alien conifer invasions. *Biological Invasions* **10**: 1389–1398.

Nunney L (2003) Managing captive populations for release: a population-genetic perspective. In *Quality Control and Production of Biological Control Agents: Theory and Testing Procedures* (Van Lenteren JC, ed.), Wallingford, UK, CABI Publishing, pp. 73–87.

Oaten A & Murdoch WW (1975) Functional response and stability in predator–prey systems. *American Naturalist* **109**: 289–298.

Obrycki JJ (2000) Coccinellid introductions: potential for and evaluation of nontarget effects. In *Nontarget Effects of Biological Control* (Follett PA & Duan JJ, eds.), Dordrecht, The Netherlands, Kluwer, pp. 127–146.

Obrycki JJ, Giles KL & Ormord AM (1998) Interactions between an introduced and indigenous coccinellid species at different prey densities. *Oecologia* **117**: 279–285.

Ochanda J (2010) Potential areas of collaboration and communication strategies between researchers and potential GMO vector user communities. In *Progress and Prospects for the Use of Genetically Modified Mosquitoes to Inhibit Disease Transmission* (Crawford VL & Reza JN, eds.), Geneva, Switzerland, WHO, pp. 37–39.

Ode PJ, Antolin MF & Strand MR (1995) Brood-mate avoidance in the parasitic wasp *Bracon hebetor* say. *Animal Behaviour* **49**: 1239–1248.

Ode PJ, Antolin MF & Strand MR (1998) Differential dispersal and female-biased sex allocation in a parasitic wasp. *Ecological Entomology* **23**: 314–318.

Oelrichs PB, MacLeod JK, Seawright AA & Grace PB (2001) Isolation and identification of the toxic peptides from *Lophyrotoma zonalis* (Pergidae) sawfly larvae. *Toxicon* **39**: 1933–1936.

Offenberg J (2011) *Oecophylla smaragdina* food conversion efficiency: prospects for ant farming. *Journal of Applied Entomology* **135**: 575–581.

Olden JD, Kennard MJ & Pusey BJ (2008) Species invasions and the changing biogeography of Australian freshwater fishes. *Global Ecology and Biogeography* **17**: 25–37.

Olivain C, Humbert C, Nahalkova J, Fatehi J, L'Haridon F & Alabouvette C (2006) Colonization of tomato root by pathogenic and nonpathogenic *Fusarium oxysporum* strains inoculated together and separately into the soil. *Applied and Environmental Microbiology* **72**: 1523–1531.

Oliver KM, Campos J, Moran NA & Hunter MS (2008) Population dynamics of defensive symbionts in aphids. *Proceedings of the Royal Society B* **275**: 293–299.

Oliver KM, Moran NA & Hunter MS (2005) Variation in resistance to parasitism in aphids is due to symbionts not host genotype. *Proceedings of the National Academy of Sciences, USA* **102**: 12795–12800.

Oliver KM, Moran NA & Hunter MS (2006) Costs and benefits of a superinfection of facultative symbionts in aphids. *Proceedings of the Royal Society of London, B* **273**: 1273–1280.

Oliver KM, Russell JA, Moran NA & Hunter MS (2003) Facultative bacterial symbionts in aphids confer resistance to parasitic wasps. *Proceedings of the National Academy of Sciences, USA* **100**: 1803–1807.

Oliver KM, Smith AH & Russell JA (2014) Defensive symbiosis in the real world – advancing ecological studies of heritable, protective bacteria and beyond. *Functional Ecology* **28**: 341–355.

Olkowski W & Zhang A (1998) Habitat management for biological control, examples from China. In *Enhancing Biological Control* (Pickett CH & Bugg RL, eds.), Berkeley, University of California Press, pp. 255–270.

O'Neil RJ, Giles KL, Obrycki JJ, Mahr DL, Legaspi JC & Katovich K (1998) Evaluation of the quality of four commercially available natural enemies. *Biological Control* **11**: 1–8.

O'Neil RJ, Nagarajan K, Wiedenmann RN & Legaspi JC (1996) A simulation model of *Podisus maculiventris* (Say) (Heteroptera: Pentatomidae) and Mexican bean beetle, *Epilachna varivestis* (Mulsant) (Coleoptera: Coccinellidae), population dynamics in soybean, *Glycine max* (L). *Biological Control* **6**: 330–339.

Onstad DW, Liu XX, Chen M, Roush R & Shelton AM (2013) Modeling the integration of parasitoid, insecticide, and transgenic insecticidal crop for the long-term control of an insect pest. *Journal of Economic Entomology* **106**: 1103–1111.

Onstad DW & McManus DP (1996) Risks of host range expansion by parasites of insects. *BioScience* **46**: 430–435.

Opit GP, Nechols JR & Margolies DC (2004) Biological control of twospotted spider mites, *Tetranychus urticae* Koch (Acari: Tetranychidae), using *Phytoseiulus persimilis* Athias-Henriot (Acari: Phytoseidae) on ivy geranium: assessment of predator release ratios. *Biological Control* **29**: 445–452.

Opit GP, Nechols JR, Margolies DC & Williams KA (2005) Survival, horizontal distribution, and economics of releasing predatory mites (Acari: Phytoseiidae) using mechanical blowers. *Biological Control* **33**: 344–351.

Oppenheim SJ, Gould F & Hopper KR (2012) The genetic architecture of a complex ecological trait: host plant use in the specialist moth, *Heliothis subflexa*. *Evolution* **66**: 3336–3351.

Orr DB & Suh CP-C (2000) Parasitoids and predators. In *Biological and Biotechnological Control of Insect Pests* (Rechcigl JE & Rechcigl NA, eds.), Boca Raton, FL, CRC Press, pp. 3–34.

Orr DB, Garcia-Salazar C & Landis DA (2000) *Trichogramma* nontarget impacts: a method for biological control risk assessment. In *Nontarget Effects of Biological Control* (Follett PA & Duan JJ, eds.), Dordrecht, The Netherlands, Kluwer, pp. 111–126.

Orr DB, Landis DA, Mutch DR, Manley GV, Stuby SA & King RL (1997) Ground cover influence on microclimate and *Trichogramma* (Hymenoptera: Trichogrammatidae) augmentation in seed corn production. *Environmental Entomology* **26**: 433–438.

Orrock JL, Preisser EL, Grabowski JH & Trussel GC (2013) The cost of safety: refuges increase the impact of predation risk in aquatic systems. *Ecology* **94**: 573–579.

Ortega YK & Pearson DE (2005) Weak vs. strong invaders of natural plant communities: Assessing invasibility and impact. *Ecological Applications* **15**: 651–661.

Ortega YK, Pearson DE & McKelvey KS (2004) Effects of biological control agents and exotic plant invasion on deer mouse populations. *Ecological Applications* **14**: 241–253.

Ortega YK, Pearson DE, Waller LP, Sturdevant NJ & Maron JL (2012) Population-level compensation impedes biological control of an invasive forb and indirect release of a native grass. *Ecology* **93**: 783–792.

Ouedraoga RM, Goettel MS & Brodeur J (2004) Behavioral thermoregulation in the migratory locust: a therapy to overcome fungal infection. *Oecologia* **138**: 312–319.

Pacala SW, Hassell MP & May RM (1990) Host-parasitoid associations in patchy environments. *Nature* **344**: 150–153.

Pachepsky E, Nisbet RM & Murdoch WW (2008) Between discrete and continuous: consumer-resource dynamics with synchronized reproduction. *Ecology* **89**: 280–288.

Panaccione DG, Beaulieu WT & Cook D (2014) Bioactive alkaloids in vertically transmitted fungal endophytes. *Functional Ecology* **28**: 299–314.

Paraiso O, Hight SD, Kairo MTK & Bloem S (2013a) Host specificity and risk assessment of *Trichogramma fuentesi* (Hymenoptera: Trichogrammatidae), a potential biological control agent of *Cactoblastis cactorum* (Lepidoptera: Pyralidae). *Florida Entomologist* **96**: 1305–1310.

Paraiso O, Kairo MTK, Hight SD, Leppla NC, Cuda JP, Owens M & Olexa MT (2013b) Opportunities for improving risk communication during the permitting process for entomophagous biological control agents: a review of current systems. *BioControl* **58**: 1–15.

Parker IM (2000) Invasion dynamics of *Cytisus scoparius*: a matrix model approach. *Ecological Applications* **10**: 726–743.

Parker JD, Burkepile DE & Hay ME (2006) Opposing effects of native and exotic herbivores on plant invasions. *Science* **311**: 1459–1461.

Parker JD & Hay ME (2005) Biotic resistance to plant invasions? Native herbivores prefer non-native plants. *Ecology Letters* **8**: 959–967.

Parker IM, Simberloff D, Lonsdale WM, Goodell K, Wonham M, Kareiva PM, Williamson MH, Von Holle B, Moyle PB, Byers JE & Goldwasser L (1999) Impact: toward a framework for understanding the ecological effects of invaders. *Biological Invasions* **1**: 3–19.

Parkes JP, Norbury GL, Heyward RP & Sullivan G (2002) Epidemiology of rabbit haemorrhagic disease (RHD) in the South Island, New Zealand, 1997–2001. *Wildlife Research* **29**: 543–555.

Parolin P, Bresch C, Poncet C & Desneux N (2014) Introducing the term 'biocontrol plants' for integrated pest management. *Scientia Agricola* **71**: 77–80.

Parry D (2009) Beyond Pandora's box: quantitatively evaluating non-target effects of parasitoids in classical biological control. *Biological Invasions* **11**: 47–58.

Parry D, Spence JR & Volney JA (1997) Responses of natural enemies to experimentally increased populations of the forest tent caterpillar, *Malacosoma distra*. *Ecological Entomology* **22**: 97–108.

Pasteur L (1874) Observations sure la coescistence der phyllocera et du mycelium constate a cully. *Comptes Rendus de l'Académie des Sciences Paris* **79**: 1233.

Paynter Q, Fowler SV, Gourlay AH, Groenteman R, Peterson PG, Smith L & Winks CJ (2010) Predicting parasitoid accumulation on biological control agents of weeds. *Journal of Applied Ecology* **47**: 575–582.

Paynter Q, Fowler SV, Gourlay AH, Peterson PG, Smith LA & Winks CJ (2015) Relative performance on test and target plants in laboratory tests predicts the risk of non-target attack in the field for arthropod weed biocontrol agents. *Biological Control* **80**: 133–142.

Paynter Q, Gourlay AH, Oboyski PT, Fowler SV, Hill RL, Withers TM, Parish H & Hona S (2008a) Why did specificity testing fail to predict the field host-range of the gorse pod moth in New Zealand? *Biological Control* **46**: 453–462.

Paynter Q, Martin N, Berry J, Hona S, Peterson P, Gourlay AH, Wilson-Davey J, Smith L, Winks C & Fowler SV (2008b) Non-target impacts of *Phytomyza vitalbae* a biological control agent of the European weed *Clematis vitalba* in New Zealand. *Biological Control* **44**: 248–258.

Paynter Q, Overton JM, Hill RL, Bellgard SE & Dawson MI (2012) Plant traits predict the success of weed biocontrol. *Journal of Applied Ecology* **49**: 1140–1148.

Paz A, Jareno D, Arroyo L, Viñuela J, Arroyo B, Mougeot F, Luque-Larenac JJ & Fargallo JA (2012) Avian predators as a biological control system of common vole (*Microtus arvalis*) populations in north-western Spain: experimental set-up and preliminary results. *Pest Management Science* **69**: 444–450.

Pearson DE & Callaway RM (2003) Indirect effects of host-specific biological control agents. *Trends in Ecology & Evolution* **18**: 456–461.

Pearson DE & Callaway RM (2004) Response to Thomas et al.: biocontrol and indirect effects. *Trends in Ecology & Evolution* **19**: 62–63.

Pearson DE & Callaway RM (2005) Indirect nontarget effects of host-specific biological control agents: implications for biological control. *Biological Control* **35**: 288–298.

Pearson DE & Callaway RM (2006) Biological control agents elevate hantavirus by subsidizing deer mouse populations. *Ecology Letters* **9**: 443–450.

Pearson DE, McKelvey KS & Ruggiero LF (2000) Non-target effects of an introduced biological control agent on deer mouse ecology. *Oecologia* **122**: 121–128.

Pearson DE, Potter T & Maron JL (2012) Biotic resistance: exclusion of native rodent consumers releases populations of a weak invader. *Journal of Ecology* **100**: 1383–1390.

Pech P, Fric Z & Konvicka M (2007) Species-specificity of the *Phengaris (Maculinea)-Myrmica* host system: fact or myth? (Lepidoptera: Lycaenidae; Hymenoptera: Formicidae). *Sociobiology* **50**: 983–1003.

Pedersen BS & Mills NJ (2004) Single vs. multiple introduction in biological control: the roles of parasitoid efficiency, antagonism and niche overlap. *Journal of Applied Ecology* **41**: 973–984.

Peever TL, Liu YC, Cortesi P & Milgroom MG (2000) Variation in tolerance and virulence in the chestnut blight fungus-hypovirus interaction. *Applied and Environmental Microbiology* **66**: 4863–4869.

Pekar S & Zd'Arkova E (2004) A model of the biological control of *Acarus siro* by *Cheyletus eruditus* (Acari: Acaridae, Cheyletidae) on grain. *Journal of Pest Science* **77**: 1–10.

Pels B & Sabelis MW (1999) Local dynamics, overexploitation and predator dispersal in an acarine predator–prey system. *Oikos* **86**: 573–583.

Pemberton RW (2000) Predictable risks to native plants in weed biological control. *Oecologia* **125**: 489–494.

Pemberton RW & Cordo HA (2001) Potential and risks of biological control of *Cactoblastis cactorum* (Lepidoptera: Pyralidae) in North America. *Florida Entomologist* **84**: 513–526.

Pemberton RW & Liu H (2007) Control and persistence of native *Opuntia* on Nevis and St. Kitts 50 years after the introduction of *Cactoblastis cactorum*. *Biological Control* **41**: 272–282.

Peng RK & Christian K (2006) Effective control of Jarvis's fruit fly, *Bactrocera jarvisi* (Diptera: Tephritidae), by the weaver ant, *Oecophylla smaragdina* (Hymenoptera: Formicidae), in mango orchards in the Northern Territory of Australia. *International Journal of Pest Management* **52**: 275–282.

Peng RK, Christian K & Reilly D (2011) The effect of weaver ants *Oecophylla smaragdina* on the shoot borer *Hypsipyla robusta* on African mahoganies in Australia. *Agricultural and Forest Entomology* **13**: 165–171.

Pennacchio F & Strand MR (2006) Evolution of developmental strategies in parasitic Hymenoptera. *Annual Review of Entomology* **51**: 233–258.

Perez C, Dill-Macky R & Kinkel LL (2008) Management of soil microbial communities to enhance populations of *Fusarium graminearum* antagonists in soil. *Plant and Soil* **302**: 53–69.

Perfecto I & Castiñeras A (1998) Deployment of the predaceous ants and their conservation in agroecosystems. In *Conservation Biological Control* (Barbosa P, ed.), San Diego, CA, Academic Press, pp. 269–290.

Perfecto I & Snelling R (1995) Biodiversity and the transformation of a tropical agroecosystem – ants in coffee plantations. *Ecological Applications* **5**: 1084–1097.

Perfecto I, Vandermeer JH, Bautista GL, Nunez GI, Greenberg R, Bichier P & Langridge S (2004) Greater predation in shaded coffee farms: the role of resident neotropical birds. *Ecology* **85**: 2677–2681.

Periquet G, Hedderwick MP, El Agoze M & Poirie M (1993) Sex determination in the hymenopteran *Diadromus pulchellus* (Ichneumonoidea): validation of the one-locus multi-allele model. *Heredity* **70**, 420–427.

Perkins RCL (1897) The introduction of beneficial insects into the Hawaiian islands. *Nature* **55**: 499–500.

Perlman SJ, Kelly SE, Zchori-Fein E & Hunter MS (2006) Cytoplasmic incompatibility and multiple symbiont infection in the ash whitefly parasitoid, *Encarsia inaron*. *Biological Control* **39**: 474–480.

Perrin RM (1975) Role of perennial stinging nettle, *Urtica dioica*, as a reservoir of beneficial natural enemies. *Annals of Applied Biology* **81**: 289–297.

Persello-Cartieaux F, Nussaume L & Robaglia C (2003) Tales from the underground: molecular plant-rhizobacteria interactions. *Plant Cell and Environment* **26**: 189–199.

Peschken DP & McClay A (1995) Picking the target: a revision of McClay's scoring system to determine the suitability of a weed for classical biological control. In *Proceedings of the VIII International Symposium on Biological Control of Weeds* (Delfosse ES & Scott RR, eds.), Melbourne, Australia, DSIR/CSIRO, pp. 137–143.

Peterson AT, Soberon J, Pearson RG, Anderson RP, Martínez-Meyer E, Nakamura M & Araújo MB (2011) *Ecological Niches and Geographic Distributions*. Princeton, NJ, Princeton University Press.

Petitpierre B, Kueffer C, Broennimann O, Randin C, Daehler C & Guisan A (2012) Climatic niche shifts are rare among terrestrial plant invaders. *Science* **335**: 1344–1348.

Pfeiffer DG (1986) Effects of field applications of paraquat on densities of *Panonychus ulmi* (Koch) and *Neoseiulus fallacis* (Garman). *Journal of Agricultural Entomology* **3**: 322–325.

Pfeiffer DG (2000) Selective insecticides. In *Insect Pest Management: Techniques for Environmental Protection* (Rechcigl JE & Rechcigl NA, eds.), Boca Raton, FL, CRC Press, pp. 131–144.

Phatak S, Callaway MB & Vavrina CS (1987) Biological control and its integration in weed management systems for purple and yellow nutsedge *Cyperus rotundus* and *Cyperus esculentus*. *Weed Technology* **1**: 84–91.

Phillips CB, Baird DB, Iline II, McNeill MR, Proffitt JR, Goldson SL & Kean JM (2008) East meets west: adaptive evolution of an insect introduced for biological control. *Journal of Applied Ecology* **45**: 948–956.

Philpott SM, Perfecto I & Vandermeer J (2008) Effects of predatory ants on lower trophic levels across a gradient of coffee management complexity. *Journal of Animal Ecology* **77**: 505–511.

Philpott SM, Soong O, Lowenstein JH, Pulido AL, Lopez DT, Flynn DFB & DeClerck F (2009) Functional richness and ecosystem services: bird predation on arthropods in tropical agroecosystems. *Ecological Applications* **19**: 1858–1867.

Pickett CH & Bugg RL (1998) *Enhancing Biological Control: Habitat Management to Promote Natural Enemies of Agricultural Pests*. Berkeley, University of California Press.

Pickett CH, Roltsch W & Corbett A (2004) The role of a rubidium marked natural enemy refuge in the establishment and movement of *Bemisia* parasitoids. *International Journal of Pest Management* **50**: 183–191.

Pierson EA & Mack RN (1990) The population biology of *Bromus tectorum* in forests: effect of disturbance, grazing, and litter on seedling establishment and reproduction. *Oecologia* **84**: 526–533.

Pierson EA & Weller DM (1994) Use of mixtures of fluorescent pseudomonads to suppress take-all and improve the growth of wheat. *Phytopathology* **84**: 940–947.

Pike KS & Burkhardt CC (1974) Hyperparasites of *Bathyplectes curculionis* in Wyoming. *Environmental Entomology* **3**: 953–956.

Pilz C, Wegensteiner R & Keller S (2007) Selection of entomopathogenic fungi for the control of the western corn rootworm *Diabrotica virgifera virgifera*. *Journal of Applied Entomology* **131**: 426–431.

Pimentel D (1980) Environmental risks associated with biological controls. In *Environmental Protection and Biological Forms of Control of Pest Organisms* (Lundholm B & Stackerud M, eds.), Stockholm, Sweden, Swedish Natural Science Research Council, pp. 11–24.

Pimentel D, Glenister C, Fast S & Gallahan D (1984) Environmental risks of biological pest controls. *Oikos* **42**: 283–290.

Pimentel D, Lach L, Zuniga R & Morrison D (2000) Environmental and economic costs of nonindigenous species in the United States. *BioScience* **50**: 53–65.

Pineda A & Marcos-Garcia MA (2008) Use of selected flowering plants in greenhouses to enhance aphidophagous hoverfly populations (Diptera: Syrphidae). *Annales De La Societe Entomologique De France* **44**: 487–492.

Piper GL, Coombs EM, Markin GP & Joley DB (2004) Rush skeletonweed. In *Biological Control of Invasive Plants in the United States* (Coombs EM, Coombs JK, Clark GL & Cofrancesco Jr. AF, eds.), Corvallis, Oregon State University Press, pp. 293–303.

Pizzatto L & Shine R (2012) Typhoid Mary in the frogpond: can we use native frogs to disseminate a lungworm biocontrol for invasive cane toads? *Animal Conservation* **15**: 545–552.

Plećaš M, Gagic V, Jankovic M, Petrovic-Obradovic O., Kavallieratos NG, Tomanovic Z, Thies C, Tscharntke T & Cetkovic A (2014) Landscape composition and configuration influence cereal aphid-parasitoid-hyperparasitoid interactions and biological control differentially across years. *Agriculture Ecosystems & Environment* **183**: 1–10.

Poehling HM (1989) Selective application strategies for insecticides in agricultural crops. In *Pesticides and Non-target Invertebrates* (Jepson PC, ed.), Andover, UK, Intercept, pp. 151–175.

Pointier J-P, David P & Jarne P (2011) The biological control of the snail hosts of schistosomes: the role of competitor snails and biological invasions. *Biomphalaria Snails and Larval Trematodes* (Toledo R & Fried B, eds.), New York, Springer, pp. 215–238.

Polis GA & Strong DR (1996) Food web complexity and community dynamics. *American Naturalist* **147**: 813–846.

Pomerinke MA, Thompson DC & Clason DL (1995) Bionomics of *Cleonidius trivittatus* (Coleoptera: Curculionidae): native biological control of purple locoweed (Rosales: Fabaceae). *Environmental Entomology* **24**: 1696–1702.

Port GR, Glen DM & Symondson WOC (2000) Success in biological control of terrestrial molluscs. In *Biological Control: Measures of Success* (Gurr GM & Wratten S, eds.), Dordrecht, The Netherlands, Kluwer, pp. 133–158.

Postma J, Montanari M & Van den Boogert P (2003) Microbial enrichment to enhance the disease suppressive activity of compost. *European Journal of Soil Biology* **39**: 157–163.

Powell W (1989) Enhancing parasitoid activity in crops. In *Insect Parasitoids* (Waage J & Greathead D, eds.), London, UK, Academic Press, pp. 319–340.

Power M & McCarty LS (2002) Trends in the development of ecological risk assessment and management frameworks. *Human and Ecological Risk Assessment* **8**: 7–18.

Power AG & Mitchell CE (2004) Pathogen spillover in disease epidemics. *American Naturalist* **164**: S79–S89.

Pozo MJ, Baek J-M, Garcia JM & Kenerley CM (2004) Functional analysis of tvsp1, a serine protease-encoding gene in the biocontrol agent *Trichoderma virens*. *Fungal Genetics and Biology* **41**: 336–348.

Prasad RP & Snyder WE (2006) Polyphagy complicates conservation biological control that targets generalist predators. *Journal of Applied Ecology* **43**: 343–352.

Pratt PD, Rayamajhi MB, Van TK & Center TD (2004) Modeling the influence of resource availability on population densities of *Oxyops vitiosa* (Coleoptera: Curculionidae), a biological control agent of the invasive tree *Melaleuca quinquenervia*. *Biocontrol Science and Technology* **14**: 51–61.

Pratt PD, Slone DH, Rayamajhi MB, Van TK & Center TD (2003) Geographic distribution and dispersal rate of *Oxyops vitiosa* (Coleoptera: Curculionidae), a biological control agent of the invasive tree *Melaleuca quinquenervia* in south Florida. *Environmental Entomology* **32**: 397–406.

Preisser EL, Bolnick DI & Benard MF (2005) Scared to death? The effects of intimidation and consumption in predator–prey interactions. *Ecology* **86**: 501–509.

Price RE, Müller EJ, Brown HD, D'Uamba P & Jone AA (1999) The first trial of a *Metarhizium anisopliae* var. *acridum* mycoinsecticide for the control of the red locust in a recognised outbreak area. *Insect Science and Its Application* **19**: 323–331.

Prior C, Jollands P & le Patourel G (1988) Infectivity of oil and water formulations of *Beauveria bassiana* (Deuteromycotina: Hyphomycetes) to the cotton weevil pest *Pantorhytes plutus* (Coleoptera: Curculionidae). *Journal of Invertebrate Pathology* **52**: 66–72.

Prischmann DA, James DG, Storm CP, Wright LC & Snyder WE (2007) Identity, abundance, and phenology of *Anagrus* spp. (Hymenoptera: Mymaridae) and leafhoppers (Homoptera: Cicadellidae) associated with grape, blackberry, and wild rose in Washington State. *Annals of the Entomological Society of America* **100**: 41–52.

Pyke GH & White AW (2000) Factors influencing predation on eggs and tadpoles of the endangered green and golden bell from *Litora aurea* by the introduced plague minnow *Gambusia holbrooki*. *Australian Zoologist* **31**: 496–505.

Pysek P & Richardson DM (2006) The biogeography of naturalization in alien plants. *Journal of Biogeography* **33**: 2040–2050.

Quicke DLJ (1997) *Parasitic Wasps*. London, UK, Chapman & Hall.

Quimby PC, Gras S, Widmer T, Meikle W & Sands D (2004) Formulation of *Selerotinia sclerotiorum* for use against *Cirsium arvense*. *Zeitschrift Fur Pflanzenkrankheiten Und Pflanzenschutz* **19**: 491–495.

Raaijmakers JM, Vlami M & de Souza JT (2002) Antibiotic production by bacterial biocontrol agents. *Antonie van Leeuwenhoek International Journal of General and Molecular Microbiology* **81**: 537–547.

Raaijmakers JM & Weller DM (1998) Natural plant protection by 2,4-diacetylphloroglucinol – producing *Pseudomonas* spp. in take-all decline soils. *Molecular Plant-Microbe Interactions* **11**: 144–152.

Raak-van den Berg CL, De Lange HJ & Van Lenteren JC (2012) Intraguild predation behaviour of ladybirds in semi-field experiments explains invasion success of *Harmonia axyridis*. *PLoS One* **7**: e40681.

Rabb RL (1971) Naturally-occurring biological control in the eastern United States, with particular reference to tobacco insects. In *Biological Control* (Huffaker CB, eds.), New York, Plenum, pp. 294–311.

Radcliffe EB & Flanders KL (1998) Biological control of alfalfa weevil in North America. *Integrated Pest Management Reviews* **3**: 225–242.

Rafter MA, Wilson AJ, Senaratne KADW & Dhileepan K (2008) Climatic-requirements models of cat's claw creeper *Macfadyena unguis-cati* (Bignoniaceae) to prioritise areas for exploration and release of biological control agents. *Biological Control* **44**: 169–179.

Raghu S, Dhileepan K & Scanlan JC (2007) Predicting risk and benefit a priori in biological control of invasive plant species: a systems modelling approach. *Ecological Modelling* **208**: 247–262.

Raghu S, Purcell MF & Wright AD (2013) Catchment context and the bottom-up regulation of the abundance of specialist semi-aquatic weevils on water hyacinth. *Ecological Entomology* **38**: 117–122.

Raghu S & Van Klinken RD (2006) Refining the ecological basis for agent selection in weed biological control. *Australian Journal of Entomology* **45**: 251–252.

Raghu S & Walton C (2007) Understanding the ghost of *Cactoblastis* past: historical clarifications on a poster child of classical biological control. *BioScience* **57**: 699–705.

Ramsell J, Malloch AJC & Whittaker JB (1993) When grazed by *Tipula paludosa, Lolium perenne* is a stronger competitor of *Rumex obtusifolius*. *Journal of Ecology* **81**: 777–786.

Ramula S, Knight TM, Burns JH & Buckley YM (2008) General guidelines for invasive plant management based on comparative demography of invasive and native plant populations. *Journal of Applied Ecology* **45**: 1124–1133.

Rand TA & Louda SM (2004) Exotic weed invasion increases the susceptibility of native plants to attack by a biocontrol herbivore. *Ecology* **86**: 1548–1554.

Rand TA & Louda SM (2006) Invasive insect abundance varies across the biogeographic distribution of a native host plant. *Ecological Applications* **16**: 877–890.

Rand TA, Van Veen FJF & Tscharnkte T (2012) Landscape complexity differentially benefits generalized fourth, over specialized third, trophic level natural enemies. *Ecography* **35**: 97–104.

Randlkofer B, Obermaier E, Casas J & Meiners T (2010) Connectivity counts: disentangling effects of vegetation structure elements on the searching movement of a parasitoid. *Ecological Entomology* **35**: 446–455.

Randlkofer B, Obermaier E & Meiners T (2007) Mother's choice of the oviposition site: balancing risk of egg parasitism and need of food supply for the progeny with an infochemical shelter? *Chemoecology* **17**: 177–186.

Raoult D, Audic S, Robert C, Abergel C, Renesto P, Ogata H, La Scola B, Suzan M & Claverie JM (2004) The 1.2 megabase genome sequence of mimivirus. *Science* **306**: 1344–1350.

Rapp G & Salum MS (1995) Ant fauna, pest damage and yield in relation to the density of weeds in coconut sites in Zanzibar, Tanzania. *Journal of Applied Entomology* **119**: 45–48.

Rasgon JL, Stryer LM & Scott TW (2003) *Wolbachia*-induced mortality as a mechanism to modulate pathogen transmission by vector arthropods. *Journal of Medical Entomology* **40**: 125–132.

Ravensberg WJ (2011) *A Roadmap to the Successful Development and Commercialization of Microbial Pest Control Products for Control of Arthropods.* Dordrecht, The Netherlands, Springer.

Rechcigl JE & Rechcigl NA (1998) *Biological and Biotechnological Control of Insect Pests.* Boca Raton, FL, Lewis.

Rees M & Hill RL (2001) Large-scale disturbances, biological control and the dynamics of gorse populations. *Journal of Applied Ecology* **38**: 364–377.

Rees M & Paynter Q (1997) Biological control of Scotch broom: modelling the determinants of abundance and the potential impact of introduced insect herbivores. *Journal of Applied Ecology* **34**: 1203–1221.

Reeve JD (1988) Environmental variability, migration, and persistence in host-parasitoid systems. *American Naturalist* **132**: 810–836.

Reeves JL & Lorch PD (2012) Biological control of invasive aquatic and wetland plants by arthropods: a meta-analysis of data from the last three decades. *BioControl* **57**: 103–116.

Rehage JS & Sih A (2004) Dispersal behavior, boldness, and the link to invasiveness: a comparison of four *Gambusia* species. *Biological Invasions* **6**: 379–391.

Reid AM, Morin L, Downey PO, French K & Virtue JG (2009) Does invasive plant management aid the restoration of natural ecosystems? *Biological Conservation* **142**: 2342–2349.

Reitz SR & Trumble JT (2002) Competitive displacement among insects and arachnids. *Annual Review of Entomology* **47**: 435–466.

Rejmanek M (1996) Species richness and resistance to invasions. In *Biodiversity and Ecosystem Processes in Tropical Forests* (Orians GH, Dirzo R & Cushman JH, eds.), Heidelberg, Germany, Springer, pp. 153–172.

Remington CL (1968) The population genetics of insect introduction. *Annual Review of Entomology* **13**: 415–426.

Rethwisch MD & Manglitz GR (1986) Parasitoids of *Bathyplectes curculionis* (Hymenoptera: Ichneumonidae) in southeastern Nebraska. *Journal of the Kansas Entomological Society* **59**: 648–652.

Reznick DN & Ghalambor CK (2001) The population ecology of contemporary adaptations: what empirical studies reveal about the conditions that promote adaptive evolution. *Genetica* **112–113**: 183–198.

Rhodes EM & Liburd OE (2006) Evaluation of predatory mites and Acramite for control of twospotted spider mites in strawberries in north central Florida. *Journal of Economic Entomology* **99**: 1291–1298.

Ricciardi A & Cohen J (2007) The invasiveness of an introduced species does not predict its impact. *Biological Invasions* **9**: 309–315.

Ricciardi A, Hoopes MF, Marchetti E & Lockwood JL (2013) Progress toward understanding the ecological impacts of nonnative species. *Ecological Monographs* **83**: 263–282.

Ricciardi A & Simberloff D (2009) Assisted colonization is not a viable conservation strategy. *Trends in Ecology & Evolution* **24**: 248–253.

Richardson DM (2000) Naturalization and invasion of alien plants: concepts and definitions. *Diversity and Distributions* **6**: 93–107.

Richardson DM (2011) *Fifty Years of Invasion Ecology. The Legacy of Charles Elton.* Oxford, UK, Wiley-Blackwell.

Richardson DM, Allsopp N, D'Antonio CM, Milton SJ & Rejmanek M (2000) Plant invasions – the role of mutualisms. *Biological Reviews* **75**: 65–93.

Richardson DM & Pyšek P (2012) Naturalization of introduced plants: ecological drivers of biogeographical patterns. *New Phytologist* **196**: 383–396.

Ridenour WM & Callaway RM (2001) The relative importance of allelopathy in interference: the effects of an invasive weed on a native bunchgrass. *Oecologia* **126**: 444–450.

Ridenour WL & Callaway RM (2003) Root herbivores, pathogenic fungi, and competition between *Centaurea maculosa* and *Festuca idahoensis. Plant Ecology* **169**: 161–170.

Riechert SE (1998) The role of spiders and their conservation in agroecosystems. In *Enhancing Biological Control: Habitat Management to Promote Natural Enemies of Agricultural Pests* (Pickett CH & Bugg RL, eds.), Berkeley, University of California Press, pp. 211–237.

Rincon C, Bordat D, Lohr B & Dupas S (2006) Reproductive isolation and differentiation between five populations of *Cotesia plutellae* (Hymenoptera: Braconidae), parasitoid of *Plutella xylostella* (Lepidoptera: Plutellidae). *Biological Control* **36**: 171–182.

Ripper WE, Greenslade RM & Hartley GS (1951) Selective insecticides and biological control. *Journal of Economic Entomology* **44**: 448–459.

Risch SJ, Andow DA & Altieri MA (1983) Agroecosystem diversity and pest control: data, tentative conclusions, and new research directions. *Environmental Entomology* **12**: 625–629.

Rishbeth J (1963) Stump protection against *Fomes annosus*: III. Inoculation with *Peniophora gigantea. Annals of Applied Biology* **52**: 63–77.

R'Kha S, Capy P & David JR (1991) Host-plant specialization in the *Drosophila melanogaster* species complex: a physiological, behavioral and genetical analysis. *Proceedings of the National Academy of Sciences, USA* **88**: 1835–1939.

Robinson AS, Franz G & Atkinson PW (2004) Insect transgenesis and its potential role in agriculture and human health. *Insect Biochemistry and Molecular Biology* **34**: 113–120.

Robinson KA, Jonsson M, Wratten SD, Wade MR & Buckley HL (2008) Implications of floral resources for predation by an omnivorous lacewing. *Basic and Applied Ecology* **9**: 172–181.

Rochat J & Gutierrez AP (2001) Weather-mediated regulation of olive scale by two parasitoids. *Journal of Animal Ecology* **70**: 476–490.

Roderick GK (1992) Postcolonization evolution of natural enemies. In *Selection Criteria and Ecological Consequences of Importing Natural Enemies* (Kauffman WC & Nechols JR, eds.), Lanham, MD, Entomological Society of America, pp. 71–86.

Roderick GK (2004) Tracing the origins of pests and natural enemies: genetic and statistical approaches. In *Genetics, Evolution and Biological Control* (Ehler LE, Sforza R & Mateille T, eds.), Wallingford, UK, CABI Publishing, pp. 97–112.

Roderick GK, Hufbauer R & Navajas M (2012) Evolution and biological control. *Evolutionary Applications* **5**: 419–423.

Roderick GK & Navajas M (2003) Genes in new environments: genetics and evolution in biological control. *Nature Reviews Genetics* **4**: 889–899.

Roderick GK & Navajas M (2008) The primacy of evolution in biological control. In *Proceedings of the XII International Symposium on Weed Biological Control* (Julien M, ed.), 403–409. Wallingford, UK, CABI Publishing, pp. 403–409.

Rogers DJ (1972) Random search and insect population models. *Journal of Animal Ecology* **41**: 369–383.

Rogers DJ & Hassell MP (1974) General models for insect parasite and predator searching behaviour: interference. *Journal of Animal Ecology* **43**: 239–253.

Rogers MA, Krischik VA & Martin LA (2007) Effect of soil application of imidacloprid on survival of adult green lacewing, *Chrysoperla carnea* (Neuroptera: Chrysopidae), used for biological control in greenhouses. *Biological Control* **42**: 172–177.

Rogers WE & Siemann E (2004) Invasive ecotypes tolerate herbivory more effectively than native ecotypes of the Chinese tallow tree *Sapium sebiferum*. *Journal of Applied Ecology* **41**: 561–570.

Rohani P, Godfray HJC & Hassell MP (1994) Aggregation and the dynamics of host-parasitoid systems: a discrete-generation model with within-generation redistribution. *American Naturalist* **144**: 491–509.

Roitberg BD, Sircom J, Roitberg C, Van Alphen JJM & Mangel M (1992) Seasonal dynamic shifs in patch exploitation by parasitic wasps. *Behavioral Ecology* **3**: 156–165.

Roitberg BD, Sircom J, Roitberg C, Van Alphen JJM & Mangel M (1993) Life expectancy and reproduction. *Nature* **364**: 108.

Roland J & Taylor PD (1997) Insect parasitoid species respond to forest structure at different spatial scales. *Nature* **386**: 710–713.

Romeis J, Babendreier D, Wäckers FL & Shanower TG (2005) Habitat and plant specificity of *Trichogramma* egg parasitoids – underlying mechanisms and implications. *Basic and Applied Ecology* **6**: 215–236.

Romeis J & Wäckers FL (2000) Feeding responses by female *Pieris brassicae* butterflies to carbohydrates and amino acids. *Physiological Entomology* **25**: 247–253.

Room PM (1990) Ecology of a simple plant-herbivore system: biological control of *Salvinia*. *Trends in Ecology & Evolution* **5**: 74–79.

Room PM, Harley KLS, Forno IW & Sands DPA (1981) Successful biological control of the floating weed salvinia. *Nature* **394**: 78–80.

Rondelaud D, Vingoles P, Dreyfuss G & Mage M (2006) The control of *Galba truncula*ta (Gastropoda: Lymnaeidae) by the terrestrial snail *Zonitoides nitidus* on acid soils. *Biological Control* **39**: 290–299.

Root RB (1973) Organization of a plant-arthropod association in simple and diverse habitats: the fauna of collards (*Brassica oleracea*). *Ecological Monographs* **43**: 95–124.

Rose KE, Louda SM & Rees M (2005) Demographic and evolutionary impacts of native and invasive insect herbivores on *Cirsium canescens*. *Ecology* **86**: 453–465.

Rose M & DeBach P (1992) Biological control of *Parabemisia myricae* (Kuwana) (Homoptera: Aleyrodidae) in California. *Israel Journal of Entomology* **25–26**: 73–95.

Rose MR, Nusbaum TJ & Chippendale AK (1996) Laboratory evolution: the experimental wonderland and the Cheshire Cat Syndrome. In *Adaptation* (Rose MR & Lauder GV, eds.), San Diego, CA, Academic Press, pp. 221–224.

Rosen D (1994) *Advances in the Study of Aphytis*. Andover, UK, Intercept.

Rosen D & DeBach P (1977) Use of scale-insect parasites in Coccoidea systematics. *Virginia Polytechnic Institute and State University Research Division Bulletin* **127**: 5–21.

Rosen D & DeBach P (1979) *Species of Aphytis of the World*. The Hague, The Netherlands, W. Junk.

Rosenheim JA (1998) Higher-order predators and the regulation of insect herbivore populations. *Annual Review of Entomology* **43**: 421–448.

Rosenheim JA & Corbett A (2003) Omnivory and the indeterminacy of predator function: can a knowledge of foraging behavior help? *Ecology* **84**: 2538–2548.

Rosenheim JA & Harmon J (2006) The influence of intraguild predation on the suppression of a shared prey population: an empirical assessment. In *Trophic and Guild Interactions in Biological Control* (Brodeur J & Boivin G, eds.), Dordrecht, The Netherlands, Springer, pp. 1–20.

Rosenheim JA, Kaya HK, Ehler LE, Marois JJ & Jaffee BA (1995) Intraguild predation among biological-control agents: theory and evidence. *Biological Control* **5**: 303–335.

Rosenheim JA, Wilhoit LR & Armer CA (1993) Influence of intraguild predation among generalist insect predators on the suppression of an herbivore population. *Oecologia* **96**: 439–449.

Rosenheim JA, Wilhoit LR, Goodell PB, Grafton-Cardwell EE & Leigh TF (1997) Plant compensation, natural biological control, and herbivory by *Aphis gossypii* on pre-reproductive cotton: the anatomy of a non-pest. *Entomologia Experimentalis et Applicata* **85**: 45–63.

Rouchet R & Vorburger C (2012) Strong specificity in the interaction between parasitoids and symbiont-protected hosts. *Journal of Evolutionary Biology* **25**: 2369–2375.

Roush RT (1990) Genetic variation in natural enemies: critical issues for colonization in biological control. In *Critical Issues in Biological Control* (Mackauer M, Ehler LE & Roland J, eds.), Andover, UK, Intercept, pp. 265–288.

Roush RT & Hopper KR (1995) Use of single family lines to preserve genetic variation in laboratory colonies. *Annals of the Entomological Society of America* **88**: 713–717.

Roy HE, Adriaens T, Isaac NJB, Kenis M, Onkelinx T, San Martin G, Brown PMJ, Hautier L, Poland R, Roy DB, Comont R, Eschen R, Frost R, Zindel R, Van Vlaenderen J, Nedved O, Ravn HP, Gregoire JC, de Biseau J-C & Maes D (2012) Invasive alien predator causes rapid declines of native European ladybirds. *Diversity and Distributions* **18**: 717–725.

Roy HE, Lawson-Handley L-J, Schonrogge K, Poland RL & Purse BV (2011) Can the enemy release hypothesis explain the success of invasive alien predators and parasitoids? *BioControl* **56**: 451–468.

Roy HE, Pell JK, Clark SJ & Alderson PG (1998) Implications of predator foraging on aphid pathogen dynamics. *Journal of Invertebrate Pathology* **71**: 236–247.

Royama T (1984) Population dynamics of the spruce budworm *Choristoneura fumiferana*. *Ecological Monographs* **54**: 429–462.

Ruiz GM & Carlton JT (2003) *Invasive Species: Vectors and Management Practices*. Washington, DC, Island Press.

Ruiz L, Flores S, Cancino J, Arredondo J, Valle J, Diaz-Fleischer F & Williams T (2008) Lethal and sublethal effects of spinosad-based GF-120 bait on the tephritid parasitoid *Diachasmimorpha longicaudata* (Hymenoptera: Braconidae). *Biological Control* **44**: 296–304.

Russell EP (1989) Enemies hypothesis: a review of the effect of vegetational diversity on predatory insects and parasitoids. *Environmental Entomology* **18**: 590–599.

Russell K & Bessin R (2009) Integration of *Trichogramma ostriniae* releases and habitat modification for suppression of European corn borer (*Ostrinia nubilalis* Hubner) in bell peppers. *Renewable Agriculture and Food Systems* **24**: 19–24.

Russell M (2015) A meta-analysis of physiological and behavioral responses of parasitoid wasps to flowers of individual plant species. *Biological Control* **82**: 96–103.

Ruxton GD, Gurney WSC & de Roos AM (1992) Interference and generation cycles. *Theoretical Population Biology* **42**: 235–253.

Ryan PR, Dessaux Y, Thomashow LS & Weller DM (2009) Rhizosphere engineering and management for sustainable agriculture. *Plant and Soil* **321**: 363–383.

Ryan RB (1990) Evaluation of biological control: introduced parasites of larch casebearer (Lepidoptera: Coleophoridae) in Oregon. *Environmental Entomology* **19**: 1873–1881.

Sailer RI (1978) Our immigrant insect fauna. *ESA Bulletin* **24**: 3–11.

Saint KM, French N & Kerr P (2001). Genetic variation in Australian isolates of myxoma virus: an evolutionary and epidemiological study. *Archives of Virology* **146**: 1105–1123.

Saito Y, Urano S, Nakao H, Amimoto K & Mori H (1996) A simulation model for predicting the efficiency of biological control of spider mites by phytoseiid predators 2. Validity tests and data necessary for practical usage. *Japanese Journal of Applied Entomology and Zoology* **40**: 113–120.

Sakai AK, Allendorf FW, Holt JS, Lodge DM, Molofsky J, With KA, Baughman S, Cabin RJ, Cohen JE, Ellstrand NC, McCauley DE, O'Neil P, Parker IM, Thompson JN & Weller SG (2001) The population biology of invasive species. *Annual Review of Ecology and Systematics* **32**: 305–332.

Sakuratani Y, Matsumoto Y, Oka M, Kubo T, Fujii A, Uotani M & Teraguchi T (2000) Life history of *Adalia bipunctata* (Coleoptera: Coccinellidae) in Japan. *European Journal of Entomology* **97**: 555–558.

Salin C, Deprez B, Van Bockstaele DR, Mahillon J & Hance T (2004) Sex determination mechanism in the hymenopteran parasitoid *Aphidius rhopalosiphi* De Stefani-Peres (Braconidae: Aphidiinae). *Belgian Journal of Zoology* **134**: 15–21.

Salt G & Van den Bosch R (1967) The defense reactions of three species of *Hypera* (Coleoptera: Curculionidae) to an ichneumon wasp. *Journal of Invertebrate Pathology* **9**: 164–177.

Samac DA & Kinkel LL (2001) Suppression of the root-lesion nematode (*Pratylenchus penetrans*) in alfalfa (*Medicago sativa*) by *Streptomyces* spp. *Plant and Soil* **235**: 35–44.

Samac DA, Willert AM, McBride MJ & Kinkel LL (2003) Effect of antibiotic-producing *Streptomyces* on nodulation and leaf spot in alfalfa. *Applied Soil Ecology* **22**: 55–66.

Sandrock C & Vorburger C (2011) Single-locus recessive inheritance of asexual reproduction in a parasitoid wasp. *Current Biology* **21**: 433–437.

Sands DPA (1997) The 'safety' of biological control agents: assessing their impact on beneficial and other non-target hosts. *Memoirs of the Museum of Victoria* **56**: 611–615.

Sands DPA & Van Driesche RG (2004) Using the scientific literature to estimate the host range of a biological control agent. In *Assessing Host Ranges for Parasitoids and Predators Used for Classical Biological Control: A Guide to Best Practice* (Van Driesche RG & Reardon R, eds.), Morgantown, WV, U.S. Forest Service, pp. 15–23.

Sant D, Casanova E, Segarra G, Aviles M, Reis M & Trillas MI (2010) Effect of *Trichoderma asperellum* strain T34 on *Fusarium* wilt and water usage in carnation grown on compost-based growth medium. *Biological Control* **53**: 291–296.

Santoyo G, Orozco-Mosqueda MdC & Govindappa M (2012) Mechanisms of biocontrol and plant-growth promoting activity in soil bacterial species of *Bacillus* and *Pseudomonas*: a review. *Biocontrol Science and Technology* **22**: 855–872.

Sato S & Dixon AFG (2004) Effect of intraguild predation on the survival and development of three species of aphidophagous ladybirds: consequences for invasive species. *Agricultural and Forest Entomology* **6**: 21–24.

Sato Y & Ohsaki N (2004) Response of the wasp (*Cotesia glomerata*) to larvae of the large white butterfly. *Ecological Research* **19**: 445–449.

Sato S, Yasuda H, Evans EW & Dixon AFG (2009) Vulnerability of larvae of two species of aphidophagous ladybirds, *Adalia bipunctata* Linnaeus and *Harmonia axyridis* Pallas, to cannibalism and intraguild predation. *Entomological Science* **12**: 111–115.

Saunders G, Cooke B, McColl K, Shine R & Peacock T (2010) Modern approaches for the biological control of vertebrate pests: an Australian perspective. *Biological Control* **52**: 288–295.

Sawyer RC (1996) *To Make a Spotless Orange*. Ames, University of Iowa Press.

Sax DF (2001) Latitudinal gradients and geographic ranges of exotic species: implications for biogeography. *Journal of Biogeography* **28**: 139–150.

Scanlan, JC, Grant WE, Hunter DM & Milner RJ (2001) Habitat and environmental factors influencing the control of migratory locusts (*Locusta migratoria*) with an entomopathogenic fungus (*Metarhizium anisopliae*). *Ecological Modelling* **136**: 223–236.

Scarborough CL, Ferrari J & Godfray HCJ (2005) Aphid protected from pathogen by endosymbiont. *Science* **310**: 1781.

Schellhorn NA, Bianchi FJJA & Hsu CL (2014) Movement of entomophagous arthropods in agricultural landscapes: links to pest suppression. *Annual Review of Entomology* **59**: 559–581.

Schellhorn NA, Glatz RV & Wood GM (2010) The risk of exotic and native plants as hosts for four pest thrips (Thysanoptera: Thripinae). *Bulletin of Entomological Research* **100**: 501–510.

Schellhorn NA, Kuhman TR, Olson AC & Ives AR (2002) Competition between native and introduced parasitoids of aphids: nontarget effects and biological control. *Ecology* **83**: 2745–2757.

Schemske DW, Mittelbach GG, Cornell HV, Sobel JM & Roy K (2009) Is there a latitudinal gradient in the importance of biotic interactions? *Annual Review of Ecology, Evolution and Systematics* **40**: 245–269.

Schisler DA, Slininger PJ, Behle RW & Jackson MA (2004) Formulation of *Bacillus* spp. for biological control of plant diseases. *Phytopathology* **94**: 1267–1271.

Schlatter D, Fubuh A, Xiao K, Hernandez D, Hobbie S & Kinkel L (2009) Resource amendments influence density and competitive phenotypes of *Streptomyces* in soil. *Microbial Ecology* **57**: 413–420.

Schmidl L (1972) Studies on the control of ragwort, *Senecio jacobaea* L. with the cinnabar moth, *Callimorpha jacobaeae* (L.) (Arctiidae: Lepidoptera), in Victoria. *Weed Research* **12**: 46–57.

Schmidt MH, Thies C, Nentwig W & Tscharntke T (2008) Contrasting responses of arable spiders to the landscape matrix at different spatial scales. *Journal of Biogeography* **35**: 157–166.

Schmidt MH & Tscharntke T (2005) Landscape context of sheetweb spider (Araneae: Linyphiidae) abundance in cereal fields. *Journal of Biogeography* **32**: 467–473.

Schmidt NP, O'Neal ME & Singer JW (2007) Alfalfa living mulch advances biological control of soybean aphid. *Environmental Entomology* **36**: 416–424.

Schmitz OJ, Krival V & Ovadia O (2004) Trophic cascades: the primacy of trait-mediated indirect interactions. *Ecology Letters* **7**: 153–163.

Schoener TW (1983) Field experiments on interspecific competition. *American Naturalist* **122**: 240–285.

Schooler SS, De Barro P & Ives AR (2011) The potential for hyperparasitism to compromise biological control: why don't hyperparasitoids drive their primary parasitoid hosts extinct? *Biological Control* **58**: 167–173.

Schreiner I (1989) Biological control introductions in the Caroline and Marshall Islands. *Proceedings of the Hawaiian Entomological Society* **29**: 57–69.

Schrey SD & Tarkka MT (2008) Friends and foes: streptomycetes as modulators of plant disease and symbiosis. *Antonie van Leeuwenhoek International Journal of General and Molecular Microbiology* **94**: 11–19.

Seaman GA & Randall JE (1962) The mongoose as a predator in the Virgin Islands. *Journal of Mammalogy* **43**: 544–546.

Seamark RF (2001) Biotech prospects for the control of introduced mammals in Australia. *Reproduction, Fertility and Development* **13**: 705–711.

Seastedt TR (2015) Biological control of invasive plant species: a reassessment for the Anthropocene. *New Phytologist* **205**: 490–502.

Sebolt DC & Landis DA (2002) Neonate *Galerucella calmariensis* (Coleoptera: Chrysomelidae) behavior on purple loosestrife (*Lythrum salicaria*) contributes to reduced predation. *Environmental Entomology* **31**: 880–886.

Sebolt DC & Landis DA (2004) Arthropod predators of *Galerucella calmariensis* L. (Coleoptera: Chrysomelidae): an assessment of biotic interference. *Environmental Entomology* **33**: 356–461.

Secord D (2003) Biological control of marine invasive species: cautionary tales and land-based lessons. *Biological Invasions* **5**: 117–131.

Segal D & Glazer I (2000) Genetics for improving biological control agents: the case of entomopathogenic nematodes. *Crop Protection* **19**: 685–689.

Segoli M & Rosenheim JA (2013) The link between host density and egg production in a parasitoid insect: comparison between agricultural and natural habitats. *Functional Ecology* **27**: 1224–1232.

Seguni ZSK, Way MJ & Van Mele P (2011) The effect of ground vegetation management on competition between the ants *Oecophylla longinoda* and *Pheidole megacephala* and implications for conservation biological control. *Crop Protection* **30**: 713–717.

Settle WH, Ariawan H, Astuti ET, Cahyana W, Hakim AL, Hindayana D, Lestari AS & Pajarningsih (1996) Managing tropical rice pests through conservation of generalist natural enemies and alternative prey. *Ecology* **77**: 1975–1988.

Shabana YM (2005) The use of oil emulsions for improving the efficacy of *Alternaria eichhorniae* as a mycoherbicide for waterhyacinth (*Eichhornia crassipes*). *Biological Control* **32**: 78–89.

Shah FA, Wang CS & Butt TM (2005) Nutrition influences growth and virulence of the insect-pathogenic fungus *Metarhizium anisopliae*. *FEMS Microbiology Letters* **251**: 259–266.

Shah PA, Gbongboui C, Godonou I, Hossou A & Lomer CJ (1998) Natural incidence of *Metarhizium flavoviride* infection in two grasshopper communities in northern Benin. *Biocontrol Science and Technology* **8**: 335–344.

Shanmuganathan T, Pallister J, Doody S, McCallum H, Robinson T, Sheppard A, Hardy C, Halliday D, Venables D, Voysey R, Strive T, Hinds L & Hyatt A (2010) Biological control of the cane toad in Australia: a review. *Animal Conservation* **13**: 16–23.

Shapiro DI, Glazer I & Segal D (1997) Genetic improvement of heat tolerance in *Heterorhabditis bacteriophora* through hybridization. *Biological Control* **8**: 153–159.

Shapiro-Ilan DI, Stuart RJ & McCoy CW (2005) Targeted improvement of *Steinernema carpocapsae* for control of pecan weevil, *Curculio caryae* (Horn) (Coleoptera: Curculionidae) through hybridization and bacterial transfer. *Biological Control* **34**: 215–221.

Shaw MR (1994) Parasitoid host ranges. In *Parasitoid Community Ecology* (Hawkins BA & Sheehan W, eds.), Oxford, UK, Oxford University Press, pp. 111–144.

Shaw PB (1984) Simulation model of a predator–prey system comprised of *Phytoseiulus persimilis* (Acari: Phytoseiidae) and *Tetranychus urticae* (Acari: Tetranychidae). 1. Structure and validation of the model. *Researches on Population Ecology* **26**: 235–260.

Shea K & Kelly D (1998) Estimating biocontrol agent impact with matrix models: *Carduus nutans* in New Zealand. *Ecological Applications* **8**: 824–832.

Shea K, Jongejans E, Skarpaas O, Kelly D & Sheppard AW (2010) Optimal management strategies to control local population growth or population spread may not be the same. *Ecological Applications* **20**: 1148–1161.

Shea K, Kelly D, Sheppard AW & Woodburn TL (2005) Context-dependent biological control of an invasive thistle. *Ecology* **86**: 3174–3181.

Shea K & Possingham HP (2000) Optimal release strategies for biological control agents: an application of stochastic dynamic programming to population management. *Journal of Applied Ecology* **37**: 77–86.

Shea K, Possingham HP, Murdoch WW & Roush R (2002) Active adaptive management in insect pest and weed control: intervention with a plan for learning. *Ecological Applications* **12**: 927–936.

Shepherd RL, Otvos IS, Chorney RJ & Cunningham JC (1984) Pest management of Douglas-fir tussock moth (Lepidoptera: Lymantriidae): prevention of an outbreak through early treatment with a nuclear polyhedrosis virus by ground and aerial applications. *Canadian Entomologist* **116**: 1533–1542.

Shearer PW & Jones VP (1998) Suitability of selected weeds and ground covers as host plants of *Nezara viridula* (L.) (Hemiptera: Pentatomidae). *Proceedings of the Hawaiian Entomological Society* **33**: 75–82.

Sheehan W (1986) Response by specialist and generalist natural enemies to agroecosystem diversification: a selective review. *Environmental Entomology* **15**: 456–461.

Sheldon SP & Creed RP (2003) The effect of a native biological control agent for Eurasian watermilfoil on six North American watermilfoils. *Aquatic Botany* **76**: 259–265.

Sheppard AW, Hill R, DeClerck-Floate RA, McClay A, Olckers T, Quimby PCJ & Zimmermann HG (2003) A global review of risk-benefit-cost analysis for the introcuction of classical biological control agents against weeds: a crisis in the making? *Biocontrol News and Information* **24**: 91 N-108 N.

Sheppard AW & Raghu S (2005) Working at the interface of art and science: how best to select an agent for classical biological control? *Biological Control* **34**: 233–235.

Sheppard AW, Rees M, Smyth M, Grigulis K & Buckley YM (2002) Modelling the population interactions between *Mogulones larvatus* and its host plant *Echium plantagineum*. In *Procceedings of the 13th Australian Weed Conference* (Spafford Jacob H, Dodd J & Moore JH, eds.), Perth, Australia, Plant Protectin Society of Western Australia, Pages 248–251.

Sheppard AW, Smyth MJ & Swirepik A (2001) The impact of a root-crown weevil and pasture competition on the winter annual *Echium plantagineum*. *Journal of Applied Ecology* **38**: 291–300.

Sheppard AW, Van Klinken RD & Heard TA (2005) Scientific advances in the analysis of direct risks of weed biological control agents to nontarget plants. *Biological Control* **35**: 215–226.

Sherratt TN & Jepson PC (1993) A metapopulation approach to modeling the long-term impact of pesticides on invertebrates. *Journal of Applied Ecology* **30**: 696–705.

Shigesada N & Kawasaki K (1997) *Biological Invasions: Theory and Practice*. Oxford, UK, Oxford University Press.

Shishido M, Miwa C, Usami T, Amemiya Y & Johnson KB (2005) Biological control efficiency of *Fusarium* wilt of tomato by nonpathogenic *Fusarium oxysporum* Fo-B2 in different environments. *Phytopathology* **95**: 1072–1080.

Siddiqui IA & Shaukat SS (2002) Mixtures of plant disease suppressive bacteria enhance biological control of multiple tomato pathogens. *Biology and Fertility of Soils* **36**: 260–268.

Siddiqui Y, Meon S, Ismail MR & Ali A (2008) *Trichoderma*-fortified compost extracts for the control of choanephora wet rot in okra production. *Crop Protection* **27**: 385–390.

Siemann E, Haarstad J & Tilman D (1999) Dynamics of plant and arthropod diversity during old field succession. *Ecography* **22**: 406–414.

Siemann E & Rogers WE (2003a) Increased competitive ability of an invasive tree may be limited by an invasive beetle. *Ecological Applications* **13**: 1503–1507.

Siemann E & Rogers WE (2003b) Reduced resistance of invasive varieties of the alien tree *Sapium sebiferum* to a generalist herbivore. *Oecologia* **135**: 451–457.

Siemann E, Tilman D, Haarstad J & Ritchie M (1998) Experimental tests of the dependence of arthropod diversity on plant diversity. *American Naturalist* **152**: 738–750.

Silva I, Van Meer MMM, Roskam MM, Hoogenboom A, Gort G & Stouthamer R (2000) Biological control potential of *Wolbachia*-infected versus uninfected wasps: laboratory and greenhouse evaluation of *Trichogramma cordubensis* and *T. deion* strains. *Biocontrol Science and Technology* **10**: 223–238.

Simberloff D (1986) Introduced insects: a biogeographic and systematic perspective. In *Ecology of Biological Invasions of North America and Hawaii* (Mooney HA & Drake JA, ed.), New York, Springer, pp. 3–26.

Simberloff D (1995) Why do introduced species appear to devastate islands more than mainland areas? *Pacific Science* **49**: 87–97.

Simberloff D (2009) The role of propagule pressure in biological invasions. *Annual Review of Ecology Evolution and Systematics* **40**: 81–102.

Simberloff D (2012) Risks of biological control for conservation purposes. *Biocontrol* **57**: 263–276.

Simberloff D (2013) *Invasive Species: What Everyone Needs to Know*. Oxford, UK, Oxford University Press.

Simberloff D & Gibbons L (2004) Now you see them, now you don't! – population crashes of established introduced species. *Biological Invasions* **6**: 161–172.

Simberloff D & Stiling P (1996a) How risky is biological control? *Ecology* **77**: 1965–1974.

Simberloff D & Stiling P (1996b) The risks of species introduced for biological control. *Biological Conservation* **78**: 185–192.

Simberloff D & Von Holle B (1999) Positive interactions of nonindigenous species: invasional meltdown? *Biological Invasions* **1**: 21–32.

Simmonds FJ (1958) The effect of lizards on the biological control of scale insects in Bermuda. *Bulletin of Entomological Research* **49**: 601–612.

Simmonds FJ (1963) Genetics and biological control. *Canadian Entomologist* **95**: 561–567.

Simmonds FJ & Bennett FD (1966) Biological control of *Opuntia* spp. by *Cactoblastis cactorum* in the Leeward Islands (West Indies). *Entomophaga* **11**: 183–189.

Simmonds FJ, Franz JM & Sailer RI (1976) History of biological control. In *Theory and Practice of Biological Control* (Huffaker CB & Messenger PS, eds.), New York, Academic Press, pp. 17–41.

Simoes N & Rosa JS (1996) Pathogenicity and host specificity of entomopathogenic nematodes. *Biocontrol Science and Technology* **6**: 403–411.

Simon JC, Carre S, Boutin M, Prunier-Leterne N, Sabater-Munoz B, Latorre A & Bournoville R (2003) Host-based divergence in populations of the pea aphid: insights from nuclear markers and the prevalence of facultative symbionts. *Proceedings of the Royal Society of London B* **270**: 1703–1712.

Simpson RG, Cross JE, Best RL, Ellsbury MM & Coseglia AF (1979) Effects of hyperparasites on population levels of *Bathyplectes curculionis* in Colorado. *Environmental Entomology* **8**: 96–100.

Singh A & Nisbet RM (2007) Semi-discrete host-parasitoid models. *Journal of Theoretical Biology* **247**: 733–742.

Sinzogan AAC, Van Mele P & Vayssieres J-F (2008) Implications of on-farm research for local knowledge related to fruit flies and the weaver ant *Oecophylla longinoda* in mango production. *International Journal of Pest Management* **54**: 241–246.

Sisterson MS, Biggs RW, Manhardt NM, Carriere Y, Dennehy T & Tabashnik BE (2007) Effects of transgenic Bt cotton on insecticide use and abundance of two generalist predators. *Entomologia Experimentalis et Applicata* **124**: 305–311.

Sisterson M & Tabashnik BE (2005) Effects of transgenic Bt crops on specialist parasitoids of target pests. *Environmental Entomology* **34**: 733–742.

Sivakoff FS, Rosenheim JA & Hagler JR (2012) Relative dispersal ability of a key agricultural pest and its predators in an annual agroecosystem. *Biological Control* **63**: 296–303.

Skalski GT & Gilliam JF (2001) Functional responses with predator interference: viable alternatives to the Holling Type II model. *Ecology* **82**: 3083–3092.

Skirvin DJ & Fenlon JS (2003) The effect of temperature on the functional response of *Phytoseiulus persimilis* (Acari: Phytoseiidae). *Experimental and Applied Acarology* **31**: 37–49.

Skirvin DJ, Perry JN & Harrington R (1997) A model describing the population dynamics of *Sitobion avenae* and *Coccinella septempunctata*. *Ecological Modelling* **96**: 29–39.

Skirvin DJ, Williams MED, Fenlon JS & Sunderland KD (2002) Modelling the effects of plant species on biocontrol effectiveness in ornamental nursery crops. *Journal of Applied Ecology* **39**: 469–480.

Skovgard H & Nachma G (2015a) Temperature dependent functional response of *Spalangia cameroni* (Perkins) (Hymenoptera: Pteromalidae), a parasitoid of *Stomoxys calcitrans* (L.) (Diptera: Muscidae). *Environmental Entomology* **44**: 90–99.

Skovgard H & Nachma G (2015b) Effect of mutual interference on the ability of *Spalangia cameroni* (Hymenoptera: Pteromalidae) to attack and parasitize pupae of *Stomoxys calcitrans* (Diptera: Muscidae). *Environmental Entomology* **44**: 1076–1084.

Smith CA & Gardiner MM (2013) Biodiversity loss following the introduction of exotic competitors: does intraguild predation explain the decline of native lady beetles? *PLoS One* **8**: e84448.

Smith KP, Handelsman J & Goodman RM (1997) Modeling dose-response relationships in biological control: partitioning host responses to the pathogen and biocontrol agent. *Phytopathology* **87**: 720–729.

Smith L (2014) Prediction of the geographic distribution of the psyllid, *Arytinnis hakani* (Homoptera: Psyllidae), a prospective biological control agent of *Genista monspessulana*, based on the effect of temperature on development, fecundity, and survival. *Environmental Entomology* **43**: 1389–1398.

Smith MC, Reeder RH & Thomas MB (1997) A model to determine the potential for biological control of *Rottboellia cochinchinensis* with the head smut *Sporisorium ophiuri*. *Journal of Applied Ecology* **34**:388–398.

Smith RF & Van den Bosch R (1967) Integrated control. In *Pest Control: Biological, Physical, and Selected Chemical Methods* (Kilgore WW & Doutt RL, eds.), New York, Academic Press, pp. 295–340.

Smith RH & Mead R (1974) Age structure and stability in models of prey predator systems. *Theoretical Population Biology* **6**: 308–322.

Smith SF & Krischik VA (1999) Effects of systemic imidacloprid on *Coleomegilla maculata* (Coleoptera: Coccinellidae). *Environmental Entomology* **28**: 1189–1195.

Smith SM (1996) Biological control with *Trichogramma*: Advances, successes, and potential of their use. *Annual Review of Entomology* **41**: 375–406.

Smith SM & You M (1990) A life system simulation model for improving inundative releases of the egg parasite, *Trichogramma minutum* against the spruce budworm. *Ecological Modelling* **51**: 123–142.

Snow AA, Andow DA, Gepts P, Hallerman EM, Power A, Tiedje JM & Wolfenbarger LL (2005) Genetically engineered organisms and the environment: current status and recommendations. *Ecological Applications* **15**: 377–404.

Snyder WE, Chang GC & Prasad RP (2005) Conservation biological control: biodiversity influences the effectiveness of predators. In *Ecology of Predator–Prey Interactions* (Barbosa P & Castellanos I, eds.), Oxford, UK, Oxford University Press, pp. 324–343.

Snyder WE, Clevenger GM & Eigenbrode SD (2004) Intraguild predation and successful invasion by introduced ladybird beetles. *Oecologia* **140**: 559–565.

Sobey WR (1969) Selection for resistance to myxomatosis in domestic rabbits (*Oryctolagus cuniculus*). *Journal of Hygiene* **67**: 743–754.

Sobey WR & Conolly D (1986) Myxomatosis: non-genetic aspects of resistance to myxomatosis in rabbits, *Oryctolagus cuniculus*. *Australian Wildlife Research* **13**: 177–187.

Solarz SL & Newman RM (2001) Variation in host plant preference and performance by the milfoil weevil, *Euhrychiopsis lecontei* Dietz, exposed to native and exotic watermilfoils. *Oecologia* **126**: 66–75.

Soldaat LL & Auge H (1998) Interactions between an invasive plant, *Mahonia aquifolium*, and a native phytophagous insect, *Rhagoletis meigenii*. In *Plant Invasions: Ecological Mechanisms and Human Responses* (Starfinger U, Edwards K, Kowarik I & Williamson M, eds.), Leiden, The Netherlands, Backhuys, pp. 347–360.

Sotherton NW (1985) The distribution and abundance of predatory Coleoptera overwintering in field boundaries. *Annals of Applied Biology* **106**: 17–21.

Sotherton NW (1995) Beetle banks – helping nature to control pests. *Pesticide Outlook* **6**: 13–17.

Soti PG & Volin JC (2010) Does water hyacinth (*Eichhornia crassipes*) compensate for simulated defoliation? *Biological Control* **54**: 35–40.

Spadaro D & Gullino ML (2005) Improving the efficacy of biocontrol agents against soilborne pathogens. *Crop Protection* **24**: 601–613.

Speyer ER (1927) An important parasite of the greenhouse whitefly (*Trialeurodes vaporariorum* Westwood). *Bulletin of Entomological Research* **17**: 301–308.

St. Leger RJ (2007) Genetic modification for improvement of virulence of *Metarhizium anisopliae* as a microbial insecticide. In *Biological Control: A Global Perspective* (Vincent C, Goettel MS & Lazarovotis G, eds.), Wallingford, UK, CABI Publishing, pp. 328–335.

St. Leger RJ, Joshi L, Bidochka MJ & Roberts DW (1996) Construction of an improved mycoinsecticide overexpressing a toxic protease. *Proceedings of the National Academy of Science USA* **93**: 6349–6354.

St. Leger RJ & Wang C (2010) Genetic engineering of fungal biocontrol agents to achieve greater efficacy against insect pests. *Applied Microbiology and Biotechnology* **85**: 901–907.

St. Martin CCG & Brathwaite RAI (2012) Compost and compost tea: principles and prospects as substrates and soil-borne disease management strategies in soil-less vegetable production. *Biological Agriculture & Horticulture* **28**: 1–33.

Stadler B & Dixon AFG (2005) Ecology and evolution of aphid-ant interactions. *Annual Review of Ecology Evolution and Systematics* **36**: 345–372.

Stamp NE & Bowers MD (1990) Parasitism of New England buckmoth caterpillars (*Hemileuca lucina*: Saturniidae) by tachinid flies. *Journal of the Lepidopterists' Society* **44**: 199–200.

Stanford MM, Werden SJ & McFadden G (2007) Myxoma virus in the European rabbit: interactions between the virus and its susceptible host. *Veterinary Research* **38**: 299–318.

Stanley JG (1976) Production of hybrid, androgenetic, and gynogenetic grass carp and carp. *Transactions of the American Fisheries Society* **105**: 10–16.

Stapel JO, Cortesero AM, De Moraes CM, Tumlinson JH & Lewis WJ (1997) Extrafloral nectar, honeydew, and sucrose effects on searching behavior and efficiency of *Microplitis*. *Environmental Entomology* **26**: 617–623.

Stark JD, Vargas R & Banks JE (2007) Incorporating ecologically relevant measures of pesticide effect for estimating the compatibility of pesticides and biocontrol agents. *Journal of Economic Entomology* **100**: 1027–1032.

Stark JD, Vargas R & Miller N (2004) Toxicity of spinosad in protein bait to three economically important tephritid fruit fly species (Diptera: Tephritidae) and their parasitoids (Hymenoptera: Braconidae). *Journal of Economic Entomology* **97**: 911–915.

Stary P (1974) Taxonomy, origin, distribution and host range of *Aphidius* species (Hym., Aphidiidae) in relation to biological control of the pea aphid in Europe and North America. *Zeitschrift fuer Angewandte Entomologie* **77**: 141–171.

Stary P, Lyon JP & Leclant F (1988) Biocontrol of aphids by the introduced *Lysiphlebus testaceipes* (Cress.)

(Hym., Aphidiidae) in Mediterranean France. *Journal of Applied Entomology* **105**: 74–87.

Steiner WWM & Teig DA (1989) *Microplitis croceipes* (Cresson): genetic characterization and developing insecticide resistant biotypes. *Southwestern Entomologist* **12**: 71–87.

Steinger T & Müller-Schärer H (1992) Physiological and growth responses of *Centaurea maculosa* (Asteraceae) to root herbivory under varying levels of interspecific plant competition and soil nitrogen availability. *Oecologia* **91**: 141–149.

Steinhaus EA (1956) Microbial control – the emergence of an idea: a brief history of insect pathology through the nineteenth century. *Hilgardia* **26**: 107–160.

Stelzl M & Devetak D (1999) Neuroptera in agricultural ecosystems. *Agriculture Ecosystems & Environment* **74**: 305–321.

Stephens AE, Srivastava DS & Myers JH (2013) Strength in numbers? Effects of multiple natural enemy species on plant performance. *Proceedings of the Royal Society of London B* **280**: 20122756.

Stephens MJ, France CM, Wratten SD & Frampton C (1998) Enhancing biological control of leafrollers (Lepidoptera: Tortricidae) by sowing buckwheat (*Fagopyrum esculentum*) in an orchard. *Biocontrol Science and Technology* **8**: 547–558.

Sterk G, Hassan SA, Baillod M, Bakker F, Bigler F, Blumel S, Bogenschutz H, Boller E, Bromand B, Brun J, Calis JNM, Coremans-Pelseneer J, Duso C, Garrido A, Grove A, Heimbach U, Hokkanen H, Jacas J, Lewis G, Moreth L, Polgar L, Roversti L, Samsoe-Peterson L, Sauphanor B, Schaub L, Staubli A, Tuset JJ, Vainio A, Van de Veire M, Viggiani G, Vinuela E & Vogt H (1999) Results of the seventh joint pesticide testing programme carried out by the IOBC/WPRS-Working Group 'Pesticides and Beneficial Organisms'. *BioControl* **44**: 99–117.

Stern VM & Van den Bosch R (1959) Field experiments on the effects of insecticides. *Hilgardia* **29**: 103–130.

Steyaert JM, Ridgway HJ, Elad Y & Stewart A (2003) Genetic basis of mycoparasitism: a mechanism of biological control by species of *Trichoderma*. *New Zealand Journal of Crop and Horticultural Science* **31**: 281–291.

Stiling P (1990) Calculating the establishment rates of parasitoids in classical biological control. *American Entomologist* **36**: 225–230.

Stiling P (1993) Why do natural enemies fail in classical biological control programs? *American Entomologist* **39**: 31–37.

Stiling P (2002) Potential non-target impacts of a biological control agent, prickly pear moth, *Cactoblastis cactorum* (Berg) (Lepidoptera: Pyralidae) in North America, and possible management actions. *Biological Invasions* **4**: 273–281.

Stiling P & Cornelissen T (2005) What makes a successful biocontrol agent? A meta-analysis of biological control agent performance. *Biological Control* **34**: 236–246.

Stiling P, Moon D & Gordon D (2004) Endangered cactus restoration: mitigating the non-target effects of a biological control agent (*Cactoblastis cactorum*) in Florida. *Restoration Ecology* **12**: 605–610.

Stiling P & Simberloff D (2000) The frequency and strength of nontarget effects of invertebrate biological control agents of plant pests and weeds. In *Nontarget Effects of Biological Control* (Follett PA & Duan JJ, eds.), Dordrecht, The Netherlands, Kluwer, pp. 31–44.

Stinner RE (1977) Efficacy of inundative releases. *Annual Review of Entomology* **22**: 515–531.

Stireman JO, Dyer LA, Janzen DH, Singer MS, Lill JT, Marquis RJ, Ricklefs RE, Hallwachs W, Coley PD, Barone JA, Greeney HF, Connahs H, Barbosa P & Morais HC (2005) Climatic unpredictability and parasitism of caterpillars: implications of global warming. *Proceedings of the National Academy of Sciences of the United States of America* **102**: 17384–17387.

Stohlgren TJ, Barnett DT & Kartesz J (2003) The rich get richer: patterns of plant invasions in the United States. *Frontiers in Ecology and the Environment* **1**: 11–14.

Stokes JFG (1917) Notes on the Hawaiian rat. *Occasional Papers of the Bishop Museum* **3**: 11–21.

Story G, Berman D, Palmer R & Scanlan J (2004) The impact of rabbit haemorrhagic disease on wild rabbit (*Oryctolagus cuniculus*) populations in Queensland. *Wildlife Research* **31**: 183–193.

Story JM (1995) Spotted knapweed. In *Biological Control in the Western United States* (Nechols JR, ed.), Oakland, University of California Press, pp. 258–263.

Story JM, Good KM, Harris P & Nowierskyi WR (1991) *Metzneria paucipunctella* Zeller Lepidoptera: Gelechidae), a moth introduced against spotted knapweed: its feeding strategy and impact on two introduced *Urophora* spp. (Diptera: Tephritidae). *Canadian Entomologist* **123**: 1001–1007.

Story JM & Nowierski RM (1984) Increase and dispersal of *Urophora affinis* (Diptera: Tephritidae) on spotted knapweed in western Montana. *Environmental Entomology* **13**: 1151–1156.

Story JM, Smith L, Corn JG & White LJ (2008) Influence of seed head-attacking biological control agents on spotted knapweed reproductive potential in western Montana over a 30-year period. *Environmental Entomology* **37**: 510–519.

Story JM & Stougaard RN (2006) Compatibility of two herbicides with *Cyphocleonus achates* (Coleoptera: Curculionidae) and *Agapeta zoegana* (Lepidoptera: Tortricidae), two root insects introduced for biological control of spotted knapweed. *Environmental Entomology* **35**: 373–378.

Stouthamer R (1993) The use of sexual versus asexual wasps in biological control. *Entomophaga* **38**: 3–6.

Stouthamer R (2004) Sex-ratio distorters and other selfish genetic elements: implications for biological control. In *Genetics, Evolution and Biological Control* (Ehler LE, Sforza R & Mateille T, eds.), Wallingford, UK, CABI Publishing, pp. 235–252.

Stouthamer R, Jochemsen P, Platner GR & Pinto JD (2000) Crossing incompatibility between *Trichogramma minutum* and *T. platneri* (Hymenoptera: Trichogrammatidae): implications for application in biological control. *Environmental Entomology* **29**: 832–837.

Stouthamer R, Luck RF & Werren JH (1992). Genetics of sex determination and improvement of biological control using parasitoids. *Environmental Entomology* **21**: 427–435.

Stouthamer R, Russell JE, Vavre F & Nunney L (2010) Intragenomic conflict in populations infected by parthenogenesis inducing *Wolbachia* ends with irreversible loss of sexual reproduction. *BMC Evolutionary Biology* **10**: 229.

Strand MR & Vinson SB (1983) Factors affecting host recognition and acceptance in the egg parasitoid *Telenomus heliothidis* (Hymenoptera, Scelionidae). *Environmental Entomology* **12**: 1114–1119.

Straub CS, Finke DL & Snyder WE (2008) Are the conservation of natural enemy biodiversity and biological control compatible goals? *Biological Control* **45**: 225–237.

Strauss SY & Agrawal AA (1999) The ecology and evolution of plant tolerance to herbivory. *Trends in Ecology and Evolution* **14**: 179–185.

Strecker RL, Marshall Jr. JT, Jackson WB, Barbehenn KR & Johnson DH (1962) *Pacific Island Rat Ecology*. Honolulu, HI, Bishop Museum.

Strong DR (1997) Fear no weevil? *Science* **277**: 1058–1059.

Strong DR, Lawton JH & Southwood TRE (1984) *Insects on Plants*. Cambridge, MA, Harvard University Press.

Strong DR & Pemberton RW (2000) Biological control of invading species-risk and reform. *Science* **288**: 1969–1970.

Strong DR & Pemberton RW (2001) Food webs, risks of alien enemies and reform of biological control. In *Evaluating Indirect Ecological Effects of Biological Control* (Wajnberg E, Scott JK & Quimby PC, eds.), Wallingford, UK, CABI Publishing, pp. 57–80.

Stuart RJ, Barbercheck ME, Grewal PS, Taylor RAJ & Hoy CW (2006) Population biology of entomopathogenic nematodes: concepts, issues, and models. *Biological Control* **38**: 80–102.

Stuart RJ & Gaugler R (1996) Genetic adaptation and founder effect in laboratory populations of the entomopathogenic nematode *Steinernema glaseri*. *Canadian Journal of Zoology* **74**: 164–170.

Styrsky JD & Eubanks MD (2010) A facultative mutualism between aphids and an invasive ant increases plant reproduction. *Ecological Entomology* **35**: 190–199.

Subramaniam TSS, Lee HL, Ahmad NW & Murad S (2012) Genetically modified mosquito: the Malaysian public engagement experience. *Biotechnology Journal* **7**: 1323–1327.

Suckling DM (2013) Benefits from biological control of weeds in New Zealand range from negligible to massive: a retrospective analysis. *Biological Control* **66**: 27–32.

Sullivan DJ & Völkl W (1999) Hyperparasitism: multitrophic ecology and behaviour. *Annual Review of Entomology* **44**: 291–315.

Summers CG (1976) Population fluctuations of selected arthropods in alfalfa: influence of two harvesting practices. *Environmental Entomology* **5**: 103–110.

Sun XL, W. Van der Werf W, Bianchi FJJA, Hu ZH & Vlak JM (2006) Modelling biological control with wild-type and genetically modified baculoviruses in the *Helicoverpa armigera*-cotton system. *Ecological Modelling* **198**: 387–398.

Sun XL, Wang HL, Sun XC, Chen XW, Peng CM, Pan DM, Jehle JA, Van der Werf W, Vlak JM & Hu ZH (2004) Biological activity and field efficacy of a genetically modified *Helicoverpa armigera* SNPV expressing an insect-selective toxin from a chimeric promoter. *Biological Control* **29**: 124–137.

Sutherst RW (2005) Prediction of species geographical ranges. *Journal of Biogeography* **30**: 805–816.

Sutton JC, Li D-W, Peng G, Yu H, Zhang P & Valdebenito-Sanhueza RM (1997) *Gliocladium roseum*: a versatile adversary of *Botrytis cinerea* in crops. *Plant Disease* **81**: 316–328.

Swift MJ, Vandermeer J, Ramakrishnan JM, Anderson C, Ong CK & Hawkins BA (1996) Biodiversity and agroecosystem function. In *Functional Roles of Biodiversity: A Global Perspective* (Mooney HA, Cushman JH, Medina E, Sala OE & Schulze E-D, eds.), Chichester, UK, Wiley, pp. 261–298.

Swope SM (2014) Biocontrol attack increases pollen limitation under some circumstances in the invasive plant *Centaurea solstitialis*. *Oecologia* **174**: 205–215.

Swope SM & Satterthwaite WH (2012) Variable effects of a generalist parasitoid on a biocontrol seed predator and its target weed. *Ecological Applications* **20**: 22–34.

Swope SM & Stein IR (2012) Soil type mediates indirect interactions between *Centaurea solstitialis* and its biocontrol agents. *Biological Invasions* **14**: 1697–1710.

Symondson WOC, Sunderland KD & Greenstone MH (2002) Can generalist predators be effective biological control agents? *Annual Review of Entomology* **47**: 561–594.

Symstad AJ, Siemann E & Haarstad J (2000) An experimental test of the effect of plant functional group diversity on arthropod diversity. *Oikos* **89**: 243–253.

Szendrei Z & Weber DC (2009) Response of predators to habitat manipulation in potato fields. *Biological Control* **50**: 123–128.

Szűcs M, Eigenbrode SD, Schwarzlander M & Schaffner U (2012a) Hybrid vigor in the biological control agent, *Longitarsus jacobaeae*. *Evolutionary Applications* **5**: 489–497.

Szűcs M, Schaffner U, Price WJ & Schwarzlaender M (2012b) Post-introduction evolution in the biological control agent *Longitarsus jacobaeae* (Coleoptera: Chrysomelidae). *Evolutionary Applications* **5**: 858–868.

Tabashnik BE (1994) Evolution of resistance to *Bacillus thuringiensis*. *Annual Review of Entomology* **39**: 47–79.

Tabashnik BE & Johnson MW (1999) Evolution of pesticide resistance in natural enemies. In *Handbook of Biological Control* (Bellows TS & Fisher TW, eds.), San Diego, CA, Academic Press, pp. 673–689.

Takabayashi J & Dicke M (1996) Plant-carnivore mutualism through herbivore-induced carnivore attractants. *Trends in Plant Science* **1**: 109–113.

Takano-Lee M & Hoddle MS (2001) Biological control of *Oligonychus perseae* (Acari: Tetranychidae) on avocado: IV. Evaluating the efficacy of a modified mistblower to mechanically dispense *Neoseiulus californicus* (Acari: Phytoseiidae). *International Journal of Acarology* **27**: 157–169.

Tallamy DW (2004) Do alien plants reduce insect biomass? *Conservation Biology* **18**: 1689–1692.

Tanada Y & Kaya HK (1993) *Insect Pathology*. San Diego, CA, Academic Press.

Tanaka S, Nishida T & Ohsaki N (2007) Sequential rapid adaptation of indigenous parasitoid wasps to the invasive butterfly *Pieris brassicae*. *Evolution* **61**: 1791–1802.

Tarmann GM (2004) *Zygaenid Moths of Australia*. Collingwood, Australia, CSIRO.

Tate CD, Carpenter JE & Bloem S (2007) Influence of radiation dose on the level of F-1 sterility in the cactus moth, *Cactoblastis cactorum* (Lepidoptera: Pyralidae). *Florida Entomologist* **90**: 537–544.

Tauber MJ, Tauber CA, Daane KM & Hagen KS (2000) Commercialization of predators: recent lessons from green lacewings (Neuroptera: Chrysopidae: *Chrysoperla*). *American Entomologist* **46**: 26–38.

Taylor LAV & Cruzon MB (2015) Propagule pressure and disturbance drive the invasion of perennial false-brome (*Brachypodium sylvaticum*). *Invasive Plant Science and Management* **8**: 169–180.

Taylor THC (1955) Biological control of insect pests. *Annals of Applied Biology* **42**: 190–196.

TeBeest DO & Templeton GE (1985) Mycoherbicides: progress in the biological control of weeds. *Plant Disease* **69**: 6–10.

Tedders WL & Schaefer PW (1994) Release and establishment of *Harmonia axyridis* (Coleoptera: Coccinellidae) in the southeastern United States. *Entomological News* **105**: 228–243.

Templeton GE, TeBeest DO & Smith RJ (1979) Biological weed control with mycoherbicides. *Annual Review of Phytopathology* **17**: 301–310.

Teplitski M & Ritchie K (2009) How feasible is the biological control of coral diseases? *Trends in Ecology & Evolution* **24**: 378–385.

Thébaud C & Simberloff D (2001) Are plants really larger in their introduced ranges? *American Naturalist* **157**: 231–236.

Thiel A & Weeda AC (2014) Haploid, diploid, and triploid-discrimination ability against polyploid mating partner in the parasitic wasp, *Bracon brevicornis* (Hymenoptera: Braconidae). *Journal of Insect Science* **14**: 291.

Thiem SM (1997) Prospects for altering host range for baculovirus bioinsecticides. *Current Opinion in Biotechnology* **8**: 317–322.

Thiem SM, Du X, Quentin ME & Berner MM (1996) Identification of a baculovirus gene that promotes *Autographa californica* nuclear polyhedrosis virus replication in a nonpermissive insect cell line. *Journal of Virology* **70**: 2221–2229.

Thies C, Steffan-Dewenter I & Tscharntke T (2003) Effects of landscape context on herbivory and parasitism at different spatial scales. *Oikos* **101**: 18–25.

Thies C & Tscharntke T (1999) Landscape structure and biological control in agroecosystems. *Science* **285**: 893–895.

Thomas CD & Kunin WE (1999) The spatial structure of populations. *Journal of Animal Ecology* **57**: 536–546.

Thomas CFG, Parkinson L, Griffiths GJK, Garcia AF & Marshall EJP (2001a) Aggregation and temporal stability of carabid beetle distributions in field and hedgerow habitats. *Journal of Applied Ecology* **38**: 100–116.

Thomas J (1980) Why did the large blue become extinct in Britain? *Oryx* **15** 243–247.

Thomas JA (1995) The ecology and conservation of *Maculinea arion* and other European species of large blue butterfly. In *Ecology and Conservation of Butterflies* (Pullin AS, ed.) London, UK, Chapman & Hall, pp. 180–197.

Thomas MB, Arthurs SP & Watson EL. 2006. Trophic and guild interactions and the influence of multiple species on disease. In *Trophic and Guild Interactions in Biological Control* (Brodeur J & Boivin G, eds.), Dordrecht, The Netherlands, Springer, pp. 101–122.

Thomas MB, Casula P & Wilby A (2004) Biological control and indirect effects. *Trends in Ecology & Evolution* **19**: 61.

Thomas MB & Reid AM (2007) Are exotic natural enemies an effective way of controlling invasive plants? *Trends in Ecology & Evolution* **22**: 447–453.

Thomas MB, Wratten SD & Sotherton NW (1991) Creation of 'island' habitats in farmland to manipulate populations of beneficial arthropods: predator densities and emigration. *Journal of Applied Ecology* **28**: 906–917.

Thomas MB, Wratten SD & Sotherton NW (1992) Creation of 'island' habitats in farmland to manipulate populations of beneficial arthropods: predator densities and species composition. *Journal of Applied Ecology* **29**: 524–531.

Thomas SR, Goulson D & Holland JM (2001b) Resource provision for farmland gamebirds: the value of beetle banks. *Annals of Applied Biology* **139**: 111–118.

Thomashow LS & Weller DM (1988) Role of a phenazine antibiotic from *Pseudomonas fluorescens* in biological control of *Gaeumannomyces graminis* var. *tritici*. *Journal of Bacteriology* **170**: 3499–3508.

Thompson JN (1998) Rapid evolution as an ecological process. *Trends in Ecology & Evolution* **13**: 329–332.

Thompson JN & Burdon JJ (1992) Gene-for-gene coevolution between plants and parasites. *Nature* **360**: 121–125.

Thomson LJ & Hoffmann AA (2007) Ecologically sustainable chemical recommendations for agricultural pest control? *Journal of Economic Entomology* **100**: 1741–1750.

Thorbek P & Bilde T (2004) Reduced numbers of generalist arthropod predators after crop management. *Journal of Applied Ecology* **41**: 526–538.

Thorpe KW (1985) Effects of height and habitat type on egg parasitism by *Trichgramma minutum* and *T. pretiosum* (Hymenoptera: Trichogrammatidae). *Agriculture, Ecosystems & Environment* **12**: 117–126.

Thrall PH & Burdon JJ (2004) Host-pathogen life history interactions affect biological control success. *Weed Technology (Supplement)* **18**: 1269–1274.

Tilman D (1997) Community invasibility, recruitment limitation, and grassland biodiversity. *Ecology* **78**: 81–92.

Tilman D (2004) Niche tradeoffs, neutrality, and community structure: a stochastic theory of resource competition, invasion, and community assembly. *Proceedings of the National Academy of Sciences, USA* **101**: 10854–10861.

Tilman D, Hill J & Lehman C (2006) Carbon-negative biofuels from low-input high-diversity grassland biomass. *Science* **314**: 1598–1600.

Tilman D, Knops J, Wedin D, Reich P, Ritchie M & Siemann E (1997) The influence of functional diversity and composition on ecosystem processes. *Science* **277**: 1300–1302.

Tilman D, Reich PB, Knops J, Wedin D, Mielke T & Lehman C (2001) Diversity and productivity in a long-term grassland experiment. *Science* **294**: 843–845.

Tilman D, Wedin D & Knops J (1996) Productivity and sustainability influenced by biodiversity in grassland ecosystems. *Nature* **379**: 718–720.

Timper P (2014) Conserving and enhancing biological control of nematodes. *Journal of Nematology* **46**: 75–89.

Tipping PW, Martin MR, Nimmo KR, Pierce RM, Smart MD, White E, Madeira PT & Center TD (2009) Invasion of a West Everglades wetland by *Melaleuca quinquenervia* countered by classical biological control. *Biological Control* **48**: 73–78.

Tisdell CA (1990) Economic impact of biological control of weeds and insects. In *Critical Issues in Biological Control* (Mackauer M, Ehler LE & Roland J, eds.), Andover, UK, Intercept, pp. 301–316.

Tisdell CA & Auld BA (1990) Evaluation of biological control projects. In *Proceedings of the VII International Symposium on Biological Control of Weeds* (Delfosse ES, ed.), Rome, Italy, Istituto Sperimentale per la Patalogia Vegetale, pp. 93–100.

Toda Y & Sakuratani Y (2006) Expansion of the geographical distribution of an exotic ladybird beetle, *Adalia bipunctata* (Coleoptera: Coccinellidae), and its interspecific relationships with native ladybird beetles in Japan. *Ecological Research* **21**: 292–300.

Tomasino SF, Leister RT, Dimock MB, Beach RM & Kelly JL (1995) Field performance of *Clavibacter xyli* subsp. *cynodontis* expressing the insecticidal protein gene *cryia(c)* of *Bacillus thuringiensis* against European corn borer in field corn. *Biological Control* **5**: 442–448.

Tonhasca A & Byrne DN (1994) The effects of crop diversification on herbivorous insects: a meta-analysis approach. *Ecological Entomology* **19**: 239–244.

Torchin ME, Lafferty KD, Dobson AP, McKenzie VJ & Kuris AM (2003) Introduced species and their missing parasites. *Nature* **421**: 628–630.

Tothill JD, Taylor THC & Paine RW (1930) *The Coconut Moth in Fiji: A History of Its Control by Means of Parasites*. London, UK, Imperial Bureau of Entomology.

Toyoshima S & Amano H (1998) Effect of prey density on sex ratio of two predacious mites, *Phytoseiulus persimilis* and *Amblyseius womersleyi* (Acari: Phytoseiidae). *Experimental and Applied Acarology* **22**: 709–723.

Trenbath BR (1993) Intercropping for the management of pests and diseases. *Field Crops Research* **34**: 381–405.

Trillas MI, Casanova E, Cotxarrera L, Ordovas J, Borrero C & Aviles M (2006) Composts from agricultural waste and the *Trichoderma asperellum* strain T-34 suppress *Rhizoctonia solani* in cucumber seedlings. *Biological Control* **39**: 32–38.

Trumble JT & Kok LT (1982) Integrated pest management techniques in thistle suppression in pastures of North America. *Weed Research* **22**: 345–359.

Trumble JT & Morse JP (1993) Economics of integrating the predaceous mite *Phytoseiulus persimilis* (Acari: Phytoseiidae) with pesticides in strawberries. *Journal of Economic Entomology* **86**: 879–885.

Trumper EV & Holt J (1998) Modelling pest population resurgence due to recolonization of fields following an insecticide application. *Journal of Applied Ecology* **35**: 273–285.

Tscharntke T, Bommarco R, Clough Y, Crist TO, Kleijn D, Rand TA, Tylianakis J, van Nouhuys S & Vidal S (2007) Conservation biological control and enemy diversity on a landscape scale. *Biological Control* **43**: 294–309.

Tscharntke T, Klein AM, Kruess A, Steffan-Dewenter I & Thies C (2005a) Landscape perspectives on agricultural intensification and biodiversity – ecosystem service management. *Ecology Letters* **8**: 857–874.

Tscharntke T, Rand TA & Bianchi FJJA (2005b) The landscape context of trophic interactions: insect spillover across the crop-noncrop interface. *Annales Zoologici Fennici* **42**: 421–432.

Tscharntke T, Sekercioglu CH, Dietsch TV, Sodhi NS, Hoehn P & Tylianakis JM (2008) Landscape constraints on functional diversity of birds and insects in tropical agroecosystems. *Ecology* **89**: 944–951.

Tsuchida T, Koga R & Fukatsu T (2004) Host plant specialization governed by facultative symbiont. *Science* **303**: 1989.

Tullett AG, Hart AJ, Worland MR & Bale JS (2004) Assessing the effects of low temperature on the establishment potential in Britain of the non-native biological control agent *Eretmocerus eremicus*. *Physiological Entomology* **29**: 363–371.

Turchin P (1995) Population regulation: old arguments and a new synthesis. In *Population Dynamics: New Approaches and Synthesis* (Cappuccino N. & Price PW, eds.), San Diego, CA, Academic Press, pp. 19–40.

Turelli M (2010) Cytoplasmic incompatibility in populations with overlapping generations. *Evolution* **64**: 232–241.

Turnbull LA, Crawley MJ & Rees M (2000) Are plant populations seed limited? A review of seed sowing experiments. *Oikos* **88**: 225–238.

Turner CE (1985) Conflicting interests in biological control of weeds. In *Proceedings of the VI International Symposium of Biological Control of Weeds* (Delfosse ES, ed.), Ottawa, Canada, Agriculture Canada, pp. 203–225.

Turnock WJ, Wise IL & Matheson FO (2003) Abundance of some native coccinellids (Coleoptera: Coccinellidae) before and after the appearance of *Coccinella septempunctata*. *Canadian Entomologist* **135**: 391–404.

Twohey MB, Heinrich JW, Seelye JG, Fredricks KT, Bergstedt RA, Kaye CA, Scholefield RJ, McDonald RB & Christie CG (2003) The sterile-male-release technique in Great Lakes sea lamprey management. *Journal of Great Lakes Research* **29**: 410–423.

Tylianakis JM, Didham R & Wratten SD (2004) Improved fitness of aphid parasitoids receiving resource subsidies. *Ecology* **85**: 658–666.

Tyndale-Biscoe CH (1994) Virus-vectored immunocontraception of feral mammals. *Reproduction Fertility and Development* **6**: 281–287.

Tyndale-Biscoe M & Vogt WG (1996) Population status of the bush fly, *Musca vetustissima* (Diptera: Muscidae), and native dung beetles (Coleoptera: Scarabaeinae) in south-eastern Australia in relation to establishment of exotic dung beetles. *Bulletin of Entomological Research* **86**: 183–192.

Unruh T, White W, Gonzalez D, Gordh G & Luck RF (1983) Heterozygosity and effective size in laboratory populations of *Aphidius ervi* (Hym.: Aphidiidae). *Entomophaga* **28**: 245–258.

Urano S, Shima K, Hongo K & Suzuki Y (2003) A simple criterion for successful biological control on annual crops. *Population Ecology* **45**: 97–103.

USDA (2003) *Reviewer's Manual for the Technical Advisory Group for Biological Control Agents of Weeds: Guidelines for Evaluating the Safety of Candidate Biological Control Agents.* Washington, DC, USDA APHIS.

US EPA (1998) *Guidelines for Ecological Risk Assessment.* Washington, DC, Science Advisory Board.

US National Academy of Sciences (1988) *Report of the Research Briefing on Biological Control in Managed Ecosystems*, Washington, DC, National Academy of Sciences.

Usher MB (1988) Biological invasions of nature reserves – a search for generalization. *Biological Conservation* **44**: 119–135.

Van Bael SA, Philpott SM, Greenberg R, Bichier P, Barber NA, Mooney KA & Gruner DS (2008) Birds as predators in tropical agroforestry systems. *Ecology* **89**: 928–934.

Van den Bosch R (1971) Biological control of insects. *Annual Review of Ecology and Systematics* **2**: 45–66.

Van den Bosch R & Dietrick EJ (1959) The interrelationships of *Hypera brunneipennis* (Coleoptera: Curculionidae) and *Bathyplectes curculionis* Hymenoptera: Ichneumonidae) in Southern California. *Annals of the Entomological Society of America* **52**: 609–616.

Van den Bosch R, Hom R, Matteson P, Frazer BD, Messenger PS & Davis CS (1979) Biological control of the walnut aphid in California: impact of the parasite, *Trioxys pallidus*. *Hilgardia* **47**: 1–13.

Van den Bosch R, Messenger PS & Gutierrez AP (1982) *An Introduction to Biological Control*. New York, Plenum.

Van den Bosch R & Stern VM (1969) The effect of harvesting practices on insect populations in alfalfa. *Proceedings of the Tall Timbers Conference on Ecological Animal Control by Habitat Management* **3**: 47–54.

Van den Bosch R & Telford AD (1964) Environmental modification and biological control. In *Biological Control of Insect Pests and Weeds* (DeBach P, ed.), London, UK, Chapman & Hall, pp. 459–488.

Van Driesche RG & Bellows TS (1993) *Steps in Classical Arthropod Biological Control*. Annapolis, MD, Entomological Society of America.

Van Driesche RG & Bellows TS (1996) *Biological Control*. New York, Chapman & Hall.

Van Driesche RG, Carruthers RI, Center T, Hoddle MS, Hough-Goldstein J, Morin L, Smith L, Wagner DL, Blossey B, Brancatini V, Casagrande R, Causton CE, Coetzee JA, Cudam J, Ding J, Fowler SV, Frankm JH, Fuester R, Goolsby J, Grodowitz M, Heard TA, Hill MP, Hoffmann JH, Huber J, Julien M, Kairo MTK, Kenis M, Mason P, Medalm J, Messing R, Miller R, Moore A, Neuenschwander P, Newman R, Norambuena H, Palmer WA, Pemberton R, Perez Panduro A, Pratt PD, Rayamajhi M, Salom S, Sands D, Schooler S, Schwarzländer M, Sheppard A, Shaw R, Tipping PW & van Klinken RD (2010) Classical biological control for the protection of natural ecosystems. *Biological Control* **54**: S2–S33.

Van Driesche RG & Hoddle M (1997) Should arthropod parasitoids and predators be subject of host range testing when used as biological control agents? *Agriculture and Human Values* **14**: 211–226.

Van Driesche RG, Hoddle M & Center TD (2008) *Control of Pests and Weeds by Natural Enemies: An Introduction to Biological Control*. Hoboken, NJ, Wiley-Blackwell.

Van Driesche RG & Murray TJ (2004) Overview of testing schemes and designs used to estimate host ranges. In *Assessing Host Ranges for Parasitoids and Predators Used for Classical Biological Control: A Guide to Best Practice* (Van Driesche RG & Reardon R, eds.), Morgantown, WV, USDA Forest Service, pp. 68–89.

Van Driesche RG, Nunn C, Kreke N, Goldstein B & Benson J (2003) Laboratory and field host preferences of introduced *Cotesia* spp. parasitoids (Hymenoptera: Braconidae) between native and invasive *Pieris* butterflies. *Biological Control* **28**: 214–221.

Van Driesche RG, Nunn C & Pasquale A (2004) Life history pattern, host plants, and habitat as determinants of population survival of *Pieris napi oleracea* interacting with an introduced braconid parasitoid. *Biological Control* **29**: 278–287.

Van Driesche R & Reardon R (2004) *Assessing Host Ranges for Parasitoids and Predators Used for Classical Biological Control: A Guide to Best Practice*. Morgantown, WV, USDA Forest Service.

Van Eenennaam JP, Stocker RK, Thiery RG, Hagstrom NT & Doroshov SI (1990) Egg fertility, early development and survival from crosses of diploid female x triploid male grass carp (*Ctenopharyngodon idella*). *Aquaculture* **86**: 111–125.

Van Emden HF (1965) The role of uncultivated land in the biology of crop pests and beneficial insects. *Scientific Horticulture* **17**: 121–136.

Van Emden HF (1990) Plant diversity and natural enemy efficiency in agroecosystems. In *Critical Issues in Biological Control* (Mackauer M, Ehler LE & Roland J, eds.), Andover, UK, Intercept, pp. 63–80.

Van Emden HF & Williams GF (1974) Insect stability and diversity in agro-ecosystems. *Annual Review of Entomology* **19**: 455–475.

Van Hamburg H & Hassell MP (1984) Density dependence and the augmentative release of egg parasitoids against graminaceous stalk borers. *Ecological Entomology* **9**: 101–108.

Van Kleunen M & Schmid B (2003) No evidence for an evolutionary increased competitive ability in an invasive plant. *Ecology* **84**: 2816–2823.

Van Klinken RD & Edwards OR (2002) Is host-specificity of weed biological control agents likely to evolve rapidly following establishment? *Ecology Letters* **5**: 590–596.

Van Klinken RD, Fichera G & Cordo H (2003) Targeting biological control across diverse landscapes: the release, establishment, and early success of two insects on mesquite (*Prosopis* spp.) insects in Australian rangelands. *Biological Control* **26**: 8–20.

Van Lenteren JC (2000) Success in biological control of arthropods by augmentation of natural enemies. In *Biological Control: Measures of Success* (Gurr G & Wratten SD, eds.), Dordrecht, The Netherlands, Kluwer, pp. 77–103.

Van Lenteren JC (2003) *Quality Control and Production of Biological Control Agents: Theory and Testing Procedures*. Wallingford, UK, CABI Publishing.

Van Lenteren JC (2006) How not to evaluate augmentative biological control. *Biological Control* **39**: 115–118.

Van Lenteren JC (2012) The state of commercial augmentative biological control: plenty of natural enemies, but a frustrating lack of uptake. *BioControl* **57**: 1–20.

Van Lenteren JC, Babendreier D, Bigler F, Burgio G, Hokkanen HMT, Kuske S, Loomans AJM, Menzler-Hokkanen I, Van Rijn PCJ, Thomas MB, Tommasini MG & Zeng QQ (2003) Environmental risk assessment of exotic natural enemies used in inundative biological control. *BioControl* **48**: 3–38.

Van Lenteren JC, Bale J, Bigler F & Hokkanen HMT (2006a) Assessing risks of releasing exotic biological control agents of arthropod pests. *Annual Review of Entomology* **51**: 609–634.

Van Lenteren JC, Cock MJW, Hoffmeister TS & Sands DPA (2006b) Host specificity in arthropod biological control, methods for testing and interpretation of the data. In *Environmental Impact of Invertebrates for Biological Control of Arthropods: Methods and Risk Assessment* (Bigler F, Babendreier D & Kuhlmann U, eds.), Wallingford, UK, CABI Publishing, pp. 38–63.

Van Lenteren JC & Loomans AJM (2006) Environmental risk assessment: methods for comprehensive evaluation and quick scan. In *Environmental Impact of Invertebrates for Biological Control of Arthropods: Methods and Risk Assessment* (Bigler F, Babendreier D & Kuhlmann U, eds.), Wallingford, UK, CABI Publishing, pp. 254–272.

Van Lenteren JC & Tommasini MG (2003) Mass production, storage, shipment and release of natural enemies. In *Quality Control and Production of Biological Control Agents: Theory and Testing Procedures* (van Lenteren JC, ed.), Wallingford, UK, CABI Publishing, pp. 181–189.

Van Lenteren JC & Woets J (1988) Biological and integrated pest control in greenhouses. *Annual Review of Entomology* **33**: 239–270.

Van Mele P (2008) A historical review of research on the weaver ant *Oecophylla* in biological control. *Agricultural and Forest Entomology* **10**: 13–22.

Van Mele P & Chien HV (2004) Farmers, biodiversity and plant protection: developing a learning environment for sustainable tree cropping systems. *International Journal of Agricultural Sustainability* **2**: 67–76.

Van Mele P & Cuc NTT (2000) Evolution and status of *Oecophylla smaragdina* (Fabricius) as a pest control agent in citrus in the Mekong Delta, Vietnam. *International Journal of Pest Management* **46**: 295–301.

Van Mele P, Vayssieres J-F, Van Tellingen E & Vrolijks J (2007) Effects of an African weaver ant, *Oecophylla longinoda*, in controlling mango fruit flies (Diptera: Tephritidae) in Benin. *Journal of Economic Entomology* **100**: 695–701.

Van Rijn PCJ & Sabelis MW (2005) Impact of plant-provided food on herbivore-carnivore dynamics.

In *Plant-Provided Food for Carnivorous Insects: A Protective Mutualism and Its Applications* (Wäckers FL, Van Rijn PCJ & Bruin J, eds.), Cambridge, UK, Cambridge University Press, pp. 223–266.

Van Valen L (1973) A new evolutionary law. *Evolutionary Theory* **1**: 1–30.

Van Veen FJF, Memmott J & Godfray HCJ (2006) Indirect effects, apparent competition and biological control. In *Trophic and Guild Interactions in Biological Control* (Brodeur J & Boivin G, eds.), Dordrecht, The Netherlands, Springer, pp. 145–170.

Van Wilgenberg E, Driessen G & Beukeboom LW (2006) Single locus complementary sex determination in Hymenoptera: an "unintelligent" design? *Frontiers in Zoology* **3**:1.

Vandergheynst J, Scher H, Guo H-Y & Schultz D (2007) Water-in-oil emulsions that improve the storage and delivery of the biolarvacide *Lagenidium giganteum*. *BioControl* **52**: 207–229.

Vanderplank FL (1960) The bionomics and ecology of the red tree ant, *Oecophylla* sp. and its relationship to the coconut bug, *Pseudotheraptus wayi* (Coreidae). *Journal of Animal Ecology* **29**: 15–33.

Vargas RI, Miller NW & Prokopy RJ (2002) Attraction and feeding responses of Mediterranean fruit fly and a natural enemy to protein baits laced with two novel toxins, phloxine B and spinosad. *Entomologia Experimentalis et Applicata* **102**: 273–282.

Vargas RI, Peck SL, McQuate GT, Jackson CG, Stark JD & Armstrong JW (2001) Potential for areawide integrated management of Mediterranean fruit fly (Diptera: Tephritidae) with a braconid parasitoid and a novel bait spray. *Journal of Economic Entomology* **94**: 817–825.

Varley GC (1959) The biological control of agricultural pests. *Journal of the Royal Society of Arts* **107**: 475–490.

Varley GC, Gradwell GR & Hassel MP (1973) *Insect Population Ecology: An Analytical Approach*. Berkeley, University of California Press.

Vasquez CJ, Stouthamer R, Jeong G & Morse JG (2011) Discovery of a CI-inducing *Wolbachia* and its associated fitness costs in the biological control agent *Aphytis melinus* DeBach (Hymenoptera: Aphelinidae). *Biological Control* **58**: 192–198.

Vega FE, Dowd PF, Lacey LA, Pell JK, Jackson DM & Klein MG (2007) Dissemination of beneficial microbial agents by insects. In *Field Manual of Techniques in Invertebrate Pathology: Application and Evaluation of Pathogens for Control of Insects and Other Invertebrate Pests*, 2nd edition (Lacey LA & Kaya HK, eds.), Dordrecht, The Netherlands, Springer, pp. 127–146.

Veldtman R, Lado TF, Botes A, Proches S, Timm AE, Geertsema H & Chown SL (2011) Creating novel food webs on introduced Australian acacias: indirect effects of galling biological control agents. *Diversity and Distributions* **17**: 958–967.

Venette RC, Kriticos DJ, Magarey RD, Koch FH, Baker RHA, Worner SP, Raboteaux NNG, McKenney DW, Dobesberger EJ, Yemshanov D, De Barro PJ, Hutchison WD, Fowler G, Kalaris TM & Pedlar J (2010) Pest risk maps for invasive alien species: a roadmap for improvement. *BioScience* **60**: 349–362.

Verhulst PF (1838) Notice sur la loi que la population suit dans son accroissement. *Correspondance Mathématique et Physique* **10**: 113–121.

Vetukhiv M (1956) Fecundity of hybrids between geographic populations of *Drosophila pseudoobscura*. *Evolution* **10**: 139–146.

Vetukhiv M (1957) Longevity of hybrids between geographic populations of *Drosophila subobscura*. *Evolution* **11**: 348–360.

Vietch CR & Clout MN (2001) Human dimensions in the management of invasive species in New Zealand. In *The Great Reshuffling: Human Dimensions of Invasive Alien Species* (McNeely JA, ed.), Gland, Switzerland, pp. 63–74.

Vigueras AL & Portillo L (2001) Uses of *Opuntia* species and the potential impact of *Cactoblastis cactorum* (Lepidoptera: Pyralidae) in Mexico. *Florida Entomologist* **84**: 493–498.

Vilá M & Gimeno C (2003) Seed predation of two alien *Opuntia* species invading Mediterranean communities. *Plant Ecology* **167**: 1–8.

Vilá M, Gomez A & Maron JL (2003) Are alien plants more competitive than their native conspecifics? A test using *Hypericum perforatum* L. *Oecologia* **137**: 211–215.

Villareal LP (2004) Are viruses alive? *Scientific American* **291**: 100–105.

Vitousek PM (1990) Biological invasions and ecosystem processes: towards an integration of population biology and ecosystem studies. *Oikos* **57**: 7–13.

Volterra V (1926) Fluctuations in the abundance of a species considered mathematically. *Nature* **118**: 558–560.

Vorley WT & Wratten SD (1987) Migration of parasitoids (Hymenoptera: Braconidae) of cereal aphids (Hemiptera: Aphididae) between grassland, early-sown cereals and late-sown cereals in southern England. *Bulletin of Entomological Research* **77**: 555–568.

Vorsino AE, Wieczorek AM, Wright MG & Messing RH (2012) Using evolutionary tools to facilitate the prediction and prevention of host-based differentiation in biological control: a review and perspective. *Annals of Applied Biology* **160**: 204–216.

Vos M & Hemerik L (2003) Linking foraging behavior to lifetime reproductive succes for an insect parasitoid: adaptation to host distributions. *Behavioral Ecology* **14**: 236–245.

Vos M & Vet LEM (2004) Geographic variation in host acceptance by an insect parasitoid: genotype versus experience. *Evolutionary Ecology Research* **6**: 1021–1035.

Waage JK (1990) Ecological theory and the selection of biological control agents. *Critical Issues in Biological Control* (Ehler LE & Roland J, eds.), Andover, UK, Intercept, pp. 135–157.

Waage JK & Greathead DJ (1988) Biological control: challenges and opportunities. *Philosophical Transactions of the Royal Society B* **318**: 111–128.

Waage JK, Hassell MP & Godfray HCJ (1985) The dynamics of pest parasitoid insecticide interactions. *Journal of Applied Ecology* **22**: 825–838.

Wäckers FL, Van Rijn PCJ & Bruin J (2005) *Plant-Provided Food for Carnivorous Insects*. Cambridge, UK, Cambridge University Press.

Wäckers F, Van Rijn PCJ & Heimpel GE (2008) Honeydew as a food source for natural enemies: making the best of a bad meal? *Biological Control* **45**: 176–184.

Wade MR, Gurr GM & Wratten SD (2008a) Ecological restoration of farmland: progress and prospects. *Philosophical Transactions of the Royal Society B* **363**: 831–847.

Wade MR, Zalucki MP, Wratten SD & Robinson KA (2008b) Conservation biological control of arthropods using artificial food sprays: current status and future challenges. *Biological Control* **45**: 185–199.

Waite GK (1988) Integrated control of *Tetranychus urticae* in strawberries in southeast Queensland, Australia. *Experimental and Applied Acarology* **5**: 23–32.

Wajnberg E (2004) Measuring genetic variation in natural enemies used for biological control: why and how? In *Genetics, Evolution and Biological Control* (Ehler LE, Sforza R & Mateille T, eds.), Wallingford, UK, CABI Publishing, pp. 19–38.

Wajnberg E, Roitberg BD & Boivin G (2016) Using optimality models to improve the efficacy of parasitoids in biological control programmes. *Entomologia Experimentalis et Applicata* **158**: 2–16.

Wajnberg E, Scott JK & Quimby PC (2001) *Evaluating Indirect Ecological Effects of Biological Control*. Wallingford, UK, CABI Publishing.

Walde SJ & Nachman G (1999) Dynamics of spatially structured spider mite populations. In *Theoretical Approaches to Biological Control* (Hawkins BA & Cornell HV, eds.), Cambridge, UK, Cambridge University Press, pp. 163–189.

Walker GP, Herman TJB, Kale AJ & Wallace AR (2010) An adjustable action threshold using larval parasitism of *Helicoverpa armigera* (Lepidoptera: Noctuidae) in IPM for processing tomatoes. *Biological Control* **52**: 30–36.

Wang B, Ferro DN & Hosmer DW (1999) Effectiveness of *Trichogramma ostriniae* and *T. nubilale* for controlling the European corn borer *Ostrinia nubilalis* in corn. *Entomologia Experimentalis et Applicata* **91**: 297–303.

Wang CS & St. Leger RJ (2007) A scorpion neurotoxin increases the potency of a fungal insecticide. *Nature Biotechnology* **25**: 1455–1456.

Wang X & Grewal PS (2002) Rapid genetic deterioration of environmental tolerance and reproductive potential of an entomopathogenic nematode during laboratory maintenance. *Biological Control*, **23**: 71–78.

Wang X-G, Duff J, Keller MA, Zalucki MP, Liu S-S & Bailey P (2004) Role of *Diadegma semiclausum*

(Hymenoptera: Ichneumonidae) in controlling *Plutella xylostella* (Lepidoptera: Plutellidae): Cage exclusion experiments and direct observation. *Biocontrol Science and Technology* **14**: 571–586.

Wang Y, Van Oers MM, Crawford AM, Vlak JM & Jehle JA (2007) Genomic analysis of *Oryctes rhinocerus* virus reveals genetic relatedness to *Heliothis zea* virus 1. *Archives of Virology* **152**: 519–531.

Wang YH & Gutierrez AP (1980) An assessment of the use of stability analyses in population ecology. *Journal of Animal Ecology* **49**: 435–452.

Wanger TC, Wielgoss AC, Motzke I, Clough Y, Brook BW, Sodhi NS & Tscharnkte T (2011) Endemic predators, invasive prey and native diversity. *Proceedings of the Royal Society B-Biological Sciences* **278**: 690–694.

Wanner H, Gu H, Hattendorf B, Gunther D & Dorn S (2006) Using the stable isotope marker ^{44}Ca to study dispersal and host-foraging activity in parasitoids. *Journal of Applied Ecology* **43**: 1031–1039.

Wapshere AJ (1974) A strategy for evaluating the safety of organisms for biological weed control. *Annals of Applied Biology* **77**: 201–211.

Wapshere AJ (1989) A testing sequence for reducing rejection of potential biological control agents for weeds. *Annals of Applied Biology* **114**: 515–526.

Ward-Fear G, Brown GP & Shine R (2010) Using a native predator (the meat ant, *Iridomyrmex reburrus*) to reduce the abundance of an invasive species (the cane toad, *Bufo marinus*) in tropical Australia. *Journal of Applied Ecology* **47**: 273–280.

Wardle DA (1987) The ecology of ragwort (*Senecio jacobaea* L.) – A review. *New Zealand Journal of Ecology* **10**: 67–76.

Warner KD (2012) Fighting pathophobia: how to construct constructive public engagement with biocontrol for nature without augmenting public fears. *BioControl* **57**: 307–317.

Watanabe S, Kato H, Kumakura K, Ishibashi E & Nagayama K (2006) Properties and biological control activities of aerial and submerged spores in *Trichoderma asperellum* SKT-1. *Journal of Pesticide Science* **31**: 375–379.

Waterhouse DF (1974) The biological control of dung. *Scientific American* **230**: 100–109.

Waterhouse DF & Norris KR (1987) *Biological Control: Pacific Prospects*. Melbourne, Australia, Inkata.

Watkinson AR (1980) Density dependence in single species populations of plants. *Journal of Theoretical Biology* **83**: 345–358.

Watt KEF (1959) A mathematical model for the effect of densities of attacked and attacking species on the number attacked. *Canadian Entomologist* **91**: 129–144.

Watt MS, Whitehead D, Kriticos DJ, Gous SF & Richardson B (2007) Using a process-based model to analyse compensatory growth in response to defoliation: simulating herbivory by a biological control agent. *Biological Control* **43**: 119–129.

Way MJ (1951) An insect pest of coconuts and its relationship to certain ant species. *Nature* **168**: 302.

Way MJ (1966) The natural environment and integrated methods of pest control. *Journal of Applied Ecology* **3 (supplement)**: 29–32.

Way MJ, Cammell ME & Paiva MS (1992) Studies on egg predation by ants (Hymenoptera: Formicidae) especially on the eucalyptus borer *Phoracantha semipunctata* (Coleoptera: Cerambycidae). *Bulletin of Entomological Research* **82**: 425–432.

Way MJ, Paiva MR & Cammell ME (1999) Natural biological control of the pine processionary moth *Thaumetopoea pityocampa* (Den. & Schiff.) in the Argentine ant *Linepithema humile* (Mayr) in Portugal. *Agricultural and Forest Entomology* **1**: 27–31.

Weed AS & Schwarzlander M (2014) Density dependence, precipitation and biological control agent herbivory influence landscape-scale dynamics of the invasive Eurasian plant *Linaria dalmatica*. *Journal of Applied Ecology* **51**: 825–834.

Weisser WW, Jansen VAA & Hassell MP (1997) The effects of a pool of dispersers on host-parasitoid systems. *Journal of Theoretical Biology* **189**: 413–425.

Welch KD & Harwood JD (2014) Temporal dynamics of natural enemy–pest interactions in a changing environment. *Biological Control* **75**: 18–27.

Weller DM (1988) Biological control of soilborne pathogens in the rhizosphere with bacteria. *Annual Review of Phytopathology* **26**: 379–407.

Weller DM & Cook RJ (1983) Suppression of take-all of wheat by seed treatments with fluorescent pseudomonads. *Phytopathology* **73**: 463–469.

Weller DM, Landa BB, Mavrodi OV, Schroeder KL, De La Fuente L, Bankhead SB, Molar RA, Bonsall RF, Mavrodi DV & Thomashow LS (2007) Role of 2,4-diacetylphloroglucinol-producing fluorescent *Pseudomonas* spp. in the defense of plant roots. *Plant Biology* **9**: 4–20.

Weller DM, Raaijmakers JM, Gardener BBM & Thomashow LS (2002) Microbial populations responsible for specific soil suppressiveness to plant pathogens. *Annual Review of Phytopathology* **40**: 309–348.

Werling BP & Gratton C (2008) Influence of field margins and landscape context on ground beetle diversity in Wisconsin (USA) potato fields. *Agriculture Ecosystems & Environment* **128**: 104–108.

Weseloh RM (1982) Implications of tree microhabitat preferences of *Compsilura concinnata* (Diptera: Tachinidae) for its effectiveness as a gypsy moth parasitoid. *The Canadian Entomologist* **114**: 617–622.

Weseloh RM (1984) Effect of size, stress, and ligation of gypsy moth (Lepidoptera: Lymantriidae) larvae on development of the tachinid parasite *Compsilura concinnata* Meigen (Diptera: Tachinidae). *Annals of the Entomological Society of America* **77**: 423–428.

Wheeler AG & Stoops CA (1996) Status and spread of the Palearctic lady beetles *Hippodamia variegate* and *Propylea quatordecimpunctata* (Coleoptera: Coccinellidae) in Pennsylvania, 1993–1995. *Entomological News* **107**: 291–298.

Whipps JM (2001) Microbial interactions and biocontrol in the rhizosphere. *Journal of Experimental Botany* **52**: 487–511.

Whipps JM (2004) Prospects and limitations for mycorrhizas in biocontrol of root pathogens. *Canadian Journal of Botany* **82**: 1198–1227.

Whipps JM & McQuilken MP (2009) Biological control agents in plant disease control. In *Disease Control in Crops: Biological and Environmentally Friendly Approaches* (Walters D, ed.), Oxford, UK, Wiley, pp. 27–61.

White PJ, Norman RA & Hudson PJ (2002) Epidemiological consequences of a pathogen having both virulent and avirulent modes of transmission: the case of rabbit haemorrhagic disease virus. *Epidemiology and Infection* **129**: 665–677.

Whitham TG, Maschinski J, Larson KC & Paige KN (1991) Plant responses to herbivory: the continuum from negative to positive and underlying physiological mechanisms. In *Plant-Animal Interactions: Evolutionary Ecology in Tropican and Temperate Regions* (Price PW, Lewinsohn TM, Fernandes GW & Benson WW, eds.), Hoboken, NJ, Wiley, pp. 227–255.

Whitham TG, Young WP, Martinsen GD, Gehring CA, Schweitzer JA, Shuster SM, Wimp GM, Fischer DG, Bailey JK, Lindroth RL, Woolbright S & Kuske CR (2003) Community and ecosystem genetics: a consequence of the extended phenotype. *Ecology* **84**: 559–573.

Wiebe AP & Obrycki JJ (2002) Prey suitability of *Galerucella pusilla* eggs for two generalist predators, *Coleomegilla maculata* and *Chrysoperla carnea*. *Biological Control* **23**: 143–148.

Wiebe AP & Obrycki JJ (2004) Quantitative assessment of predation of eggs and larvae of *Galerucella pusilla* in Iowa. *Biological Control* **31**: 16–28.

Wiedenmann RN & Smith Jr. JW (1997) Novel assocations and importation biological control: the need for ecological and physiological equivalencies. *International Journal of Tropical Insect Science* **17**: 51–60.

Wiens JJ & Graham CH (2005) Niche conservatism: integrating evolution, ecology, and conservation biology. *Annual Review of Ecology, Evolution and Systematics* **36**: 519–539.

Wiggins BE & Kinkel LL (2005a) Green manures and crop sequences influence alfalfa root rot and pathogen inhibitory activity among soil-borne *Streptomycetes*. *Plant and Soil* **268**: 271–283.

Wiggins BE & Kinkel LL (2005b) Green manures and crop sequences influence potato diseases and pathogen inhibitory activity of indigenous streptomycetes. *Phytopathology* **95**: 178–185.

Wiklund C & Friberg M (2008) Enemy-free space and habitat-specific host specialization. *Oecologia* **157**: 287–294.

Wilby A & Thomas MB (2002) Natural enemy diversity and pest control: patterns of pest emergence with agricultural intensification. *Ecology Letters* **5**: 353–360.

Wilkinson ATS (1965) Releases of cinnabar moth, *Hypocrita jacobaeae* (L.) (Lepidoptera: Arctiidae) on tansy ragwort in British Columbia. *Proceedings of the Entomological Society of British Columbia* **62**: 10–15.

Wilkinson TK & Landis DA (2005) Habitat diversification in biological control: the role of plant resources. In *Plant-Provided Food for Carnivorous Insects: A Protective Mutualism and Its Application* (Wäckers FL, Van Rijn PCJ & Bruin J, eds.), Cambridge, UK, Cambridge University Press, pp. 305–325.

Williams L & Martinson TE (2000) Colonization of New York vineyards by *Anagrus* spp. (Hymenoptera: Mymaridae): Overwintering biology, within-vineyard distribution of wasps, and parasitism of grape leafhopper, *Erythroneura* spp. (Homoptera: Cicadellidae), eggs. *Biological Control* **18**: 136–146.

Williams RT, Fullagar PJ, Kogon C & Davey C (1973) Observations on a naturally-occurring winter epizootic of myxomatosis at Canberra, Australia, in the presences of rabbit fleas (*Spilopsyllus cuniculi* Dale) and virulent myxoma virus. *Journal of Applied Ecology* **10**: 417–427.

Williamson M (1991) Biocontrol risks. *Nature* **353**: 394.

Williamson M (1996) *Biological Invasions*. London, UK, Chapman & Hall.

Williamson M & Fitter A (1996) The varying success of invaders. *Ecology* **77**: 1661–1666.

Willis AJ & Blossey B (1999) Benign environments do not explain the increased vigour of non-indigenous plants: a cross-continental transplant experiment. *Biocontrol Science and Technology* **9**: 567–577.

Willis AJ & Memmott J (2005) The potential for indirect effects between a weed, one of its biocontrol agents and native herbivores: a food web approach. *Biological Control* **35**: 299–306.

Willis AJ, Memmott J & Forrester RI (2000) Is there evidence for the post-invasion evolution of increased size among invasive plant species? *Ecology Letters* **3**: 275–283.

Willis AJ, Thomas MB & Lawton JH (1999) Is the increased vigour of invasive weeds explained by a trade-off between growth and herbivore resistance? *Oecologia* **120**: 632–640.

Willoquet JMR (2010) Collaboration and communication strategies between researchers and potential GMO vector user communities. In *Progress and Prospects for the Use of Genetically Modified Mosquitoes to Inhibit Disease Transmission* (Crawford VL & Reza JN, eds.), Geneva, Switzerland, WHO, pp. 34–36.

Wilson CL (1969) Use of plant pathogens in weed control. *Annual Review of Phytopathology* **7**: 411–434.

Wilson CL (1998) Conserving epiphytic microorganisms on fruits and vegetables for biological control. In *Conservation Biological Control* (Barbosa P, ed.), San Diego, CA, Academic Press, pp. 335–350.

Wilson F (1965) Biological control and the genetics of colonizing species. In *The Genetics of Colonizing Species* (Baker HG & Stebbins GL, eds.), New York, Academic Press, pp. 307–325.

Wilson JRU, Ajuonu O, Center TD, Hill MP, Julien MH, Katagira FF, Neuenschwander P, Njoka SW, Ogwang J, Reeder RH & Van T (2007) The decline of water hyacinth on Lake Victoria was due to biological control by *Neochetina* spp. *Aquatic Botany* **87**: 90–93.

Wilson K, Thomas MB, Blanford S, Doggett M, Simpson SJ & Moore SL (2002) Coping with crowds: density-dependent disease resistance in desert locusts. *Proceedings of the National Academy of Sciences USA* **99**: 5471–5475.

Wilson M (1997) Biocontrol of aerial plant diseases in agriculture and horticulture: current approaches and future prospects. *Journal of Industrial Microbiology & Biotechnology* **19**: 188–191.

Wilson M, Xin WM, Hashmi S & Gaugler R (1999) Risk assessment and fitness of a transgenic entomopathogenic nematode. *Biological Control* **15**: 81–87.

Wilson MJ, Glen DM, Hamacher GM & Smith JU (2004) A model to optimise biological control of slugs using nematode parasites. *Applied Soil Ecology* **26**: 179–191.

Windels CE (1997) Altering community balance: organic ammendments, selection pressures, and biocontrol. In *Ecological Interactions and Biological Control* (Andow DA, Ragsdale DW & Nyvall RF, eds.), Boulder, CO, Westview, pp. 282–300.

Winkler K, Wäckers FL, Kaufman LV, Larraz V & Van Lenteren JC (2009a) Nectar exploitation by herbivores and their parasitoids is a function of flower species and relative humidity. *Biological Control* **50**: 299–306.

Winkler K, Wäckers F & Pinto DM (2009b) Nectar-providing plants enhance the energetic state of herbivores as well as their parasitoids under field conditions. *Ecological Entomology* **34**: 221–227.

Winston RL, Schwarzländer M, Hinz HL, Day MD, Cock MJW & Julien MH (2014) *Biological Control of Weeds: A World Catalogue of Agents and Their Target Weeds*, 5th edition. USDA Forest Service, FHTET-2014-04, Morgantown, WV.

Wissinger SA (1997) Cyclic colonization in predictably ephemeral habitats: a template for biological control in annual crop systems. *Biological Control* **10**: 4–15.

Withers TM & Browne BL (2004) Behavioral and physiological processes affecting outcomes of host range testing. In *Assessing Host Ranges for Parasitoids and Predators Used for Classical Biological Control: A Guide to Best Practice* (Van Driesche RG & Reardon R, eds.), Morgantown, WV, USDA Forest Service, pp. 40–55.

Withers TM, Carlson CA & Gresham BA (2013) Statistical tools to interpret risks that arise from rare events in host specificity testing. *Biological Control* **64**: 177–185.

Wodzicki K (1981) Some nature conservation problems in the South Pacific. *Biological Conservation* **21**: 5–18.

Woets J & Van Lenteren JC (1976) The parasite-host relationship between *Encarsia formosa* (Hym., Aphelinidae) and *Trialeurodes vaporariorum* (Hom., Aleyrodidae). VI. Influence of the host plant on the greenhouse whitefly and its parasite *Encarsia formosa*. *IOBC/WPRS Bulletin* **4**: 151–164.

Wolfe LM (2002) Why alien invaders succeed: support for the escape-from-enemy hypothesis. *American Naturalist* **160**: 705–711.

Wolfe LM, Elzinga JA & Biere A (2004) Increased susceptibility to enemies following introduction in the invasive plant *Silene latifolia*. *Ecology Letters* **7**: 813–820.

Wolfenbarger LL, Naranjo SE, Lundgren JG, Bitzer RJ & Watrud LS (2008) Bt crop effects on functional guilds of non-target arthropods: a meta-analysis. *PLoS One* **3**: e2118.

Wood AR, Crous PW & Lennox CL (2004) Predicting the distribution of *Endophyllum osteospermi* (Uredinales, Pucciniaceae) in Australia based on its climatic requirements and distribution in South Africa. *Australian Plant Pathology* **33**: 549–558.

Wooton JT (1994) The nature and consequences of indirect effects in ecological communities. *Annual Review of Ecology and Systematics* **25**: 443–466.

Wraight SP & Ramos ME (2005) Synergistic interaction between *Beauveria bassiana*- and *Bacillus thuringiensis tenebrionis*-based biopesticides applied against field populations of Colorado potato beetle larvae. *Journal of Invertebrate Pathology* **90**: 139–150.

Wratten SD, Bowie MH, Hickman JM, Evans AM, Sedcole JR & Tylianakis JM (2003) Field boundaries as barriers to movement of hover flies (Diptera: Syrphidae) in cultivated land. *Oecologia* **134**: 605–611.

Wratten SD, Van Emden HF & Thomas MB (1998) Within-field and border refugia for the enhancement of natural enemies. In *Enhancing Biological Control: Habitat Management to Promote Natural Enemies of Agricultural Pests* (Pickett CH & Bugg RL, eds.), Berkeley, University of California Press, pp. 375–403.

Wright DJ & Verkerk RHJ (1995) Integration of chemical and biological control systems for arthropods – evaluation in a multitrophic context. *Pesticide Science* **44**: 207–218.

Wright MG, Hoffmann MP, Kuhar TP, Gardner J & Pitcher SA (2005) Evaluating risks of biological control introductions: a probabilistic risk-assessment approach. *Biological Control* **35**: 338–347.

Wu Z, Hopper KR, Ode PJ, Fuester R, Tuda M & Heimpel GE (2005). Single-locus complementary sex determination absent in *Heterospilus prosopidis* (Hymenoptera: Braconidae). *Heredity* **95**: 228–234.

Wu Z, Hopper KR, O'Neil RJ, Voegtlin DJ, Prokrym DR & Heimpel GE (2004) Reproductive compatibility and genetic variation between two strains of *Aphelinus albipodus* (Hymenoptera: Aphelinidae), a parasitoid of the soybean aphid, *Aphis glycines* (Homoptera: Aphididae). *Biological Control* **31**: 311–319.

Wyckhuys KAG & Heimpel GE (2007) Response of the soybean aphid parasitoid *Binodoxys communis* (Gahan) (Hymenoptera: Braconidae) to olfactory cues from target and non-target host-plant complexes. *Entomologia Experimentalis et Applicata* 123: 149–158.

Wyckhuys KAG, Hopper KR, Wu K-M, Straub C, Gratton C & Heimpel GE (2007a) Predicting potential ecological impact of soybean aphid biological control introductions. *Biocontrol News and Information* 28: 30 N–34 N.

Wyckhuys KAG, Koch RL & Heimpel GE (2007b) Physical and ant-mediated refuges from parasitism: implications for non-target effects in biological control. *Biological Control* **40**: 306–313.

Wyckhuys KAG, Koch RL, Kula RR & Heimpel GE (2009) Potential exposure of a classical biological control agent of the soybean aphid, *Aphis glycines*, on non-target aphids in North America. *Biological Invasions* 11: 857–871.

Wyckhuys KAG, Lu Y, Morales H, Vazquez LL, Legaspi JC, Eliopoulos PA & Hernandez LM (2013) Current status and potential of conservation biological control for agriculture in the developing world. *Biological Control* 65: 152–167.

Wyss E, Niggli U & Nentwig W (1995) The impact of spiders on aphid populations in a strip-managed apple orchard. *Journal of Applied Entomology* **119**: 473–478.

Xi Z, Khoo CCH & Dobson SL (2005) *Wolbachia* establishment and invasion in an *Aedes aegypti* laboratory population. *Science* **310**: 326–328.

Xiao K, Kinkel LL & Samac DA (2002) Biological control of *Phytophthora* root rots on alfalfa and soybean with *Streptomyces*. *Biological Control* 23: 285–295.

Xu X-M & Jeger MJ (2013) Theoretical modeling suggests that synergy may result from combined use of two biocontrol agents for controlling foliar pathogens under spatial heterogeneous conditions. *Phytopathology* 103: 768–775.

Xu X-M, Jeffries P, Pautasso M & Jeger MJ (2011) Combined use of biocontrol agents to manage plant diseases in theory and practice. *Phytopathology* **101**: 1024–1031.

Xu X-M, Salama N, Jeffries P & Jeger MJ (2010) Numerical studies of biocontrol efficacies of foliar plant pathogens in relation to the characteristics of a biocontrol agent. *Phytopathology* **100**: 814–821.

Yachi S & Loreau M (1999) Biodiversity and ecosystem productivity in a fluctuating environment: the insurance hypothesis. *Proceedings of the National Academy of Sciences, USA* **96**: 1463–1468.

Yano E (1989) A simulation study of population interaction between the greenhouse whitefly *Trialeurodes vaporariorum* Westwood (Homoptera: Aleyrodidae) and the parasitoid *Encarsia formosa* Gahan (Hymenoptera: Aphelinidae). II. Simulation analysis of population dynamics and strategy of biological control. *Researches on Population Ecology* 31: 89–104.

Yano E (2006) Ecological considerations for biological control of aphids in protected culture. *Population Ecology* 48: 333–339.

Yara K, Sasawaki T & Kunimi Y (2010) Hybridization between introduced *Torymus sinensis* (Hymenoptera: Torymidae) and indigenous *T. beneficus* (late-spring strain), parasitoids of the Asian chestnut gall wasp *Dryocosmus kuriphilus* (Hymenoptera: Cynipidae). *Biological Control* **54**: 14–18.

Yasuda H, Evans EW, Kajita Y, Urakawa K & Takizawa T (2004) Asymmetric larval interactions between introduced and indigenous ladybirds in North America. *Oecologia* **141**: 722–731.

Ye SD, Ying SH, Chen C, and M. G. & Feng MG (2006) New solid-state fermentation chamber for bulk production of aerial conidia of fungal biocontrol agents on rice. *Biotechnology Letters* 28: 799–804.

Yeates AG, Schooler SS, Garono RJ & Buckley YM (2012) Biological control as an invasion process: disturbance and propagule pressure affect the invasion success of *Lythrum salicaria* biological control agents. *Biological Invasions* **14**: 255–271.

Yeates GW & Wardle DA (1996) Nematodes and predators and prey: relationships to biological and soil processes. *Pedobiologia* **40**: 43–50.

Yong TH & Hoffmann MP (2006) Habitat selection by the introduced biological control agent *Trichogramma ostriniae* (Hymenoptera: Trichogrammatidae) and implications for nontarget effects. *Environmental Entomology* 35: 725–732.

Yong TH, Pitcher S, Gardner J & Hoffmann MP (2007) Odor specificity testing in the assessment of efficacy and non-target risk for *Trichogramma ostriniae* (Hymenoptera: Trichogrammatidae). *Biocontrol Science and Technology* **17**: 135–153.

Yu H & Sutton JC (1997) Effectiveness of bumblebees and honeybees for delivering inoculum of *Gliocladium roseum* to raspberry flowers to control *Botrytis cinerea. Biological Control* **10**: 113–122.

Zayed A & Packer L (2005) Complementary sex determination substantially increases extinction proneness of haplodiploid populations. *Proceedings of the National Academy of Sciences, USA* **102**: 10742–10746.

Zchori-Fein E, Perlman SJ, Kelly SE, Katzir N & Hunter MS (2004) Characterization of a 'Bacteroidetes' symbiont in *Encarsia* wasps (Hymenoptera: Aphelinidae): proposal of 'Candidatus *Cardinium hertigii*'. *International Journal of Systematic and Evolutionary Microbiology* **54**: 961–968.

Zelazny B, Alfiler AR & Lolong A (1989) Possibility of resistance to a baculovirus in populations of the coconut beetle (*Oryctes rhinoceros*). *FAO Plant Protection Bulletin* **37**: 77–82.

Zera AJ & Harshman LG (2001) The physiology of life history trade-offs in animals. *Annual Review of Ecology and Systematics* **32**: 95–126.

Zhou Y, Gu H & Dorn S (2006) Single-locus sex determination in the parasitoid wasp *Cotesia glomerata*(Hymenoptera: Braconidae). *Heredity* **96**: 487–492.

Zimmerman EC (1948) *Insects of Hawaii. Volume 1: Introduction*. Honolulu, University of Hawaii Press.

Zimmermann HG, Moran VC & Hoffmann JH (2000) The renowned cactus moth, *Cactoblastis cactorum*: its natural history and threat to native *Opuntia* floras in Mexico and the United States. *Diversity and Distributions* **6**: 259–269.

Zuniga MC (2002). A pox on thee! Manipulation of the host immune system by myxoma virus and implications for viral-host co-adaptation. *Virus Research* **88**: 17–33.

Zwölfer H & Harris P (1984) Biology and host specificity of *Rhinocyllys conicus* (Froel.) (Col., Curculionidae), a successful agent for biocontrol of the thistle, *Carduus nutans* L. *Zeitschrift für Angewandte Entomologie* **97**: 36–62.

INDEX